• 青阳非金属矿研究院
• 浙江工业大学化学工程学院工业催化学科 组织编写

第二届世界非金属矿科技和产业论坛论文集

PROCEEDINGS OF THE SECOND WORLD FORUM ON INDUSTRIAL MINERALS

非金属矿
科学前沿和绿色高新技术

Industrial Minerals
Cutting-edge Science and Green High-tech Industry

主 编
周春晖 边 亮 储茂泉
刘海波 刘明贤 欧阳静
王文波 周岩民 朱润良

中国农业科学技术出版社

图书在版编目（CIP）数据

非金属矿科学前沿和绿色高新技术 / 周春晖等主编. —北京：中国农业科学技术出版社，2018. 9

ISBN 978-7-5116-3491-7

Ⅰ. ①非…　Ⅱ. ①周…　Ⅲ. ①非金属矿—研究　Ⅳ. ①TD163

中国版本图书馆 CIP 数据核字（2018）第 224290 号

责任编辑　李　雪　乔利利　徐定娜
责任校对　马广洋

出版发行　中国农业科学技术出版社
北京市中关村南大街 12 号　　邮编：100081
电　　话　(010)82109707　82105169(编辑室)　(010)82109702(发行部)
(010)82109709(读者服务部)
传　　真　(010)82106626
网　　址　http://www.castp.cn
经　　销　各地新华书店
印　　刷　北京建宏印刷有限公司
开　　本　787mm×1 092mm　1/16
印　　张　24.75
字　　数　606 千字
版　　次　2018 年 9 月第 1 版　2018 年 9 月第 1 次印刷
定　　价　88.00 元

前　言

第二届世界非金属矿科技和产业论坛（The Second World Forum on Industrial Minerals，WFIM－2）于2018年10月20—23日在中国安徽省青阳县召开。本次论坛的主题是“非金属矿科学前沿和绿色高新技术”，旨在交流世界范围内的非金属矿及相关领域的学术研究动态、面临的科学问题和新的技术进展，切磋非金属矿产业的实际难题、挑战和科技需求，探讨产学研合作和多学科交叉融合解决机制，推进学术交流、技术创新和产业升级，助力科技引领发展绿色的、高值化的非金属矿加工业和产业链；论坛融合展示学术基础理论研究的突破和工程技术的创新成果，面向世界非金属矿和相关领域的前沿科技、新兴产业、经济和社会的可持续发展，交流和推动政、企、商、民、专家学者之间的广泛交流与合作，提供相关政府部门、企业、投融资、科技专家学者等多界共商非金属矿科技、经济、社会、环境及其他相关问题的信息交流和对话平台，增强国内业界与世界其他国家地区的非金属矿科技和经济对话、联系和合作，拓展非金属矿科技成果跨区域转移扩散、转化和产业化空间。

本次论坛收到中外科研人员投来的会议摘要稿件170余份，参会的国际非金属矿及相关领域科研人员来自希腊、法国、意大利、澳大利亚、美国、韩国、印度、波兰、赞比亚等国家，国内非金属矿及相关领域的科研人员来自众多的高校和科研院所，参会的还有众多的企业人员和政府有关部门人员等。

本世界非金属矿科技和产业论坛系全公益性质。首届论坛由安徽省青阳县经济和信息化委员会、青阳县科学技术局、酉华镇人民政府、青阳非金属矿研究院、中国矿物岩石地球化学学会第9届矿物物理矿物结构专业委员会共同发起，于2017年10月28—31日在青阳县“五溪山色”召开，标志着论坛的正式启动和运行。首届论坛主题是“非金属矿科技与产业的联动”，首届论坛主席是中国科学院广州地球化学研究所何宏平研究员和浙江工业大学化学工程学院周春晖研究员。

论坛选址安徽省青阳县为论坛常规所在地，亦称“青阳论坛”。安徽省青阳县非金属矿产资源非常丰富，其中以“三石”（石灰石、方解石、白云石）资源储量巨大，特别是酉华镇及周边地域的石灰石以质优量大闻名于国内外，资源储量位居全国前列。青阳县酉华镇是安徽省政府认定的“安徽省非金属矿采选及深加工产业集群镇”。青阳县境内及周边区域，其他非金属矿产如花岗岩、黏土矿物、页岩等储量也十分丰富，各类非金属矿及制品的生产企业和从业人员众多。

非金属矿是与人类生产、生活密切相关的重要矿产资源之一，在当前的工业生产和社会生活中占有重要地位，广泛应用于化工、石油、造纸、冶金、建筑、机械、农业、环保、医药、保健等行业，正越来越多地被用于能源、国防、航天、通信、智能材料等高科技领域。据称当今世界非金属矿物用量和产值增长速度已超过了金属矿产，其开发利用水平已成为衡量一个国家科学技术发展水平和人民生活水平的重要标志之一。非金属矿科学技术不仅在工业生产中十分重要，而且非金属矿的科学研究涉及地球和矿物形成、古气候和环境演化、外星球探索分析、生命起源等重大科技前沿领域，具有重大科学意义和作用。

青阳非金属矿研究院系正式登记注册的、非企业性质的、公益性的科技机构，理念和宗旨为“Qing Yang Institute for Industrial Minerals（QYIM）：A non-profit research and development（R&D）

institution dedicated to the advancement and promotion of science and technology of industrial minerals and related areas for making the world green，sustainable，and better.” 青阳非金属矿研究院办院和运行坚持定位作为非营利性、开放式、协同创新的学术、技术的科学研究和交流合作平台，携手非金属矿领域及相关的地质、矿物、化工、材料、生物医药、农业、环保、日化等领域的科学研究人员、学者、企业家、企业技术人员等，共同促进和提升非金属矿及相关领域的科学研究、技术开发、成果转化、产品开发、应用推广、科普、教育和文化建设等。在此恳请大家给予青阳非金属矿研究院更多支持并经常性开展合作、交流，共同投身和见证非金属矿产学研的发展及为地方、国家和社会可持续发展做出应有的贡献。

本次论坛设置了八个议题：非金属矿物和相关材料的结构化学和模拟；非金属矿地质、采选和分析表征科学；非金属矿/聚合物复合功能材料；非金属矿和光、电、磁、声、热等先进材料；非金属矿和新型化工、轻工、建材、能源、冶金；非金属矿和现代农林牧渔业；非金属矿、纳米技术与医药健康；非金属矿与星球探索、地球环境、生态保护和尾矿、矿渣、固体废弃物利用。旨在寻求在一定程度上较全面地研讨当前非金属矿物基础科学和应用开发的研究水平和前沿动态。相信辅以本论文摘要集的正式出版，既有利论坛现场交流，也有利于非金属矿开采、生产和应用的生产企业、检测分析部门、地矿部门、科学研究、教学、商贸等相关人员的后续交流。

本次论坛的论文摘要汇集出版总体设计、分章规划和整理由周春晖完成。各章组稿人员（议题分会主席）为：第一章，边亮（西南科技大学环境与资源学院）；第二章，刘海波（合肥工业大学资源与环境工程学院）；第三章，王文波（中国科学院兰州化学物理研究所）；第四章，欧阳静（中南大学资源加工与生物工程学院）；第五章，周春晖（浙江工业大学化学工程学院）；第六章，周岩民（南京农业大学动物科技学院）、边亮、朱润良、周春晖；第七章，储茂泉（同济大学生命科学与技术学院）和刘明贤（暨南大学化学与材料学院）；第八章，朱润良（中国科学院广州地球化学研究所）。

本次论坛得到了安徽省青阳县经济和信息化委员会、青阳县科学技术局、酉华镇人民政府、青阳县经济开发区酉华工业园（筹）、浙江工业大学等单位和部门的大力支持，青阳非金属矿研究院的国际和国内学术委员会成员、理事等给予了诸多关怀、指导和帮助，在此深表谢意。

出版社编辑人员对本书的出版给予了辛勤的、高效的文稿编辑工作，在此表示感谢。

由于时间仓促、水平有限，在各章下有些内容不能完全贴切对应议题，这一方面是由于科学研究交叉性或依据作者投稿所选的议题所造成，另一方面也可能是因本书统筹整理过程中存在一些疏忽和编者水平有限，恳望作者、读者和同行谅解和不吝指正。

特别感谢各位非金属矿及相关领域科研人员的积极投稿、参与、支持和协助。

周春晖
青阳非金属矿研究院
池州非金属矿产业技术创新战略联盟
浙江工业大学化学工程学院工业催化学科
绿色化学合成技术国家重点实验室培育基地
2018 年 10月

目　录

第一章　非金属矿矿物学基础、结构和理论

第二章 非金属矿地质、采选和分析表征

第三章　非金属矿—聚合物复合功能材料

第四章　非金属矿光、电、磁、声、热材料

第五章 非金属矿和新型化工、建材、能源

第六章　非金属矿与养殖、农业和土壤

第七章 非金属矿纳米技术与生物医药健康

第八章 非金属矿与地球化学、生态环境

附 录

第一章

非金属矿矿物学基础、结构和理论

癸酸/硅藻土/碳纳米管复合相变材料的制备及性能

刘　鹏　顾晓滨*　庞增瑞　宋宪伟
（河北地质大学宝石与材料工艺学院，石家庄 050031）

经济社会的快速发展使得能源短缺问题日益突出，目前人类利用的能源以化石能源为主，然而化石能源使用过程中会对环境产生严重的污染，因此，开发利用可代替化石能源的新型能源迫在眉睫。而新能源存在不稳定、热转化效率低等问题，所以储能技术的研究成为全球学者的研究热点，相变储能材料（phase change materials，简称 PCM）作为储能材料的典型代表也因此，得到了人们的广泛关注。与传统显热储能材料相比，PCM 可储存 5～14 倍的热量，且在能量存储过程中保持近似恒温，故在可再生能源开发利用领域应用前景广阔[1]。

在几大类相变材料中，有机 PCM 具有很多优良的性能成为相变储能材料研究的热点之一，而其导热性差的问题阻碍了其实际应用进程，因此，探索强化有机 PCM 导热性的方法，对 PCM 的应用具有重要意义。具有众多优势特点的硅藻土（Diatomite，简称 DT）[2-5]作为典型的多孔矿物可成为有机相变材料良好的封装载体[1]。该研究以硅藻土为封装载体，碳纳米管（Carbon nanotubes，简称 CNTs）为强化导热剂，癸酸（Capric acid，简称 CA）为相变材料，采用熔融浸渍法制备了癸酸/硅藻土/碳纳米管复合 PCM，并采用 XRD、SEM、DSC、TG 对相变材料的结构、形貌和热性能等进行了系统分析。

通过 CA、DT、CA＋DT、CA＋DT＋CNTs 的 SEM 观察可知，癸酸或癸酸与碳纳米管的混合物在加热搅拌的过程中，分散到了硅藻土空间结构表面及内部孔隙结构之中。图 1 为 CA、DT、CA＋DT、CA＋DT ＋CNTs 的 XRD 图，从图 1 可知，癸酸、硅藻土及碳纳米管在复合过程中无化学反应，只进行了物理混合，具有较好的化学相容性。对 CA、CA＋DT、CA＋DT ＋CNTs 进行热重分析可知，复合相变材料的热稳定性得到很大程度的改善，这为实际应用提供了可能。通过步冷曲线测试分析发现，碳纳米管的添加能够改善癸酸的传热效率，且随着碳纳米管添加质量分数的增加，癸酸/硅藻土/碳纳米管复合相变材料的传热效果先变好而后变差。利用 DSC 对泄漏实验中样品分析可知，DT 含量越大，CA＋DT 相变材料的相变焓越小，然而 DT 添加过量后，虽然相变过程中不发生泄漏，但相变材料储能下降，亦会影响材料性能。图 2 为 CA、CA＋DT、CA＋DT＋CNTs 的 DSC 对比图，从图 2 可知，硅藻土可作为有机酸的封装基体，碳纳米管可改善癸酸的传热效率。

较优癸酸/硅藻土/碳纳米管复合相变材料，其熔化、凝固的相变温度和相变潜热分别为 34.05 ℃、28.07 ℃、79.03 J/g 和 79.17 J/g。

---

*通信作者：顾晓滨；E-mail：XBGu@hgu. edu. cn；手机号：18813159087。河北省科技支撑计划项目（17214016）、环境友好能源材料国家重点实验室开放基金资助（17kffk13）。

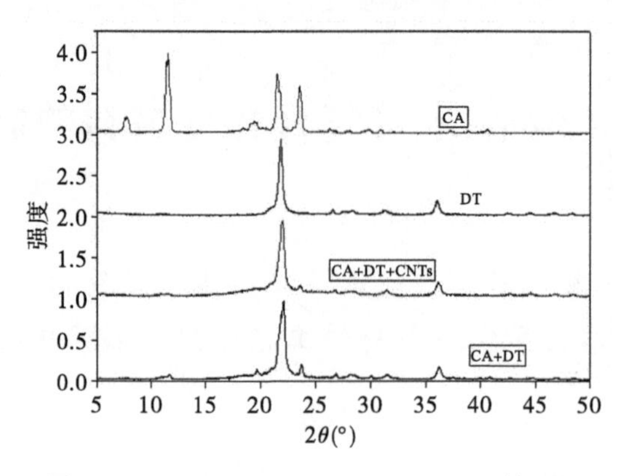

图 1　CA、CA+DT、CA+DT+CNTs 的 XRD

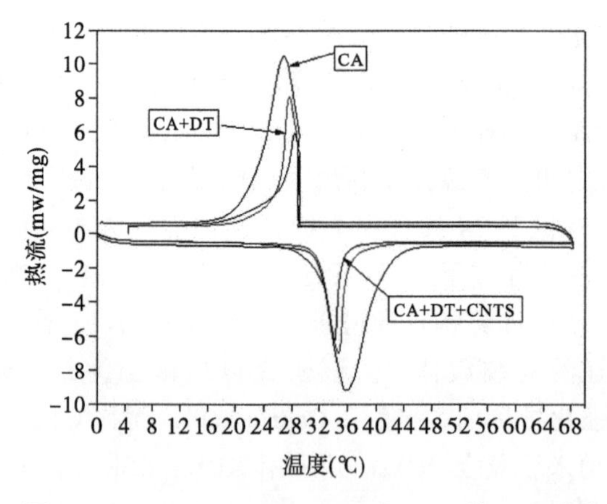

图 2　CA、CA+DT、CA+DT+CNTs 的 DSC 曲线

参考文献

[1] Gu X B，Qi S，Niu J J. 2014. Research status and prospect on phase change mineral materials [J]. Bulletin of mineralogy，Petrology and Geochemistry，33：932－940.

[2] Lv P Z，Liu C Z，Rao Z H. 2017. Review on clay mineral-based form-stable phase change materials：Preparation，characterization and applications [J]. Renewable & Sustainable Energy Reviews，68：707－726.

[3] Jeong S G，Jeon J，Chung O，et al. 2013. Evaluation of PCM/diatomite composites using exfoliated graphite nanoplatelets (xGnP) to improve thermal properties [J]. Journal of Thermal Analysis & Calorimetry，114：689－698.

[4] Yong D，Li J H，Qian T T，et al. 2016. Preparation and characterization of KNO_3/diatomite shape-stabilized composite phase change material for high temperature thermal energy storage [J]. Journal of Materials Science & Technology，33：198－203.

[5] Liu Z Y，Hu D，Lv H L，et al. 2017. Mixed mill-heating fabrication and thermal energy storage of diatomite/paraffin phase change composite incorporated gypsum-based materials [J]. Applied Thermal Engineering，118：703－713.

水滑石超族矿物多型结构的计算机模拟

陈　锰* 朱润良　朱建喜　何宏平

（中国科学院矿物学与成矿学重点实验室/广东省矿物物理与材料研究开发重点实验室，中国科学院广州地球化学研究所，广州 510640）

层状矿物是由相同或相似的片层在一维方向上堆垛而组成的。片层间堆垛方式的差异反映为片层间的平移、旋转或倒转。多型指的是具有相同或相似片层组成的矿物在一维方向上堆垛差异而形成的同质多像现象。水滑石超族矿物（Hydrotalcite Supergroup 或 LDHs），是一类具有与水镁石相似片层结构的矿物（图 1）。其与水镁石的差异是存在阳离子替代，常见为 Al^{3+} 对 Mg^{2+} 的替代使片层带正电荷。片层与片层间由阴离子和水分子柱撑，体系保持电中性。片层堆垛方式的差异形成一系列多型（图 2），如 $3R_1$（原胞由 3 个片层组成，三方晶系，菱面体格子）和 1T（原胞为单个片层，三方晶系，六方格子）。

研究表明，土壤中的黏土矿物和氧化物矿物可以转化为 LDHs。该族矿物在土壤中的稳定存在，对阴离子（CO_3^{2-}、SO_4^{2-}、NO_3^-、Cl^- 等）和二价过渡金属离子的封存起到重要作用。LDHs 可以作为催化剂、催化剂载体、吸附剂、电极材料而广泛应用。其稳定性主要由其结构决定，尤其是反映层堆垛差异的多型结构。而影响多型结构有很多因素，如金属阳离子组成、层间阴离子类型和水含量等。前人使用 X 射线衍射方法研究了不同天然或合成 LDHs 的多型，但仍缺乏系统性的认识。另一方面，X 射线衍射的结果具有一定多解性。关于 LDHs 多型结构的基础问题是：多型是如何由阳离子、阴离子和水分子的组成所决定？层堆垛如何与层间离子、水分子的络合结构相关联？

笔者使用计算机模拟方法，系统分析了金属阳离子组成、层间阴离子类型和水含量等因素对 LDHs 多型的影响。计算机模拟研究，是基于牛顿力学原理，使用经验力场描述原子间相互作用，模拟晶体结构与运动获取物理认识。笔者使用了广受验证的 ClayFF 力场 LDHs 组成原子间的相互作用力，使用退火方法来得到稳定的结构构型。

模拟研究揭示：层间离子为 NO_3^- 时，水含量上升使 LDHs 的结构由 $3R_1$ 向 1T 转变（图 3）。伴随多型转变，层间距上升，NO_3^- 离子的构型由 D_{3h} 向 C_{2v} 对称性转变。当三价金属离子替代量更高时，多型转变出现于更低的水含量。而当层间离子为 SO_4^{2-} 时（图 3），水含量的变化造成三阶段的多型转变。第一和第三阶段的多型均为 $3R_1$ 多型，而中间阶段的多型与阳离子比例相关，为 1T 多型或出现随机混层现象。多型结构与 SO_4^{2-} 离子结构耦合，中间阶段以出现较高比例的 C_s 构型 SO_4^{2-} 离子为标志，而在其余情况，SO_4^{2-} 离子大多为 C_{3v} 构型。对于层间为 SO_4^{2-} 离子的情况，多型转变时的水含量几乎与阳离子比例无关。层间为 CO_3^{2-} 离子或 Cl^- 离子时，水含量的变

* 通信作者：陈锰；E-mail：chenmeng@gig. ac. cn；手机号：13710224156。国家自然科学基金（41602034）、中国科学院前沿科学重点研究项目（QYZDJ-SSW-DQC023）资助。

化不会造成多型的改变，始终为 $3R_1$ 多型。层间阴离子的构型反映了局域的氢键作用，而片层的堆垛影响长程静电作用。局域的氢键作用和长程的静电作用共同维系水滑石超族矿物的结构与稳定性。

计算机模拟方法系统揭示了金属阳离子组成、层间阴离子类型和水含量等因素对水滑石超族矿物多型的影响。这为揭示该族矿物在地球化学环境中的稳定性提供了重要的结构认识，另一方面也为其在催化剂、吸附剂、电极等材料上的应用提供了重要认识。

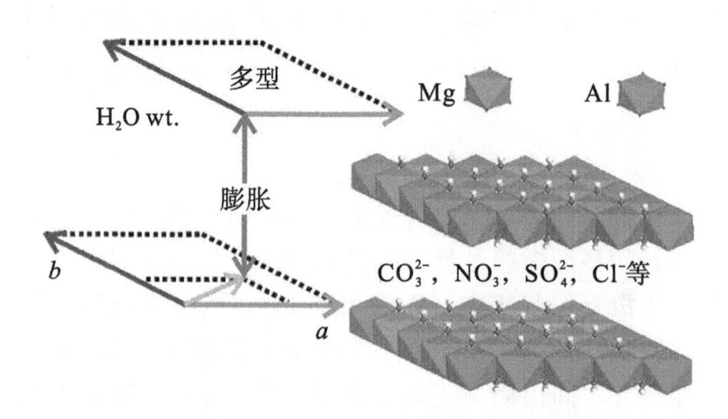

图 1　水滑石超族矿物的层结构与多型影响因素图解

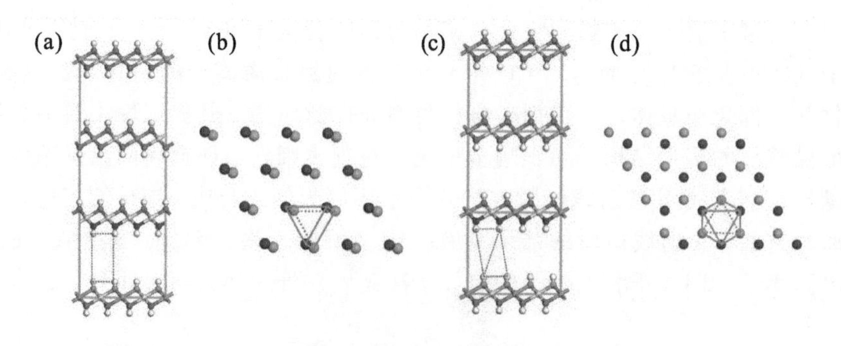

图 2　$3R_1$ 和 $1T$ 多型

注：a. $3R_1$ 侧视图；b. $3R_1$ 顶视图；c. 1T 侧视图；d. 1T 顶视图

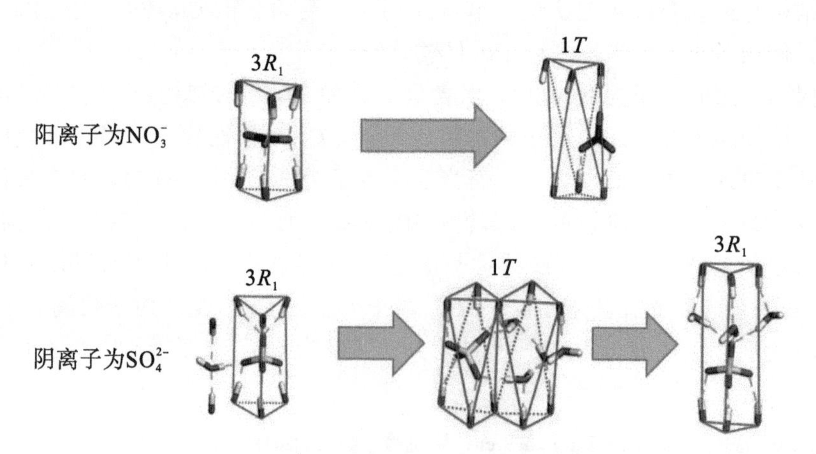

图 3　层间离子为 NO_3^-、SO_4^{2-} 时的多型转变

蒙脱石—碳酸盐矿化菌对 Pb^{2+} 的联合滞固行为

代群威[1,2*]　许凤琴[1]　董　净[1,2]　赵玉连[1]　党　政[1]　黄云碧[1]
（1. 西南科技大学环境与资源学院，绵阳 621010；
2. 西南科技大学固体废物处理与资源化教育部重点实验室，绵阳 621010）

土壤重金属污染日益严重，治理重金属污染的任务迫在眉睫。目前蒙脱石等黏土吸附重金属的研究众多，碳酸盐矿化菌固化重金属的研究也有报道。许多国内外研究学者利用环保型材料蒙脱石或改性蒙脱石对 Pb^{2+}、Zn^{2+}、Cu^{2+}、Cd^{2+} 等重金属离子进行批量吸附，研究发现蒙脱石对重金属有一定的吸附效果，改性后效果更好[1-3]。也有学者利用微生物对 Pb^{2+}、Sr^{2+}、Cd^{2+}、Cu^{2+} 和 Zn^{2+} 等重金属离子进行诱导矿化或控制矿化，研究发现重金属离子的去除率均在 90%以上[4]。但上述研究多集中在单一微生物体系或矿物体系修复重金属污染方面。笔者则是以 Pb^{2+} 为常见重金属离子代表，基于蒙脱石对重金属离子的快速吸附和碳酸盐矿化菌的高效矿化，将二者结合起来构建缓存滞留和矿化固定功能的蒙脱石—碳酸盐矿化菌体系，开展该体系对 Pb^{2+} 的吸附—矿化行为过程及滞/固效果研究。

实验过程中，采用培养液相 Pb^{2+} 浓度为 372.4 mg/L、pH＝7.0。在 250 mL 锥形瓶中封装 150 mL 培养基溶液，灭菌备用。利用微孔滤膜可以过滤除菌的特性，将尿素溶液（20 g/L）过滤加入上述培养基。加入一定量的蒙脱石，2 h 后按 1∶100 体积比接种碳酸盐矿化菌，在 30 ℃、150 r/min 条件下振荡培养。在一定作用时间内，通过分析蒙脱石—碳酸盐矿化菌体系的液相 Pb^{2+} 浓度等，得出该体系对 Pb^{2+} 的去除效果。取样时间分别为 0 d、2 h、1 d、2 d、3 d、5 d、7 d、15 d 和 30 d。取样后，将样品在 4 000 r/min 下离心 10 min，取上清液稀释后用 ICP 测试 Pb^{2+} 浓度。沉淀物用去离子水洗涤 3 次，60 ℃下烘干后，放置干燥皿中作为 SEM/EDS、FTIR 等样品备用。

对蒙脱石—碳酸盐矿化菌对 Pb^{2+} 滞/固实验液相 pH 值变化进行分析（图 1）。其中蒙脱石对照组的液相 pH 值上升并稳定在 7.8 左右，实验组的蒙脱石—碳酸盐矿化菌体系明显高于对照组，30 d 时 pH 值为 8.5。分析认为，碳酸盐矿化菌分解尿素产生碳酸根，使得体系中 pH 值逐渐升高。蒙脱石—碳酸盐矿化菌体系碳酸根情况进行分析，在 30 d 时，碳酸根浓度约为 120.0 mg/L，远低于纯菌液相下的 2 200.0 mg/L。这主要源自碳酸盐矿化菌能分解尿素，并产生碳酸根，这将与液相中 Pb^{2+} 作用，具体反应过程如下：尿素→HCO_3^-→CO_3^{2-}＋Pb^{2+}→$PbCO_3$↓，可知体系菌体代谢产生碳酸根与 Pb^{2+} 作用。

对蒙脱石—碳酸盐矿化菌体系滞/固实验中 Pb^{2+} 的赋存特征进行分析（图 2）。可以看出，蒙脱石—碳酸盐矿化菌体系对 Pb^{2+} 的去除效果较好，第 7 天后该体系对 Pb^{2+} 的去除率基本稳定，最终可达 90.0%；其中，7 d 时碳酸盐矿化菌的固化率为 85.3%，占总去除量的 94.8%。可见，

* 通信作者：代群威；E-mail：qw_dai@163. com；手机号：13550824213。国家自然科学基金（41102212）、四川省科技厅应用基础重点研究项目（2016JY0213）。

碳酸盐矿化菌对该体系的整体滞/固效果基本一直占据主导作用。

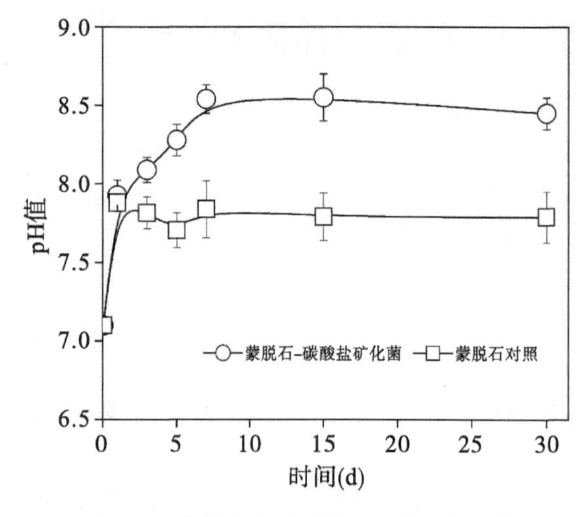

图 1　Pb^{2+} 滞/固实验液相 pH 值变化曲线

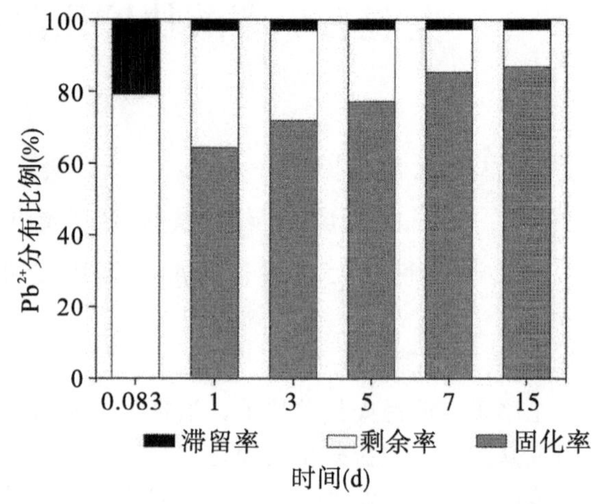

图 2　蒙脱石—矿化菌体系中 Pb^{2+} 的分布比例

蒙脱石—碳酸盐矿化菌体系作用 30 d 后，对实验样品进行 SEM、FTIR 测试分析。有棒状晶体和颗粒状晶体产生，粒径范围为 1～3 μm。1 339、681 cm^{-1} 处吸收峰分别对应 CO_3^{2-} 的对称伸缩振动吸收峰及 CO_3^{2-} 的面外弯曲振动吸收峰。因此，可知样品中含有碱式碳酸铅，说明碳酸盐矿化菌矿化 Pb^{2+} 形成结晶。

该研究中所用的蒙脱石—碳酸盐矿化菌体系，蒙脱石起到对 Pb^{2+} 的快速缓存作用，碳酸盐矿化菌在此基础上对 Pb^{2+} 进行矿化固定，最终实现该体系的联合滞/固效果，7 d 即可达 90％的去除率，其中固化率占 85％以上。该技术主要针对水体、土壤重金属污染的治理提供新思路，特别是对高浓度、突发性重金属污染具有更好效果，在当今环保产业快速发展的大背景下，具有良好应用市场和产业化前景。

参考文献

［1］Adraa K E，Georgelin T，Lambert J F，et al. 2017. Cysteine-montmorillonite composites for heavy metal cation complexation：a combined experimental and theoretical study ［J］. Chemical Engineering Journal，314：406－417.

［2］Akpomie K G，Dawodu F A. 2015. Treatment of an automobile effluent from heavy metals contamination by an eco-friendly montmorillonite ［J］. Journal of Advanced Research，6（6）：1003－1013.

［3］张媛，王亚娥，康峰，等. 2015. 钠基蒙脱石对水体中 Cd^{2+} 的大容量吸附实验研究 ［J］. 非金属矿，38（2）：74－76.

［4］王茂林，吴世军，杨永强，等. 2018. 微生物诱导碳酸盐沉淀及其在固定重金属领域的应用进展 ［J］. 环境科学研究，31（2）：206－214.

黏土矿物基活性SiO_2纳米片及功能化应用

王　兰[1*]　顾士庆[1]　边　亮[2*]　王传义[3]

（1. 中国科学院新疆理化技术研究所环境科学与技术研究室，乌鲁木齐 830011；

2. 西南科技大学固体废物处理与资源化教育部重点实验室，绵阳 621010；

3. 陕西科技大学环境科学与工程学院，西安 710021）

二维纳米材料独特的空间结构、大的比表面积等优良特征赋予了其较多的优异性质，充分利用这些特性有利于实现新材料的制备和新产品的开发。在众多制备二维纳米材料的各类材料中，层状黏土矿物因其储量丰富且具有独特的硅酸盐层结构而备受关注。天然层状黏土矿物一般由Si—O四面体片和Al—O八面体片彼此连结成结构层[1]。四面体片与八面体片组合形式不同，可将结构层分为1∶1型和2∶1型两种基本类型，而高岭石和蛭石是黏土矿物的主要代表性矿物种类。其中，高岭石是1∶1型的层状硅酸盐，其结构单元层由一层［SiO_4］四面体片和一层［AlO_6］八面体片组成；蛭石为2∶1型的层状硅酸盐，结构单元层由层状硅氧骨架通过氢氧镁铝层结合而成的双层［SiO_4］四面体，且四面体中的Si^{4+}可以被Al^{3+}或Mg^{2+}等阳离子同晶置换。利用无机方法处理此类矿物[2,3]，制备活性二氧化硅纳米片，不仅可以得到纳米尺寸的二维空间效应，而且可赋予它们新的表面物理化学特性，并应用于环境污染物吸附、纳米结构调控、表面固载等功能化应用中。

通过对蛭石进行简单的酸处理，获得了SiO_2百分含量达82%、比表面积为764 m^2/g、Zeta电位为−38.9 mV，表面富含Si—OH反应活性基团的二氧化硅纳米片；高岭石经过热活化—酸处理的改性方式，最终获得了SiO_2百分含量达81.7%、比表面积为270 m^2/g、Zeta电位为−35.9 mV，表面富含Si—OH反应活性基团的二氧化硅纳米片（图1）。通过分析酸/酸—热条件对矿物组成、结构和形态的影响，深入了解了从硅酸盐结构层到活性二氧化硅纳米片的演变规律。

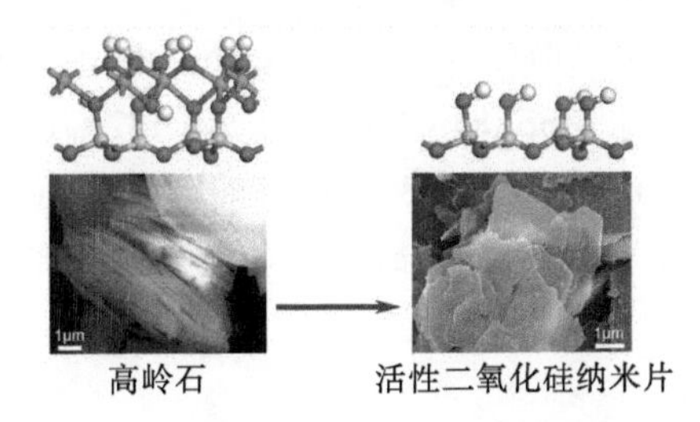

图1　高岭石及其活性二氧化硅纳米片的扫描电镜图

注：插图为相应材料的原子结构图，白色：H原子；红色：O原子；粉色：Al原子；黄色：Si原子

*通信作者：王兰，E-mail：wanglan@ms. xjb. ac. cn；手机号：15199133731；边亮，E-mail：bianliang@swust. edu. cn。国家自然科学基金（U1703129和U1403295）、973项目（2014CB8460003）、四川省“千人计划”项目、河北省青年拔尖人才资助项目。

在此基础上，以蛭石基活性二氧化硅纳米片为基体，以带有不同官能团的硅烷偶联剂为修饰剂，通过偶联反应在其表面连接了对 Pb（Ⅱ）具有吸附作用的官能团，分别为 $-NH_2$、$-COOH$、$-SH$，所得吸附剂简单定义为 Verm - NH_2、Verm - COOH 和 Verm - SH。SEM、EDS、热分析、FTIR 和 BET 分析结果表明，$-NH_2$、$-COOH$ 和 $-SH$ 基团被成功地连接在二氧化硅的表面；通过吸附性能对比发现，吸附量大小顺序为 Verm - NH_2＞Verm - SH＞Verm - COOH，表明 $-NH_2$ 基团对 Pb（Ⅱ）的结合能力更强（图 2）；准二级动力学模型和 Langmuir 模型分别可以很好地拟合 Verm - NH_2 吸附 Pb（Ⅱ）的动力学和热力学行为，理论最大吸附量为 265.8 mg/g。

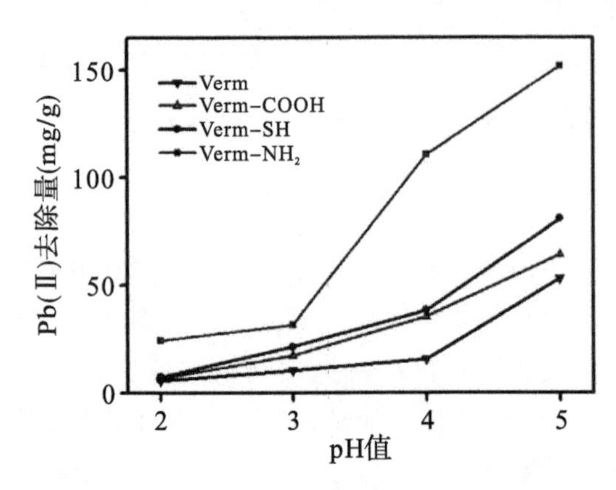

图 2　溶液 pH 值对 4 种吸附剂性能的影响

注：反应条件：初始浓度＝400 mg/L，吸附时间 t＝120 min，吸附剂剂量＝0.05 g

因此，基于黏土矿物获得活性二氧化硅纳米片，在得到不同矿物的结构和物理化学特性演变规律的同时，对其进一步有机功能化修饰并应用于重金属离子吸附，可为黏土矿物高附加值利用提供理论指导和依据。

参考文献

［1］Brigatti M F，Galan E，Theng B K G. 2006. Structures and mineralogy of clay minerals，in：Bergaya F，Theng B K G，Lagaly G.（Eds.），Handbook of Clay Science，Developments in Clay Science，vol. 1. Elsevier，Amsterdam，chap. 2.

［2］Wang L，Wang X，Yin J，et al. 2016. Insights into the physicochemical characteristics from vermiculite to silica nanosheets［J］. Applied Clay Science，132：17 - 23.

［3］Zhang Q，Yan Z L，Ouyang J，et al. 2018. Chemically modified kaolinite nanolayers for the removal of organic pollutants［J］. Applied Clay Science，157：283 - 290.

$CaCO_3$的生物诱导合成及潜在应用

刘仁绿　连　宾*
（南京师范大学生命科学学院，南京 210023）

$CaCO_3$的生物沉积在自然界非常普遍[1,2]。然而，不同微生物诱导的$CaCO_3$的形态和晶体形状差异很大，有关微生物诱导$CaCO_3$合成的调控机制还很不完善，亟待进一步研究。此外，已有研究表明生物源$CaCO_3$对重金属离子具有较好的吸附效果[3]。故而，有关生物$CaCO_3$的合成及其应用备受地质生物学研究者的关注。枯草芽孢杆菌（*Bacillus subtilis*）是我国常见的微生物肥料菌种[4]，研究该菌对$CaCO_3$的诱导矿化机理对解决土壤污染的环境问题具有理论和实践价值。

B. subtilis（GenBank 登录序列号为 KT343639），源自我国国家微生物肥料技术研究推广中心，是我国微生物肥料用的功能菌株。本文采用 FESEM－EDS、AAS、XRD 以及 TEM－SAED 等方法，研究了该菌在不同$CaCl_2$添加量的实验条件（置于 30 ℃、180 r/min 的摇床中培养 7 d）下对$CaCO_3$的诱导合成机制并探讨了其对 Cd（Ⅱ）的吸附作用。

对培养 7 d 后的沉淀物进行 FESEM 及 TEM 分析，可观察到方块状、球状以及不规则状和鳞片状团聚体等形态（图 1a/d），其表面多孔，边角不完整。能谱分析表明，该沉淀物主要含 C、O、Ca 三种元素，用 TEM－SAED 观察添加 0.2% $CaCl_2$（m/V）培养体系所形成的沉淀物，未出现明显的衍射环或斑点（图 1b），XRD 结果也未见明显衍射峰（图 1c），表明该沉淀物为无定形$CaCO_3$[5]，进一步研究认为这些无定形的$CaCO_3$是结晶程度较好的球霰石的前躯体[6]。TEM－SAED 观察添加 0.8% $CaCl_2$（m/V）培养体系形成的沉淀物，可见明显的衍射环（图 1e），结晶程度较好，XRD 分析表明为球霰石（图 1f）。综上所述，微生物不仅能调控$CaCO_3$的晶体形貌，还能根据Ca^{2+}浓度的高低调控$CaCO_3$的结晶程度和晶体结构。微生物分泌的碳酸酐酶（CA）可以催化CO_2与HCO_3^-之间的转换反应[7]，本研究发现通过添加 CA 抑制剂能显著减弱该菌对$CaCO_3$的诱导矿化能力，说明这些$CaCO_3$的形成与该菌分泌的 CA 密切相关。

有关生物源$CaCO_3$的形成机理包括：微生物通过分泌 CA，同时以自身细胞外壁为模板在液体培养条件下诱导$CaCO_3$的形成，通常情况下以无定形的$CaCO_3$为主，并有少量球霰石形成，其可能的形成过程如图 2 所示。对 Cd（Ⅱ）吸附研究发现，当 Cd（Ⅱ）浓度为 49.83 mg/L 时，该菌诱导合成的无定形$CaCO_3$及部分球霰石对 Cd（Ⅱ）的吸附效率均在 97%以上，进一步研究表明，当Cd（Ⅱ）为 476.59 mg/L 时，该菌诱导合成的球霰石对 Cd（Ⅱ）的吸附效率（80.09%）显著高于无定形$CaCO_3$的吸附效率（75.64%），XRD 分析结果表明生物球霰石吸附重金属前后晶体结构并无显著变化，暗示其吸附过程为物理吸附过程。本研究结果提高了对有关$CaCO_3$生物诱导成矿机理的认识，为微生物肥料在土壤固持重金属中的潜在贡献提供了理论依据。

* 通信作者：连宾；E-mail：bin2368@vip. 163. com。国家自然科学基金（41772360）资助项目。

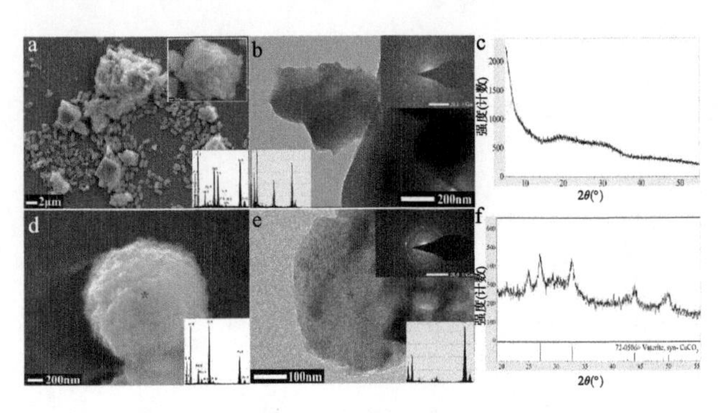

图 1 生物源 $CaCO_3$ 的形态和矿物学分析

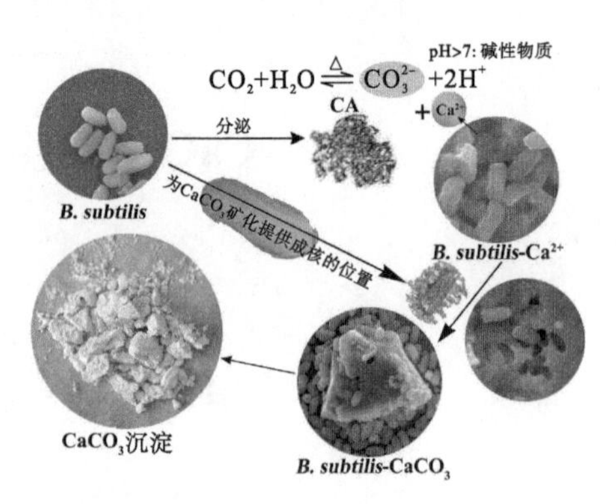

图 2 生物源 $CaCO_3$ 诱导矿化的机制模式

参考文献

[1] Phillips A J, Lauchnor E, Eldring J, et al. 2013. Potential CO_2 leakage reduction through biofilm-induced calcium carbonate precipitation [J]. Environmental Science & Technology, 47 (1): 142-149.

[2] Mitchell A C, Dideriksen K, Spangler L H, et al. 2010. Microbially enhanced carbon capture and storage by mineral-trapping and solubility-trapping [J]. Environmental Science & Technology, 44 (13): 5270-5276.

[3] Liu R L, Guan Y, Chen L, et al. 2018. Adsorption and desorption characteristics of Cd^{2+} and Pb^{2+} by micro and nano-sized biogenic $CaCO_3$ [J]. Frontiers in Microbiology, 9 (41).

[4] Kim Y S, Kim S H, Cho S H, et al. 2014. Effects of two formulation types of a microbial fertilizer containing Bacillus subtilis on growth of creeping bentgrass [J]. Journal of International Medical Research, 43 (3): 3-8.

[5] Rodriguez-navarro C, Kudłacz K, Cizer Ö, et al. 2015. Formation of amorphous calcium carbonate and its transformation into mesostructured calcite [J]. Crystengcomm, 17 (1): 58-72.

[6] Addadi L, Raz S, Weiner S. 2003. Taking advantage of disorder: amorphous calcium carbonate and its roles in biomineralization [J]. Advanced Materials, 15 (12): 959-970.

[7] Power I M, Harrison A L, Dipple G M. 2016. Accelerating mineral carbonation using carbonic anhydrase [J]. Environmental Science & Technology, 50 (5): 2610-2618.

高岭土及金属掺杂高岭土表面固碳的第一性原理研究

张红平[1*]　边　亮[2]　张亚萍[1]　何辉超[2]　董发勤[2]

（1. 西南科技大学环境友好能源材料国家重点实验室，
材料科学与工程学院，绵阳 621010；
2. 西南科技大学固体废物处理与资源化教育部重点实验室，绵阳 621010）

黏土矿物因其独特的物理结构和较大的比表面积而引起了人们的广泛关注。高岭土是自然界中非常丰富的天然黏土矿物之一。高岭土除了在陶瓷、橡胶、橡胶填料、耐火材料原料等工业应用之外，在环境领域也具有巨大的潜在应用价值。研究表明，高岭石可用于气体分离，同时对重金属离子有很好的去除能力[1,2]。由于温室效应，CO_2的捕获和转化的科学问题是非常重要的科学问题。研究表明，高岭土对 CO_2具有良好的吸附能力，通常 CO_2在高岭土表面是线性吸附过程，但是到超临界区域后 CO_2的吸附量急剧上升[3,4]。同时，研究表明金属掺杂对材料的结构和性能影响较大，Cr、Mn、Fe 等元素掺杂能显著改变高岭土的电子结构及晶胞体积。虽然针对金属元素掺杂高岭土的电子结构及电子态等相关的理论研究已有报道，但是 CO_2在天然高岭土，金属元素掺杂的高岭土表面的吸附行为却鲜见相关研究。明确 CO_2在天然或金属元素掺杂的高岭土表面的吸附行为，对开发基于高岭土的 CO_2固化、存储材料具有重要的理论及现实意义。

计算模型及参数：通过构建 CO_2与高岭土相互作用的模型，在 CO_2上方添加厚度为 20Å 的真空层规避周期性边界效应。本研究涉及的高岭土固化 CO_2的球棍模型见图 1。采用第一性原理计算程序 $Dmol^3$，该程序基于密度泛函理论，其中交换关联能部分采用数值基组进行近似。

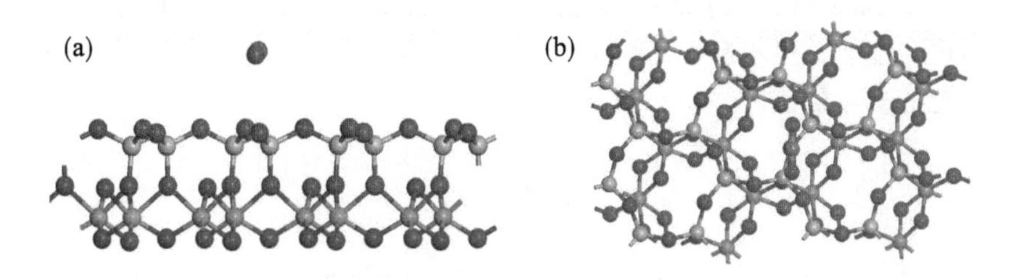

图 1　CO_2分子与高岭土表面相互作用的俯视（a）及侧视（b）图

本研究采用的基于 DNP 极化函数的双数值基组与 Gaussian 软件的 6－31G* * 基组相媲美。计算体系的原子核的核电子采用密度泛函的半核赝势进行处理。电子交换关联能采用 PBE 和 GGA 方法处理。

第一性原理计算表明，CO_2在天然的及金属元素（Ni、Co、Fe 和 Mn）掺杂的高岭土表面都

* 通信作者：张红平；E-mail：zhp1006@126.com；手机号：18781121580。西南科技大学龙山人才计划（18lzx447）、973 项目（2014CB8460003）、四川省“千人计划”项目、河北省青年拔尖人才项目资助。

表现出良好的吸附行为。由表 1 可知，单个 CO_2 分子在不同的高岭土表面的吸附能都维持在 −0.32 eV 左右，说明金属元素掺杂不会改变高岭土固化 CO_2 的能力。

表 1　不同种类高岭土与 CO_2 分子间的吸附能

项目	不同种类高岭土				
	纯的	Ni 掺杂	Co 掺杂	Fe 掺杂	Mn 掺杂
总能	−13 994.365 68	−14 114.881 51	−14 065.952 07	−14 039.326 46	−14 015.758 75
表面能	−13 805.874 79	−13 926.390 83	−13 877.461 11	−13 850.835 54	−13 827.267 84
CO_2 能量	−188.479 113	−188.479 011	−188.479 12	−188.479 121	−188.479 116
E_{ads}（eV）	−0.32	−0.32	−0.32	−0.32	−0.32

图 2 为 CO_2 在天然高岭土表面吸附的电子差分密度等值面图，图 2 可直观地看到 CO_2 在高岭土表面的吸附相对比较牢固，体系稳定后电子主要从高岭土表面向二氧化碳分子上转移。

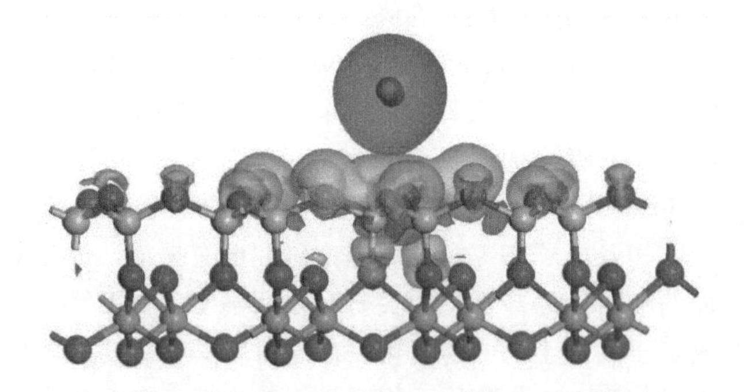

图 2　CO_2 固化在高岭土表面的电子差分密度等值面图

注：蓝色代表电子富集，黄色代表电子流失，等值面值为 0.005

综上所述，CO_2 在高岭土表面能够稳定地吸附，过渡金属（Ni、Co、Fe 和 Mn）掺杂能有效调控高岭土的结构和性能，但不影响高岭土对 CO_2 的吸附能力，该研究对设计基于高岭土的 CO_2 固化材料具有一定的参考价值。

参考文献

[1] Lvov Y M，Devilliers M M，Fakhrullin R F. 2016. The application of halloysite tubule nanoclay in drug delivery [J]. Expert Option on Drug Delivery，13：977 - 986.

[2] Golestani B，Nam B H，Nejad F M，et al. 2015. Nanoclay application to asphalt concrete：Characterization of polymer and linear nanocomposite-modified asphalt binder and mixture [J]. Construction and Building Materials，91：32 - 38.

[3] Schaef H T，Glezakou V A，Owen A T，et al. 2014. Surface condensation of CO_2 onto kaolinite [J]. Environmental Science Technology Letters，1：142 - 145.

[4] Chen Y H，Lu D L. 2015. CO_2 capture by kaolinite and its adsorption mechanism [J]. Applied Clay Science，104：221 - 228.

高温下地球深部矿物中的缺陷氢

杨　燕* 夏群科

（浙江大学地球科学学院，杭州 310008）

表层地球和深部地球之间的相互影响和耦合关系是当今地球系统科学研究的前沿。地球深部是挥发分的主要储库，挥发分在深部地球和地表层圈之间的循环影响着星球的气候、宜居性及其演化。因此，研究高温高压下地球深部硅酸盐矿物中的缺陷氢对理解其在地球深部的循环机制，以及对认识表生环境的演化和生命起源有着重要的意义。

天然和合成样品的研究表明名义上无水矿物（NAMs）是地球深部最主要的水储库[1]。这些以缺陷氢（H）形式存在的水，显著影响着矿物的结构和性质，以及地球深部的物理化学性质、重大地质现象和过程[2,3]，因此，NAMs 中水的发现是 20 世纪 90 年代以来地球科学领域很重要的进展之一。尽管 NAMs 中水的含量及其宏观效应方面取得了丰硕成果，但是水在 NAMs 中的结合机制尚未得到清晰的认识，而这是理解水的宏观效应的基本依据[4,5]。目前对 NAMs 中的缺陷氢的认识，少见原位高温观察。然而，地球深部是高温高压的环境，并且氢具有极高的活动性，因此，为了认识高温下地球深部矿物中的缺陷氢，该研究对长石、单斜辉石、斜方辉石、金红石等矿物进行了原位高温红外光谱分析[6]。

实验用红外光谱仪为尼高力 iS50FTIR，配接红外显微镜。测量范围是 4 000～400 cm^{-1}。原位高温分析使用的是配接在红外显微镜上的 Linkam FTIR600 冷热台和 Instec HS 系列高热台。从 －100 ℃每隔 50～100 ℃升到 1 000 ℃，升温速率是 10～15 ℃/min。每一个温度下都测一次背景。详细观察矿物中同一个分析区域在不同温度下红外谱图的变化。所有实验在中国科学技术大学地球和空间科学学院红外光谱实验室以及浙江大学地球科学学院的红外光谱实验室完成。

实验结果表明，500 ℃以下，未见显著脱氢，但是长石和金红石中的缺陷氢在晶格内部发生质子迁移（图 1）。相对于长石和金红石，单斜辉石和斜方辉石中的缺陷氢主要存在于 Si 空位或 M 空位。随着温度升高至 1 000 ℃，未见 H 质子在不同位置之间的迁移，但是发生了脱氢，不同占位的 H 的热稳定性也不同，并且，在高温脱氢的过程中会产生新的占位的缺陷氢（图 2）。因此，地球深部矿物中的缺陷氢在高温下的状态不同于常温，在解释缺陷氢对矿物乃至地球深部宏观效应时，需要考虑其在高温下的状态。

以上工作利用高温红外光谱技术，原位观察地球深部矿物中的缺陷氢在高温下的状态，为在原子尺度上认识地球深部挥发分的宏观效应和循环机制提供了新的制约。

* 通信作者：杨燕；E-mail：yanyang2005@zju. edu. cn；手机号：18334357859。浙江省自然科学基金（LY18D020001）资助项目。

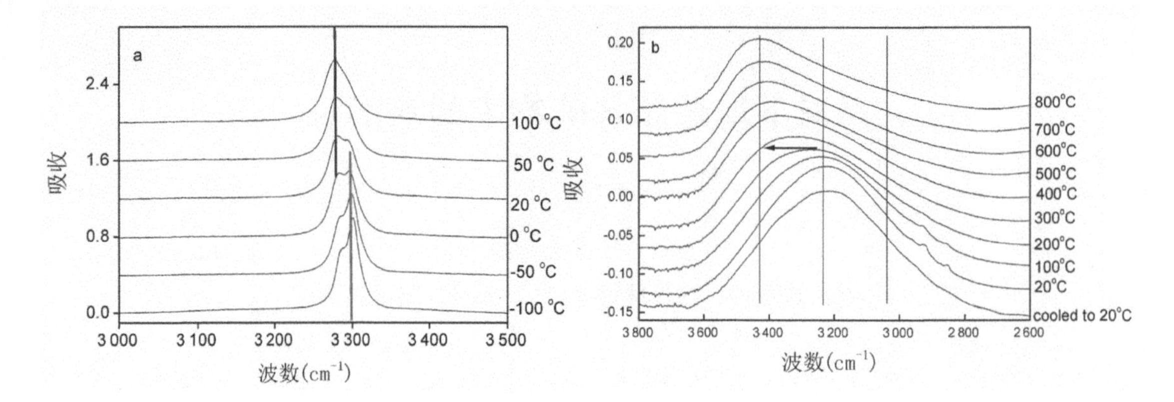

图 1　金红石（a）和歪长石（b）的原位高温红外光谱（修改已发表工作[7,8]）

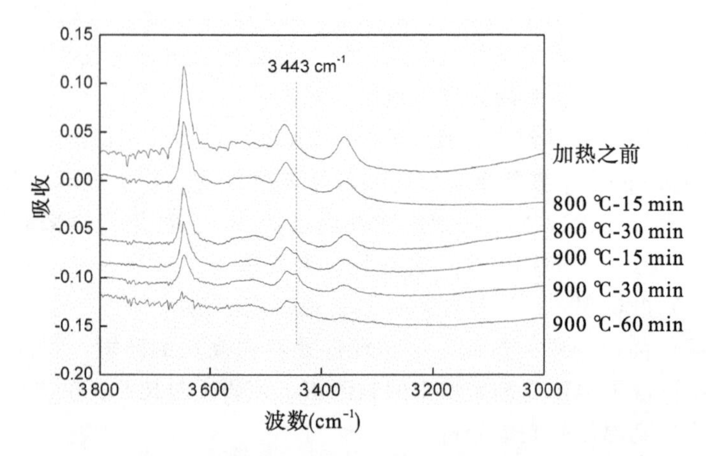

图 2　单斜辉石的等热红外光谱（高温下脱氢—产生氢的过程）

参考文献

[1] Peslier A H，Scönbächler M，Busemann H，et al. 2017. Water in the Earth’s interior：Distribution and Origin [J]. Space Science Reviews，212：743 - 810.

[2] 杨晓志，李岩. 2016. 高温高压实验和硅酸盐地幔中的水 [J]. 中国科学，46：287 - 300.

[3] 夏群科. 2017. 地幔中的水与重大地质现象和过程 [J]. 自然杂志，39：1 - 4.

[4] Karato S. 2015. Some notes on hydrogen-related point defects and their role in the isotope exchange and electrical conductivity in olivine [J]. Physics of the Earth and Planetary Interiors，248：94 - 98.

[5] Jones A. 2016. Proton conduction and hydrogen diffusion in olivine：an attempt to reconcile laboratory and field observations and implications for the role of grain boundary diffusion in enhancing conductivity [J]. Physics and Chemistry of Minerals，43：237 - 265.

[6] 杨燕. 2017. 名义上无水矿物的原位高温分子光谱 [J]. 矿物岩石地球化学通报，36：48 - 58.

[7] Yang Y，Xia Q，Feng M，et al. 2011. In situ varying temperature FTIR experiment study of rutile [J]. American Mineralogist，96：1851 - 1855.

[8] Liu W D，Yang Y，Zhu K Y，et al. 2018. Temperature dependences of hydrous species in feldspars [J]. Physics and Chemistry of Minerals，45：609 - 620.

铀酰—重金属离子在铁酸盐—蒙脱石界面的竞争吸附

蒋小强[1]　聂嘉男[1]　顾晓滨[2]　董发勤[1]　宋绵新[1]　尚丽平[1]　邓　琥[1]
李伟民[1]　王　彬[1]　何辉超[1]　霍婷婷[1]　邹　浩[1]　边　亮[1,2*]
（1. 西南科技大学固体废物处理与资源化教育部重点实验室，绵阳 621010；
2. 河北地质大学宝石与材料工艺学院，石家庄 050000）

随着铀矿石开发和核废物的处理技术的发展，铀盐的循环和可持续性发展逐渐引起社会关注。在酸性氧化环境下，铀酰离子（UO_2^{2+}）是铀的最稳定形态，而吸附则被认为是从水溶液中提取铀酰的最主要方法[1]。其中，影响铀酰吸附的因素较多，例如 pH 值、腐殖酸等。据报道，核裂变产物[2]、矿物开采或铀矿床中铀酰离子往往伴随重金属（Cd、Cu、Hg、Pb、Zn 和 As）释放到环境中[3]。此外，天然水和海水中也含有低浓度的金属阳离子（Sr^{2+}、Co^{2+}、Cr^{3+}、Cd^{2+}、Zn^{2+}、Ni^{2+}）与铀酰离子[4]。研究表明，共存重金属阳离子与铀酰之间会发生竞争吸附反应，基于铀超标会破坏水体酸碱平衡，污染土壤和危害人体健康。因此，迫切需要研究环境中重金属和铀酰的竞争吸附机制。

前期研究表明，蒙脱石作为传统吸附剂具有优异的吸附能力，适用于去除水中的金属离子，而磁性纳米粒子尤其是 Fe_3O_4 具有成本低、通过磁场易分离的特性。因此，该研究采用原位共沉淀法制备 MFe_2O_4（M＝Mn，Fe，Zn，Co，Ni）-蒙脱石粉体，即将一定量蒙脱石加入 $MCl_2\cdot nH_2O$（M＝Fe，Mn，Cu，Zn，Ni）和 $FeCl_3\cdot 6H_2O$（摩尔比 1∶2）混合溶液中，80 ℃条件下，滴加 NaOH 溶液至碱性分离得样品。吸附实验通过称取定量的产物分别加入一定体积的 $UO_2^{2+}-X^{n+}$（X＝Rb^+，Sr^{2+}，Cr^{3+}，Mn^{2+}，Ni^{2+}，Zn^{2+} 和 Cd^{2+}）混合溶液中，持续振荡 24 h 后测试各离子浓度。

通过 XRD、FT-IR、拉曼光谱对所得材料进行表征，结果显示 MFe_2O_4 成功负载到蒙脱石，并通过 Si—O—O—M 和 Al—O—O—M 化合键相连接。采用 SEM、HR-TEM 分析材料形貌，发现 MFe_2O_4 主要结合在蒙脱石的边缘位置，介于 Fe_3O_4 晶格能较大，其在蒙脱石表面出现明显的团聚现象，而 M^{2+} 作为介质，其团聚现象减弱但 MFe_2O_4 粒子粒径增大。XPS、EDX 图谱显示，M^{2+} 均匀地分布在蒙脱石的表面，具有更强的表面效应，并通过电位测试和太赫兹光谱加以说明 MFe_2O_4 在蒙脱石边缘的高分散性和电子转移特性。

吸附试验结果显示（图 1），X^{n+} 随着初始浓度的增加平衡吸附量增加。通过 Langmuir、Freundlich 和 Dubinin-Radushkevich（DR）方程对吸附数据拟合，X^{n+} 的 Langmuir、Freundlich 线性拟合较好（R^2＝0.775～0.994），说明重金属离子主要发生单层吸附在 MFe_2O_4-蒙脱石的表面，而 UO_2^{2+} 和 Cr^{3+} 更加符合 DR 方程，说明其稳定吸附在蒙脱石的层间区域，因为 X^{n+} 与·OH 结合的

* 通信作者：边亮；E-mail：bianliang@swust. edu. cn；手机号：13795778099。973 项目（2014CB8460003）、西南科技大学龙山基金项目（17QR004）、四川省“千人计划”项目、河北省青年拔尖人才项目。

离子单层阻止 UO_2^{2+} 被吸收，说明 X^{n+} 和 UO_2^{2+} 在材料表面发生竞争吸附。同时 EDX 结果显示，UO_2^{2+} 稳定吸附在 MFe_2O_4-蒙脱石界面区域。通过循环伏安法研究吸附过程间的电荷转变，表明 UO^{2+} 与 Cr^{3+} 存在活性电子转移，UO_2^{2+} 的选择性吸附和电荷转移主要与 MFe_2O_4-蒙脱石界面有关。

综上所述，本文研究了水溶液中重金属和铀酰离子在 MFe_2O_4-蒙脱石表面的竞争吸附。AlO－OM和 SiO－OM 键的存在导致界面的活性空穴增强，引起活性电子转移到界面，改善了铀酰的选择性去除。其中，UO_2^{2+} - X^{n+} 混合水溶液中铀酰吸附量可达 89.25 mg/g，证实了该材料从 X^{n+} 水溶液中选择性去除 UO_2^{2+} 的潜在价值，该技术有望应用于矿山铀废水处理、海水提铀等方面。

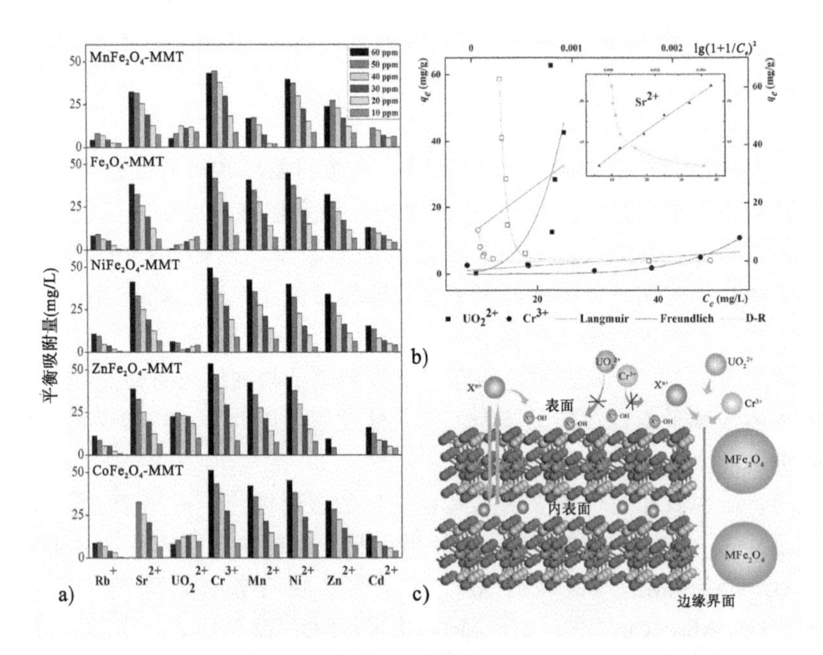

图 1 铁酸盐—蒙脱石吸附铀酰—重金属离子的平衡吸附量图（a）、$ZnFe_2O_4$-蒙脱石对于 UO_2^{2+}、Cr^{3+} 和 Sr^{2+} 线性拟合曲线（b）、MFe_2O_4-蒙脱石吸附位点图（c）

参考文献

［1］Zhang X，Wang J，Li R，et al. 2013. Removal of uranium（Ⅵ）from aqueous solutions by surface modified magnetic Fe_3O_4 particles［J］. New Journal of Chemistry，37：3914－3919.

［2］Burns P C，Ewing R C，Navrotsky A. 2012. Nuclear fuel in a reactor accident［J］. Science，335：1184.

［3］Wang F，Liu Q，Li R，et al. 2016. Selective adsorption of uranium（Ⅵ）onto prismatic sulfides from aqueous solution［J］. Colloids & Surfaces A Physicochemical & Engineering Aspects，490：215－221.

［4］Voegelin A，Vulava V M，Kretzschmar R. 2001. Reaction-based model describing competitive sorption and transport of Cd，Zn，and Ni in an acidic soil［J］. Environmental Science & Technology，35：1651－1657.

钾长石水热碱法分解及制备沸石粉体

刘昶江[1,2]　马鸿文[1*]

（1. 中国地质大学（北京）材料科学与工程学院，北京 100083；

2. 河北地质大学宝石与材料工艺学院，石家庄 050031）

中国的水溶性钾资源（KCl、K_2SO_4）资源短缺且分布不均，钾肥生产长期依赖进口，严重威胁中国农业的可持续发展。以钾长石为主要矿物组成的富钾岩石资源在我国储量丰富且分布广泛，开发这类非水溶性钾资源对于解决上述问题意义重大[1]。

钾长石（$KAlSi_3O_8$）属架状硅酸盐，钾元素被固封于骨架中，其加工技术的关键是设法破坏其硅铝骨架，使钾以离子形式溶出，以进一步制备可溶性钾盐。钾长石加工方法多样，其中水热碱法具有能耗低、资源利用率高、生产过程清洁等特点[2,3]。钾长石的理论化学组成中 K_2O 仅占16.9%，其加工过程应兼顾剩余 Al_2O_3、SiO_2 组分的综合利用。钾长石的水热碱法分解体系与传统沸石分子筛的合成体系相近，故该研究旨在探索 $NaOH-H_2O$ 水热体系中钾长石中钾的溶出行为，同时综合利用其硅铝组分制备沸石粉体。

首先对反应体系不同条件下的相平衡产物进行模拟计算，再根据模拟结果，配制混合料浆、选定优化反应条件进行验证。实验在容积为 1 L 的高压反应釜中进行，反应温度为 200～280 ℃。

钾长石水热分解—沸石合成属溶解—沉淀过程，涉及到体系内原矿物的溶解和新化合物的形成，待体系达到固—液化学平衡时，可通过相平衡计算预测其最终产物[4]。结果表明，不同条件下可能出现的产物分别为方沸石、羟钙霞石或二者混合物（图 1c）。据此，控制相应水热反应条件，可分别制备羟钙霞石（图 1a）及方沸石（图 1d）两种沸石粉体。羟钙霞石粒度分布为：D（10）=3.0 μm，D（50）=21.9 μm，D（90）=45.7 μm；方沸石粒度分布为：D（10）=3.0 μm，D（50）=18.1 μm，D（90）=84.5 μm。二者理论晶体化学组成分别为 $Na_8Al_6Si_6O_{24}(OH)_2 \cdot H_2O$ 和 $NaAlSi_2O_6 \cdot H_2O$，结构中均不含 K 元素，故通过一步水热过程可同时实现钾的溶出及沸石化合物的制备。增加碱液浓度有助于 K^+ 的快速溶出，当 NaOH 浓度为 3.0 mol/kg 时，240 ℃下反应 4 h 即可实现 K^+ 的近乎全部溶出（图 1b）。

动力学分析表明，钾长石水热分解反应表观活化能为 46.61 kJ/mol，反应级数为 2.61。自然界中钾长石的风化可为土壤提供少量 K^+，但其过程漫长，反应级数约为 0.5。通过水热碱法处理可将该过程缩短至数小时内，同时副产沸石粉体，对于非水溶性钾资源的综合、集约利用与工程实践具有指导意义。

作者简介：刘昶江；E-mail：liucj2017@hgu. edu. cn；手机号：18332391133。*通信作者，马鸿文，E-mail：mahw@cugb. edu. cn。国家科技支撑计划课题（2006BAD10B04）、中地质调查项目（12120113087700）、中央高校基本科研业务费（2652015371）资助。

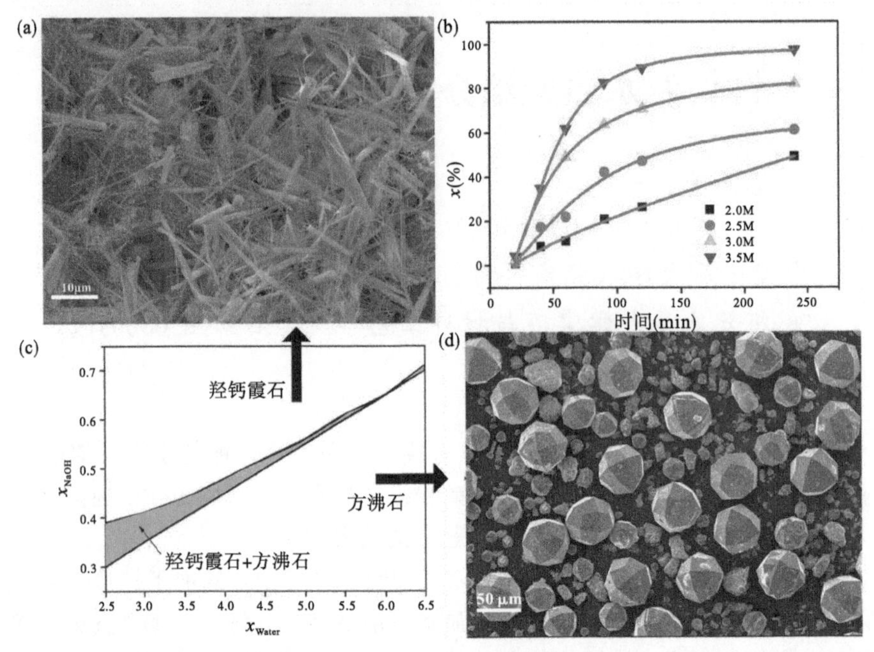

图 1　羟钙霞石的 SEM 照片（a）；不同浓度 NaOH 溶液中 K^+ 的溶出率随时间变化（M——mol/kg）（b）；钾长石-NaOH-H_2O 体系相平衡模拟结果（x_{NaOH}——NaOH/钾长石质量比，x_{Water}——H_2O/钾长石质量比）（c）；方沸石的 SEM 照片（d）

参考文献

[1] 马鸿文，苏双青，刘浩，等. 2010. 中国钾资源与钾盐工业可持续发展［J］. 地学前缘，17：294-310.

[2] Ma H W，Yang J，Su S Q，et al. 2015. 20 years advances in preparation of potassium salts from potassic rocks：a review［J］. Acta Geologica Sinica，89：2058-2071.

[3] 马鸿文，杨静，张盼，等. 2017. 中国富钾正长岩资源与水热碱法制取钾盐反应原理［J］. 地学前缘，doi：10. 13745/j. esf. yx. 2017-7-27.

[4] 刘昶江，马鸿文，张盼. 2018. 富钾正长岩水热分解生成沸石反应热力学［J］. 物理化学学报 34，168-176.

蒙脱石吸附重金属离子的解吸附行为与机制

蔡元峰* 张晓科 朱 强 白利娟 麦地娜·努尔太
麦尔旦·图尔荪 艾米尔丁·艾尔肯
（内生金属成矿机制研究国家重点实验室，
南京大学地球科学与工程学院，南京 210023）

重金属离子污染是困扰当前农业生产的重大的生态安全问题。重金属离子在土壤圈和水圈的迁移受诸如金属离子溶解度、液态水的 pH 值和 Eh 以及土壤的物理化学条件影响和制约。重金属离子在土壤中主要与黏土矿物发生相互作用而导致吸附和解吸附，其行为和机制已经被广泛研究。前人的研究多以受污染的水体和土壤为研究对象，尝试解决环境污染问题，而对吸附了重金属离子的黏土矿物在水圈中的解吸附行为不很关注，导致学者们忽视解吸附行为和机制的主要原因可能源于重金属离子在水体中的浓度极低，即便是严重污染的水体或土壤其重金属离子浓度依然很低。针对吸附了重金属离子的黏土矿物的解吸附问题，笔者们遴选了铬、锰、镍等很普遍的且易于表征的重金属离子，以蒙脱石这一具有极强的离子交换能力和吸附能力的黏土矿物为介质，开展了不同浓度、不同 pH 值的解吸附实验，以期探明其解吸附行为和解吸附机制。

吸附了铬、镍和锰的蒙脱石在 pH 值为 3 和 11 的代表性酸性和碱性条件下的解吸附曲线分别如图 1 和图 2 所示。3 种离子的解吸附行为几乎完全相同，均表现出在短时间内溶液中的离子浓度急剧增加，在 20 h 左右达到解吸附的峰值浓度，随后离子浓度缓慢降低，直至稳定在某一平衡浓度，如铬、锰和镍分别稳定在 25、21 和 19.5 mg/kg 左右。这也与大多数的吸附极低浓度重金属离子的蒙脱石的解吸附重金属离子的行为一致[1]。在碱性条件下，溶液中的 3 种金属离子的浓度经历短暂的迅速增加（1 h 内）至 1.62 mg/kg 左右，然后其浓度逐渐降低，最终稳定在 0.1 mg/kg 左右(图 2)。

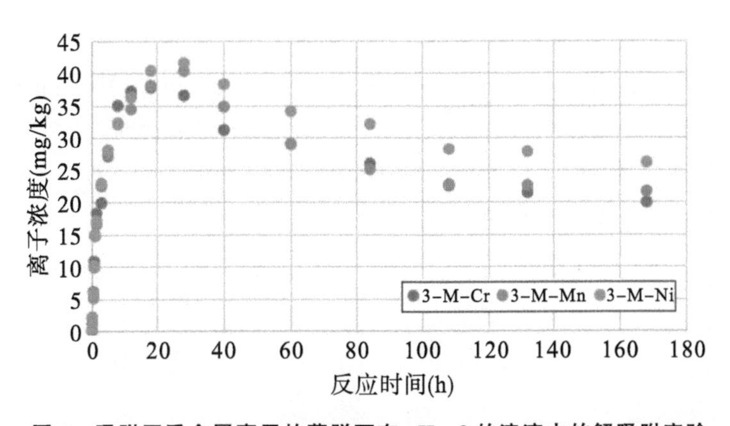

图 1　吸附了重金属离子的蒙脱石在 pH=3 的溶液中的解吸附实验

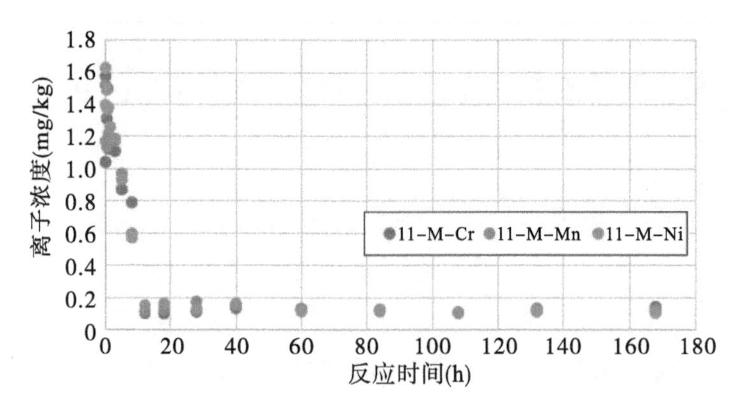

图 2　吸附了重金属离子的蒙脱石在 pH=11 的溶液中的解吸附实验

在解吸附达到平衡后，测得溶液的 pH 值分别为 3.40±0.05 和 4.80±0.05，而溶液中的重金属离子的浓度却相差近 200 倍。

pH=3　$Me-Mmt+nH^{+} \Leftrightarrow H^{+}-Mmt+Me^{n+1}$ ……………………………………（1）

pH=11　$Me-Mmt+nOH^{-} \Leftrightarrow Me(OH)_{n}\downarrow+[W-Mmt]^{-}$
$Me-Mmt+nH_2O \Leftrightarrow Me(OH)_{l}\downarrow+xH^{+}-Mmt+(n-l-x)H^{+}$ ………………（2）

式（1）给出了吸附了重金属离子的蒙脱石在酸性溶液中 H^{+} 交换吸附在颗粒表面和部分层间域中阳离子的过程，该过程是一个耗酸的过程，因而导致最终溶液的 pH 值升高至 3.40 左右。而在碱性条件下，OH^{-} 首先夺取了吸附在蒙脱石表面的阳离子并形成氢氧化物沉淀，同时溶液因为消耗了 OH^{-} 而向酸性转化，也引起了水的水解，形成的 H^{+} 又置换了吸附在表面的和部分层间域的重金属阳离子，也以氢氧化物沉淀的形式存在于蒙脱石颗粒的表面或粒间，最终溶液的 pH 值稳定在 4.80 左右的弱酸性。在酸性或碱性条件下的解吸附受特定条件下重金属离子水解反应发生的难易程度，可由它们的溶度积常数（Ksp）大小所反映。据有关资料[2]，Cr^{3+}、Ni^{2+} 和 Mn^{2+} 的溶度积常数分别为 6.3×10^{-30}、6.0×10^{-16} 和 1.6×10^{-13}。根据上述溶度积常数可计算出相应的氢氧化物的溶解度分别为 0.005 15、12.7、3.04 mg/kg。实验结果表明在碱性条件下，除 Cr^{3+} 外，Ni^{2+}、Mn^{2+} 均未达到平衡，也说明蒙脱石对这 2 种离子的吸附能力要强于 Cr^{3+}。

参考文献

［1］Zhu J，Cozzolino V，Pigna M，et al. 2011. Sorption of Cu，Pb and Cr on Na-montmorillonite：competition and effect of major elements ［J］. Chemosphere，84：484－489.

［2］Sillen L G，Martell A E. 1964. Stability constants of metal-ion complexes. The Chemical Society，London，Lange's Handbook，pps. 8-6 to 8－11.

Al^{3+}对混层六方水钠锰矿结构和性质的影响——与Fe^{3+}对比

冯雄汉[1*]　殷　辉[1]　Kideok D. Kwon[2]　王小明[1]
严玉鹏[1]　谭文峰[1]　刘　凡[1]　张　静[3]
（1. 华中农业大学，资源与环境学院，农业部长江中下游
耕地保育重点实验室，武汉 430070；
2. Department of Geology，Kangwon National
University，Chuncheon 24347，Korea；
3. 中国科学院高能物理研究所，北京 100039）

氧化锰矿物与过渡金属的相互作用是自然界普遍发生的现象，土壤、沉积物和大洋锰结核中常常富集多种过渡金属元素。六方水钠锰矿是环境中广泛存在、活性最强的氧化锰矿物。过渡金属进入水钠锰矿结构中，既影响这些元素的存在形态，也会引起矿物结构和性质的显著变化，进而改变其对环境中的重金属和有机污染物地球化学行为的影响。铝、铁是地壳中含量最为丰富的两种金属元素。铁氧化物常常与锰氧化物紧密伴生，且较高含量的Fe^{3+}可以进入六方水钠锰矿结构中[1]，但是据文献报道仅有少量Al^{3+}能进入水钠锰矿结构中，其机制尚不清楚。

该研究在采用浓盐酸还原煮沸高锰酸钾制备混层六方水钠锰矿过程中加入Al^{3+}或Fe^{3+}，合成了不同Al^{3+}或Fe^{3+}含量的混层六方水钠锰矿。应用粉末X射线衍射（XRD）、化学分析、傅里叶转换红外吸收光谱（FTIR）、X射线光电子能谱（XPS）、X射线吸收精细结构光谱（又称X射线吸收近边结构光谱，XANES）、扩展X射线吸收精细结构光谱（EXAFS）等手段表征产物晶体结构、微观形貌和化学组成，并与过渡金属离子Fe对水钠锰矿结构和性质的影响进行了对比。使用密度函数理论（DFT）对Al、Fe在水钠锰矿结构中的可能配位、自旋态、晶胞参数等进行了计算。并考察了Al掺杂水钠锰矿对重金属离子（Pb^{2+}、Zn^{2+}）等吸附行为。

粉晶X射线衍射分析表明，样品均为单相混层六方水钠锰矿。尽管初始Al/（Al＋Mn）和Fe/（Fe＋Mn)摩尔比均高达0.20，但是最终Al/（Al＋Mn）和Fe/（Fe＋Mn）摩尔比分别为～0.07和～0.20。Al^{3+}或Fe^{3+}的引入显著改变了水钠锰矿的物理化学特征：（1）矿物结晶度减弱，如沿c轴方向晶体的厚度和a－b方向上的相干散射尺寸（CSD）均减小，比表面积和总羟基含量增加，但是空位处羟基含量减少，边面位点增多。（2）Mn K边XANES的Combo拟合分析表明，Al或Fe的引入使样品中锰平均氧化度降低；Mn K边EXAFS分析表明，随着Al或Fe含量的增加，水钠锰矿锰氧八面体层内共边Mn－Mn距离减小，而共角Mn－Mn距离随Al含量增加而减小，而随Fe含量增加而增加（图1左）。（3）粉末XRD模拟和DFT计算均表明，a－b方向上晶胞参数随Al含量增加而减小，而随Fe含量先减小后增加。结合对高、低自旋Fe的总能量和相

* 通信作者：冯雄汉，E-mail：fxh73@mail. hzau. edu. cn。

对于锰氧八面体层内 Mn 位置的计算，认为 Fe 可能以高自旋态存在于水钠锰矿结构中。（4）Fe K边 EXAFS 分析表明，在含 Fe 水钠锰矿结构中，32%～50%的 Fe^{3+} 位于锰氧八面体层内（图 1 右），而其余 Fe^{3+} 则吸附于空位上下方或边面位点。（5）Al^{3+} 主要以三齿共角（TCS）络合物的形式吸附于空位上下方。所得水钠锰矿样品对 Pb^{2+}、Zn^{2+} 的等温吸附实验表明，含 Al 或 Fe 水钠锰矿对 Pb^{2+} 的吸附量显著增加，而对 Zn^{2+} 的吸附量降低。研究加深了对过渡金属与氧化锰矿物相互作用机制和氧化锰矿物环境地球化学行为的理解，亦为开发大洋多金属锰结核资源提供参考[2]。

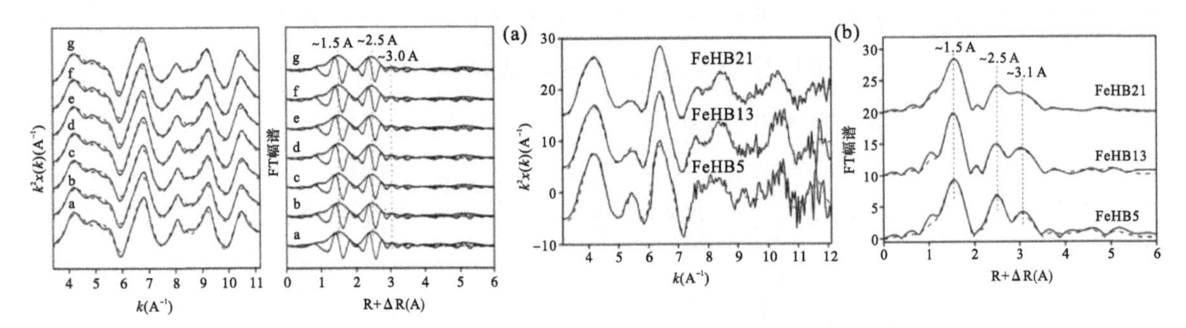

图 1　Mn（左）与 Fe（右）的 K 边 EXAFS 谱及其傅里叶变换谱

注：蓝线是实验谱，红色实线或虚线是拟合谱。a. HBir；b. AlHB4；c. AlHB5；d. AlHB7；e. FeHB5；f. FeHB13；g. FeHB21

参考文献

[1] Yin H，Liu F，Feng X H，et al. 2013. Effects of Fe doping on thestructures and properties of hexagonal birnessites-comparisonwith Co and Ni doping [J]. Geochim. Cosmochim. Acta，117：1－15.

[2] Yin H，Kwon K D，Lee J Y，et al. 2017. Distinct effects of Al^{3+} doping on the structure and properties of hexagonal turbostraticbirnessite：a comparison with Fe^{3+} doping [J]. Geochim. Cosmochim. Acta，208：268－284.

钙—蒙脱石吸附氨基酸的第一性原理研究

李海龙[1]　边　亮[1*]　顾晓滨[2]　董发勤[1]　李伟民[1]　宋绵新[1]　邹　浩[1]　霍婷婷[1]
（1. 西南科技大学固体废物处理与资源化教育部重点实验室，绵阳 621010；
2. 河北地质大学宝石与材料工艺学院，石家庄 050000）

氨基酸的解吸、吸附、选择和缩合等作用是自然环境中的重要过程，影响蛋白质在土壤和沉积物中的转运过程。蒙脱石作为黏土矿物的主要成分具有较高的阳离子交换容量（88～90 Meq/100 g）和表面积（～800 M^2/g），对氨基酸具有很强的吸附能力[1]。蒙脱石与氨基酸的相互作用主要包含阳离子交换作用、静电/范德华力相互作用、氢键、疏水/亲水作用、水桥/阳离子桥等[2]。前期研究表明，氨基酸以中性形式吸附在矿物表面上时，既可以作为中性的分子（H_2N－CH－COOH），也可以作为两性离子（^+H_3N－CH－COO^-）与矿物表面作用。在自然水环境中，研究人员证实了蒙脱石吸附甘氨酸除了阳离子交换作用，还包含静电相互作用（－COO^-基团和Al－O八面体的Al（OH）$_2^+$）和氢键作用（NH_3^+基团和表面氧）[2,3]。尽管已有大量的实验研究黏土矿物表面与氨基酸的吸附作用机制，但是对于氨基酸在水—黏土矿物界面上产生的相互作用机制还处于基础的认知阶段，尤其是在电子级水平上的微观机制研究较为欠缺[4]。因此，系统地研究氨基酸与蒙脱石表面的相互作用机制，有助于更好地理解土壤生态系统安全、地球的生物化学的进化和生命起源。

该研究借助密度泛函理论计算了蒙脱石吸附氨基酸（中性：甘氨酸和丝氨酸；负电性：谷氨酸；正电性：精氨酸）的作用过程。同时，针对随时间的变化的氨基酸与蒙脱石表面轨道耦合转变过程，利用二维相关分析技术（2D－CA技术）定量处理了在能量转变点附近电子轨道的简并与劈裂机制。对氨基酸与蒙脱石表面间可能存在的阳离子交换作用、静电相互作用、水合作用进行了定性和定量的分析。计算结果表明：Ca^{2+}作为电子受体，八面体中的O和Al（Mg或Fe）、氨基酸中的COO^-（H）和NH_3^+作为电子供体。主要的相互作用包含铝氧八面体的离子交换、氨基酸—氧—钙的静电（或范德华力）相互作用、氨基酸—水—氧的亲水作用。结合2D－CA定量技术（图1）分析得到：（1）Al－O八面体通过层间Ca离子与氨基酸发生阳离子交换作用。钙—蒙脱石表面的Al－O八面体作为电子的供体，在价带区域提供了部分电子—空穴（e－h^+）对。由于短程的范德华作用力，部分Ca－s轨道捕获氨基酸中的H－s（－COO－（H）或－NH_3^+）电子占据导带中的空Ca－d^0轨道，使得Al－O八面体优先与－COO－（H）中的OH^-（甘氨酸和丝氨酸）或－NH_3^+中的H^+（谷氨酸和精氨酸）发生相互作用，使得部分Ca^{2+}离子被还原而浸出。（2）H_2O的s（或p）轨道与－NH_3^+（和－COO^-）的p轨道发生sp（或pp）轨道杂化形成水合离子团簇（－NH_3·（H_2O）$^+$和－HOCO·（H_2O）$^-$），通过长程静电（－NH_3^+－

---

* 通信作者：边亮；E-mail：bianliang@swust. edu. cn；手机号：13795778099。973项目（2014CB8460003）、西南科技大学龙山基金项目（17QR004）、四川省“千人计划”项目、河北省青年拔尖人才项目。

H_2O—O或者—COO^-—H_2O—O）作用与Al—O八面体的表面氧相互作用。(3）由于电荷补偿机制，Al—O八面体中的Mg^{2+}杂质抑制氨基酸和钙—蒙脱石间部分电子—空穴（$e-h^+$）的湮灭；同时由于电荷歧化机制，Al—O八面体中的Fe^{2+}杂质增强了部分电子—空穴（$e-h^+$）的湮灭，从而实现对氨基酸与钙—蒙脱石间的静电相互（或范德华力）作用的调控。因此，本研究可为理解氨基酸与黏土矿物界面间相互作用提供一种定量的理论分析技术。

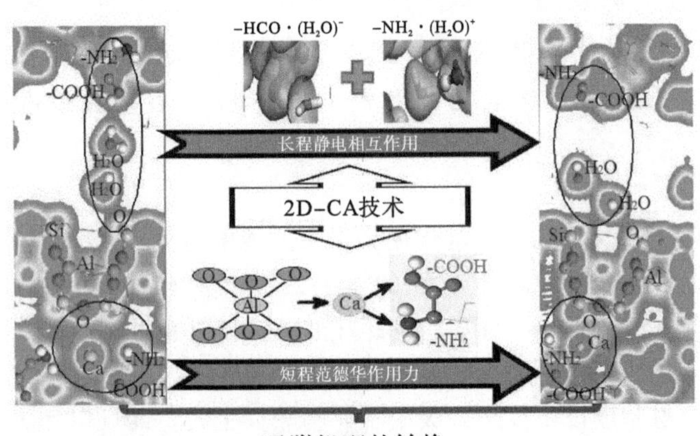

图1　Ca-蒙脱石表面与氨基酸相互作用和研究方法的示意

参考文献

[1] Yu W H, Li N, Tong D S, et al. 2013. Adsorption of proteins and nucleic acids on clay minerals and their interactions: A review [J]. Applied Clay Science, 80-81: 443-452.

[2] Lambert J F. 2008. Adsorption and Polymerization of Amino Acids on Mineral Surfaces: A Review [J]. Orig Life Evol Biosph, 38: 211-242.

[3] Ramos M E, Huertas F J. 2013. Adsorption of glycine on montmorillonite in aqueous solutions [J]. Applied Clay Science, 80-81: 10-17.

[4] Zhao H X, Yang Y, Shu X, et al. 2018. Adsorption of organic molecules on mineral surfaces studied by first-principle calculations: A review [J]. Advances in Colloid and Interface Science, 256: 230-241.

蒙脱石与A549细胞膜磷脂分子的表界面作用

霍婷婷[1,2]　董发勤[1,2*]　边　亮[1,2]　李海龙[1]　宋绵新[1]　贺小春[2]　邓建军[3]

（1. 西南科技大学环境与资源学院，绵阳 621010；
2. 固体废物处理与资源化教育部重点实验室，绵阳 621010；
3. 四川绵阳四〇四医院，绵阳 621000）

黏土矿物，如蒙脱石是风化程度较高、在土壤中分布很广的重要活性组分，通常具有分散程度高、粒度小、比表面大等特点，也是最容易迁移至大气环境中的矿物种类，在大气矿物质中所占比重很大。蒙脱石颗粒在大气中主要以集合体存在，且由于特殊的理化性质，极易吸附大气环境中各种重金属和有机污染。因此，开展关于蒙脱石矿物粉尘，尤其是微纳米尺度的蒙脱石矿尘对人体肺细胞的毒性研究对全面了解 PM2.5 的污染特征及其对人体健康和大气环境的影响具有重要的意义。本文在开展蒙脱石粉尘对人肺上皮细胞 A549 细胞毒性效应研究的前提下，通过实验和模拟计算探讨了蒙脱石引起 A549 细胞膜磷脂结构破坏方式，揭示蒙脱石对细胞膜损伤的分子机理。

实验所用蒙脱石矿物采自新疆阿尔泰地区，经湿法分选提纯，D_{50}为 0.273 μm。受试细胞株为 A549 人肺腺癌上皮细胞株，Ⅱ型肺泡上皮细胞表型。采用 MTT 法辅以激光共聚焦显微镜 Calcein- AM/PI 双荧光标记和 SEM 观察 A549 细胞存活率和形貌变化。采用 DiI 对细胞膜磷脂分子进行荧光标记，FTIR 和 Peakfit 对细胞膜磷脂分子的振动变化进行解析。并对蒙脱石和磷酸甘油酯（$C_{19}H_{37}O_8P$）进行建模优化，蒙特卡洛和分子动力学模拟计算蒙脱石与磷脂分子之间的吸附能、作用基团及作用方式。

实验结果表明，蒙脱石粉尘浓度为 25. 0 μg/mL 作用 3 h 时即对 A549 细胞活性产生显著影响（85.21%），且随着作用浓度升高，细胞死亡增加，呈现明显的剂量和时间效应。蒙脱石紧紧包覆在细胞膜周围（图 1－A、1－C），引起细胞膜磷脂分子荧光分散度变化（图 1－B），表现为磷脂分子局部凝集反应，细胞膜结构破坏，胞核碎片化（25 μg/mL，24 h）和显著的 DNA 损伤[1,2]。蒙脱石粉尘在细胞膜周围的大量堆积直接导致约 90%细胞膜起泡，出现膜凸起和凹陷的动态变化（图 1－D）。FTIR 分析发现与蒙脱石粉尘作用后 1 037 cm^{-1}处 ν_{as}（PO_2^-）蓝移；磷脂头部的 ν_s（P＝O）的 1 063 和 1 091 cm^{-1}峰分裂为 1 052、1 078 和 1 095 cm^{-1}三个峰，且峰位置和相对强度发生显著变化；1 233 cm^{-1}处振动峰蓝移，说明蒙脱石的作用引起磷脂分子极性头部磷酯酰基团的振动变化[3]。

蒙特卡洛计算结果表明，蒙脱石与磷脂分子之间发生较为强烈的吸附作用，吸附能为

---

* 通信作者：董发勤；E-mail：fqdong@swust. edu. cn；联系电话：0816-6089013。国家自然科学基金（41602033、41130746、41572025、41472046），四川省应用基础研究项目重点项目（2018JY0426），西南科技大学龙山人才计划（18lzx651）。

522.40 kcal/mol，解离能为 490.57 kcal/mol。利用分子动力学对体系结构进行弛豫和优化，计算得到硅氧四面体与磷脂分子的作用距离为 3.30～4.50Å，铝氧八面体与磷脂分子的作用距离为 2.90～4.10Å（图 1－E)。表明蒙脱石与磷脂分子之间以短程范德华力作用为主。蒙脱石与磷脂分子间的主要作用发生在极性亲水端，磷脂分子通过磷酸甘油酯基（－COOH）和磷酸头部的羟基（－OH）中裸露 H 原子直接和蒙脱石中硅氧四面体和铝氧八面体的发生共用氧作用。

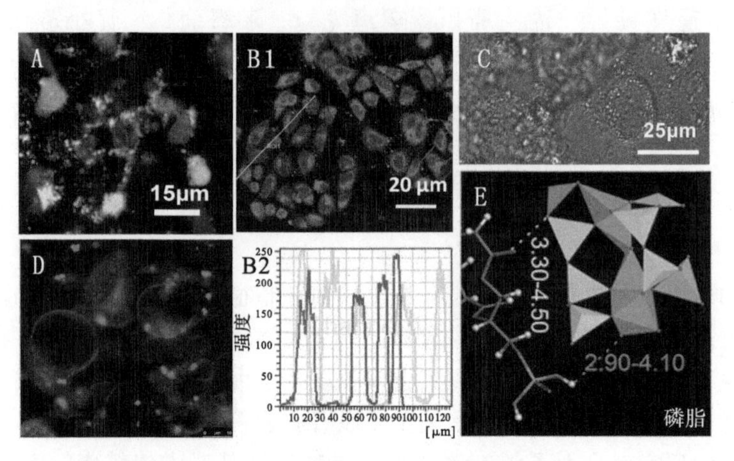

图 1　A549 细胞与蒙脱石粉尘作用后的形貌

注：A. Calcein－AM/PI 标记后的 A549 细胞荧光图片；B1. DiI 标记的 A549 细胞形态；B2. 膜磷脂荧光变化；C. DIC 图片；D. DiI 标记细胞膜，细胞膜表面凸起；E. 蒙脱石与磷脂分子的作用距离

综合实验和模拟计算结果，蒙脱石硅氧四面体和铝氧八面体中的氧原子主要与 A549 细胞膜磷脂分子极性头部的－COOH 和－OH 之间发生共用氧作用，引起磷脂分子振动的变化，使细胞膜本身由磷脂分子双亲性控制的双层膜结构受到影响，出现膜磷脂分子的翻转或凝集，细胞膜流动性变差，甚至膜磷脂分子结构重组等现象，最终诱使细胞膜功能丧失、细胞凋亡或坏死。

参考文献

［1］Shi B，Yun K S，Hassanali A A，et al. 2015. DNA binding to the silica surface［J］. Journal of Physical Chemistry B，119（34）：11030－11040.

［2］Huo T，Dong F，Wang M，et al. 2015. Cytotoxicity of quartz and montmorillonite in human lung epithelial cells（A549）［J］. Proceedings of the 11th International Congress for Applied Mineralogy（ICAM），Springer International Publishing，159－171.

［3］Zenobi M C，Luengo C V，Avena M J，et al. 2010. An ATR-FTIR study of different phosphonic acids adsorbed onto boehmite［J］. Spectrochimica Acta Part A Molecular and Biomolecular Spectroscopy，75：1283－1288.

机械力作用下石英均裂反应产生活性氧（ROS）

吴 逍[1,2] 鲜海洋[1,2] 唐红梅[1,2] 魏景明[1] 朱建喜[1*] 何宏平[1]
（1. 中国科学院矿物学与成矿学重点实验室/
广东省矿物物理与材料研究开发重点实验室，广州 510640；
2. 中国科学院大学，北京 100049）

石英化学成分为 SiO_2，晶体结构为硅氧四面体共角顶联接成的架状结构，一般表现出稳定的物理化学性质。但在机械破碎加工过程中，石英颗粒发生形变、摩擦和破裂，新鲜断面上出现的高活性悬键将可能引发自由基化学过程[1]。

本研究通过机械球磨在石英表面构造缺陷，分析石英-水界面产生活性氧（ROS）的情况。实验采用微型球磨机在一定频率下对 1 g 石英砂（d＝0.84～2 mm）球磨 0、1、5、10、15 min，分别采用 DMPO/H_2O（10 mM）溶液捕获·OH，DMPO/DMSO（10 mM）溶液捕获·O_2^-，TEMP/H_2O（10 mM)溶液捕获$^1O_2$。悬浮液过滤得到透明液体样品，采用德国 Bruker A300 - 10 - 12 型电子顺磁自旋波谱仪（ESR）测试样品中的自由基；悬浮液离心干燥得到石英粉末，采用 Micromeritics ASAP2020 比表面积分析仪测试样品比表面积值变化。

图 1 是样品的 ESR 谱，(a) 谱线是干燥状态下球磨得到石英粉末的信号，归属为≡SiOO·；(b) 谱线中 4 个峰的强度比值近于 1∶2∶2∶1，这种特征信号归属于 DMPO - OH·加合物；(c) 谱线中峰强较弱的 6 折峰应归属于DMPO -·O_2^-；(d) 谱线中的峰强比近于 1∶1∶1，这是$^1O_2$的特征峰形。图 2 是不同时间的球磨作用下羟基自由基电子自旋共振谱强度与石英比表面积变化。随着球磨时间增加，石英粉末的比表面积从 0.005 m^2/g 逐渐增加到 1.44 m^2/g，羟基自由基的 ESR 峰强也逐渐升高，二者的变化趋势较为相似。

结合 ESR 谱和石英比表面积值变化情况分析发现，石英在机械球磨下破裂形成了表面缺陷，由均裂形成的≡Si·和≡SiO·与界面水发生自由基化学反应，产生了·OH、·O_2^- 和$^1O_2$等活性氧组分。≡SiO·与 H_2O 反应生成≡SiOH 和·OH，而·O_2^- 和$^1O_2$的产生与溶解在水中的 O_2有关，O_2可以被石英破裂过程中发射的少量电子还原形成·O_2^-，而$^1O_2$可能与≡SiOO·和 H_2O 进一步反应生成的 H_2O_2有关[2]。

石英机械破碎过程中形成自由基活性位点，与界面水反应生成活性氧组分，在环境中可进一步引发链式自由基化学反应，形成的这些活性氧组分（ROS）给人体健康和环境带来危害。

*通信作者：朱建喜；E-mail：zhujx@gig. ac. cn；手机号：13428862003。

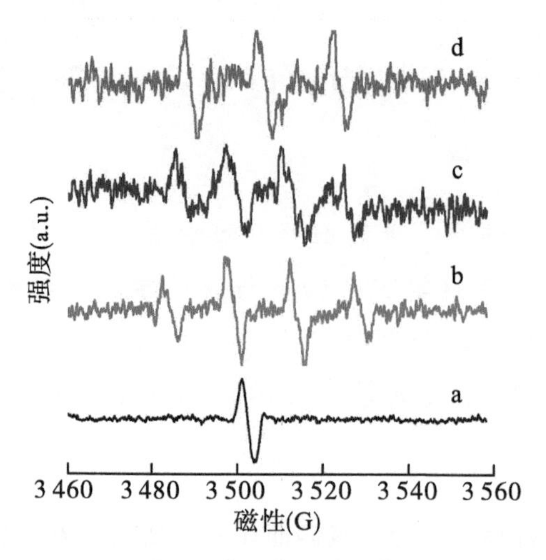

图 1　电子自旋共振谱

注：a. 石英粉末；b. DMPO 捕获的羟基自由基加合物；c. DMPO/DMSO 捕获的超氧自由基加合物；d. TEMP 捕获的单线态氧

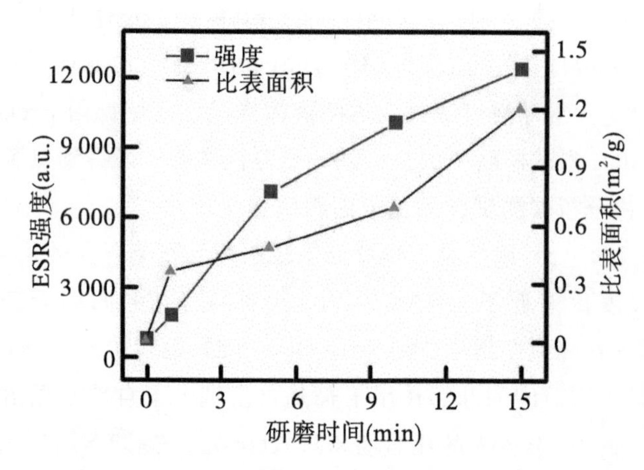

图 2　不同研磨时间下羟基自由基电子自旋共振谱强度与石英比表面积变化

参考文献

[1] Kita I，Matsuo S，Wakita H. 1982. H_2 generation by reaction between H_2O and crushed rock：An experimental study on H_2 degassing from the active fault zone [J]. Journal of Geophysical Research：Solid Earth，87：10789 - 10795.

[2] Shi X L，Dalal N S，Vallyathan V. 1988. ESR evidence for the hydroxyl radical formation in aqueous suspension of quartz particles and its possible significance to lipid peroxidation in silicosis [J]. J Toxicol Environ Health，25（2）：237 - 245.

碳纳米管负载钴系氧族化合物及电催化氧还原性能

李雨鑫　马帅飞　吕国诚*　廖立兵*
（中国地质大学（北京），材料科学与工程学院，北京 100083）

随着化石能源的日益衰竭，可持续能源的前景也越来越光明。由电化学水分解产生清洁氢能为未来的可持续能源提供了保障，然而制氢效率受到析氧反应（OER）的限制。之所以受到限制，是因为这个反应具有很大的过电位，反应很难进行。电解催化剂的应用克服了这个难题，并使整个反应过程更加高效。贵金属催化剂在析氢反应和析氧反应中具有极高的活性。但这些材料资源匮乏且昂贵，阻碍了其广泛应用。近年来人们致力于寻找高效、非贵金属的电解催化剂用于电解水，并且过去十年已取得很大的进展[1-5]。

本研究采用冷凝回流的方法制备了 CoO/MWCNTs、Co_9S_8/MWCNTs、CoSe/MWCNTs 三种电化学催化剂，并探索三种电化学催化剂对电解水析氧反应的催化性能。碳纳米管具有良好的力学性能、稳定性、导电性和大比表面积，钴系氧族化合物具有良好的催化性能，两者结合形成一种新型的电催化剂。研究结果（图 1～3）表明，CoO、CoS、CoSe 纳米颗粒均匀分布在 MWCNTs 上；CoO/MWCNTs、CoS/MWCNTs、CoSe/MWCNTs 材料具有良好的电解水析氧催化性能，从测试结果可知，当电流密度为 10 mA/cm^2 时，CoO/MWCNTs、CoS/MWCNTs、CoSe/MWCNTs 三种复合材料的阳极析氧过电位分别为 429、364、293 mV。可以看出：碳材料的纳米结构和电催化剂两种组分的协同效应导致增强的表面积和导电性，降低了水分解的超电势。

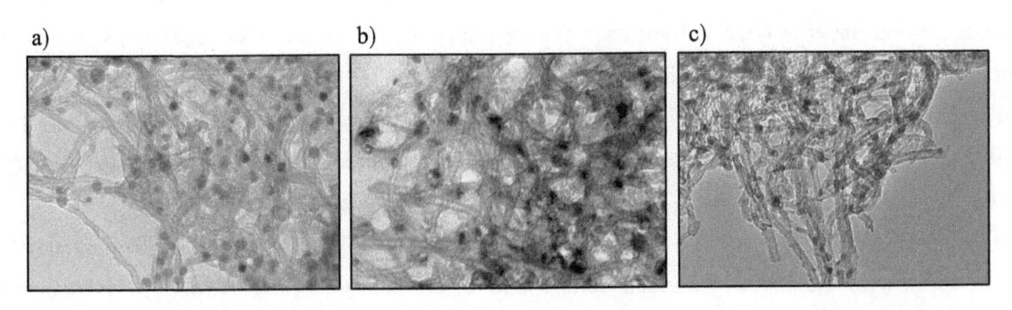

图 1　CoO/MWCNTs (a)、Co_9S_8/MWCNTs (b)、CoSe /MWCNTs (c) 的 TEM 图像

* 通信作者：吕国诚，E-mail：guochenglv@cugb. edu. cn；廖立兵，E-mail：lbliao@cugb. edu. cn。国家重点科技攻关计划（2017YFB0310704）、国家自然科学基金青年基金（51604248）资助项目。

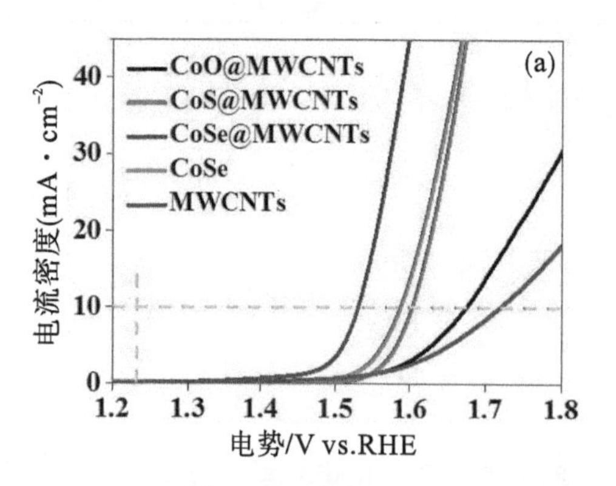

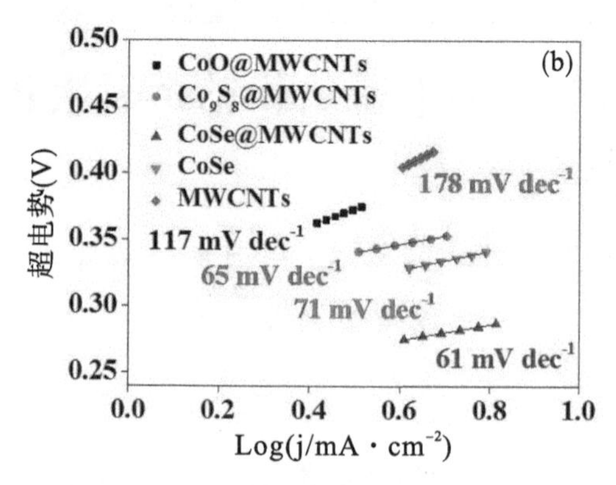

图 2　CoO/MWCNTs、Co_9S_8/MWCNTs、CoSe /MWCNTs 在 10 mA · cm^{-2} 电流密度下的电势（a）和塔菲尔斜率（b）

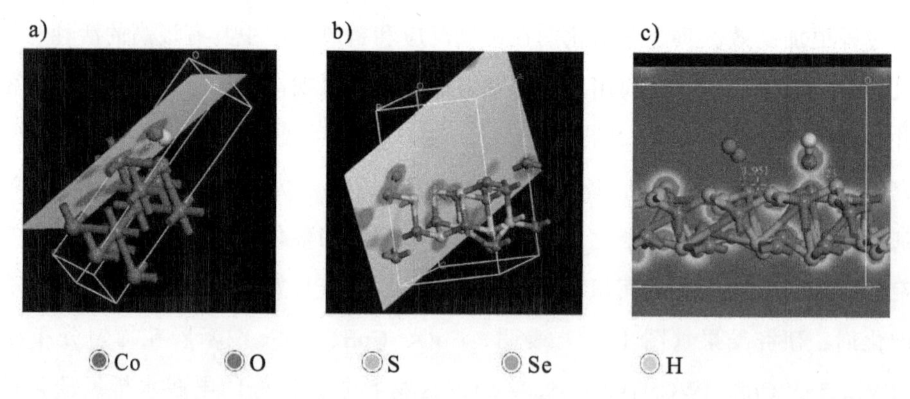

图 3　CoO/MWCNTs（a）、Co_9S_8/MWCNTs（b）、CoSe /MWCNTs（c）的模拟计算

参考文献

［1］Li X，Fang Y，Lin X，et al. 2015. MOF derived Co_3O_4 nanoparticles embedded in N-doped mesoporous carbon layer/MWCNT hybrids：extraordinary bi-functional electrocatalysts for OER and ORR［J］. J. Mater. Chem. A，3：17392 – 17402.

［2］Hu H，Han L，Yu M，et al. 2016. Metal-organic-framework-engaged formation of Co nanoparticle-embedded carbon@Co9S8 double-shelled nanocages for efficient oxygen reduction［J］. Energy Environ. Sci，9：107 – 111.

［3］Yu X，Yu L，Wu H，et al. 2015. Formation of nickel sulfide nanoframes from metal-organic frameworks with enhanced pseudocapacitive and electrocatalytic properties［J］. Nanostructures，54：5331 – 5335.

［4］Choi W，Yang G，Kim S，et al. 2016. One-step synthesis of nitrogen-iron coordinated carbon nanotube catalysts for oxygen reduction reaction［J］. Journal of Power Sources，313：128 – 133.

［5］Weng Z，Liu W，Yin L，et al. 2015. Metal/oxide interface nanostructures generated by surface segregation for electrocatalysis［J］. Nano Lett.，11：7704 – 7710.

石墨烯基复合矿物相变材料的导热强化

顾晓滨[1,2*]　刘　鹏[2]　边　亮[3]　刘昶江[1]
（1. 河北地质大学宝石与材料工艺学院，石家庄 050031；
2. 环境友好能源材料国家重点实验室，绵阳 621010；
3. 西南科技大学固体废物处理与资源化教育部重点实验室，绵阳 621010）

随着世界经济的发展和人口的增长，能源危机的问题日益凸显，然而目前传统能源利用的效率和可再生能源的开发都还处于较低的水平，因此全球的学者都将目光聚焦到能源存储技术的研发和应用问题上[1]。在各种能源存在形式中，热能是人们日常生活中接触最多也使用最多的能源，其他能源形式通常要转化成热能才能更好地被人们所利用，因此热能存储问题就成为问题的焦点。热能存储分为潜热存储、显热存储和化学反应存储，其中潜热存储，即利用相变材料来进行能量存储，因其储热密度大且在吸放热过程中近似恒温因而受到了人们的广泛关注。然而目前相变材料本身存在的导热率低的问题严重影响了传热效率，从而大大限制了其应用进程。

不少学者通过向相变材料中添加高导热率的添加剂，试图提高复合材料的整体热导率，其中部分研究取得了较好效果。石墨烯，是一种二维蜂窝网状的规则六方晶体，与 C60、碳纳米管相似，是石墨的同素异形体。石墨烯的特殊结构，使其具有优良的力学、电学、热学性能。石墨烯的传热主要是靠声子传热，导热系数可以达到 5 000 W/（m·K），因而可以作为提高相变材料热导率的优良添加剂。然而以石墨烯为添加剂来提高棕榈酸（PA）/高岭土相变复合材料的研究并未见文献报道，本文就以该体系为研究对象，分析讨论石墨烯含量对复合材料的相变潜热、熔化和凝固时间、导热性能等相变特性的影响，并探究其作用机理。

实验中使用的棕榈酸、高岭土均来自国药集团上海有限公司，而石墨烯是由北京德科岛津科技有限公司提供。其中，高岭石作为复合材料的支撑材料，棕榈酸为相变材料，而石墨烯则用于提高相变复合材料整体热导率。PA/高岭土定型相变材料是通过直接熔融法制备而成的[2]。图 1a 为高岭石分子结构模式图，由图可知高岭石是典型的 1∶1 型黏土矿物，层与层之间的空隙为相变材料提供了赋存空间；而高岭石层间分子与相变材料间的相互作用力又使相变材料即使在液体状态也不会从中泄露。如图 1b 的红外光谱测试结果表明，复合材料的红外光谱是其中各物质组分红外光谱的叠加，因此可知制备的定型相变材料中 PA、高岭土和石墨烯有很好的相容性。

*通信作者：顾晓滨；E-mail：XBGu@hgu. edu. com；手机号：18810518219。河北省科技支撑计划项目（17214016）、河北省高等学校科学技术研究项目-青年基金（QN2018124）、环境友好能源材料国家重点实验室开放基金（17kffk13）、河北地质大学博士启动基金（BQ2017021）资助项目。

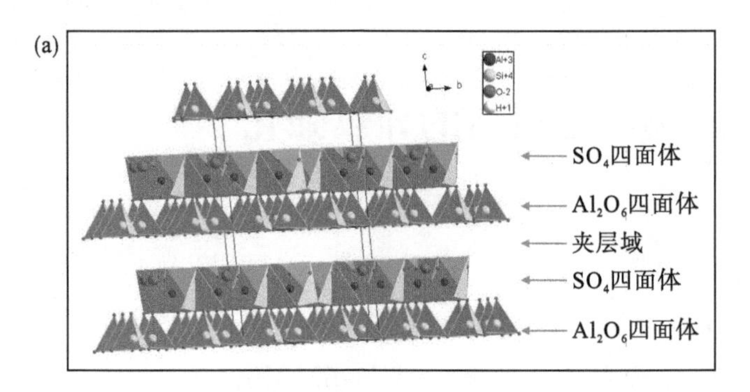

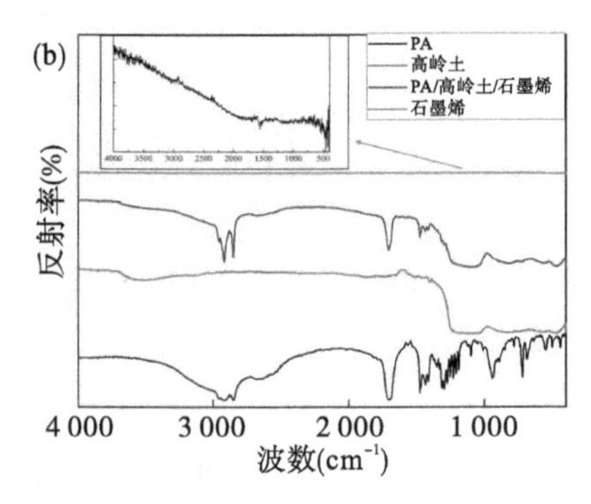

图 1　高岭石分子结构模式图（a）和相变复合材料红外光谱图（b）

图 2 为不同比例相变复合材料的 DSC 测试图，从图中可以看出，随着复合材料中 PA 质量分数的降低，复合材料升降温过程中的相变潜热均逐步降低，潜热值处于 213.10～54.7 J/g。而相变点在升降温的过程中则变化不大，最高为 66.11 ℃，最低为 64.75 ℃。

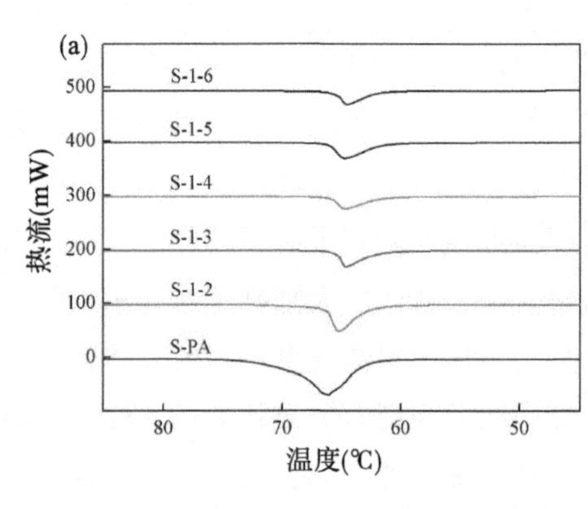

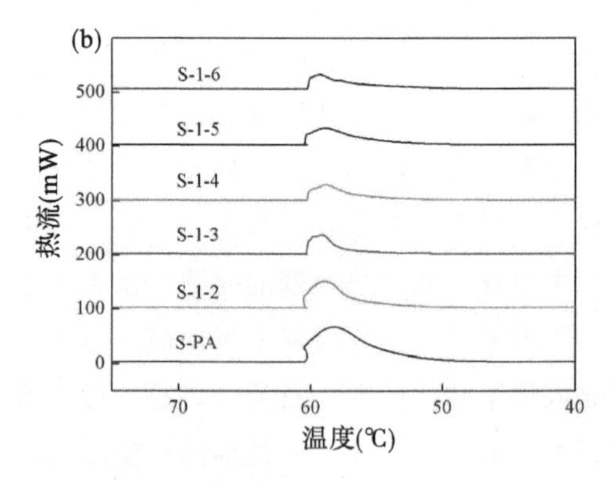

图 2　不同比例相变复合材料的 DSC 测试图

注：a. 吸热过程；b. 放热过程

综合泄露测试、扫描电镜、热重测试、步冷曲线和寿命测试的结果可知，制备的定型相变复合材料具有良好的热物理性能。高岭土、棕榈酸和石墨烯间通过氢键和毛细管力结合，具有良好的相容性。石墨烯的添加使相变复合材料的热导率得到了很大的提升。因此，本实验制备的定型相变复合材料具有良好的性能，不仅为相变材料的导热强化提供了思路，而且为其在室内热水等领域的应用提供了理论依据。

参考文献

［1］顾晓滨，秦善，牛菁菁. 2014. 相变储能矿物材料研究现状及其展望［J］. 矿物岩石地球化学通报，6：932－940.

［2］Gu Xiaobin，Qin Shan，Wu Xiang，et al. 2016. Preparation and thermal characterization of sodium acetate trihydrate/expanded graphite composite phase change material［J］. Journal of Thermal Analysis and Calorimetry，125：831－838.

单斜晶型 $BiVO_4$ 和 WO_3 表面态钝化及光催化活性调控

何辉超[1]　杨民基[1]　钟小辉[1]　边　亮[2]　张红平[1]　董发勤[2*]　周　勇[1*]
（1. 西南科技大学环境友好能源材料国家重点实验室，绵阳 621010；
2. 西南科技大学固体废物处理与资源化教育部重点实验室，绵阳 621010）

单斜晶型 $BiVO_4$ 和 WO_3 的禁带宽度和价带位置分别为：$BiVO_4$ $Eg\approx2.4$ eV，$E_{VB}\approx2.4$ V vs. RHE；WO_3 $Eg\approx2.7$ eV，$E_{VB}\approx3.4$ V vs. RHE，光生载流子在两种材料体相的转移距离均大于 100 nm，故两种半导体都是具有可见光响应的光催化剂。前期研究表明[1,2]，单斜晶型 $BiVO_4$ 表面存在一定数量的 Bi－O 悬空键，单斜晶型 WO_3 表面存在一定数量的 W－O 悬空键，这些悬空键会诱导两种光催化剂形成高表面态，易引起光生电子—空穴在其表面的高复合率，影响材料的光催化活性。In^{3+} 的离子半径约 0.8Å，其与 Bi^{3+} 的离子半径相近（1.03Å），如果 In^{3+} 定向掺杂取代单斜晶型 $BiVO_4$ 中的部分 Bi^{3+} 位点，可使 $BiVO_4$ 晶型结构发生扭曲，进而可能改变 $BiVO_4$ 高表面态。另一方面，无定形 TiO_2 表面含有大量的电子缺陷位点，若与单斜晶型 WO_3 发生耦合接触时，可形成 TiO_2（O^{2-}）－WO_3（W^{6+}）相互作用，进而可能改变 WO_3 高表面态。

基于上述分析，该研究首先采用密度泛函理论（DFT）计算预测了两种钝化单斜晶型 $BiVO_4$ 和 WO_3 表面态方法的作用效果。根据 DFT 计算预测，进一步制得了目标物质的薄膜材料，通过系统的表征和光催化活性测试验证了两种钝化方法有效性。

（1）DFT 计算：分别计算了 7% In^{3+} 定向掺杂取代单斜晶型 $BiVO_4$ 中 Bi^{3+} 位点前后，（110）和（121）面 $BiVO_4$ 表面能变化；无定形 TiO_2 耦合在单斜晶型 WO_3 表面前后，（002）面 WO_3 的表面能及 Ti－O 键长变化。（2）验证实验：采用滴液法制备 7% In^{3+} 掺杂和未掺杂的单斜晶型 $BiVO_4$ 薄膜；采用水热法制备 WO_3 纳米片薄膜，同时在该 WO_3 薄膜基础上，利用电化学法在 WO_3 纳米片上沉积无定形 TiO_2 薄层。通过 XRD、SEM 和 XPS 等技术表征了制得薄膜样品，利用原位 PL 光谱和系统的光电化学测试方法检测验证薄膜材料的光电催化活性。

DFT 计算发现（图 1）：7% In^{3+} 定向掺杂取代单斜晶型 $BiVO_4$ 中 Bi^{3+} 位点前后，（110）面 $BiVO_4$ 表面能由 0.283 33 J/m^2 降低到 0.144 67 J/m^2，（121）面 $BiVO_4$ 表面能由 0.617 02 J/m^2 降低到 0.358 34 J/m^2。光催化活性测试实验表明，7% In^{3+} 定向掺杂的 $BiVO_4$ 薄膜具有显著增强的光电氧化水性能和光电转化效率。原位 PL 光谱（图 2）发现光生电子—空穴在 7% In^{3+} 定向掺杂 $BiVO_4$ 薄膜表面的非辐射复合率增强，辐射复合率减弱，证实了 In^{3+} 定向掺杂钝化了 $BiVO_4$ 的表面态。

DFT 计算表明（图 1）：无定形 TiO_2 耦合在 WO_3 表面后，（002）面 WO_3 的表面能由 1.17 J/m^2 降低到 0.30 J/m^2，无定形 TiO_2 中的 Ti－O 键长由 1.929Å 变成 1.886/2.290Å，无定形 TiO_2（O^{2-}）－WO_3（W^{6+}）成键效果明显。光催化活性测试发现，无定形 TiO_2 耦合的 WO_3 薄膜具有显著增强

* 通信作者：董发勤，教授，E-mail：fqdong@swust. edu. cn；周勇，教授，E-mail：zhouyong199@nju. edu. cn，手机号：18281625592。国家自然科学基金（41702037）、四川省科技厅面上基金（2017JY0146）资助项目。

的光电氧化水性能。原位 PL 光谱（图 2）显示光生电子—空穴在无定形 TiO_2 耦合的 WO_3 薄膜表面的非辐射复合率增强，辐射复合率减弱，证实了无定形 TiO_2 耦合钝化了 WO_3 的表面态。该研究可为钝化改性半导体矿，提高其光催化活性提供理论及方法参考。

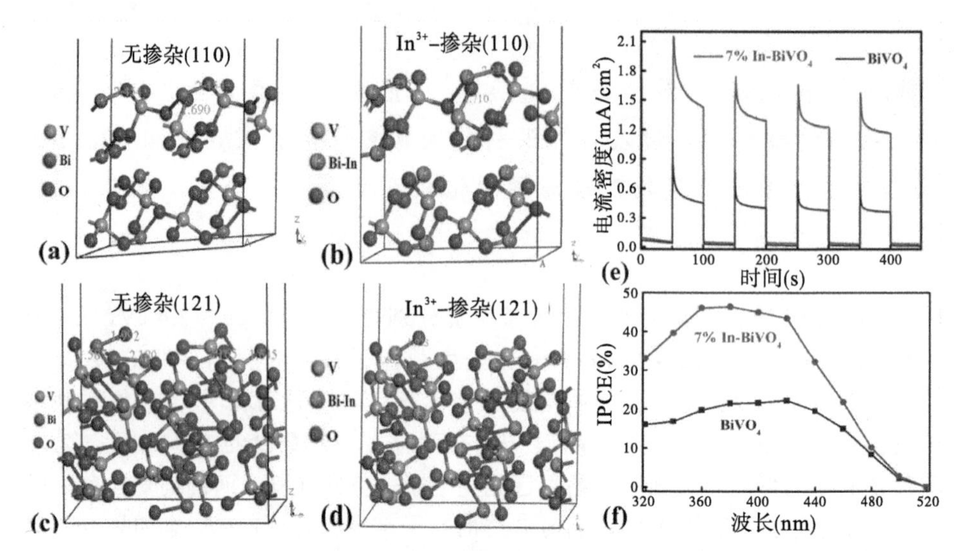

图 1　7% In^{3+} 定向掺杂单斜晶型 $BiVO_4$ 前后，(110) 和 (121) 面 $BiVO_4$ 晶型结构图，以及掺杂前后 $BiVO_4$ 薄膜的光电响应性能对比图

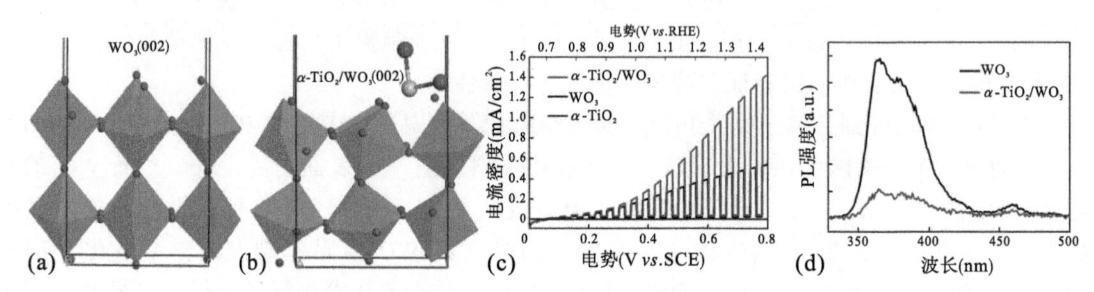

图 2　无定形 TiO_2 耦合在 WO_3 表面前后，(002) 面 WO_3 晶型结构图，以及耦合前后 WO_3 薄膜的光电响应性能和原位 PL 光谱对比图

参考文献

[1] Zhong X H，He H H，Yang M J，et al. 2018. In^{3+}-doped $BiVO_4$ photoanode with passivated surface states for photoelectrochemical water oxidation [J]. Journal of Materials Chemistry A，6：10456－10465.

[2] Yang M J，He H H，Zhang H P，et al. 2018. Enhanced photoelectrochemical water oxidation on WO_3 nanoflake films by coupling with amorphous TiO_2 [J]. Electrochimica Acta，DOI：10. 1016/j. electacta. 2018. 06. 056.

VTMS改性高岭石稳定多重乳液研究

李存军[1,2]　梁少彬[1]　王林江[1*]
（1. 桂林理工大学大学材料科学与工程学院，桂林 541004；
2. 浙江工业大学化学工程学院，杭州 310014）

乳液被广泛应用于化妆品、石油和药品制备等领域。黏土矿物作为一种固体乳化剂在Pickering乳液领域具有重要的研究和应用意义[1]。不对称的或极性Pickering颗粒（Janus颗粒）作为乳化剂同时具备Pickering颗粒和表面活性剂的优点，在环境友好型乳液研究和应用中具有广泛的前景[2]。

多重乳液在食品、药物控制释放等行业越来越受到重视[3]。本文以高岭石矿物为对象，基于高岭石表面因结构差异而具有的天然Janus特征，采用有机改性提高高岭石极性，凸显高岭石的Janus特征。Janus高岭石作为乳化剂可以降低乳液油/水界面能，从而提高颗粒在界面的稳定性并获得多重乳液。

按照不同油相体积分数（ϕ_o）量取液体石蜡为油相，将VTMS加入液体石蜡并于室温下搅拌均匀；称量一定量高岭石加入上述混合液，搅拌均匀制得高岭石分散体；向高岭石分散体中加入氯化钠溶液，分别置于不同温度下搅拌24 h，乳化后乳液静置90 d。不同温度稳定乳液的改性高岭石记为Kaol－T，T表示温度。

对于Kaol－60稳定的乳液体系（图1A），在$\phi_o=0.5\sim0.53$范围出现W/O/W（水包油包水）多重乳液；对于Kaol－67稳定的乳液体系（图1B），在$\phi_o=0.4\sim0.5$范围出现W/O/W多重乳液。Kaol－67稳定乳液体系的多重乳液范围较Kaol－60更大，且在较小的ϕ_o时已经出现多重乳液。这可能是由于VTMS对高岭石的修饰温度较Kaol－60高，高岭石表明润湿性差异所致。此外，研究发现高岭石改性温度低于60 ℃或高于67 ℃时，均无多重乳液出现，表明疏水性能太弱及太强均不能稳定油水体系获得多重乳液。

多重乳液的出现会导致乳液液滴粒径的变化，出现较大粒径的液滴。根据图2 Kaol－60稳定乳液分散相粒径分布图，乳液分散相主要液滴粒径约为200 μm，与光学显微照片中所观察到乳液液滴大小相符。当$\phi_o=0.57$时，粒度分布曲线在350～450 μm范围出现对应多重乳液曲线平台，且其粒度范围与光学显微照片中多重乳液的粒径大小接近，证明乳液中存在W/O/W型多重乳液。由此可知，通过调节修饰的温度和ϕ_o等因素，用Kaol－60和Kaol－67可以获得W/O/W多重乳液。

*通信作者：王林江；E-mail：wlinjiang@163. com；手机号：13507736656。国家自然科学基金项目（41572034）、广西有色金属及特色材料加工国家重点实验室培育基地开放基金（15KF-11）。

高岭石作为一种丰富的天然矿物资源，通过 VTMS 改性高岭石表面获得改性高岭石，将其用作乳化剂可以稳定油水体系从而获得多重乳液。该研究可以拓宽高岭石矿物应用范围，并为其他非金属矿物用作乳化剂稳定 Pickering 乳液提供参考。

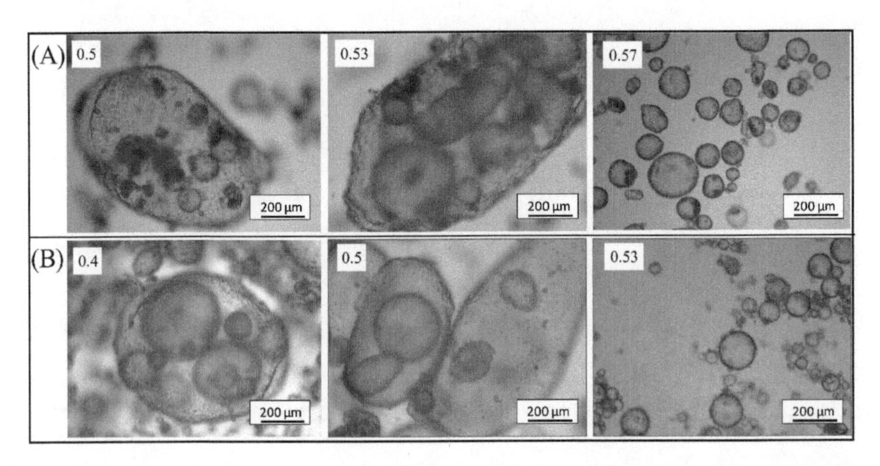

图 1　Kaol－60（A）和 Kaol－67（B）稳定的乳液在不同油相含量 ϕ_o 下的光学显微照片

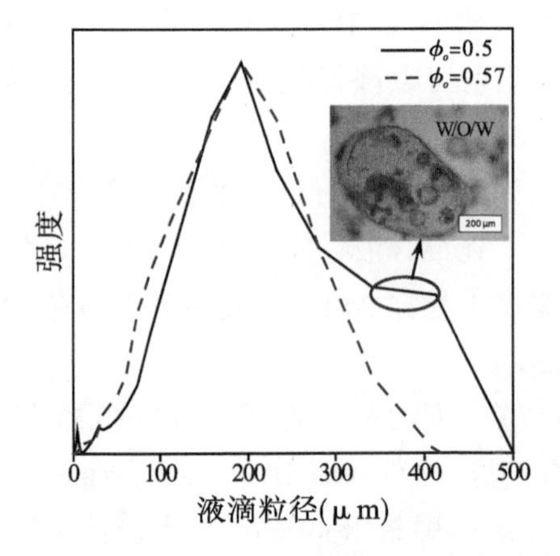

图 2　Kaol－60 稳定的乳液液滴的粒径分布

参考文献

［1］Yan N X，Masliyah J H. 1996. Effect of pH on adsorption and desorption of clay particles at oil-water interface［J］. Journal of Colloid and Interface Science，181：20－27.

［2］Liang S B，Li C J，Dai L X，et al. 2018. Selective modification of kaolinite with vinyltrimethoxysilane for stabilization of Pickering emulsions［J］. Applied Clay Science，161：282－289.

［3］Silva B F，Rodríguez-Abreu C，Vilanova N. 2016. Recent advances in multiple emulsions and their application as templates［J］. Current Opinion in Colloid & Interface Science，25：98－108.

方解石族碳酸盐矿物的热膨胀行为

王美丽[1,2*]　施光海[2]　白　清[2]
（1. 河北地质大学宝石与材料工艺学院，石家庄 050031；
2. 中国地质大学，地质过程与矿产资源国家重点实验室，北京 100083）

伴随着全球变暖等问题的出现，地球深部碳循环逐渐受到了地质学家的重视。碳酸盐矿物在含碳矿物相中所占的比重最大，种数达 100 多种，是地球圈层中最大的碳储库。前人大量实验岩石学研究也证实[1,2]，随着洋壳和陆壳的俯冲，地表系统中的碳以碳酸盐矿物的形式被带入到地球深部，其中部分碳又通过火山活动等相关作用带出地表。因此，碳酸盐矿物在地球内部环境条件下的高温高压行为等对碳的带入、碳的带出、碳在地球内部的存在形式以及运移等都有着十分重要的影响[3,4]。

其中，方解石族碳酸盐矿物是碳酸盐矿物中重要的一族。目前，仅有少数研究者对其在高温下的热行为进行了研究，尤其是其热膨胀性质。本研究主要利用原位中低温 X 射线衍射光谱技术对 5 种方解石族碳酸盐矿物（$MgCO_3$、$ZnCO_3$、$MnCO_3$、$CdCO_3$和 $CaCO_3$）在 83～618 K 之间的行为进行探索。

通过解析 XRD 数据，获得了 5 种矿物在不同温度条件下的晶胞参数和晶体结构相关参数；并发现 5 种矿物均呈现出轴向热膨胀性质的各向异性，c 轴的热膨胀性明显大于 a 轴，其晶体结构的布局可以很好地解释这一现象；而菱镉矿和方解石的 a 轴表现为负热膨胀，推测这可能是二者 CO_3基团沿 c 轴方向相对较大幅度的振动行为所引起：菱镉矿和方解石两种样品的 O_1-O_2键长随着温度的增加基本保持不变，然而菱镁矿和菱锌矿的该键长却随着温度的增加而增加。

此外，利用热状态方程对不同温度下的晶胞参数和键长数据进行拟合，获得了 5 种矿物的体积热膨胀系数、轴向热膨胀系数以及键长热膨胀系数。本研究实验数据显示：（1）常温下体积热膨胀系数和晶胞参数 a 热膨胀系数与阳离子半径呈反比线性关系（图 1），其表达式分别为：α_0（V）（10^{-5} K^{-1}）$=-8.713* r+9.919$ 和 α_0（a）（10^{-5} K^{-1}）$=-4.075* r+3.624$；然而受 CO_3基团沿 c 轴方向的振动所影响，晶胞参数 c 热膨胀系数与阳离子半径不再呈现反比线性关系；（2）电负性在控制该族矿物热膨胀系数中没有明显作用。

---

*通信作者：王美丽；E-mail：meili-028@163. com；手机号：18701647395。国家自然科学基金（41603064）、中国博士后基金（2016M591222）资助项目。本研究即将在《European Journal of Mineralogy》期刊发表。

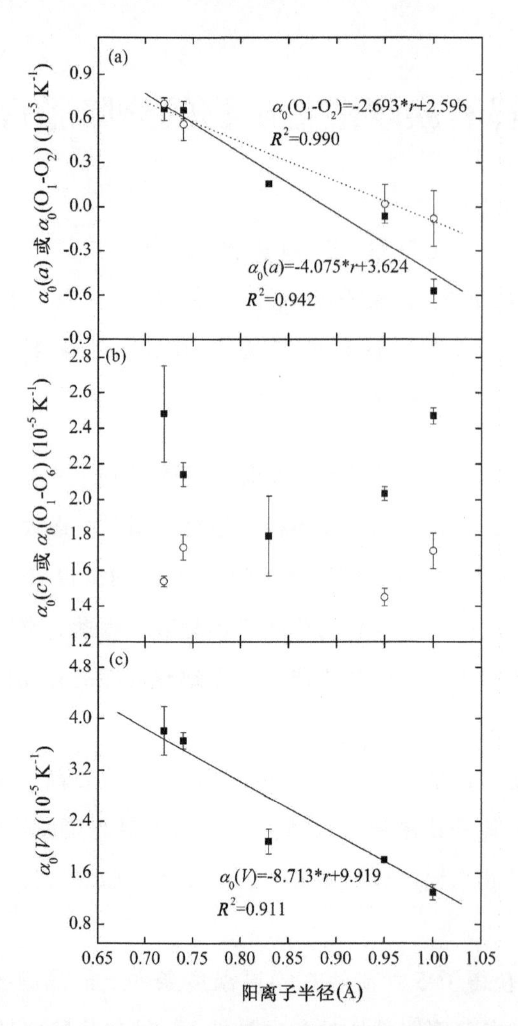

图 1　阳离子半径对方解石族碳酸盐矿物热膨胀性质
（体积热膨胀系数、轴向热膨胀系数以及键长热膨胀系数）的影响

注：实心方形符号代表晶胞参数相关的拟合结果，空心的圆形符号代表晶体结构相关的拟合结果

参考文献

[1] Dasgupta R，Hirschmann M M. 2010. The deep carbon cycle and melting in Earth's interior [J]. Earth and Planetary Science Letters，298（1）：1－13.

[2] Hazen R M，Schiffries C M. 2013. Why deep carbon [J]. Reviews in Mineralogy and Geochemistry，75：1－6.

[3] Hazen R M，Downs R T，Jones A P，et al. 2013. Carbon mineralogy and crystal chemistry [J]. Reviews in Mineralogy and Geochemistry，75：7－46.

[4] Shatskiy A F，Litasov K D，Palyanov Y N. 2015. Phase relations in carbonate systems at pressures and temperatures of lithospheric mantle：review of experimental data [J]. Russian Geology and Geophysics，56：113－142.

黑曲霉对蛇纹石的风化及草酸镁的诱导合成

孙晶晶　李　静　侯天仪　连　宾*

（南京师范大学生命科学学院，南京 210023）

蛇纹石是我国藏量丰富的含镁硅酸盐矿物，每年工业开采带来经济效益的同时，大量的边角碎料以及废弃物（尤其小于 10 mm 的尾矿）被丢弃，造成环境破坏。利用生物技术手段提取这些废料中的镁离子进行综合利用具有重要价值[1]。黑曲霉（Aspergillus niger）作为土壤中普遍存在的腐生型丝状真菌，是一种良好的浸矿菌种，研究证实其对黑云母、磷灰石、镁橄榄石等有很强的风化效果[2,3]。相关机理研究表明，在风化过程中黑曲霉可产生多种有机酸（如草酸等），增加对矿物的溶解能力。目前关于天然草酸镁石矿物结构和组成的研究虽然已有一些报道[4]，但实验室条件下由生物诱导矿化的草酸镁石却甚少涉及。

近年来，随着对微生物风化矿物作用机制的研究逐渐深入，有关生物酶及其基因表达调控在矿物风化中的作用日益受到关注[5,6]，但相关研究仍然非常缺乏。本文在前期研究黑曲霉风化硅酸盐矿物分子机理的基础上[7]，研究在不同镁源条件下碳酸酐酶（CA）和漆酶（McoA）基因表达量的差异，为进一步探讨实验条件下黑曲霉风化富镁硅酸盐矿物的调控作用机理提供新的线索，并为蛇纹石尾矿的再利用提供新思路。

采用原子吸收仪（AAS）测定黑曲霉风化添加蛇纹石发酵液中 Mg^{2+} 的浸出效率；采用重铬酸钾—次甲基蓝氧化脱色法测定发酵液中草酸的含量[8]；检测不同镁源条件下发酵液中 CA 和 McoA 活性，并通过实时荧光定量 PCR（RT-qPCR）的检测比较 McoA 和 CA 基因表达量的差异；用 SEM 观察微生物—矿物相互作用形成复合体的形貌，XRD 和 XRF 分析原矿和黑曲霉风化后残渣中的矿物和化学组成。

在黑曲霉风化蛇纹石的过程中，Mg^{2+} 释放量显著高于纯水与无菌查氏培养基对照组，且在 15 d 达到最高值［(362.04 ±40.76) mg/kg］（图 1a）。由图 1b 可知，与蛇纹石原矿 XRD 图谱相比，蛇纹石经历纯水及培养基作用 30 d 后矿物组成未发生明显变化，但是相同情况下，黑曲霉作用后的蛇纹石衍射峰发生了一定的改变，并且形成了次生矿物——草酸镁石（Glushinskite）。此外，应用 RT-qPCR 技术，研究不同镁源培养条件对黑曲霉 CA 和 McoA 基因表达量的影响。由图 2c 和图 2d 可知，添加蛇纹石（缺可溶性 Mg^{2+}）的条件下，黑曲霉风化矿物至第 5 天，CA 和 McoA 基因分别表达上调了（62.13 ±0.01）倍和（1.23 ±1.15）倍，结合两种镁源下酶活的差异（图 2a 和图 2b），显示这两个基因可能在黑曲霉风化富镁硅酸盐矿物的过程中起重要作用。

综上所述，本研究利用黑曲霉风化蛇纹石，发现其风化效果显著，并可诱导草酸镁矿物形成，这为进一步研究微生物浸提镁以及蛇纹石综合利用奠定基础。另外发现了 CA 和 McoA 两种酶在风化过程中发挥重要作用，为深入探究黑曲霉在风化富镁硅酸盐矿物中的分子机理奠定了理论基础。

---

* 通信作者：连宾；E-mail：bin2368@vip. 163. com。国家自然科学基金（41772360）资助项目。

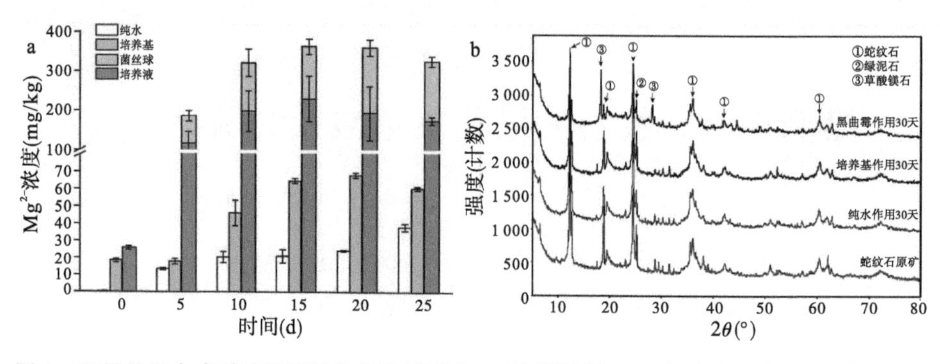

图1 不同处理方式对蛇纹石风化作用过程中 Mg^{2+} 释放量（a）与矿物组成 XRD 分析（b）

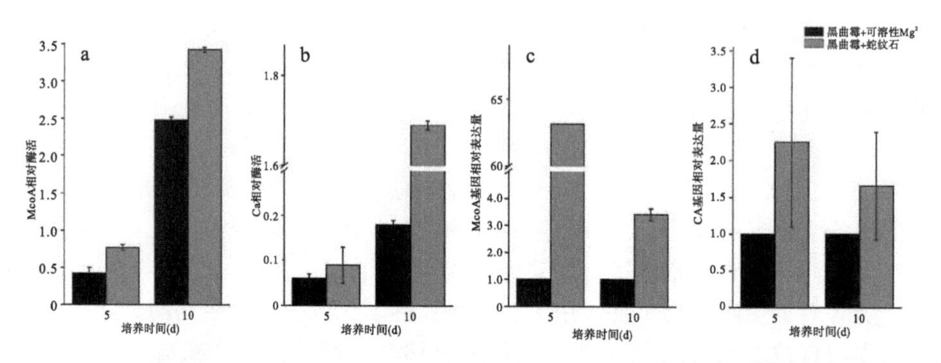

图2 不同镁源条件下黑曲霉 CA、McoA 酶活（a、b）及其相应的基因表达量（c、d）

注：图 2c 和图 2d 中以提供可溶性 Mg^{2+} 条件下 CA 基因表达量＝1 作为参照

参考文献

［1］周维卫，传秀云，周述慧. 2010. 蛇纹石及其固体废弃物固定 CO_2 的研究现状［J］. 矿物学报，1：179－180.

［2］Li Z，Liu L，Chen J，et al. 2016. Cellular dissolution at hypha-and spore-mineral interfaces revealing unrecognized mechanisms and scales of fungal weathering［J］. Geology，44：319－322.

［3］胡婕，郁建平，连宾. 2011. 黑曲霉对含钾矿物的解钾作用与机理分析［J］. 矿物岩石地球化学通报，30：277－285.

［4］蔡林，李涵，周根陶. 2012. 黑曲霉诱导下富镁硅酸盐矿物的溶解和草酸镁石的矿化研究［J］，矿物学报，s1：127－128.

［5］孙蕾蕾，肖雷雷，肖波，等. 2013. 黑曲霉风化含钾矿石过程中碳酸酐酶和半胱氨酸合成酶基因表达量的差异［J］. 中国科学：地球科学，43：1828－1833.

［6］董翠玲，连宾. 2014. 细菌与真菌对黑云母的风化作用比较：以胶质芽孢杆菌和黑曲霉为例［J］. 矿物岩石地球化学通报，33：772－777.

［7］Wang W Y，Lian B，Pan L. 2015. An RNA-sequencing study of the genes and metabolic pathways involved in Aspergillus niger weathering of potassium feldspar［J］. Geomicrobiology Journal，32：689－700.

［8］Jin Z X，Wang C，Dong W，et al. 2007. Isolation and some properties of newly isolated oxalate-degrading Pandoraea sp. OXJ-11 from soil［J］. Journal of Applied Microbiology，103：1066－1073.

高岭石/有机铵插层复合物的形成机理

王　定*

（河北地质大学宝石与材料工艺学院，石家庄 050031）

近年来，黏土矿物改性受到广泛关注。有机黏土矿物是黏土矿物改性研究中的一大热点，它是由黏土矿物与有机质在特定条件下反应生成。在黏土矿物中，蒙脱石、海泡石、锂皂石、蛭石和云母均被用于制备有机黏土矿物[1-4]。作为一种典型的 1∶1 型黏土矿物，高岭石是制备有机黏土矿物的重要原料之一。作为制备有机黏土矿物的主体材料，高岭石与各种有机质的反应主要分为插层反应和接枝反应。截至目前，与接枝反应相比，插层反应仍是制备高岭石/有机黏土矿物的最有效方法。

目前，已有多种类型的有机质被用于与高岭石反应，如极性分子、有机铵离子（主要是季铵盐和烷基胺）、不同类型的聚合物及阳离子染料和阳离子复合物。截至目前，仅有少数极性小分子可直接插入高岭石层间，其他有机质可通过置换被先前插入的分子而进入。多次置换插层反应极大地拓展了可进入高岭石层间有机质，这为研发高岭石的新应用提供了可能。

本文以张家口高岭土、二甲基亚砜、甲醇和系列季铵盐［丁基三甲基氯化铵（BTAC）、己基三甲基溴化铵（HTAB）、辛基三甲基氯化铵（OTAC）、十烷基三甲基氯化铵（DTAC）、十二烷基三甲基氯化铵（DETAC）、十四烷基三甲基氯化铵（TTAC）、十六烷基三甲基氯化铵（HTAC）、十八烷基三甲基氯化铵（STAC）］为原料，通过多次置换插层反应，分别在液相及固相条件下制备出系列高岭石/有机铵插层复合物。具体步骤如下：（1）高岭石/二甲基亚砜插层复合物的制备。10 g 高岭土原矿与 100 mL 质量分数为 90%的二甲基亚砜在 60 ℃下搅拌 8 h 后离心分离，用 50 mL 甲醇将上述固体产物清洗 5 次后自然风干备用。（2）甲氧基接枝高岭石的制备。利用甲醇对上述产物反复漂洗，置换已进入高岭石层间的二甲基亚砜分子，从而制备出甲氧基接枝高岭石。（3）液相反应。常温下，1 g甲氧基接枝高岭石与 1 mol/L 季铵盐溶液混合搅拌 24 h，获得系列高岭石/季铵盐插层复合物。（4）固相反应。1 g 甲氧基接枝高岭石与 0. 01 mol/L 季铵盐混合研磨，获得对应产物。利用 X 射线衍射、傅里叶红外光谱、热重/差示扫描量热法研究了高岭石/有机铵插层复合物产物结构和形成机理。

X 射线衍射结果显示，反应条件可显著影响插层反应的进行。不同制备条件下，季铵盐均可被插入高岭石层间，但在固相条件下仅短链季铵盐（碳链长度≤6）插层成功（表 1）。依据插层复合物层间距的变化与客体分子的立体构型，提出短链季铵盐离子在高岭石层间以单层平卧形式存在。

红外及热重分析显示，除水分含量外，插层复合物的结构、热行为等性质受反应介质影响较小。同时，对比分析发现，无论何种反应介质，季铵盐离子是通过离子—偶极反应进入高岭石层

* 通信作者：王定；E-mail：wangding0313@. com；手机号：18533269863。国家自然科学基金（51034006）。

间。但在固态反应中，水分子的存在至关重要。除离子—偶极力外，其他一些因素可能对插层反应有所影响，如空间几何约束、液相体系的 pH 值和外表面的电荷分布（溶液反应）。在固态反应条件时，离子—偶极力引起的吸引力不足以克服空间几何约束，因此长链季铵盐离子无法被插入层间。

表 1　有机铵盐改性高岭石插层复合物层间距

插层剂种类	碳链上碳原子个数	碳链长度（nm）	d_{001}（nm）			$\mid\Delta d_{001}\mid$（nm）	
			甲氧基接枝高岭石	(a)	(b)	(a)	(b)
BTAC	n=4	1.30	0.86	1.39	1.37	0.51	0.53
HTAB	n=6	1.43	0.86	1.51	1.58	0.65	0.72
OTAC	n=8	1.82	0.86	0.88(c)	3.40	0.02	2.54
DTAC	n=10	2.08	0.86	0.86(c)	3.66	0	2.80
DETAC	n=12	2.34	0.86	0.84(c)	3.50	0.02	2.64
TTAC	n=14	2.60	0.86	0.90(c)	3.80	0.04	2.94
HTAC	n=16	2.86	0.86	0.86(c)	4.09	0	3.23
STAC	n=18	3.12	0.86	0.85(c)	4.24	0.01	3.38

注：(a) 固相反应产物（实验数据）；(b) 液相反应产物（文献数据）；(c) 插层未成功

参考文献

[1] Akyuz S，Akyuz T. 2003. FT-IR spectroscopic investigations of surface and intercalated 2-aminopyrimidine adsorbed on sepiolite and montmorillonite from Anatolia [J]. Journal of Molecular Structure，651：205-210.

[2] Gorrasi G，Tortora M，Vittoria V，et al. 2003. Transport properties of organic vapors in nanocomposites of organophilic layered silicate and syndiotactic polypropylene [J]. Polymer，44：3679-3685.

[3] Klapyta Z，Fujita T，Iyi N. 2001. Adsorption of dodecyl-and octadecyltrimethylammonium ions on a smectite and synthetic micas [J]. Applied Clay Science，19：5-10.

[4] Klapyta，Gawel Z，Fujita A，et al. 2003. Structural heterogeneity of alkylammonium-exchanged，synthetic fluorotetrasilicic mica [J]. Clay Minerals，38：151-160.

对乙酰氨基酚在碳酸钙微球内的组装及缓释性

郭玉华* 潘国祥　徐敏红　伍　涛　王永亚
（湖州师范学院工学院，湖州 313000）

碳酸钙是地壳中最丰富的无机矿产资源，属于ABO_3类的多型晶体，在自然界可降解循环，无毒无害，具有生物活性，对正常细胞没有毒副作用，由于其具有良好的机械稳定性及热稳定性，在药物控释载体、基因治疗载体等生物医学领域显示巨大的应用潜力[1,2]。对乙酰氨基酚被广泛用于感冒引起的发热、头痛以及各种疼痛的治疗。该药物在人体内吸收迅速，生物半衰期短，一般为1～4 h，患者需要频繁用药[3]。碳酸钙作为辅料可以改善对乙酰氨基酚的生物利用度[4]。因此，为减少对乙酰氨基酚的给药次数，强化药效，从而降低因多次给药对患者造成的身体损害，本文研究了对乙酰氨基酚在碳酸钙微球内的组装的条件以及其在人工肠液和人工胃液中的释放。

将对乙酰氨基酚、十二烷基硫酸钠（SDS）、氯化钙加水溶解，在常温条件下搅拌，将碳酸钠溶液匀速滴入对乙酰氨基酚混悬液中，滴完后继续搅拌一定的时间，静置24 h，抽滤，产物在烘箱50 ℃条件下干燥2 h，得到对乙酰氨基酚碳酸钙微球产品。参照《中国药典》“释放度测定方法”中第一法，采用动态透析法考察碳酸钙微球的体外释药情况，分别以人工肠液和人工胃液为释放介质，测定在温度（37±0.5)℃下，不同时间其释放情况。

以载药率和包封率综合指标作为考察标准，采用正交设计对碳酸钠水溶液浓度、SDS的质量、药品质量和搅拌时间4个因素对载药微球合成影响的研究结果见表1。各因素对实验结果的影响主次顺序是搅拌时间＞SDS的质量＞碳酸钠水溶液浓度＞药品质量。由各因素K值的比较得出最佳合成条件为：搅拌时间15 min，SDS的质量3.5 g，碳酸钠的浓度1.59 g/100mL，药品质量0.5 g。该条件下的样品扫描电子显微镜照片见图1。可以看出样品呈球性较好，大部分微球的粒径为3 μm左右，微球的表面较圆整光滑。

对乙酰氨基酚碳酸钙微球在人工肠液和人工胃液中的累积释放曲线如图2所示。4 h之前对乙酰氨基酚在模拟胃液中累积释药比在模拟肠液中的快，可能由于碳酸钙在酸性介质中发生了反应，碳酸钙溶解导致对乙酰氨基酚更快地释放。4.7 h之后在模拟肠液中释药速度增加的更快，约10 h后累积释放率达到40%，而在模拟胃液中累积释放率约为30%，模拟胃液释放曲线更加平缓，在突释后的释药速率增长比较均匀。将释药数据进行动力学拟合，在人工肠液和人工胃液中的释放行为均符合一级动力学方程。

---

*通信作者：郭玉华；E-mail：guoyuhua@zjhu. edu. cn；手机号：13587936596。浙江省自然科学基金（LY14B060006）资助项目。

表 1 正交实验结果

序号	碳酸钠水溶液浓度（g/100mL）	SDS 的质量（g）	药品质量（g）	搅拌时间（min）	综合指标（载药率%+包封率%）/2
1	0.53	1.44	0.5	5	0.207
2	0.53	2.88	0.8	10	0.224
3	0.53	3.5	1.1	15	0.696
4	1.06	1.44	0.8	15	0.424
5	1.06	2.88	1.1	5	0.056
6	1.06	3.5	0.5	10	0.454
7	1.59	1.44	1.1	10	0.255
8	1.59	2.88	0.5	15	0.909
9	1.59	3.5	0.8	5	0.715
K_1	0.376	0.295	0.523	0.326	
K_2	0.311	0.396	0.454	0.311	
K_3	0.626	0.621	0.336	0.676	
R	0.315	0.326	0.187	0.365	

注：K 为 3 个水平综合指标的平均值；R 为极差

图 1 微球的扫描电子显微镜照片

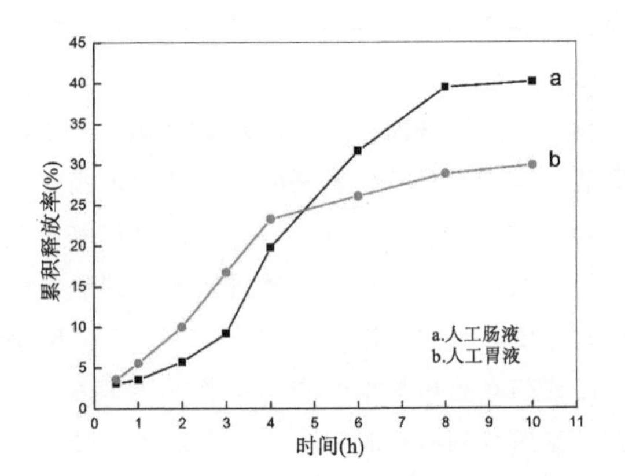

图 2 微球的累积释放曲线

参考文献

[1] Dong Z L，Feng L Z，Zhu W W，et al. 2016. $CaCO_3$ nanoparticles as an ultra-sensitive tumor-pH-responsive nanoplatform enabling real-time drug release monitoring and cancer combination therapy [J]. Biomaterials，110：60－70.

[2] Trushina D B，Bukreeva T V，Kovalchuk M V，et al. 2014. $CaCO_3$ vaterite microparticles for biomedical and personal care applications [J]. Materials Science and Engineering：C，45：644－658.

[3] 薛洪源，杨汉煜，胡玉钦，等. 2006. 对乙酰氨基酚软胶囊与片剂的人体相对生物利用度研究 [J]. 中国医药工业杂志，37（9）：621－623.

[4] 石秀芝，陈桂兰，王春民，等. 1992. 扑热息痛片辅料选择的实验研究 [J]. 中国医药工业杂志，29（3）：460－462.

非活性酿酒酵母菌生物矿化铀的机理探讨

张　伟[1,2]　董发勤[3*]　边　亮[4]　宋怀庆[4]　周　琳[3]　覃贻琳[3]

（1. 西南科技大学分析测试中心，绵阳 621010；

2. 中国工程物理研究院激光聚变研究中心，绵阳 621900；

3. 固体废物处理与资源化教育部重点实验室，绵阳 621010；

4. 西南科技大学环境与资源学院，绵阳 621010）

酿酒酵母菌作为发酵工业常用的微生物，因其具有安全价廉等优点，常用于生物吸附剂处理废水中的多种重金属离子和放射性核素[1,2]。目前研究表明，微生物吸附放射性核素根据作用方式不同可分为代谢性（活体生物）和非代谢性（无活性或死体状态）两类[3]。本文基于近年来国内外的研究成果及自身的研究工作，采用静态摇瓶吸附法考察了非活性酿酒酵母菌对U（Ⅵ）的吸附行为，通过不同pH值条件下微生物对铀的吸附特性及动力学模型的拟合，并结合FTIR和XPS等测试手段对反应前后细胞功能基团和铀赋存价态的表征，初步探讨了非活性酿酒酵母菌生物矿化铀的机理。该研究有助于更加清楚地了解生物吸附过程，为铀的富集再生应用提供基础的实验数据。

实验中，酿酒酵母菌（Saccharomyces cerevisiae）购自中国安琪酵母股份有限公司，灭活后使用。硝酸氧铀酰，分析纯，购自北京化工厂。非活性酿酒酵母菌对铀的吸附采用静态摇瓶实验法，具体方法参考文献［3］。

溶液体系pH值对非活性酿酒酵母吸附铀的影响较大。随着溶液初始pH值从1.0升高到6.0，非活性酿酒酵母菌对U（Ⅵ）的去除率出现先增加后降低的趋势。最佳吸附pH值出现在pH＝3.0左右。究其原因：pH≤3.5时，铀主要以UO_2^{2+}及少量的铀酰络合物存在，UO_2^{2+}首先与非活性酿酒酵母菌带负电的细胞表面以静电吸引作用快速接近。3.5≤pH≤6.0，UO_2^{2+}逐渐发生水解以$[(UO_2)_2(OH)_2]^{2+}$、$[(UO_2)_3(OH)_5]^+$等离子存在，离子半径较大的络阳离子与细胞表面活性位点结合的数量会减少，因此U（Ⅵ）的去除率下降。

采用准一级、准二级和Webber内扩散模型对接触时间的实验数据进行拟合分析：准二级动力学吸附模型更适合描述非活性酿酒酵母菌对U（Ⅵ）的吸附过程，揭示非活性酿酒酵母菌对U（Ⅵ）的吸附过程可能存在受速率控制的化学吸附行为。Webber内扩散模型可将拟合数据分段拟合：膜扩散部分（非活性酿酒酵母菌外表面对铀的迅速吸附过程）和吸附剂内扩散部分（铀在细胞表面与特征官能团结合的过程）。

非活性酿酒酵母菌与铀作用后沉淀物的FTIR光谱显示：细胞表面的羟基、氨基、羰基、羧基和磷酸基等活性基团参加了与铀的作用。在917 cm^{-1}附近出现了铀的特征峰[4]，证实非活性酿

* 通信作者：董发勤；E-mail：fqdong@swust.edu.cn；手机号：13980125273。国家重点基础研究发展计划（2014CB846003）资助项目。

酒酵母菌确实固化了铀。

利用 XPS 表征铀在非活性酿酒酵母菌细胞表面的赋存价态（图 1）：与铀作用前，非活性酿酒酵母菌细胞表面并未检测到铀信号。作用后，在 393 eV 和 382.2 eV 下检测到中等强度的铀信号。对 U4f 的自旋-轨道裂分峰进行分峰可知，U（Ⅵ）有部分被非活性酿酒酵母菌还原为 U（Ⅳ）。对菌体表面磷/碳比、U（Ⅳ）/U（Ⅵ）含量比分析发现，随着非活性酿酒酵母菌对铀吸附量的增加，磷/碳比例增大，U（Ⅳ）/U（Ⅵ）含量比基本相同，说明沉积在非活性酿酒酵母菌细胞表面的铀可能是磷—铀混合晶体。

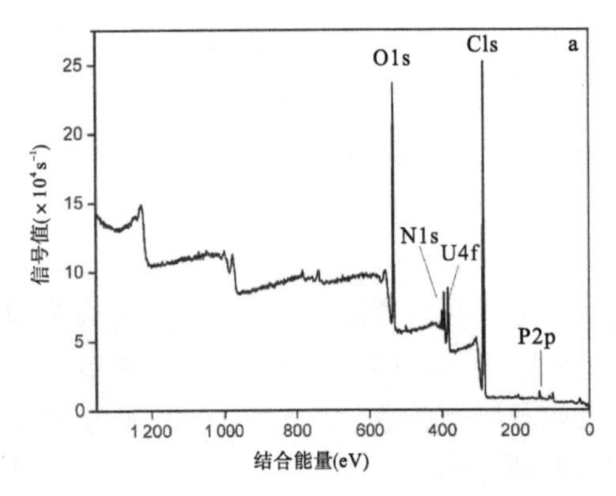

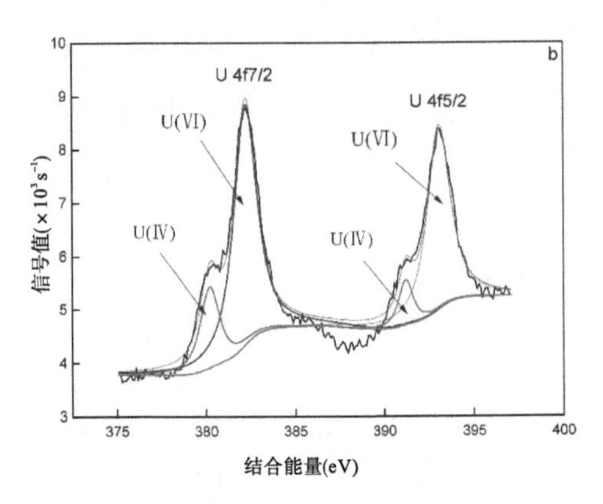

图 1　非活性酿酒酵母菌与铀作用后的 XPS 谱图

注：a. XPS 全谱；b. U4f 的高分辨分峰谱图

非活性酿酒酵母菌生物矿化铀的机理可能是：U（Ⅵ）首先通过静电引力作用快速附着在细胞表面上。然后，细胞表面上的羧基、氨基、羟基和磷酸盐等官能团与 U（Ⅵ）配位络合。同时伴随着菌体在高压蒸汽灭活过程中因细胞破裂而外泄的胞内酶、有机酸等物质将一小部分 U（Ⅵ）还原成 U（Ⅳ）。最后，不同价态的铀与磷在细胞表面逐渐形成磷-铀混合晶体。

参考文献

[1] Liu M X，Dong F Q，Yan X Y，et al. 2010. Biosorption of uranium by Saccharomyces cerevisiae and surface interactions under culture conditions [J]. Bioresource Technology，101：8573－8580.

[2] 彭国文，丁德馨，胡南，等. 2011. 化学修饰啤酒酵母菌对铀的吸附特性 [J]. 化工学报，62：3201－3206.

[3] 张伟，董发勤，覃贻琳，等. 2015. 灭活酿酒酵母菌对 U（Ⅵ）的吸附行为及减量化研究 [J]. 功能材料，46：23064－23070.

[4] Popa K，Cecal A，Drochioiu G，et al. 2003. Saccharomyces cerevisiae as uranium bioaccumulating material：the influence of contact time，pH and anion nature [J]. Nukleonika，48：121－125.

黄素及 AQS 介导光生电子还原 U（Ⅵ）

王萍萍[1,2]　刘明学[2]　王旭辉[2]　霍婷婷[2]　董发勤[2*]
（1. 西南科技大学环境与资源学院，绵阳 621010；
2. 固体废物处理与资源化教育部重点实验室，绵阳 621010）

铀矿从开采冶炼到生产应用，产生大量的含铀废弃物和废水，对水体和土壤造成严重污染，从而对人类健康具有严重影响。目前，含铀废水的处理方法主要有化学沉淀法、离子交换法、蒸发浓缩法、膜分离及吸附法等。相对于传统处理方法，光催化还原 U（Ⅵ）由于具有清洁无二次污染、耗能少、处理迅速等优点而备受关注[1]。氧化还原介体能充当电子载体可逆的参与氧化还原反应，可加速光催化还原 U（Ⅵ）这一过程[2,3]。本文系统研究分析了典型电子穿梭体——黄素（RF）、蒽醌-2-磺酸钠（AQS）在 U（Ⅵ）水溶液体系中的氧化还原特性，通过光电子直接还原 U（Ⅵ），以及添加 RF、AQS 介导光电还原 U（Ⅵ）的还原速率、还原率、电子传递动力学的影响分析，为光电化学还原处理含铀废水工艺提供理论指导。

实验结果表明（图 1），0.1 mol/L NaCl 溶液的 CV 曲线上无明显的氧化还原峰出现，负向扫描电流增大，主要是由于水中的少量的 H^+ 还原反应发生。添加 1.0 mol/L 的 RF、AQS 到 0.1 mol/L NaCl 溶液，RF 的还原峰为－0.6 V（vs. SCE），氧化峰为－0.29 V（vs. SCE）；AQS 的还原峰为－0.8 V（vs. SCE），氧化峰为－0.39 V（vs. SCE）。为进一步研究 RF、AQS 的氧化还原性能，测试了 FTO 导电玻璃电极在 RF、AQS 中不同扫速下的循环伏安曲线。RF、AQS 氧化峰 A 的峰电流值（j_{op}）和还原峰 B 的峰电流值（$-j_{rp}$）分别对扫描速度的 1/2 次方（$v^{1/2}$）作线性拟合的关系图。RF、AQS 的 j_{op} 和 $-j_{rp}$ vs. $v^{1/2}$ 线性关系较好，符合 Randles Sevcik 的控制扩散反应公式，说明 RF、AQS 的电化学转化属于扩散控制。

另外，该研究还分析了 RF 及 AQS 对光电子还原 U（Ⅵ）的还原率的影响。RF 及 AQS 均对 U（Ⅵ）的还原起到了促进作用（图 1）。反应 24 h 后，U（Ⅵ）的还原率均大于 90%，表现为 AQS 组＞RF 组＞对照组，还原率分别为 98.12%、96.12%和 88.33%。RF、AQS 及对照组反应前后阻抗有较大变化，其中 AQS 变化最为显著，对照组变化较小。反应后电极传递电子的能力有所下降，且添加 AQS 还原反应后，电极传递电子的能力最低，可能的原因是在反应后生成的 U（Ⅳ）沉淀吸附到电极上，占据反应位点，说明空白组吸附上的 U（Ⅳ）沉淀最少，而 AQS 组吸附上的 U（Ⅳ）沉淀最多。最后，该研究对 RF、AQS 对 U（Ⅵ）还原产物进行了分析。FT-IR 分析，在 472.4 cm^{-1} 与 913.9 cm^{-1} 处有特征峰出现，文献表明其对应的主要是 U（Ⅳ）与 U（Ⅵ）的特征峰，因此光电子还原铀产物主要以 U（Ⅳ）与 U（Ⅵ）晶体存在。SEM 分析，添加 RF、AQS 后对电极上均沉积了薄片状、针柱状矿物，EDS 分析有 U 元素存在。

* 通信作者：董发勤；E-mail：fqdong@swust. edu. cn；联系电话：0816-6089013。国家重点基础研究发展计划（973）项目（2014CB846003）资助。

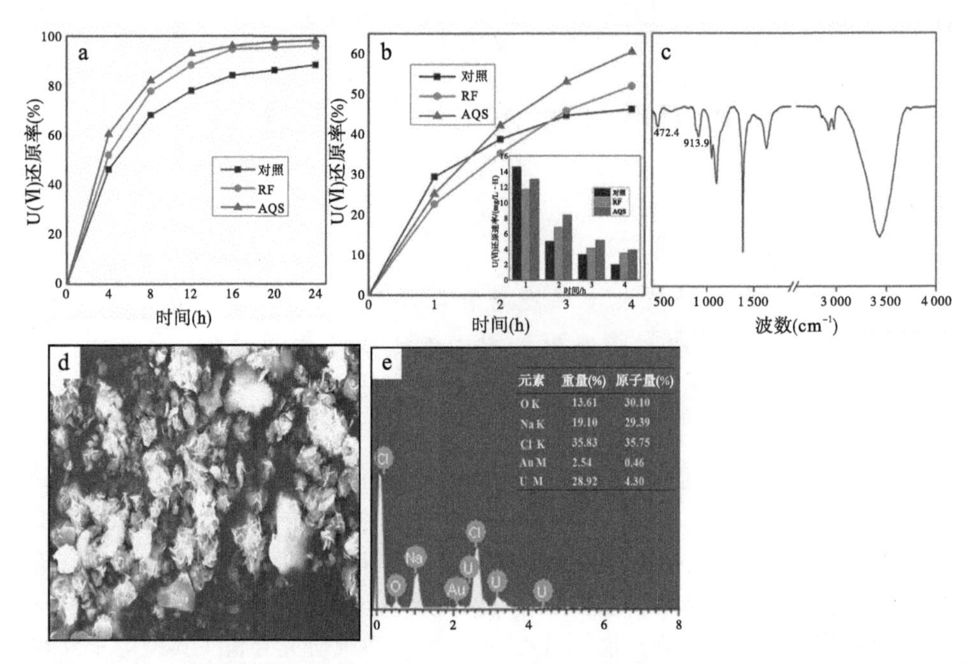

图 1　AQS 与 FMN 对 U（Ⅵ）还原率影响（a）；对 U（Ⅵ）还原率影响（b）；还原 U（Ⅵ）产物 FTIR 图谱（c）；还原 U（Ⅵ）产物 SEM 图谱（d）；还原 U（Ⅵ）产物 EDS 图谱（e）

综上所述，RF 与 AQS 可加快 U（Ⅵ）还原反应动力学速度，提高光电子对 U（Ⅵ）的还原率，该结果为光电子在还原治理重金属污染领域提供了一定的参考依据。

参考文献

[1] Kim Y K，Lee S，Ryu J，et al. 2015. Solar conversion of seawater uranium（Ⅵ）using TiO_2 electrodes［J］. Applied Catalysis B：Environmental，163：584 - 590.

[2] Wang G，Zhen J，Zhou L，et al. 2015. Adsorption and photocatalytic reduction of U（Ⅵ）in aqueous TiO_2 suspensions enhanced with sodium formate［J］. Journal of Radioanalytical and Nuclear Chemistry，304：579 - 585.

[3] Wang P，Dong F，Liu M，et al. 2018. Improving photoelectrochemical reduction of Cr（Ⅵ）ions by building α-Fe_2O_3/TiO_2 electrode［J］. Environmental Science and Pollution Research，1 - 9.

埃洛石对 PBAT 热分解行为的影响机制

李旭娟* 谭道永　孙红娟　蔡宗佐　王玉琪
（西南科技大学环境与资源学院，绵阳 621010）

埃洛石（Hal）是 1∶1 二八面体高岭土系矿物，其常见形貌为中空管状结构，与碳纳米管相似。但较碳纳米管，埃洛石具有不均匀的内外表面，其管内壁为铝氧八面体层，含 Al－OH 基团，外壁为硅氧四面体层，含 O－Si－O 基团，管边缘为 Al－OH 和 Si－OH 基团，这一独特的结构特点使其在物质吸附、储存、输运、药物缓释、催化、电化学、储能、增强、阻燃等方面具有优异性能，且其廉价易得，近年来重新引起了人们的重视和研究。

在聚合物纳米复合材料方面，埃洛石因具有较高的长径比、独特的表面化学性质、诱导结晶行为以及中空管状结构，在提升复合纳米材料热稳定性和阻燃性能方面具有独特优势。大量研究发现，埃洛石可提高聚丙烯、尼龙、聚对苯二甲酸乙二醇酯、聚己内酯、橡胶等体系的热分解温度，有助于提升复合材料的热稳定性和阻燃性[1,2]。有人将其归因于埃洛石对分解气体和热量的阻隔作用，有人总结为埃洛石中空管的吸附作用，相关机理还有待深入研究。另一方面，聚对苯二甲酸己二酸丁二醇酯（PBAT)，一种典型的全生物可降解高分子，有望替代传统聚乙烯材料，在包装、医用材料、农业覆盖薄膜领域广泛应用。研究埃洛石/PBA 复合材料的热稳定性和热分解机制，有助于理解和改善温度场下 PBAT 基复合材料的加工和使用稳定性，设计和制备出绿色材料，并拓展这类材料的应用领域。

原材料：PBAT，新疆蓝山屯河聚酯有限公司产品。熔融指数 3.8 g/10 min（2.16 kg，190 ℃），T_m＝115 ℃，T_g＝－31 ℃，密度 1.3 g/cm^3。埃洛石，产地为湖北省，使用前进行提纯，过 100 目筛，直径为（53.0±12.0） nm。

样品制备与测试：溶液浇筑法用于制备 Hal/PBAT 纳米复合薄膜。将干燥后的 PBAT 溶解于二氯甲烷溶液中，加入 2%的埃洛石及改性埃洛石，采用超声和搅拌耦合场分散埃洛石，制备出均匀的复合薄膜。采用 FTIR－TG－GC/MS 联用仪器对试样进行热分解分析。

图 1 的 DTG 曲线中，PBAT 及其复合材料均只有 1 个峰，说明埃洛石不影响 PBAT 的热分解步骤。TG 结果表明，埃洛石略微提高了 PBAT 的热分解温度（从 403 ℃提升至 406 ℃）。令人惊讶的是，与以往报道的 Hal/聚合物复合材料相比，埃洛石显著提高了 PBAT 的成碳率（从 0.88% 增加至 7.63%）。

*通信作者：李旭娟；E-mail：lixujuan2000@163. com；手机号：13550802448。西南科技大学博士基金（16zx7130）。

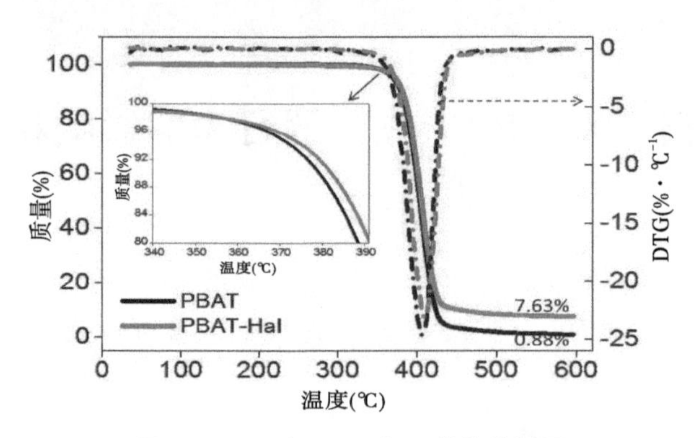

图 1 PBAT 和 PBAT/Hal 的热分解图

图 2 红外结果显示，PBAT 及其复合材料的热分解主要产物为：H_2O（4 000～3 500 cm^{-1}），CH_4（3 105、3 083 cm^{-1}），CO_2（669、2 360～2 310 cm^{-1}），CO（2 240～2 040 cm^{-1}），含有羰基和酯基的化合物（1 145、1 264、1 767 cm^{-1}）以及芳环化合物（1 605、1 507、1 457 cm^{-1}）。在相同分解温度（407 ℃）下，Hal/PBAT 试样的吸收峰强度更大，同时在 1 700～1 200 cm^{-1}出现了新的吸收峰，说明埃洛石具有催化作用，致使生成气体的浓度增加，并伴有新气体形成。这可能归因于埃洛石的二次脱氢反应，该反应促使热分解气体浓度增加，使 PBAT 生成芳环化物和残碳。

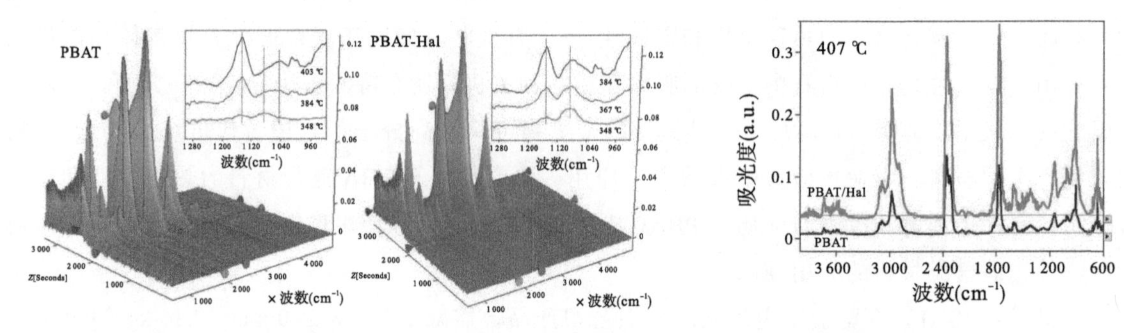

图 2 PBAT 和 PBAT/Hal 复合薄膜热分解气体的红外图谱

埃洛石独特的表面特性致使其在热分解中二次脱氢，从而提高了 PBAT 的热分解温度，显著增加了体系的芳环化物和残碳量。该结果有助于理解埃洛石/聚合物体系的热稳定机制，为绿色阻燃材料的应用开发提供理论基础。

参考文献

[1] Rooj S，Das A，Thakur V，et al. 2010. Preparation and properties of natural nanocomposites based on natural rubber and naturally occurring halloysite nanotubes [J]. Materials & Design，31（4）：2151－2156.

[2] Garcia-Garcia D，Garcia-Sanoguera D，Fombuena V，et al. 2018. Improvement of mechanical and thermal properties of poly（3-hydroxybutyrate）（PHB）blends with surface-modified halloysite nanotubes（HNT）[J]. Applied Clay Science，162：487－498.

蒙脱石基固态减水剂对水泥水化的影响及作用机理

吴丽梅* 曹诗悦　王　晴　唐　宁
（沈阳建筑大学材料科学与工程学院，沈阳 110168）

蒙脱石是典型的天然纳米层状结构矿物，由于同晶置换（如八面体层的 Al^{3+} 被 Mg^{2+} 或者 Fe^{2+} 置换，四面体层的 Si^{4+} 被 Al^{3+} 或者 Fe^{3+} 置换），使得片层中有过剩负电荷，并由于静电作用通过层间吸附 K^{+}、Na^{+}、Ca^{2+}、Mg^{2+} 等阳离子达到晶胞电荷平衡。有机改性蒙脱石体系是利用了蒙脱石的阳离子交换特性，有机溶剂通过离子交换作用或分子间吸附作用进入蒙脱石层间，蒙脱石对十六烷基三甲基溴化铵的负载量可达 1 300 mmol/kg[1]。插层后的蒙脱石的晶面间距将由于离子或分子的柱撑作用而增大，有机化程度增强，从而提高了蒙脱石的触变性、黏结性等性能。

混凝土减水剂是一种能够降低水泥用量、提高工业废渣利用率、提高混凝土强度、耐久性、工作性，改善混凝土和易性的简便而又有效的技术手段，对于现代混凝土的制备工艺来说，是必不可少的材料。相比于前两代减水剂，聚羧酸系减水剂的结构和性能具有可变性，并具有掺量低、水泥适应性好、有害成分少、环保性好、流动性保持好的优点，研究开发新型的聚羧酸系减水剂受到了国内外的广泛关注[2,3]。目前聚羧酸减水剂在市场上价格昂贵，为了降低聚羧酸减水剂在混凝土工程中的用量进而降低混凝土工程成本，并且更好地发挥其性能优势，该研究利用蒙脱石的层状结构特点，将聚羧酸减水剂与蒙脱石进行复合组装并研究其在混凝土中的应用效果。

实验原料主要有钠基蒙脱石、十六烷基三甲基溴化铵、聚羧酸减水剂、水泥等。主要实验仪器有磁力搅拌器、离心机、烘箱、X 射线衍射仪、TAM Air 微量热仪等。将钠基蒙脱石先与阳离子型有机物十六烷基三甲基溴化铵在水体系中进行插层，再与聚羧酸减水剂进行插层组装；采用 XRD 方法测定插层前后层间距的变化，以 TAM Air 微量热仪表征插层后复合减水剂体系体系的水泥水化热。

该研究利用蒙脱石和聚羧酸减水剂的各种优良性能合成了一种新型复合高效减水剂，探索层电荷以及插层组装的温度、时间、浓度等对聚羧酸在复合积水及体系中负载量的影响规律，探讨了其性能和应用情况。通过对多种不同原材料比例的试验、检测，得到蒙脱石与聚羧酸减水剂复合的最佳工艺条件。聚羧酸减水剂与蒙脱石的复合主要是插层组装和表面吸附，插层量占主要。随着聚羧酸减水剂用量的增多，插入蒙脱石层间的量逐渐增多，当减水剂用量为 6 g/L 时，蒙脱石层间距为 3.92 nm，并且插层均匀（图 1）。

将制得样品掺入硅酸盐水泥中参与水化反应，利用 TAM Air 微量热仪得到水化反应的水化热曲线，通过与聚羧酸减水剂和空白对照组进行对比从而对其参与水化反应的效果和反应机理进

*通信作者：吴丽梅；E-mail：lmwu@sjzu. edu. cn；手机号：13889325652。中国博士后科学基金（2018M631818）资助项目。

行了分析和讨论。图 2 显示，聚羧酸减水剂对水泥水化的初期和加速期有抑制作用，同时会延长其诱导期，使放热速率大幅减小，聚羧酸减水剂浓度越大，水泥水化累计放热量越小。结合蒙脱石基固态减水剂的缓释实验来看，蒙脱石基固态减水剂中的聚羧酸减水剂与蒙脱石作用力较强，导致参与到水泥水化过程中时释放量较小，但是对水泥水化反应的影响依然很大。而蒙脱石基聚羧酸复合减水剂的水泥净浆流动度要优于相同含量有效成分的聚羧酸减水剂。

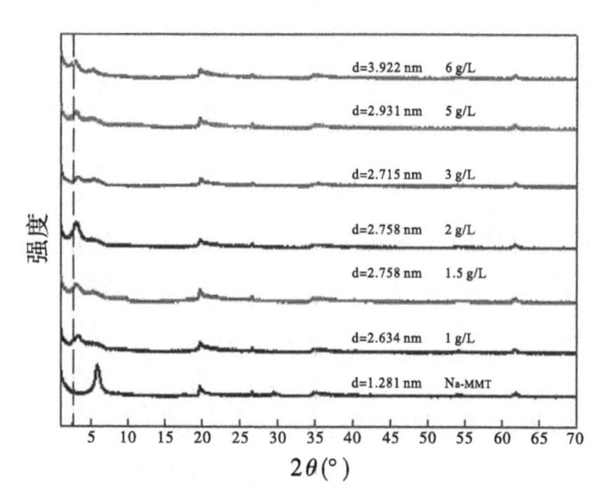

图 1 聚羧酸减水剂插层蒙脱石 XRD 图谱

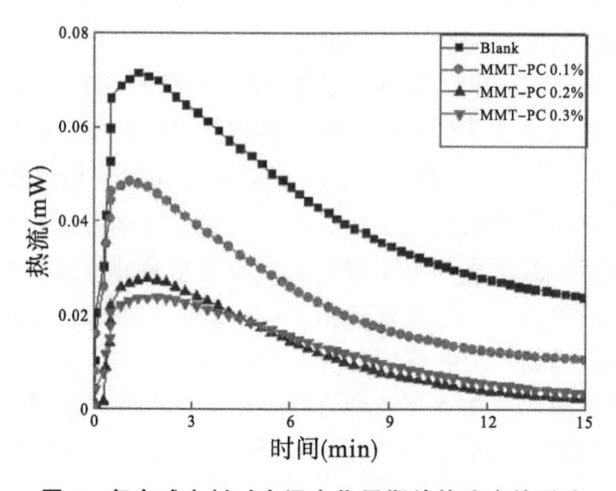

图 2 复合减水剂对水泥水化早期放热速率的影响

参考文献

[1] Zhou Q，Shen W，Zhu J X，et al. 2014. Structure and dynamic properties of water saturated CTMA-montmorillonite：molecular dynamics simulations [J]. Applied Clay Science，97：62－71.

[2] Varadwaj G B B，Rana S，Paridaand K，et al. 2014. Amulti-functionalized montmorillonite for cooperative catalysis in one-pot Henry reaction and water pollution remediation [J]. Journal of Materials Chemistry A，2（20）：7526－7534.

[3] Kagawa S，Suh S，Hubacek K，et al. 2015. CO_2 emission clusters withinglobal supply chain networks：Implications for climate change mitigatio [J]. Global Environmental Change，35：486－496.

斜硅镁石的高温振动光谱研究

叶　宇　刘　丹

（中国地质大学（武汉）地质过程与矿产资源国家重点实验室，武汉 430074）

板块俯冲是将地表水传输带入地幔深部的主要动力学过程，俯冲带中的岩石和矿物都含有一定量的水，存在形式可能是结构羟基或者水分子。俯冲板块中的蛇纹石（serpentine，13wt% H_2O）进入地幔深处后，在深度 120～180 km 的高温高压作用下将产生 Phase A（11.8wt% H_2O）、粒硅镁石（chondrodite，5.4wt% H_2O）和斜硅镁石（clinohumite，2.9%wtH_2O）。因此，这些含水矿物被作为水的搬运工，将俯冲板块中的水带入上地幔深部乃至转换带中。本文主要通过高温拉曼和红外探究天然斜硅镁石和合成斜硅镁石的高温行为，从而探究水对俯冲带的致密高含水的硅酸镁矿物（DHMS）的热力学性质的影响，以及为了解俯冲板块中的水能否带入 410 km 以下的转换带提供有力的实验约束与证据。

采用外加热装置结合显微拉曼分析技术和显微红外光谱仪，按照每次间隔 50 K，在常压下原位测量天然斜硅镁石［$Mg_{7.093(03)}Fe_{0.583}(SiO_4)_4(Mg_{0.749(83)}Ti_{0.250(17)})O_{0.5}OH_{1.3}F_{0.201(69)}$］和合成斜硅镁石［$Mg_9(SiO_4)_4(OH)_2$］的高温振动光谱。高温拉曼光谱测试的范围是 80～1 050 K，斜硅镁石的结构在实验温压范围内稳定，部分振动峰出现了弱化消失的现象，所有的拉曼振动峰随着温度的升高向低频移动。为了研究斜硅镁石中 OH 振动和温度之间的关系，本文分别进行了天然斜硅镁石从常温到 1 243 K，合成斜硅镁石从常温到 1 093 K 的原位高温红外光谱测试。合成斜硅镁石在 1 093～1 143 K 发生分解，而天然斜硅镁石在 1 243 K 晶体结构仍然保持稳定。天然斜硅镁石和合成斜硅镁石共有的吸收光谱在 3 609、3 560、3 526 cm^{-1}，由于氢氢排斥作用而随着温度的升高向低频移动。但天然斜硅镁石特有的吸收光谱在 3 309 cm^{-1}和 3 397 cm^{-1}，由于 F 的取代（$F^- = OH^-$），而随着温度的升高向高频移动。本文的实验结果也证明之前学者的研究，F 的取代使得斜硅镁石更加稳定，F 使得斜硅镁石的晶体结构发生了改变。

根据高温实验和高压拉曼实验的结果，计算了斜硅镁石的 Grüneisen 参数，并计算了以及 Phase A、粒硅镁石这类致密的高含水硅酸镁矿物的非谐参数。通过和橄榄石的对比，这类致密的高含水硅酸镁矿物的非谐参数要更小一些。在探究矿物在高温下的热力学性质时，一般采用简化的 DeBye 模型，默认分子做简谐振动。本研究支持这种近似准谐波在俯冲带的高温下对斜硅镁石是有效的，因为即使当外推到 2 000 K 时，非谐效应对于热容的贡献不超过 2%。因此，经典的 DeBye 模型可以合理地模拟这些 DHMS 相在俯冲带中的热力学性质。

Structural Alteration of Montmorillonite under Acid Activation and Heating

Meng Na Yang　Hong Juan Sun*　Tong Jiang Peng　Yan Yan Xie
(Key Laboratory of Solid Waste Treatment and Resource Recycle, Southwest University of Science and Technology, Mianyang 621010, China)

Activation is a way that can greatly improve the adsorption property of bentonite, of which montmorillonite is the main component. Previous studies have shown that activation is actually a reaction between the acid and the montmorillonite layers[1]. The activated product, also called activated clay, has strong adsorption ability that can be applied to the refining and decolorization of fats and oils[1,2]. For activated clay production industry in China, there is an urgent need to improve the properties of their products recently, so it' s necessary to explore the technological factors affecting the properties of activated clay. In the previous low-temperature calcination experiment of montmorillonite activation, we found that calcination temperature is the most influential factor for the product properties, but the mechanism of its influence is still unclear, so experiments were conducted to find out the structural alteration of montmorillonite and how it influence the properties.

Bentonite collected from JiangxiProvince of China was dry milled and then screened by 100 mesh sieve. The obtained powder (10 g) was mixed with 2.5 mL conc. H_2SO_4 (98%) (Analytical grade) in a corundum crucible and then calcinated in a tube furnace for 40 min at temperatures varying from 160 ℃ to 240 ℃. After that the samples were repeatedly centrifuged with deionized water until the pH of the supernatant reaches 5&6. The wet samples were dried at 105 ℃ in an oven and ground to pass through a 200 mesh sieve to obtain a series of samples.

The XRD result shows that there are Ca-montmorillonite, quartz and calcite in the bentonite sample. With the increase of the calcination temperature, the relative intensities of basal reflections of montmorillonite decrease obviously, indicating the damage of the structure of thelayers. Size decreasing and edge curling are found in the SEM pictures of activated clay, responding to the results of XRD. With the increasing of calcination temperature, the damage caused by acid extent rapidly. The XRF results indicate the exact damaged part of the sheet. The contents of Al_2O_3, MgO, Fe_2O_3 and CaO decrease significantly while the relative content of SiO_2 increases with the extension of calcination temperature. This is as a result of the extension and quickening removal of cations from the interlayer (Ca^{2+} and Mg^{2+}) and

* Corresponding author: Hong Juan Sun; E-mail: sunhongjuan @ swust. edu. cn; Mobile Phone No.: ++86-13550826578. Innovative Team Project of Sichuan Education Department (14TD0012).

the octahedral sheets (Al^{3+}, Mg^{2+} and Fe^{3+}) of montmorillonite[1], for the reaction intensity increases as temperature rise. Fig. 1 shows the structural alteration of montmorillonite under activation. The properties of montmorillonite change with the change of layer structure. The effect of calcination temperature has a great influence on the structure of montmorillonite, from this point of view, it influences the adsorption properties prominently. The decolorization rate tests show significant differences of samples obtained from different temperatures. AC-200 shows the best adsorption property, as the decolorization rate is 98.82%. BET test of three activated samples gives detailed explanation, with the specific surface area of 188.42 m^2/g, 192.74 m^2/g and 158.46 m^2/g, respectively. This confirmed the theory that when montmorillonite sheet is excessively damaged by acid, it will cause the structure to collapse, thereby blocking a part of the pores, resulting in a decrease in adsorption capacity[3].

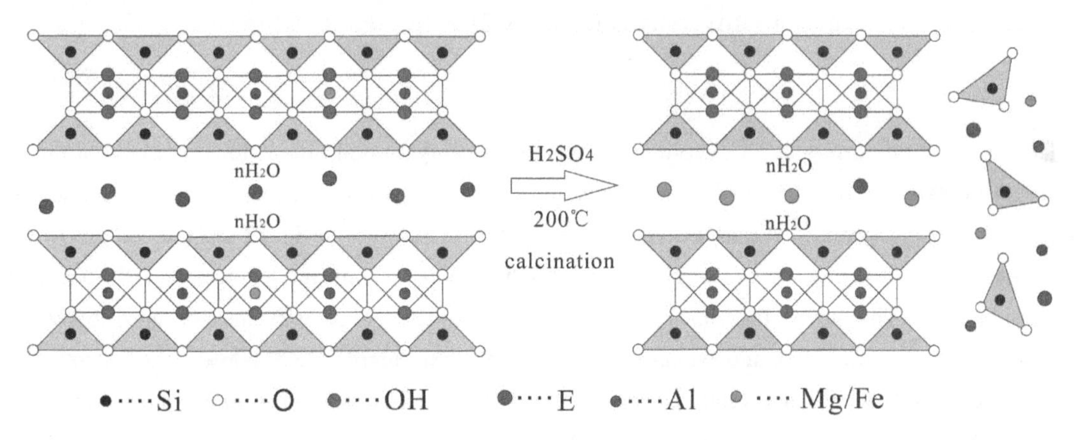

Fig. 1 Schematic diagram of the structural alteration of montmorillonite after activation

References

[1] Komadel P. 2016. Acid activated clays: Materials in continuous demand [J]. Applied Clay Science, 131: 84－99.

[2] Ajemba R O. 2013. Alteration of Bentonite From Ughelli By Nitric Acid Activation: Kinetics And Physicochemical Properties [J]. Indian Journal of Science & Technology, 6 (2): 4076.

[3] Babaki H, Salem A, Jafarizad A. 2008. Kinetic model for the isothermal activation of bentonite by sulfuric acid [J]. Materials Chemistry & Physics, 108 (2-3): 263－268.

Chloritization Simulating Starts from Montmorillonite at Hydrothermal Condition

Xiao Ke Zhang Yuan Feng Cai* Yu Guan Pan
(State Key Laboratory of Mineral Deposits Research,
School of Earth Science and Engineering,
Nanjing University, Nanjing 210046, P. R. China)

The present study provides an overview of the solution-claymineral interactions involved in the chloritization process, and focuses on the transformation processed at low-temperature in diagenetic condition or hydrothermal systems. A series of simulation experiments were designed and carried out in sealed tubes which were placed in an oven. The oven was heated to a constant temperature of 70－150 ℃ (±2 ℃) for 30 d. The parent materials used for these experiments were different montmorillonites (One of the samples was white (W) and wellcrystallized, the other sample was yellow (Y), also Y had a higher iron content than W, and relatively poorly crystallized). The montmorillonites were introduced to solutions with concentrations at different Fe/Mg/Al moral ratios with various values of pH and Eh. The products were analyzed using X-ray powder diffraction (XRD), Fourier-transform infrared (FTIR) spectrometry, scanning electron microscopy (SEM) and transmission electron microscopy (TEM).

Qualitative analysis of XRD patterns and SAD (Selected Area Diffraction) of TEM revealed that the main product phase was a chlorite group mineral. The 001 diffraction peak was present at 14.2Å, besides 001 diffraction, and the other diffractions located at 7.10, 4.73, 3.54, 2.83 and 1.537/1.499Å were assigned to 002, 003, 004 and 060 diffractions of chlorite, respectively. Products from Y had two (060) reflections at 1.537 and 1.499Å, which corresponded to 060 lattice planes of tri and dioctahedral chlorites, respectively, but those samples from W only showed the 060 refletion diffraction at 1.499Å, indicating the only presence of dioctahedral chlorite. Stretching vibrations of Fe (Ⅱ) Fe (Ⅱ) OH-AlFe (Ⅱ)OH and/or MgFe (Ⅲ) OH were observed in FTIR spectra (3 562－3 520 cm^{-1}) of the two products formed inneutral and alkaline solutions ($7<pH<9$) with redox conditions, which indicated that Fe (Ⅱ), Mg (Ⅱ) entered octahedral position. The AlFe (Ⅱ) OH-MgFe (Ⅲ) OH (3 562－3 550 cm^{-1}) vibrations were only observed in products formed in alkaline solutions ($pH>9$). SEM observations suggested a dramatically change of morphology of grains before and after reaction. Most of the montmorillonite grains were flaky, smooth surfaced before reaction. After reaction, the neoformed, ooliticchlorite material was observed on the surface of Y, and there were tiny scales of chlorite of W.

* Corresponding author: Yuan Feng Cai; E-mail: caiyf@nju. edu. cn. This study was funded by the National Natural Science Fund Project (Project number: 41672037).

The morphological change of two minerals showed that chlorite was formed by a dissolution-crystallization mechanisms（DC），and the mineral reaction that was not dominated by a solid-state transformation（SST）mechanisms which was suggested by previous study[1,2].

The simulation experiments demonstrated that the pH and redox conditions（Eh）of the environment controlled the nature of the daughter mineral species，as well as the iron content of clay itself did. Results in the present study revealed that chloritizationoccurred preferentially in a nearly neutral to alkaline and redox condition. Furtermore，trio-octaheral chlorite was apt to form when montmorillonite has a higher iron content.

References

[1] Beaufort D，Rigault C，Billon S. 2015. Chlorite and chloritization processes through mixed-layer mineral series in low-temperature geological systems-a review [J]. Clay Minerals，50：497 - 523.

[2] Inoue A，Utada M，Nagata H. 1984. Conversion of trioctahedralsmectite to interstratified chlorite/smectite in Pliocene acidic pyroclastic sediments of the Ohyu district，Akita Prefecture，Japan [J]. Clay Science，6：103 - 116.

第二章

非金属矿地质、采选和分析表征

蛋白石凹凸棒石型黏土选矿、加工及资源化应用

王 灿 刘海波 陈天虎*

（合肥工业大学资源与环境工程学院，合肥 230009）

富含蛋白石的凹凸棒石黏土称为蛋白石—凹凸棒石黏土，由于矿石中蛋白石含量过多，该类矿石的吸附性能和胶体性能均很差，尚没有得到合理利用。目前仅盱眙县黄泥山凹凸棒石矿区每年就有大约10万吨蛋白石—凹凸棒石黏土，其中有40%～50%凹凸棒石与蛋白石团块混杂在一起，由于缺少分选加工技术及相关的应用研究，巨量蛋白石—凹凸棒石黏土未得到合理利用，被丢弃矿山或者直接回填采坑，由此导致矿山废石堆积如山，不仅影响了矿山环境，而且影响矿山企业经济效益。因此，本研究针对该类矿石开发出一种低成本的干法选矿技术，实现蛋白石和凹凸棒石的有效分离。根据分选出的蛋白石和凹凸棒石不同的矿物学特性，分别进行资源化应用。利用蛋白石中SiO_2结晶度低的特性，采用蛋白石与碱反应湿法制备出水玻璃这种化工广泛应用的原料。利用凹凸棒石具有较大的比表面积、特殊的纳米棒状结构效应、物理吸附能力等特性[1,2]，用做催化剂载体，负载锰氧化物活性组分，开发出凹凸棒石—锰氧化物复合材料用于催化降解甲醛，探究其催化甲醛反应的活性及机制。

通过研究蛋白石—凹凸棒石黏土的干法选矿技术，不仅实现了蛋白石和凹凸棒石的有效分离，也为低品位的凹凸棒石和蛋白石提供了新的应用途径，实现了这类矿产资源的高值化合综合利用。

本实验所用蛋白石凹凸棒石型黏土取自江苏盱眙黄泥山矿场。整个矿层剖面从上至下可以划分成4层，依次为玄武岩层、富蛋白石凹凸棒石型黏土（硅质黏土）、凹凸棒石黏土及富蒙脱石凹凸棒石黏土（图1)。将经自然干—湿交替风化后的蛋白石凹凸棒石型黏土进行干法筛分、酸处理纯化，最终分离出蛋白石和凹凸棒石。分离出的蛋白石通过与碱水热反应制备水玻璃。分离出凹凸棒石作为催化剂载体，负载锰氧化物催化净化甲醛。

干法选矿结果如表1所示，经4 mm规格筛子筛分处理后筛上物质量比例为47.6%，蛋白石含量为71.1%，其中凹凸棒石含量约为22.2%；4 mm以下各粒级蛋白石含量在25%～30%，凹凸棒石含量在65%～75%，不同粒级间蛋白石和凹凸棒石各组分含量比较接近，因此，采用4 mm筛可以实现蛋白石和凹凸棒石的有效分离。

在单因素实验考察不同影响蛋白石碱溶因素的基础上，通过响应面分析法中Box - Behnken实验设计，对分离提纯后蛋白石碱溶制备水玻璃的工艺条件进行优化，得出在蛋白石用量为10 g时，当液固比为6∶1，碱硅比为4.05∶10，反应温度为110 ℃时，加权平均值达到最大值76.1，此时SiO_2浸出率为81.4%，水玻璃模数为2.85。

* 通信作者：陈天虎；E-mail：chentianhu@hfut. edu. cn；手机号：13956099615。中国科学院盱眙凹土应用技术研发与产业化中心开放课题资助项目（201503）。

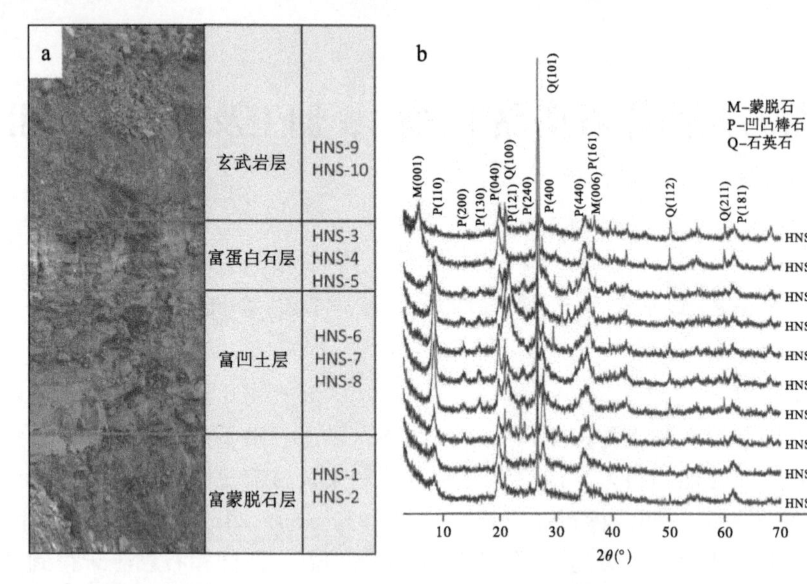

图 1　凹凸棒石黏土矿层中蛋白石的分布（a）剖面矿物组成（b）多剖点 XRD 分析

表 1　不同粒级黄泥山蛋白石—凹凸棒石黏土化学成分分析（%）

粒级（mm）	比例	SiO_2	MgO	Al_2O_3	CaO	Fe_2O_3	其他	烧失量	凹凸棒石	蛋白石
＞4	47.6	72.03	8.12	3.44	1.54	2.65	0.80	10.07	22.2	71.1
2～4	17.6	54.26	10.88	5.15	1.70	3.59	1.03	21.57	66.2	29.7
1～2	8.8	50.39	11.42	5.41	1.67	3.82	1.05	24.45	70.4	26.8
1～0.5	11.4	50.20	11.77	5.21	1.49	3.65	1.00	24.97	72.3	25.4
＜0.5	14.6	52.01	11.50	5.86	2.10	2.26	1.10	23.25	68.7	27.4

以选矿纯化后的凹凸棒石为催化剂载体，采用不同锰前驱体通过共沉淀法制备了 4 种 MnO_x/PG 催化剂，并用于催化氧化甲醛。结果表明：以高锰酸钾为前驱体的 Mn/PG－PP 催化剂中锰氧化物呈纳米化高度均匀包覆在凹凸棒石表面，可将高浓度甲醛（1 200 mg/kg）在 125 ℃完全转化为二氧化碳和水，对低浓度甲醛（1 mg/kg），在室温条件下，10 h 内甲醛转化率维持在 100％。

参考文献

［1］陈天虎，徐晓春，岳书仓. 2004. 苏皖凹凸棒石黏土纳米矿物学及地球化学［M］. 北京：科学出版社.

［2］Wang C，Liu H B，Chen T H，et al. 2018. Synthesis of palygorskite-supported $Mn_{1-x}Ce_xO_2$ clusters and their performance in catalytic oxidation of formaldehyde［J］. Applied Clay Science，159：50－59.

硫化金属氧化物柱撑黏土制备及其纤维素水解性能研究

周　扬　杨　淼　杨海燕　房　凯　童东绅* 周春晖*
（浙江工业大学化学工程学院，杭州 310014）

面对煤、石油和天然气等矿物资源急剧消耗及环境污染问题的双重危机，开发与利用生物能源成为符合可持续发展战略的必然选择。其中，纤维素占生物质组分的40%～50%，因此，从纤维素出发制备有价值的化学品成为当今研究的重要议题[1]。纤维素是由D-葡萄糖分子以β-1,4-糖苷键连接而成的大分子多糖，因此纤维素降解的关键步骤是在催化剂作用下破坏断裂糖苷键。目前，在催化水解纤维素上，液体酸催化存在腐蚀设备、不易回收等缺点，因此，寻找一种新型高效的固体酸催化剂已成为目前研究的重点和热点问题[2]。

黏土为层状结构的硅酸盐非金属纳米矿物，由于具有来源广泛、价格低廉、分散性好、膨胀性能强等优点而被广泛应用[3]。本研究利用黏土固有的膨胀性将硫化金属氧化物嵌入其层间，以沉淀硫酸化法制备硫化金属氧化物柱撑黏土，制得柱撑黏土具有可控的孔结构，更好地分散了活性相和反应物，使其具有可观的Bronsted酸性位点。

因此，本研究将制备的硫化氧化锆柱撑蒙脱土材料作为固体酸催化剂催化水解纤维素[3]。通过IR、XRD等表征技术对硫化氧化锆柱撑蒙脱土结构进行表征，利用紫外分光光度仪在520 nm波长处检测还原糖产率，以此研究其催化水解纤维素的性能。

图1是由沉淀硫酸化法制备的一系列不同硫锆比的硫化氧化锆柱撑黏土催化剂的红外谱图。由图1可知，除基团吸收强度不同外，不同硫锆比例的催化剂红外谱图与未柱撑硫化氧化锆的红外谱图基本相同。850～460 cm^{-1}范围内的吸收峰对应着蒙脱土Si—O—Si骨架伸缩振动峰和Si—O—Al骨架伸缩振动峰，说明催化剂中存在蒙脱土。

图2为不同硫锆比柱撑蒙脱土的XRD图谱。由图2可知，不同硫锆比柱撑蒙脱土的XRD图在5.68°附近基本都没有出现衍射峰，说明硫化氧化锆柱撑到了蒙脱土层间，致使蒙脱土晶面间距变大。在19.76°和22.02°附近出现的是蒙脱土的另外2个典型晶面衍射峰110晶面和010晶面；在30°附近出现的衍射峰表明催化剂中存在四方相氧化锆。结果表明，成功制备硫化氧化锆柱撑蒙脱土。

将沉淀硫酸化法制备的不同硫锆比和不同蒙脱土含量的复合材料作为固体催化剂水解纤维素，结果如图3所示，S/Zr比的变化对硫化氧化锆柱撑黏土催化剂的催化活性有明显的影响。随着S/Zr比的增大，还原糖收率基本呈现一个先增大后减小的趋势，在S/Zr比为0.3时还原糖收率最高，为17.62%；纤维素转化率也基本是先增大后减小的趋势，在S/Zr比为0.5时纤维素转化率最高，为50%。S/Zr比为0.5的催化剂纤维素转化率高，但是还原糖收率低，也就是其选择性

* 通信作者：童东绅；E-mail：tds@zjut. edu. cn；手机号：13656675377；周春晖，E-mail：clay@zjut. edu. cn。
国家自然科学基金（21506188）、浙江省自然科学基金（LY16B030010）资助项目。

不高，还原糖进一步分解成其他副产物。综上所述，当 S/Zr 比为 0.3 时，硫化氧化锆柱撑蒙脱土催化剂水解纤维素效果最佳，转化率达到 45.0%，还原糖收率达到 17.62%。

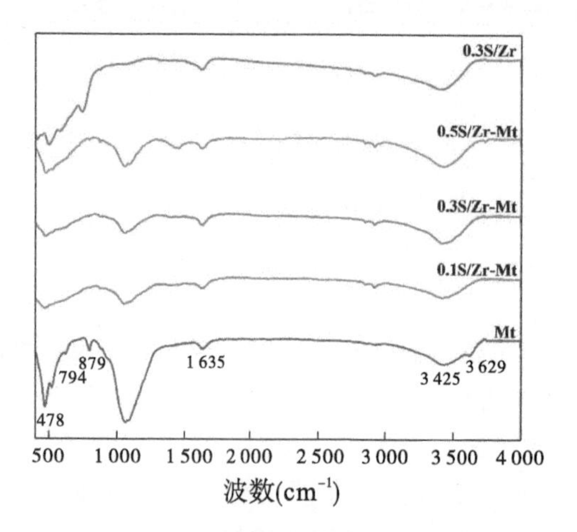

图 1　不同硫锆比柱撑蒙脱土 IR 图

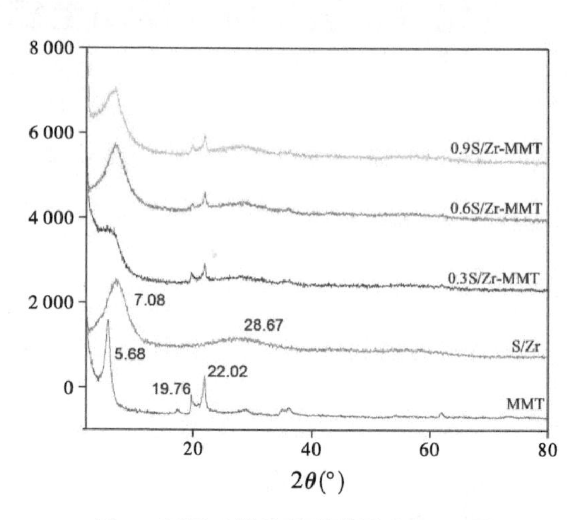

图 2　不同硫锆比柱撑蒙脱土 XRD 图

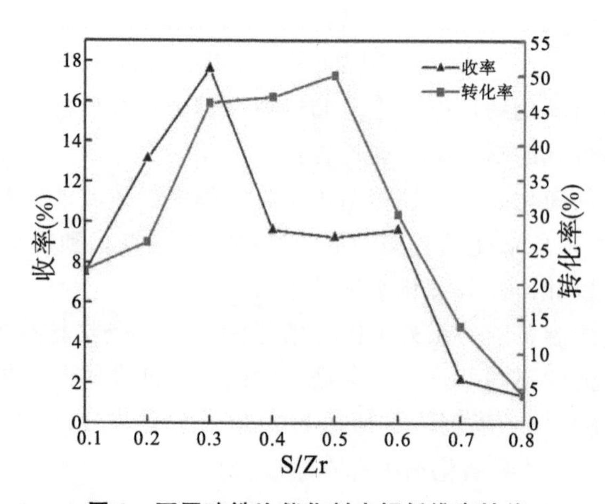

图 3　不同硫锆比催化剂水解纤维素性能

参考文献

[1] Zhou C H，Xia X，Lin C X，et al. 2011. Cheminform abstract：catalytic conversion of lignocellulosic biomass to fine chemicals and fuels [J]. Cheminform，40 (11)：5588 - 5617.

[2] Tong D S，Xia X，Luo X P，et al. 2013. Catalytic hydrolysis of cellulose to reducing sugar over acid-activated montmorillonite catalysts [J]. Applied Clay Science，74 (1)：147 - 153.

[3] Vellayan K，González B，Trujillano R，et al. 2017. Pd supported on cu-doped ti-pillared montmorillonite as catalyst for the ullmann coupling reaction [J]. Applied Clay Science，160：126 - 131.

利用凹凸棒石矿中的蛋白石岩合成4Å沸石

邬宗姗　陈天虎*　刘海波　王　灿　谢晶晶　周跃飞
（合肥工业大学资源与环境学院，合肥 230009）

4Å沸石是具有均一微孔道的架状结构铝硅酸盐结晶物质，因其结构中微孔道有效孔径为4Å（埃）而得名。理想晶体化学结构式为 $Na_{12}[(AlO_2)_{12}(SiO_2)_{12}]27H_2O$[1,2]，其中硅铝皆占据四面体位置，钠离子远程静电平衡四面体位置铝替代硅产生的结构电荷，为可交换性阳离子[3]。由于其独特的阳离子交换性、吸附选择性和催化性能，已在化工、环保、冶金、医药等领域获得广泛应用[4,5]。选择来源丰富、廉价的活性硅铝原料在较温和的条件下合成质量符合标准要求的4Å沸石是本领域研究方向之一。

苏皖地区凹凸棒石矿中普遍存在蛋白石岩透镜体或者条带，在开采过程中形成大量固体废弃物。前期研究[6]表明，凹凸棒石矿中的蛋白石岩主要是微晶蛋白石（Opal－CT），含有少量的凹凸棒石、白云石，基本不含石英、伊利石等其他原料所常见的杂质矿物。该蛋白石与强碱具有很高的反应活性，用作合成沸石分子筛的原料有望得到高品质产品。本文研究蛋白石岩合成4Å沸石的工艺条件，为苏皖地区凹凸棒石黏土矿山蛋白石岩废弃物的资源化利用提供理论支撑。

剥离出来的蛋白石—凹凸棒石黏土用推土机或者铲车堆存在矿场空地（图1左图），经过一段时间的露天放置后，蛋白石—凹凸棒石黏土团块发生了风化（图1右图）。选取经过风化后蛋白石—凹凸棒石黏土用不同粒径的筛进行筛分。XRD图谱表明，经4 mm规格筛子筛分处理后筛上物更富含蛋白石，其含量约为73％。取粒径大于4 mm筛上物料100 kg，用颚式破碎机、对辊机粗碎并缩分，再使用球磨机粉碎过200目筛。为了消除样品中凹凸棒石等杂质的影响，用10％盐酸、液固比3∶1、70 ℃酸浸2 h，过滤、洗涤至中性并干燥备用。

图1　蛋白石—凹凸棒石黏土堆放及风化现场

*通信作者：陈天虎；E-mail：chentianhu168@vip.sina.com；手机号：13956099615。国家自然科学基金（41572028）、中科院凹土工程中心（201503）资助项目。

以经过预处理凹凸棒石矿中蛋白石岩为原料，采用水热法合成 4Å 沸石，X 射线粉末衍射作为物相组成分析手段，钙离子交换量、静态吸附水量作为性能评价指标，考察了反应时间、反应温度、Na_2O/SiO_2、H_2O/Na_2O、SiO_2/Al_2O_3 对产品性能的影响，优化了利用蛋白石岩水热法合成 4Å 沸石的条件，并通过扫描电镜、红外光谱等表征最佳合成工艺下产物的结晶度、结构，分析了蛋白石转变为 4Å 沸石的机理。结果表明，以凹凸棒石矿中的蛋白石岩为主要原料，水热法可合成出性能符合国家轻工业标准要求的 4Å 沸石。最优条件为：85 ℃反应 3 h，$Na_2O/SiO_2=1.0$，$H_2O/Na_2O=40$，$SiO_2/Al_2O_3=2.0$，所获得的 4Å 沸石钙离子交换量 293 $mgCaCO_3/g$ 干 4Å 沸石，静态吸附水量 22.3%。XRD 图谱（图 2）中 4Å 沸石的特征峰符合文献报道，由 SEM 结果（图 3），合成的 4Å 沸石呈立方体晶型，完全符合文献中对 4Å 沸石晶形特征的描述。

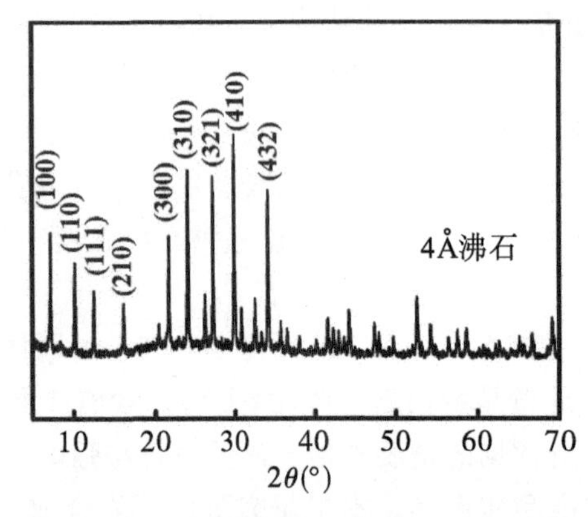

图 2　最优合成工艺下 4Å 沸石的 XRD 图

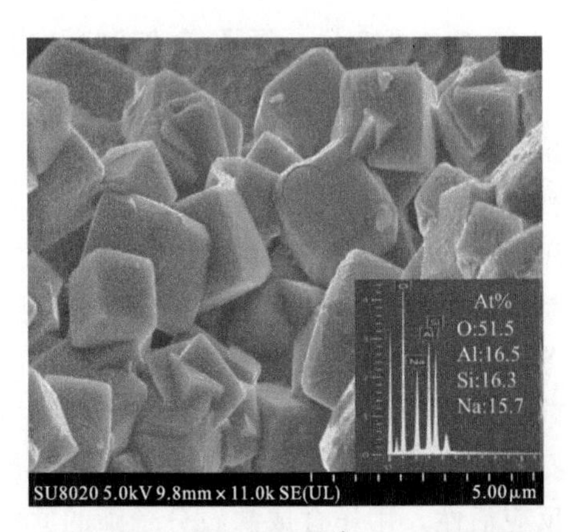

图 3　最优合成工艺下 4Å 沸石的 SEM 图

参考文献

[1] García G，Aguilar-Mamani W，Carabante I，et al. 2015. Preparation of zeolite A with excellent optical properties from clay [J]. Journal of Alloys & Compounds，619：771 - 777.

[2] 金兰淑，高湘骐，刘洋，等. 2012. 4Å 沸石对复合污染水体中 Pb^{2+}，Cu^{2+} 和 Cd^{2+} 的去除 [J]. 环境工程学报，6：1599 - 1603.

[3] Ojuva A，Järveläinen M，Bauer M，et al. 2015. Mechanical performance and CO_2，uptake of ion-exchanged zeolite A structured by freeze-casting [J]. Journal of the European Ceramic Society，35：2607 - 2618.

[4] Müller P，Russell A，Tomas J. 2015. Influence of binder and moisture content on the strength of zeolite 4Å granules [J]. Chemical Engineering Science，126：204 - 215.

[5] Ayele L，Pérez-Pariente J，Chebude Y，et al. 2015. Synthesis of zeolite A from Ethiopian kaolin [J]. Microporous&Mesoporous Materials，215：29 - 36.

[6] 陈天虎，徐晓春，Xu H F，等. 2005. 苏皖坡缕石黏土中蛋白石特征及其成因意义 [J]. 矿物学报，25：81 - 88.

固体核磁共振在矿物/水溶液界面中的应用：纳米羟基磷灰石除氟研究

任 超 苟文贤 季峻峰 王洪涛 李 伟*
（南京大学地球科学与工程学院，南京 210023）

饮用水中的氟（F）污染物被认为是目前威胁人类健康的主要问题之一。根据世界卫生组织（WHO）规定，安全饮用水中的氟含量不能超过 1.5 mg/L[1]。全世界有超过 25 个国家，超过 2 亿人口正面临饮水型氟中毒的威胁[2]。氟中毒在中国也是一种流行严重的地方病，在四种主要的地方病中，地氟病分布最广，危害最大。

羟基磷灰石（Hydroxyapatite，简称 HAP）在自然界中广泛分布，相比于其他吸附剂有吸附量大、吸附效果好、环境友好等特点。前人对 HAP 除氟也开展了一定的研究，但是由于实验条件的限制，大多数研究仅停留在吸附实验部分，缺乏谱学方面的研究，关于 HAP 吸附 F 的反应机制还存在一些争论。本研究提供了一种新型的氟固体核磁共振技术（^{19}F NMR）以查明 HAP 对 F 的吸附机制，并运用透射电子显微镜（TEM）、X 射线衍射技术（XRD）和傅里叶变换红外光谱（FTIR）等手段进行证实。

实验结果表明，HAP 吸附 F 主要受初始 F 浓度和 pH 值控制。通过^{19}F NMR 分析得出，在 pH＝7，初始 F 浓度≤50 mM 条件下，反应产物的化学位移是 $\delta=-103$ ppm，所对应的是 F－HAP（含氟羟基磷灰石），此时 HAP 对 F 的吸附机制以离子交换反应为主，即 F^- 取代了 HAP 的隧道－OH。当初始 F 浓度增大到一定程度时（＝100 mM），HAP 除氟的机制发生改变，出现了 2 种化学位移，分别为－103 和－108 ppm，所对应的产物分别是 F－HAP 和 CaF_2（氟化钙），当初始 F 浓度大于 100 mM 时，产物的化学位移 $\delta=-108$ ppm 对应 CaF_2，此时生成沉淀态 CaF_2 是主要的吸附机制（图 1）。

通过结合 TEM（图 2）和 XRD 进一步证实了 NMR 的结论并发现生成的 CaF_2 具备良好的结晶度。

研究发现在 pH＝4 条件下，初始 F 离子浓度为 5 mM 时就会产生 CaF_2 沉淀。相反，pH＝10 条件下，即使初始 F 浓度增加到 500 mM 时仍无 CaF_2 生成。本研究首次运用固体核磁共振技术系统地讨论了 HAP 吸附 F 的分子水平机制，并得到不同机制的转变边界，对将 HAP 运用于实际饮用水除氟和工业除氟具有一定的指导作用。同时本研究还通过再生实验证明了 HAP 具备很好的再生性能，并对再生机制进行了探讨。

---

* 通信作者：李伟；E-mail：liwei_ isg@nju. edu. cn；手机号：13952007316。江苏省杰青（BK20150018）、中组部青年千人计划、国家自然科学基金（No. 41473084）、南京大学登峰人才计划资助项目。

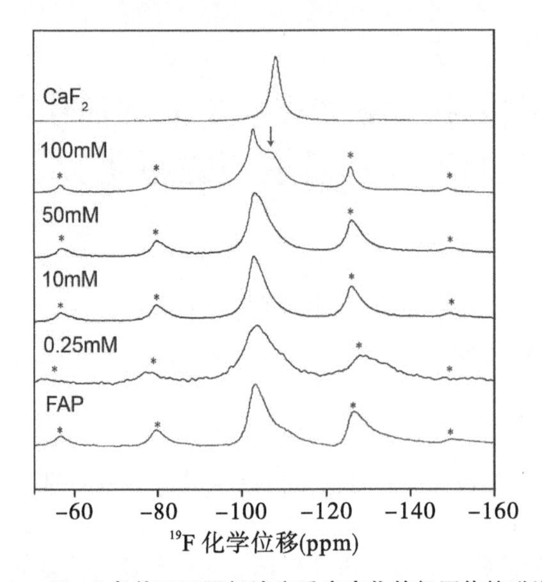

图 1　pH=7 条件下不同氟浓度反应产物的氟固体核磁谱图

注："＊"表示的是边带，红色箭头对应的肩峰化学位移是－108 ppm，对应 CaF_2 沉淀生成

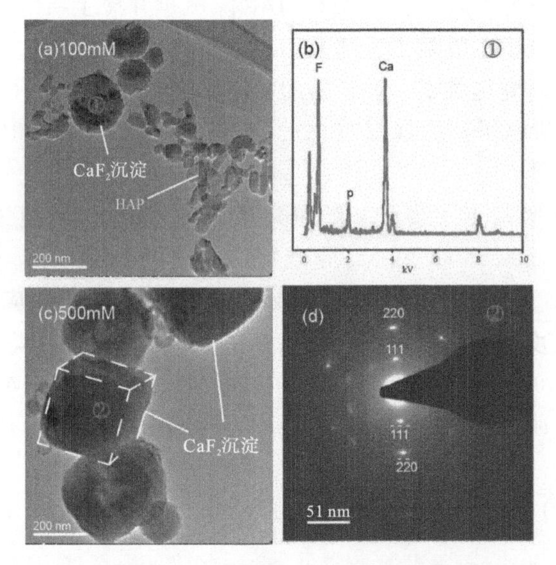

图 2　pH=7 条件下，(a) 100 mM 产物 TEM 结果显示，圆形指示新矿物的生成；(b) EDAX 能谱分析表明新生成矿物含 F、Ca；(c) 200 mM 产物的 TEM 及 (d) SEAD 分析说明 CaF_2 沉淀具有一定的结晶度

参考文献

[1] Fawell J, Bailey K, Chiliton J, et al. 2006. Fluoride in drinking water [M]. London: World Health Organization.

[2] Ayoob S, Gupta A K. 2006. Fluoride in drinking water: a review on the status and stress effects [J]. Critical Reviews in Environmental Science and Technology, 36 (6): 433 - 487.

Lap/Au 复合材料的制备及其在拉曼光谱中的应用

张　静[1,2] 张　浩[1]　吴琦琦[1]　周春晖[1,2*]
（1. 浙江工业大学化学工程学院，杭州 310014；
2. 青阳非金属矿研究院，青阳 242800）

金纳米颗粒是目前纳米材料界很常用的材料之一，其化学性质稳定、无毒性、等离子体可调、生物兼容性好、易于修饰及制备，可应用于生物成像、医药、传感器、催化、光电子器件等领域[1]。化学还原法制备金纳米颗粒成本相对较低，产物形貌、粒径可以通过改变合成条件来控制。然而在水中还原的金颗粒易团聚，有机保护胶体虽然可以控制成核，防止团聚，但其有毒性，大大限制了金纳米颗粒的生物应用。锂皂石作为人工合成的硅酸盐黏土，合成手段较成熟，可以在水中自发剥离，分散性好，生物相容性好，且具有阳离子交换性和表面反应性[2,3]。

本研究使用简单的化学还原法制备了 Lap/Au 复合光学材料，采用 UV－Vis、EDS、TEM 等手段对得到的复合材料进行表征，并将复合材料与拉曼活性染料结合，研究了复合材料的拉曼增强性质。

复合材料中含有锂皂石的特征元素与金元素（图 1）。这可能是由于氯金酸中的 Au^{3+} 与锂皂石的负电荷表面可以通过静电作用在氯金酸被还原之前就结合在一起。

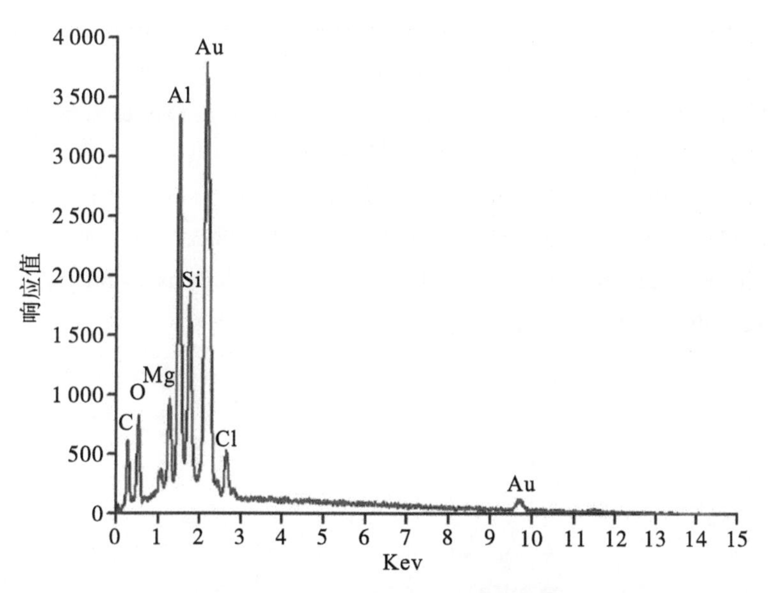

图 1　Lap/Au 复合材料的 EDS 能谱分析

* 通信作者：周春晖；E-mail：clay@zjut. edu. cn。国家自然科学基金（41672033）。

实验所得的金颗粒粒径为 50～60 nm，呈不规则的多边形状，且分散性较好，没有明显团聚；且由于锂皂石的指导作用，金颗粒的吸收波长在 600 nm 左右，与一般的金纳米颗粒相比发生了明显的红移（图 2）。这可能与颗粒尺寸的变大和非球形形状有关。这种金纳米颗粒有着很好的生物相容性，适合用作表面增强拉曼光谱的基底材料[4]。

采用化学还原法在锂皂石分散体中制备 Lap/Au 复合材料粒径较大，吸收波长较长，并且体系中没有与金颗粒结合的锂皂石可以进一步吸附阳离子拉曼活性染料，从而提高拉曼强度。

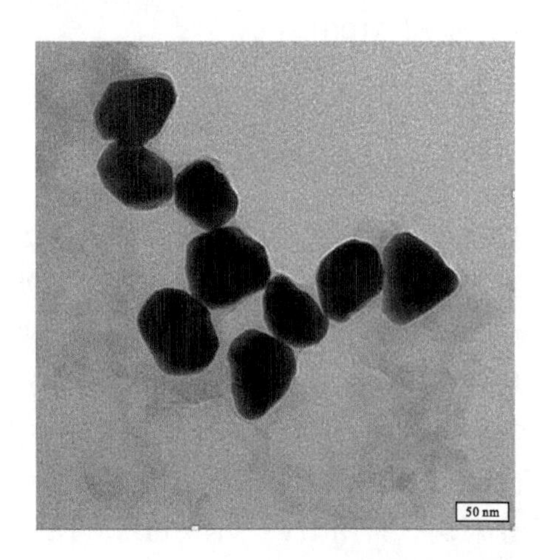

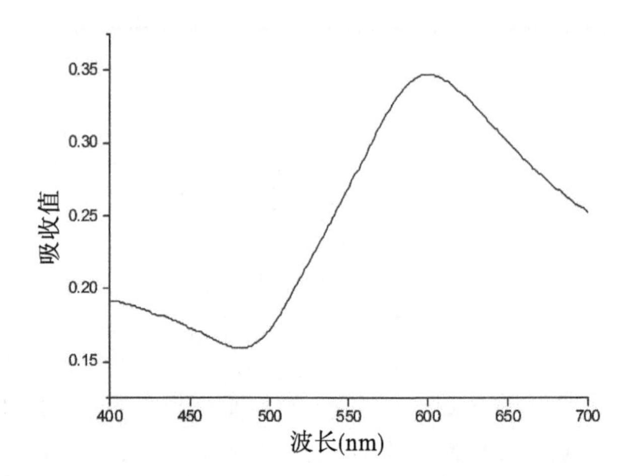

图 2　Lap/Au 样品的 TEM 图及 UV - Vis 图

参考文献

[1] Sau T K，Rogach A L，Jäckel F，et al. 2010. Properties and applications of colloidal nonspherical noble metal nanoparticles [J]. Advanced Material，22（16）：1805 - 1825.

[2] Zhao L Z，Zhou C H，Wang J，et al. 2015. Recent advances in clay mineral-containing nanocomposite hydrogels [J]. Soft Matter，11（48）：9229 - 9246.

[3] Gaharwar A K，Mihaila S M，Swami A，et al. 2013. Bioactive silicate nanoplatelets for osteogenic differentiation of human mesenchymal stem cells [J]. Adv Mater，25（24）：3329 - 3336.

[4] Hill E H，Claes N，Bals S，et al. 2016. Layered silicate clays as templates for anisotropic gold nanoparticle growth [J]. Chemistry of Materials，28（14）.

安徽明光官山凹凸棒石纯化

陈　叶　陈天虎*　王　灿　张　斌　刘海波

（合肥工业大学资源与环境工程学院，合肥 230009）

凹凸棒石又名坡缕石（palygorskite），是一种层链状结构的含水富镁铝硅酸盐黏土矿物[1]。因其具有特殊的结构、形态、物理化学性质以及潜在的应用价值，而得到各界学者的广泛研究。目前凹凸棒石已经应用到农业、食品行业、轻工业、建材行业、日用品行业等各个行业。我国凹凸棒石储量虽大但仍有限，且原矿中低品位矿较多，含有大量的杂质矿物如蒙脱石、伊利石、白云石等，严重影响了对凹凸棒石物化性质的正确认识以及凹凸棒石黏土作为材料的应用效果，还因原料的低品位导致经济效益差，资源浪费大。因此对凹凸棒石纯化以促进对凹凸棒石的深入研究与应用，减少资源化浪费很有必要[2,3]。

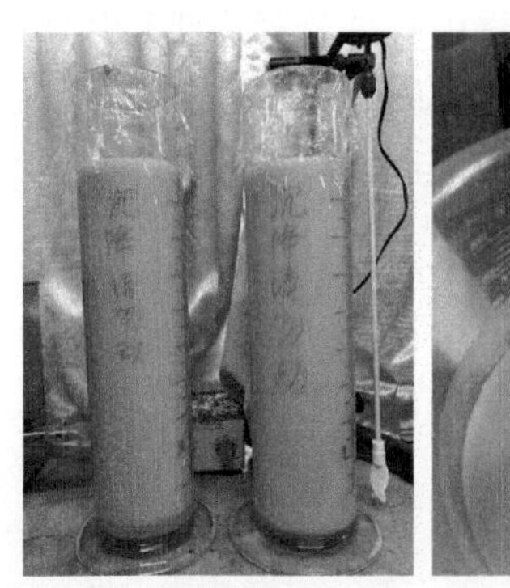

图 1　凹凸棒石沉降（左）和磁选杂质（右）

凹凸棒石的纯化实验主要分为两步，分别为沉降和酸化。所需材料和仪器主要有明光官山凹凸棒石黏土原样、六偏磷酸钠、氯化钙、磁棒、搅拌器、离心机等。实验步骤为：称取原样 200 g，放入 5 L 量筒中，加入 1 L 超纯水，超声 30 min，用磁棒吸出磁性物质（图 1 右）；加入 2 g 六偏磷酸钠和 4 L 超纯水，搅拌 30 min；静置 12 h（图 1 左），用虹吸管取 35 cm 深度以上的混合液体后补充少量水，搅拌 30 min 后，再静置 12 h，用虹吸管取 35 cm 深度以上的混合液体。加少量 $CaCl_2$后在4 000 r/min下离心分离 10 min，将固体物质于 105～110 ℃下烘干过夜。取 100 g 烘干样品放入2 L烧杯中，加入 1 000 mL 1wt%的稀盐酸溶液，于 60 ℃水浴加热以 600 r/min 搅拌 24 h 后将酸化后湿样以相同条件离心洗涤（4 000 r/min，10 min），直至 Cl^- 洗涤完全后烘干。

对提纯前后的样品进行 XRD、XRF 和 TEM 表征，观察提纯前后凹凸棒石主要成分和结构形

* 通信作者：陈天虎；E-mail：chentianhu@hfut. edu. cn；手机号：15856995526。国家自然科学基金（12345678）、安徽省自然科学基金（12345678）资助项目。

貌变化。XRD 结果（图 2）表明，原凹凸棒石中含有伴生的白云石、石英等矿物，提纯后凹凸棒石的特征衍射峰明显增强，石英的特征峰明显减弱，白云石的特征峰衍射峰消失，说明这些杂质矿物得到了有效去除。同时凹凸棒石各个特征衍射峰的位置未见明显变化，表明凹凸棒石的晶体结构在经两步法提纯后未受到破坏。XRF 结果显示，Fe_2O_3 含量从 5.9%降到 2.1%，CaO 含量从 3.7%降到 0.03%，说明杂质得到有效的去除。

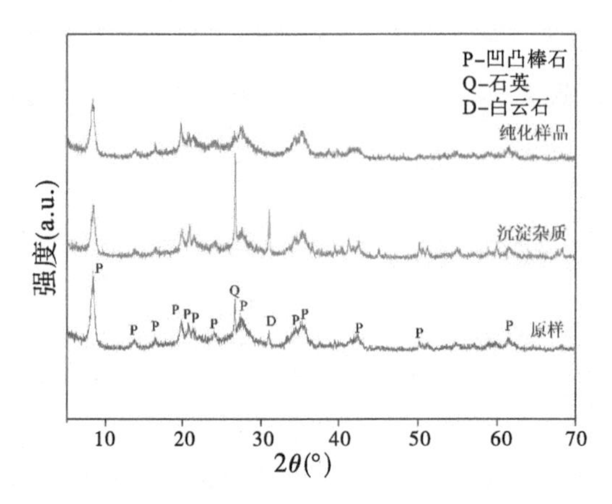

图 2　提纯前后凹凸棒石的 XRD 图

图 3 为提纯前后透射电镜，可以看到原样中有部分团块状和片状的碳酸盐杂质存在（红色虚线位置）。提纯后样品中则可以观察到清晰的凹凸棒石的棒状晶体形态，未发现块状和片状的杂质，表明两步提纯法能够有效提高凹凸棒石的纯度。

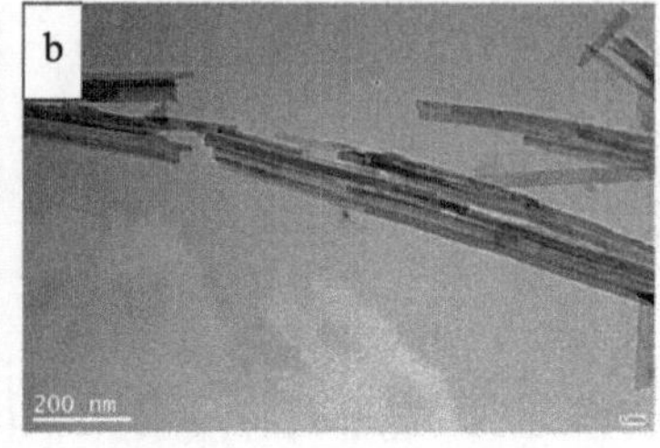

图 3　凹凸棒石提纯前后 TEM 图

注：a. 提纯前；b. 提纯后

综合以上表征结果，凹凸棒石经过静水沉降结合酸浸出的提纯方法能够有效地去除白云石和石英等杂质，可以得到纯度较高的凹凸棒石样品，可用于后续作为高效催化剂的研究。

参考文献

［1］赵善宇，王立久，程正勇. 2005. 凹凸棒石黏土提纯技术及性能研究［J］. 广东建材，61－63.

［2］陈天虎，彭书传，黄川徽，等. 2004. 从苏皖凹凸棒石黏土制备纯凹凸棒石［J］. 硅酸盐学报，32：965－969.

［3］李予晋，曾路，王娟娟，等. 2006. 江苏省盱眙凹凸棒石提纯工艺的研究［J］. 电子显微学报，344－344.

聚电解质对黏土胶体的流变性的影响

吴琦琦[1,3]　俞卫华[2,3]　张　静[1,3]　周春晖[1,3*]

（1. 浙江工业大学化学工程学院，杭州 310032；

2. 浙江工业大学之江学院，绍兴 312030；

3. 青阳非金属矿研究院，青阳 242800）

黏土矿物由于其分散体的黏度和界面性质对于许多工程和工业应用非常重要，包括陶瓷加工、化妆品、油墨和涂料、纸浆、煤炭运输和浮选矿物质和水泥加工[1]等。而聚合物分散剂具有控制胶体加工和在固-液界面保持良好吸附的良好性能，近年来越来越受关注。其在水泥和混凝土浇筑、陶瓷浆料加工、涂料、化妆品等[2]等领域都有应用。

本研究则是在水性体系中聚电解质对合成黏土矿物的流变性的影响来探究两者之间的作用方式使得两者能更好地在涂料陶瓷加工等领域中应用。通过水热合成法合成皂石黏土[3]，通过XRD、IR进行表征，并对其分散体在加入聚电解质5040（一种聚合物分散剂）后的流变性能包括黏度、触变性等的改变进行分析与讨论。

XRD图（图1）中有合成的样品有皂石001晶面的特征峰，通过IR图也证明实验成功合成出皂石。

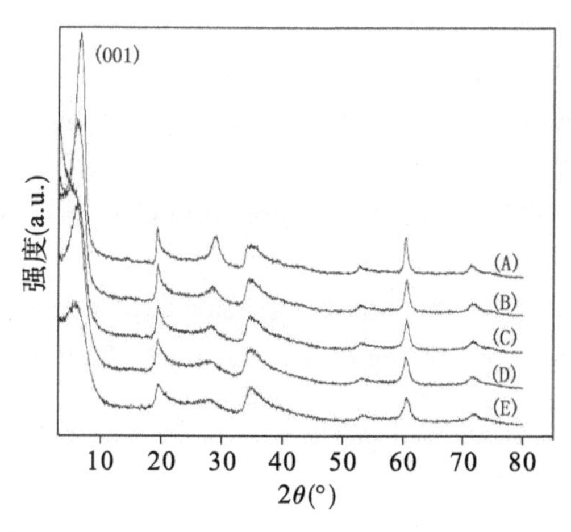

图1　硅铝比为5（A）、10（B）、15（C）、20（D）、25（E）的皂石XRD图

图2是合成的皂石样品中的其中一个流变曲线，通过流变性能的测试证明合成皂石在加入聚电解质后黏度下降由凝胶变为溶胶，触变性减弱等流变性能的改变。

---

*通信作者：周春晖；E-mail：chc. zhou@aliyun. com；手机号：13588066098。国家自然科学基金（41672033）资助项目。

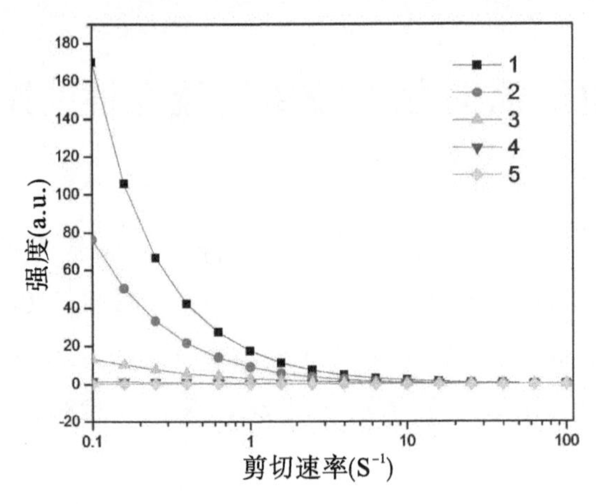

图2　1、2、3、4、5 硅铝比为10的皂石分散体分别加入0、0.25、0.5、0.75、1 mL的5040的流变曲线

本研究合成的黏土矿物的分散体可以形成胶体，并有良好触变性，但在聚合物分散剂的存在下由于其与黏土矿物的相互作用，使得黏土矿物凝胶向溶胶转变，并使得其触变性变小。聚合物分散剂的加入使得黏土矿物分散体的流变性发生改变，当不需要这种改变时，例如水性涂料中黏土矿物为作为增稠剂[4]，而聚合物分散剂作为涂料不可或缺，那么如何让黏土矿物能够保持这种流变性则是今后需要研究的内容。

参考文献

［1］Zhao L Z，Zhou C H，Wang J，et al. 2015. Recent advances in clay mineral-containing nanocomposite hydrogels［J］. Soft Matter，11（48）：9229 - 9246.

［2］Zhang L，Lu Q，Xu Z，et al. 2012. Effect of polycarboxylate ether comb-type polymer on viscosity and interfacial properties of kaolinite clay suspensions［J］. Journal of Colloid & Interface Science，378（1）：222 - 231.

［3］Zhang D，Zhou C H，Lin C X，et al. 2010. Synthesis of clay minerals［J］. Applied Clay Science，50（1）：1 - 11.

［4］Huang J，Wang L，Li T，et al. 2017. Rheological studies of mineral clay and its application in reactive dye printing of cotton［J］. Textile Research Journal，004051751668527.

新型三维有机黏土材料 LDH 制备及其对染料的去除

章　萍　曾宪哲　何　涛　马若男　黄　云*
（南昌大学资源环境与化工学院鄱阳湖环境与
资源利用教育部重点实验室，南昌 330031）

层状双金属氢氧化物（LDHs）是一类二维阴离子型黏土材料，由相互平行且带有正电荷的层板组成，层间填充有平衡层板正电荷的阴离子及水分子，其基本结构式为 $[M^{2+}_{1-x}M^{3+}_{x}(OH)_2]^{x+}(A^{n-})_{x/n}\cdot mH_2O$。其中，$M^{2+}$ 为二价金属离子（Ca^{2+}、Mg^{2+}、Zn^{2+} 等），M^{3+} 为三价金属离子（Al^{3+}、Fe^{3+}、Cr^{3+} 等），A^{n-} 代表 Cl^-、NO_3^-、PO_4^{3-} 等阴离子。因其具有层间阴离子可交换性、结构坍塌记忆效应、主层板可自组装性等特性，已广泛应用于高分子复合材料、催化材料、环境治理等领域[1]。

二维 LDHs 存在形貌单一、比表面积小、表面亲水且层片易凝聚等不足，为满足 LDHs 在实际应用方面的需求，将二维 LDHs 转变成多孔、比表面积更大的三维 LDHs 的研究近年来愈受关注[2]。本文以表面活性剂（十二烷基硫酸钠）作为软模板剂，合成有机三维 MgAl - LDH（O3D - LDH），实现了 LDHs 表面性质与形貌结构的同时改变。并以阳离子（罗丹明 B，RhB）和阴离子（甲基橙，MO）染料为模拟污染物，考察了其对阴、阳离子型染料的吸附性能。结果表明，O3D - LDH 吸附 RhB 的去除量为 73.06 mg/g，吸附 MO 的去除量为 644.84 mg/g，经有机改性的三维 LDHs 对阴、阳染料均有明显作用。

实验中采用水热法，通过调节 Mg/Al 摩尔比（2 到 1/3）、SDS 溶液浓度（0.01～0.2 mol/L）、水热反应时间（1～12 h），制备了一系列 O3D - MgAl - LDH，以探究反应条件对产物形成的影响。通过将 50 mL 镁铝盐溶液、0.48 g 尿素与 30 mL SDS 溶液混合激烈搅拌并超声 30 min，在 150 ℃下水热反应后离心、洗涤并干燥制得 O3D - LDH。

改变 O3D - LDH 样品在染料溶液中的投加量，研究了其对染料的吸附动力学，并采用 X 射线衍射仪（XRD）、红外光谱分析仪（FT - IR）、发射扫描电镜带能谱仪（SEM）、透射电子显微镜（TEM）、等离子体反射光谱仪（ICP - OES）等仪器对样品和反应固样、溶液进行表征和分析。

实验发现，Mg/Al 物质的量比、SDS 浓度和水热反应时间均会影响 3D - MgAl LDH 的形成。当 $n_{Mg}/n_{Al}=2$、SDS 浓度为 0.1 mol/L、水热反应时间为 6 h 时，有利于合成出花状形貌完整、片层厚度均一的 O3D - LDH。产物电镜扫描图如图 1 所示。

通过吸附动力学考察了 O3D - LDH 对阴、阳离子型染料的吸附性能，如图 2 所示。结果表明，O3D - LDH 吸附 RhB 符合 Freundlich 吸附模型，去除量为 73.06 mg/g；吸附 MO 符合 Lang-

---

* 通信作者：黄云；E-mail：zhangping@ncu. edu. cn；手机号：13699508783。国家自然科学基金（No. 214167014，No. 21767018）、江西省杰出青年人才资助项目（20171BCB23017）、中国博士后科学基金（No. 2017M612164）。

muir 吸附模型，去除量为 644.84 mg/g。

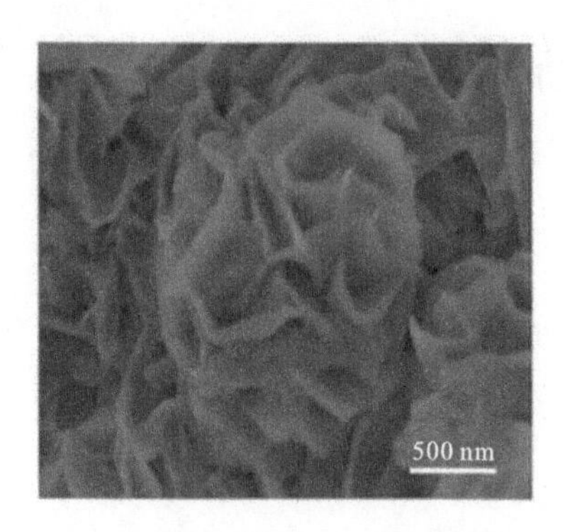

图 1 O3D－LDH 电镜扫描图

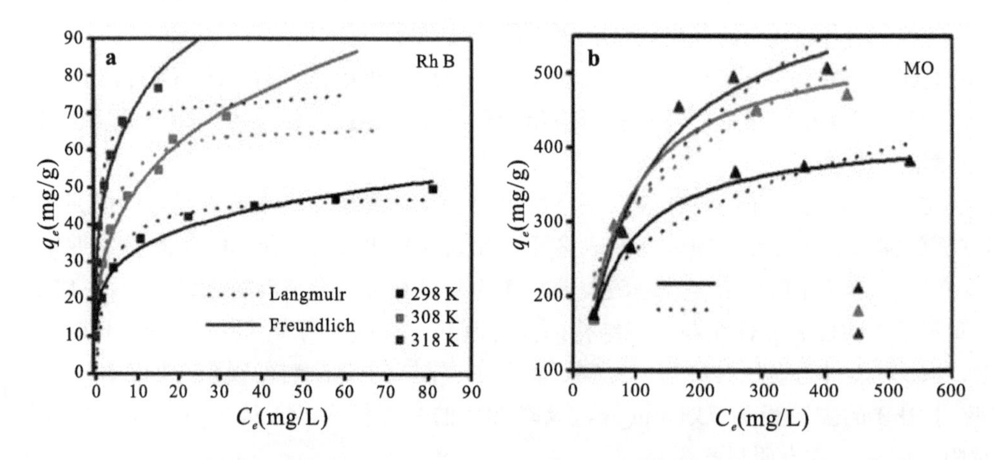

图 2 O3D－LDH 对 RhB 的吸附动力学曲线（a）以及 O3D－LDH 对 MO 的吸附动力学曲线（b）

合成 O3D－LDH 受溶液镁铝比、SDS 浓度、反应时间等影响。在最佳合成条件下制备的 O3D－LDH 在对阴离子染料具有优秀吸附性能的同时，对阳离子染料也具有明显作用。本研究结论可为三维黏土材料 LDH 的制备及其对水体染料污染物的去除提供理论依据和技术支持。

参考文献

［1］Zou Y D，Wang P Y，Yao W，et al. 2017. Synergistic immobilization of UO_2^{2+} by novel graphitic carbon nitride@ layered double hydroxide nanocomposites from wastewater［J］. Chem. Eng. J，330：573－584.

［2］Lu L，Li J，Yang P，et al. 2017. Synthesis of novel hierarchically porous Fe_3O_4@MgAl-LDH magnetic microspheres and its superb adsorption properties of dye from water［J］. Journal of Industrial and Engineering Chemistry，46：315.

硬石膏协同煅烧菱镁矿动态除磷效果

余韵青　程　鹏　刘海波　陈天虎　陈　冬*
（合肥工业大学资源与环境工程学院，合肥 230009）

根据2016年中国环境状况公报，总磷是湖泊等淡水主要的污染指标之一，尤其是在水体富营养化方面，磷是一个关键因子。因此，磷的去除成为全球性的热点问题。

目前水体除磷较为常见的方法有化学法、物理法、生物法三大类。生物除磷多数是应用在污水厂对污水的处理过程中，难以大规模应用在自然界富磷水体中。化学除磷主要是铁、铝和钙法的化学沉淀并配合有机化合物的混凝除磷，目前已有大量的研究和应用实例，主要应用于污水处理、工业废水的除磷，但是该方法易造成一定的二次污染，例如改变自然界水体的色度、pH值、增加有害离子浓度等[1]。

目前，越来越多的材料被应用到除磷当中，特别是天然矿物因其廉价易得而被广泛应用。王峰等[2]利用改性的膨润土吸附除磷，在进水浓度为10 mg/L、pH＝9、500 ℃热改性的条件下，去除率可达92.77％，但该方法对低浓度的含磷废水处理效果不理想。喻鹏辉等[3]采用方解石与硬石膏等混合处理磷废水，该方法处理后，高浓度磷废水的去除率可达95％以上，且出水pH值较为稳定，但同样对低浓度的含磷废水处理效果不佳。为解决低浓度含磷废水的处理问题，本文通过硬石膏和煅烧菱镁矿协同动态实验研究了对低浓度含磷废水的处理效果。

实验用的硬石膏矿石取自安徽省巢湖市，所取样品经过破碎、筛分等预处理获得粒径为0.9～1 mm（18～20目）的颗粒样品。实验用菱镁矿取自辽宁省大石桥市，菱镁矿矿石颗粒使用前需在800 ℃温度下空气气氛中马弗炉内煅烧2 h。

为探究硬石膏协同煅烧菱镁矿对模拟含磷（磷酸根）废水的处理能力，实验设计了动态柱装置，如图1所示。吸附层高为60 cm，柱内径为1.5 cm。实验配制了1.0 g/L（KH_2PO_4，以P计）模拟含磷废水储备液，使用时直接稀释到所对应的浓度。分别配制0.5、1.0、2.0、5.0 mg/L模拟含磷废水，模拟不同的磷负荷对动态柱除磷性能的影响。

图2显示不同的磷负荷对动态柱除磷性能的影响。由图2可以看出，当进水浓度为0.5 mg/L时，磷的去除率为100％；进水磷浓度升高为1 mg/L时，磷的去除率为88％左右；当进水磷浓度升高至2 mg/L时，磷的去除率降到80％左右；当进水磷浓度继续升高至5 mg/L时，磷的去除率降至66％。

动态除磷实验表明，硬石膏协同菱镁矿煅烧产物不但具备很好的除磷效果，进水磷浓度为1 mg/L的水体除磷后可达到地表水环境质量Ⅲ类水标准（0.2 mg/L），而且适合对低浓度含磷废水的深度处理。

*通信作者：陈冬；E-mail：cdxman@hfut. edu；手机号：13955176245。国家自然科学基金（41472047）资助项目。

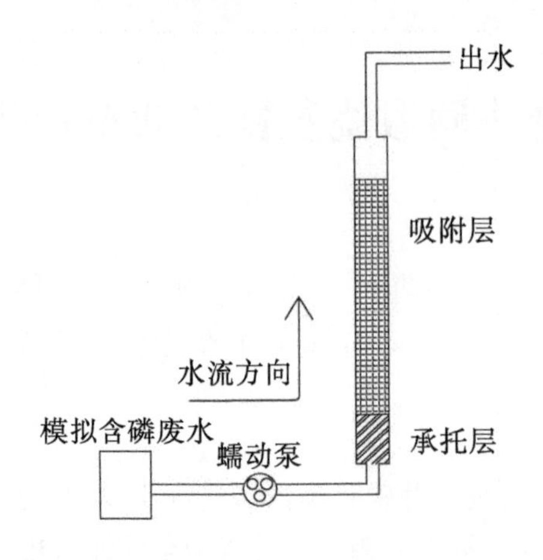

图 1　硬石膏协同菱镁矿煅烧产物除磷动态吸附装置实验

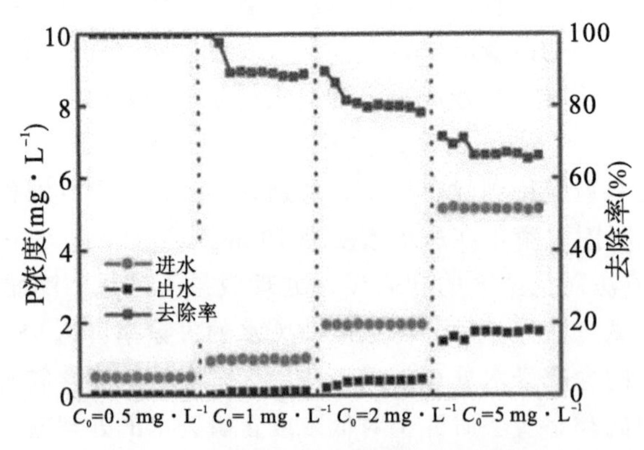

图 2　不同进水浓度对除磷性能的影响

参考文献

[1] Qayyum M F，Rehman M Z，Ali S，et al. 2017. Residual effects of monoammonium，phosphate，gypsum and elemental sulfur on cadmium phytoavailability and translocation from soil to wheat in an effluent irrigated field [J]. Chemosphere，174：515 - 523.

[2] 王峰，翟由涛，陈建林. 2011. 膨润土的改性及其对废水中磷吸附效果的研究 [J]. 安徽农业科学，39：5968 - 5970.

[3] 喻鹏辉，褚海云. 2016. 混合矿物材料处理含磷废水的研究 [J]. 江西化工，23：106 - 109.

硅藻土负载二氧化锰催化氧化甲醛

韩正严　王　灿　刘海波*　陈天虎　陈　冬
（合肥工业大学资源与环境工程学院，合肥 230009）

甲醛气体对人体的危害十分严重，潜伏期也很长，被称作“室内污染第一杀手”。主要表现为对人体的刺激作用、致敏作用、致突变作用。随着人们对环保健康问题的不断重视，如何高效去除室内甲醛引起了广泛关注。在众多去除甲醛的方法中，催化氧化法具有能耗低、效率高、方便便捷且不产生二次污染物等特点。而过渡金属催化剂由于原料低廉、成本较低且在催化氧化甲醛方面有较高的效率，具有较高的研究价值。硅藻土稳定的孔道结构和独特的化学性质使其拥有极为优异的吸附性和稳定性，其应用范围广，发展前景广，在处理室内空气污染物方面也表现出优异的性能[1-5]。

本文研究了硅藻土的提纯，并通过将锰氧化物负载在提纯后的硅藻土上制得具有催化氧化甲醛能力的催化剂，针对催化剂制备方法、锰氧化物的负载浓度等条件变量探究了最高催化效率。

本实验采用云南省宾川县开采的硅藻土。使用 HCl 对硅藻土进行酸浸法提纯，以去除低品位硅藻土中的 $CaCO_3$ 杂质。将大块硅藻土捣碎、研磨、过筛，加入去离子水充分搅拌，置于 80 ℃超声水浴箱中，缓慢倒入稀盐酸搅拌，直至不产生气泡为止，然后过滤、洗涤、干燥，即得纯净硅藻土粉末。

负载锰氧化物时使用共沉淀法。在负载过程中，使用甲醇和草酸铵 2 种还原剂将高锰酸钾还原成锰氧化物并负载于硅藻土上，分别命名为 MnO_x/Dt－MAL、MnO_x/Dt－AO。根据投加的高锰酸钾和还原剂的量来控制负载的浓度，得到 1％、5％、10％、15％、20％五种不同负载浓度的催化剂。对得到的两组共计 10 种催化剂进行 XRD、拉曼光谱、TEM、H_2－TPR、红外光谱等表征分析和催化效果评价，探究其催化氧化效率及催化剂构效关系、催化机理。

硅藻土提纯前后化学组成和构造发生了变化。粉末颜色由黄变白，XRD 衍射光谱显示提纯前硅藻土出现了归属于 $CaCO_3$ 的衍射峰，提纯后该衍射峰消失，即 $CaCO_3$ 被除去，硅藻土得到纯化。在负载了锰氧化物的催化剂中并没有发现归属于锰氧化物的特征峰，结合 TEM 图像，证明锰氧化物在硅藻土的表面及孔道中并非以结晶的形式负载，其高度分散且结晶度较差，仅以微粒或基团的形式均匀附着在硅藻土的表面。表征结果表明，负载的锰氧化物为水钠锰矿型二氧化锰，在不同条件下制得的水钠锰矿样品强度有所不同，草酸铵作还原剂降低了硅藻土孔道的稳定性，或者是降低了 MnO_2 晶型结构的稳定性。随着锰氧化物负载的增多，催化剂氧化能力也增强，使用草酸铵作还原剂制备的催化剂比使用高锰酸钾制备的还原强度更高，还原温度更低。

一般情况下，随着催化剂中锰氧化物负载浓度的提升，催化氧化的效率也随之提升。但当锰

*通信作者：刘海波；E-mail：liuhaibosky116@hfut.edu.cn；手机号：15856995526。国家自然科学基金（41672040）。

负载浓度过高时，堵塞硅藻土中天然空洞，其比表面积减小，与甲醛的接触面积降低，催化氧化能力下降。低温条件下，MnO_x/Dt－MAL 催化剂能力优于 MnO_x/Dt－AO，但 MnO_x/Dt－AO 能在较低温度下实现完全催化氧化甲醛（图 1）。总体来看，使用草酸铵作还原剂制备的 MnO_x/Dt－AO 催化剂催化效果比使用甲醇作还原剂制得的 MnO_x/Dt－MAL 催化剂好，但 2 种催化剂均能使甲醛完全催化氧化为 CO_2 和 H_2O。

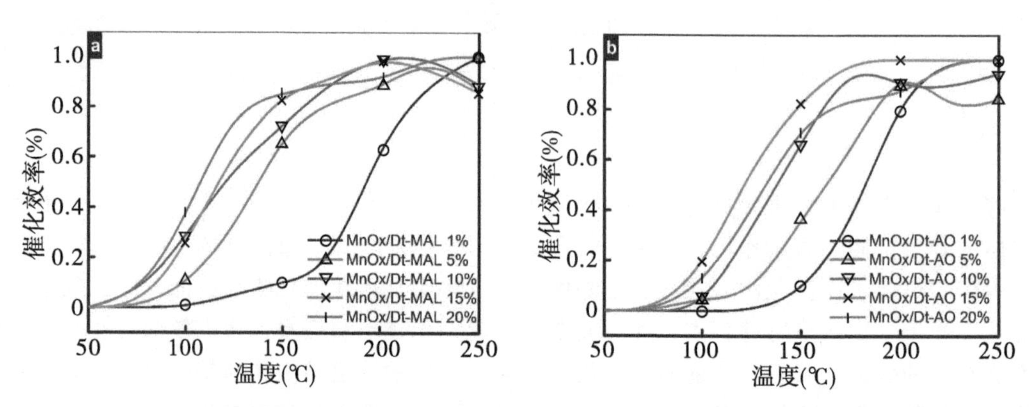

图 1　不同锰负载量的 MnO_x/Dt－MAL（a）、MnO_x/Dt－AO（b）催化剂的催化效果

参考文献

[1] 闫金萍. 2004. 甲醛及其对人体健康的危害［J］. 化学世界，（10）：558－559.

[2] 王金龙. 2016. 水钠锰矿型二氧化锰室温催化分解甲醛的研究［D］. 北京：清华大学.

[3] Meher S K，Rao G R. 2012. Enhanced activity of microwave synthesized hierarchical MnO_2 for high performance supercapacitor applications［J］. Journal of Power Sources，215（4）：317－328.

[4] Yang Y，Huang J，Zhang S Z. 2014. Catalytic removal of gaseous HCBz on Cu doped OMS：Effect of Cu location on catalytic performance［J］. Applied Catalysis B：Environmental，167－178.

[5] 任子杰，高惠民. 2013. 硅藻土提纯及制备助滤剂研究进展［J］. 矿产综合利用，5：5－12.

煅烧和碳酸化改性黏土质白云岩及其表征

孙付炜 王翰林 李宏伟 洪晓梅 项学芃 刘海波 谢晶晶 陈天虎*
（合肥工业大学资源与环境学院，合肥 230009）

黏土质白云岩是由亚微米白云石和纳米凹凸棒石共生形成的一种岩石，是一种含凹凸棒石的弱固结白云质白垩，外观为雪白色，白垩状，疏松多孔，质轻，在我国安徽明光和河南镇平有很高的探明储量[1]。凹凸棒石作为黏土质白云岩的次要组分，是一种链层状结构的含水富镁硅酸盐矿物，理想化学式为 $Mg_5Si_8O_{20}(OH)_2(OH_2)_4 \cdot 4H_2O$，晶体常呈针状、棒状的纤维集合体，且由于铝、铁对镁的类质同象替代作用，又产生了铝凹凸棒石和铁凹凸棒石的变种。田漪兮等[2]发现利用黏土质白云岩为吸附剂对铅离子有良好的去除效果，但反应速率较慢，因此，对黏土质白云岩进行改性处理来提高其反应活性，成为了新的研究方向。而黏土质白云岩经煅烧/碳酸化处理制备纳米方镁石和方解石等高活性多孔结构化复合材料（PCN），在水处理、土壤修复等方面有很大的应用潜力。

实验所用的黏土质白云岩取自安徽明光花果山，样品经破碎后筛分出粒径为 0.45～0.9 mm 的颗粒，马弗炉内 860 ℃煅烧 1 h，取出冷却后干燥器内密闭保存待用。作为对照的纯白云石取自安徽巢湖汤山，方解石由安徽砂石厂提供。将样品经过常温碳酸化、煅烧黏土质白云岩消化和高温碳酸盐消化，然后对其进行 XRD、TEM 表征。

XRD 表征结果（图 1）显示，在煅烧 1 h 后，图谱中属于白云石的特征峰被属于方钙石和方镁石的特征峰取代，说明白云石完全分解转化为 CaO 和 MgO，而煅烧黏土质白云岩在常温条件下碳酸化后，可以看到方镁石和斜硅钙石保持物相不变，方钙石的特征峰随碳酸化时间增加逐渐被方解石的特征峰替代。这说明方钙石与空气中的水和二氧化碳缓慢反应（一般 7 d）形成方解石。随着煅烧温度逐渐升高，从没有发生反应逐渐到发生碳酸化，产物为方解石，一般在 600～700 ℃时，方钙石与二氧化碳完全反应形成了方解石。而且在不同的温度下，方镁石特征峰没有明显变化，未出现菱镁矿衍射峰，说明方镁石没有参与碳酸化反应。进气加入 10%水蒸气后，碳酸化初始反应温度从 400 ℃降低到 300 ℃，完全反应最低温度从 600 ℃降低到 500 ℃。水蒸气对煅烧黏土质白云岩碳酸化反应表现出了促进作用。

对其进行 TEM 分析（图 2），由左图可以看出，相比于普通白云石，黏土质白云岩中的白云石具有相对细小的颗粒尺度，是一种亚微米粒级的白云石，白云石晶体之间相互堆叠集聚，又和凹凸棒石晶束交织镶嵌，构成了黏土质白云岩这一天然纳米—亚微米矿物堆积体。由右图可以看出，经碳酸化处理后微观形貌完全改变，白云石的菱面体结构和凹凸棒石的棒状结构基本完全消失，形成了由纳米类四面体颗粒堆叠形成的新形貌，比对可知类四面体颗粒大小在 30～50 nm，说明该材料经煅烧/碳酸化后形成了方解石和方镁石颗粒。

---

* 通信作者：陈天虎；E-mail：chentianhu@hfut. edu. cn；手机号：13956099615。国家自然科学基金（41572028）。

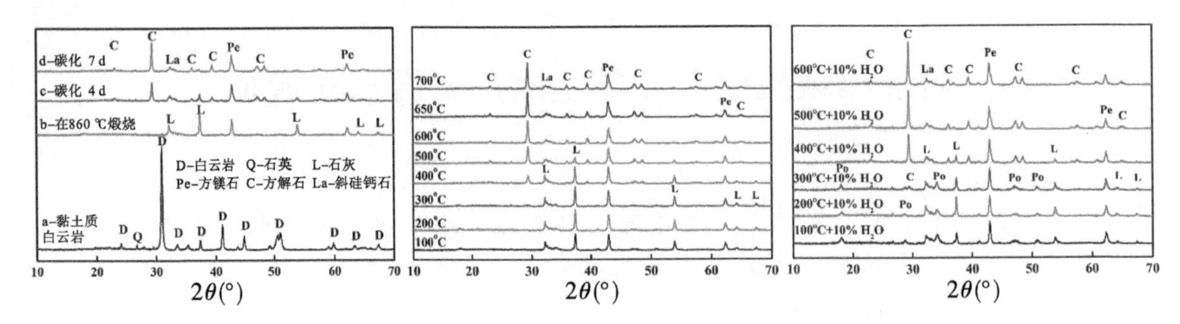

图 1 XRD 表征图谱

注：左图为黏土质白云岩煅烧/碳酸化前后；中图为不同温度煅烧/碳酸化产物；
右图为煅烧黏土质白云岩 10%水蒸气下不同温度碳酸化产物

图 2 黏土质白云岩煅烧/碳酸化前（左）后（右）TEM 图片

综上，黏土质白云岩的主要组分为白云石和凹凸棒石，高温煅烧后白云石会分解为纳米方钙石和纳米方镁石，其中方钙石在一定条件下碳酸化可完全转化为方解石，新生的方解石粒径在 30 nm 左右。此外，高温和水蒸气的存在可以有效地提升该材料的碳酸化进程，表现出促进作用。

参考文献

[1] 陈天虎，徐晓春，岳书仓. 2003. 苏皖凹凸棒石黏土纳米尺度矿物学及地球化学［M］. 北京：科学出版社.

[2] 田漪兮，李宏伟，谢晶晶，等. 2017. 黏土质白云岩去除水中铅离子作用及机理［J］. 岩石矿物学杂志，36（1）：104－109.

崩落法开采挂帮矿诱发边坡失稳研究

王大国*　韩超超
（西南科技大学环境与资源学院，绵阳 621010）

根据目前国内黏土矿山的开采现状，其大部分采用的开采方式为露天开采，如山西太湖石矿区、内蒙古杂怀沟矿区和贵州小山坝矿区等。随着露天黏土矿山开采深度不断增加，露天转地下开采逐渐成为大多数露天黏土矿山的最佳选择。在露天转地下开采的过渡期间，矿石产量的衔接问题尤为重要[1]。露天挂帮矿作为一类残留在露天境界外且靠近最终边坡面的矿产资源，在经济条件允许下，对露天挂帮矿进行安全合理的开采可以为矿山在露天转地下过渡期间提供一定的矿石产出，实现露天转地下平稳过渡。

在各类露天黏土矿山挂帮矿开采过程中，如露天铝土矿山，考虑其开采难易程度受多种因素影响，包括露天边坡形态、挂帮矿赋存条件以及边坡内部结构面。其中，露天边坡形态和挂帮矿赋存条件决定了采矿方法的选择，当最终边坡角处于某一临界值时，挂帮矿的开采极易造成露天边坡发生崩塌、滑坡等地质事件[2]；边坡内部结构面影响着露天边坡的稳定性，众多结构面中，软弱夹层所产生的影响较为显著[3]。根据黏土矿山开采实际，有学者提出采用崩落法开采挂帮矿[4]，该方法可以通过切割工程的规模、形状和放矿进度来控制边坡失稳的进程[5]。当边坡发生失稳破坏时，诱导塌落的松散岩体冲入采空区，当采空区能完全容纳塌落的松散岩体时，露天坑底的采矿作业不受挂帮矿开采的影响，实现坑底采矿作业与挂帮矿开采同时进行，解决矿山在露天转地下过渡期的矿石产量衔接问题。

本文基于岩石拉张破坏理论，建立了岩石拉张破坏有限元模型，根据贵州小山坝五龙寺铝土矿山挂帮矿的赋存形态，建立了简化后的铝土矿露天边坡模型（图 1）。模拟了露天边坡在不同坡体形态、挂帮矿及软弱夹层赋存情况影响下，坡体内第一主应力演变过程和裂纹扩展过程，分析了各因素影响下的边坡破坏过程及范围，详细讨论了松散岩体体积和采空区体积在各因素条件下的大小关系，为类似黏土矿山挂帮矿开采提供理论依据。

研究结果显示，当边坡内不含软弱夹层时，初始裂纹出现在坡面或坡顶，随着节点应力释放—转移—集中这一过程，初始裂纹始终向采空区扩展，最终形成一条贯穿采场的长裂纹。当边坡内存在软弱夹层时，初始裂纹产生于软弱夹层尾部围岩中，随着初始裂纹扩展，在坡面或坡顶出现一条向采空区扩展的长裂纹，该长裂纹最终贯穿采场并决定了边坡最终破坏范围。

---

* 通信作者：王大国；E-mail：dan_wangguo@163.com；手机号：13730717308。国家自然科学基金（51349011）、四川省教育厅科研创新团队（18TD0019）、西南科技大学龙山学术人才科研支持计划“团队支持”（18LZXT15）、西南科技大学龙山学术人才科研支持计划“重点支持”（18LZX410）资助项目。

仅考虑露天边坡形态和挂帮矿赋存条件的影响时：在边坡倾角影响下，当边坡倾角大于 40°，第二分段挂帮矿开采后形成的采空区体积小于松散岩体体积；矿体倾角影响下，当矿体倾角小于 30°，两个分段挂帮矿开采后形成的采空区体积小于松散岩体体积；矿体埋深影响下，两个分段挂帮矿开采后形成的采空区体积小于松散岩体体积。仅考虑软弱夹层赋存条件的影响时：软弱夹层长度影响下，当软弱夹层长度大于 80 m，两个分段挂帮矿开采后形成的采空区体积小于松散岩体体积；软弱夹层埋深、倾角和围岩/软弱夹层弹性模量比影响下，第二分段挂帮矿开采后形成的采空区体积小于松散岩体体积。

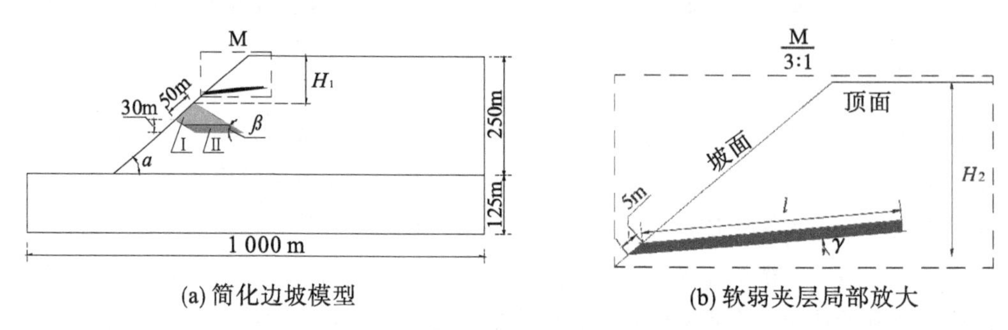

(a) 简化边坡模型　　(b) 软弱夹层局部放大

图 1　岩质边坡模型

注：Ⅰ——露天挂帮铝土矿第一分段；Ⅱ——露天挂帮铝土矿第二分段；α——边坡倾角；β——矿体倾角；H_1——矿体埋深；l——软弱夹层长度；H_2——软弱夹层埋深；γ——软弱夹层倾角；M——软弱夹层局部放大标识符

参考文献

[1] 孟繁华，赵刚. 2018. 露天转地下矿山覆盖层合理厚度及安全结构的研究 [J]. 有色矿冶，34（1）：12－15.

[2] Li L C，Tang C A，Zhao X D，et al. 2014. Block caving-induced strata movement and associated surface subsidence：a numerical study based on a demonstration model [J]. Bulletin of Engineering Geology and the Environment，73（4）：1165－1182.

[3] Lu H F，Liu Q S，Chen C X，et al. 2012. Disaster mechanism of hard and soft interbedding slope and its preventive measure [J]. Disaster Advances，5：1361－1366.

[4] 韩贞健. 2003. 崩落采矿法在小山坝铝土矿小矿体中的应用 [J]. 采矿技术，3（2）：43－45.

[5] 李海英，任凤玉，严国富，等. 2015. 露天转地下过渡期岩移危害控制方法 [J]. 东北大学学报：自然科学版，36（3）：419－422.

三聚氰酸改性 $g-C_3N_4$/高岭石复合材料及其光催化性能

袁　方[1]　孙志明[1*]　李　雪[1]　徐　洁[1]　汪　滨[2]　郑水林[1]
（1. 中国矿业大学（北京）化学与环境工程学院，北京 100083；
2. 北京服装学院服装材料研究开发与评价重点实验室，北京 100029）

类石墨相氮化碳（$g-C_3N_4$）具有禁带宽度窄、室温下稳定、不含金属元素、合成简便且成本低廉等诸多优良特性，自 2009 年被发现可以在可见光照射下催化分解水制氢以来，引起了国内外科研人员的广泛关注。但由于 $g-C_3N_4$的二维片层结构，制得的纯 $g-C_3N_4$极易团聚，导致其往往比表面积小、光生电子—空穴复合效率高、可见光利用率低，严重限制了其实际应用[1]。因此，制备高活性的 $g-C_3N_4$复合材料是目前研究的重点。本课题研究发现通过采用天然矿物负载的方式可有效提高 $g-C_3N_4$这类催化剂的吸附性能，从而提高材料的可见光光催化活性[2,3]。

如图 1 所示，本文以水洗高岭石为载体，三聚氰胺为前驱体，三聚氰酸为改性剂，采用浸渍-热聚法制备了一种 $g-C_3N_4$/高岭石复合材料。结合 XRD、SEM、BET、FT-IR、UV-Vis DRS 和 TG 等多种表征手段对样品的物相组成、微观形貌、孔结构特性、化学结构及光学性能等进行了分析，并研究了不同三聚氰酸改性剂用量对复合材料光催化性能的影响及复合材料的光催化反应机理。本研究旨在为非金属矿物负载型光催化材料的进一步实用化提供借鉴。

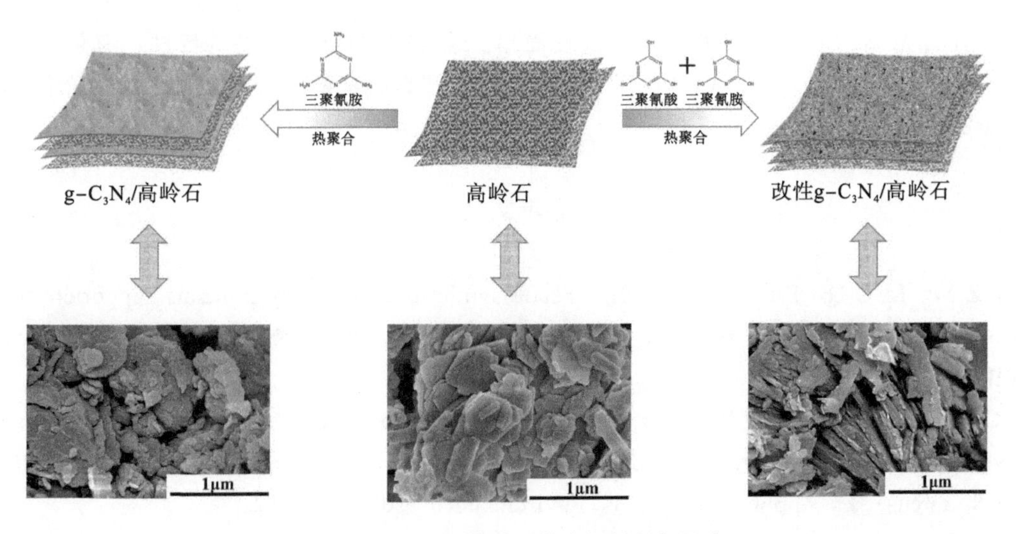

图 1　$g-C_3N_4$/高岭石复合材料制备思路

---

* 通信作者：孙志明；E-mail：zhimingsun@ cumtb. edu. cn；手机号：13466774499。国家自然科学基金（51504263）资助项目。

实验结果如图 2 所示，三聚氰酸的引入有效提高了复合材料的可见光光催化反应活性，吸附性能和光催化反应速率与纯 g－C_3N_4相比分别提高了 2.9 倍和 3.4 倍。通过 BET 分析，复合材料的比表面积为 49.522 m^2/g，与纯 g－C_3N_4相比提高近 2 倍。

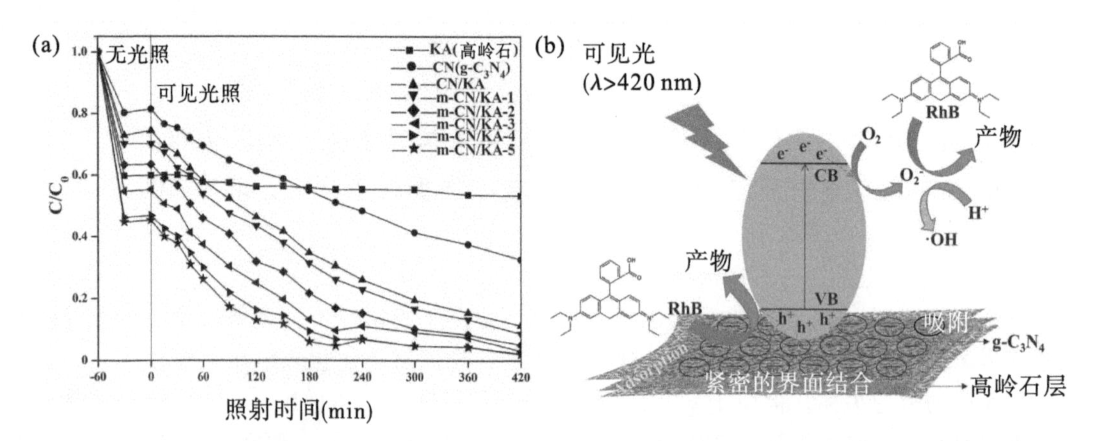

图 2　g－C_3N_4/高岭石复合材料对 RhB 净化性能（a）及复合材料光催化反应机理（b）

复合材料光催化性能的显著提高不仅是因为复合材料中 g－C_3N_4与高岭石的紧密结合，更重要的是由于改性剂三聚氰酸的引入。三聚氰酸由于与三聚氰胺具有相似结构，使二者在制备前驱体过程中易均匀混合，得到的前驱体在煅烧过程中，其中的三聚氰酸会因高温发生分解，从而在复合材料中留下丰富的孔道结构和活性位点，极大地提高了催化剂比表面积，进而有效提高了 g－C_3N_4/高岭石复合材料的吸附及光催化性能。

利用光催化技术净化工业和生活中的废水废气，是环境治理领域一种十分理想的处理方式，但要实现大规模的工业应用，目前仍存在诸多问题。本研究合成的三聚氰酸改性 g－C_3N_4/高岭石复合材料具有无毒、无金属元素、合成过程简单、合成成本低廉等优良特性，有效改善了纯 g－C_3N_4在实际应用中的不足之处，在环境污染控制领域具有良好的应用前景，可为光催化材料的进一步实用化提供借鉴。

参考文献

［1］Sun Z M，Li C Q，Du X，et al. 2018. Facile synthesis of two clay minerals supported graphitic carbon nitride composites as highly efficient visible-light-driven photocatalysts［J］. Journal of Colloid and Interface Science，511：268－276.

［2］Li C H，Sun Z M，Zhang W Z，et al. 2018. Highly efficient g-C_3N_4/TiO_2/kaolinite composite with novel three-dimensional structure and enhanced visible light responding ability towards ciprofloxacin and S. aureus［J］. Applied Catalyst B：Environment，220：272－282.

［3］Sun Z M，Yao G Y，Zhang X Y，et al. 2016. Enhanced visible-light photocatalytic activity of kaolinite/g-C_3N_4 composite synthesized via mechanochemical treatment［J］. Applied Clay Science，129：7－14.

利用电阻法测试高岭石径厚比

刘文超[1]　江发伟[1]　程宏飞[1,2]　刘钦甫[1*]
（1. 中国矿业大学（北京）地球科学与测绘工程学院，北京 100083；
2. 长安大学环境科学与工程学院，西安 710054）

高岭石的径厚比在其应用的许多领域如造纸、涂料，尤其是应用于橡胶的阻隔性能和补强性能，都是一个关键的影响因素，具有重要的研究价值[1,2]。但受其矿物形貌的特殊性及纳米级微观结构的影响，高岭石径厚比的测量一直是备受业界关注的重点和难点[3,4]。目前对于片状矿物径厚比的测量主要是基于透射电镜或扫描电镜的观察，并且仅是定性和概略性的描述，尚未有一个统一和定量检测的方法及标准。

电阻法径厚比测试的方法是在电阻法粒径方法的基础之上，对相关测试仪器原理、测试过程、测试条件等方面进行深入研究，并运用数学归纳统计知识推导出新的公式，得出的一种新的高岭石径厚比测试方法。

本文利用电阻法径厚比测试的仪器库尔特粒度分析仪来自美国贝克曼库尔特有限公司，该仪器早期主要应用于医学、生物方面，特别是用于红细胞粒径测试方面，之后随着仪器技术的发展，也常用于微细颗粒粒径的测试方面。该仪器具有准确性高、重复性好等特点。其测试原理见图 1。

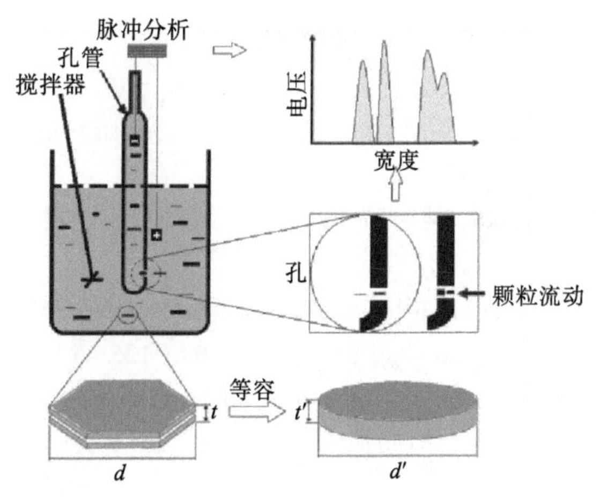

图 1　电阻法测试原理

* 通信作者：刘钦甫；E-mail：lqf@cumtb. edu. cn；手机号：13911683809。国家自然科学基金（12345678）。

3次测试结果（表1）对比发现，电阻法测试结果的重复性较好，且电阻法3次测试的结果尚未发现明显变化。通过标准样品对其进行检测，发现其准确性也较高。

表1　不同测试范围同一样品的库尔特仪测试结果

粒径	全程			0.4～1		
	1	2	3	1	2	3
D10	0.425	0.429	0.43	0.423	0.428	0.429
D50	0.54	0.554	0.563	0.529	0.546	0.556
D90	0.854	0.821	0.831	0.744	0.764	0.79
平均径	0.667	0.611	0.607	0.56	0.574	0.585

可见，因电阻法测试以记数的形式考察每一个样品颗粒，故同种样品尽管浓度不同，测得结果差别较小。然而，电阻法测试所用仪器库尔特仪尽管测算精确，但使用中仍有限制：（1）需经常使用标准样品进行仪器校准，以保证测算数据准确。（2）对于粒度分布范围相对较宽的颗粒，电阻法实用性较低。（3）电阻法测试方法的下限由最小孔径的小孔管所限制，所测样品粒度越细，所需小孔管精度要求越高，最小的小孔管比较容易被样品堵塞。

综上所述，电阻法测试对血球的粒度分析适应性较高，但对很多工业物质来说测试条件较为苛刻。测试速度虽然较快，但准备时间较长也较繁琐，故长期在矿物颗粒测试方面应用较少。然而，本研究则看重其电阻法测算原理的特殊性——可以测算颗粒的真实体积，再结合高岭石特殊的矿物形貌——片层状颗粒，对电阻法及库尔特仪进行了深入的分析和研究，发现该原理对片状高岭石颗粒的径厚比研究有着十分明显的优势。准备过程虽然较为繁琐，但是相对图像法需要手动或半自动测试很多颗粒的表面直径以及难度较大的厚度来说，工作量减少十分明显。

参考文献

[1] 刘钦甫，张玉德，陆银平. 2005. 黏土/聚合物纳米复合材料研究现状［J］. 非金属矿，28（9）：41-43.

[2] 陆银平，张玉德，刘钦甫. 2009. 纳米黏土的制备及应用研究进展［J］. 化工新型材料，（10）：8-10.

[3] Cheng H，Zhang Z，Liu Q，et al. 2014. A new method for determining platy particle aspect ratio：A kaolinite case study［J］. Applied Clay Science，97-98：125-131.

[4] Jiang F，Liu Q，Cheng H. 2015. A new method for determining the aspect ratio of kaolinite by image-resistance combination［J］. Materials Letters，159：90-93.

安徽北淮阳地区非金属矿产分布规律

方　明*　胡飞平
（安徽省地质矿产勘查局313地质队，六安 237010）

安徽北淮阳成矿带属北秦岭成矿带的东延部分，带内深大断裂发育，构造活动频繁，岩浆活动广泛，变质作用显著，为钼、铅锌、铜、金多金属矿的形成提供了有利成矿条件，同时也导致区内非金属矿产种类多、类型多。

区内主要的非金属矿种有萤石、明矾石、晶质石墨、橄榄岩、蛇纹岩、刚玉、蛭石、含钾岩石、磷矿、脉石英、水泥大理岩、熔剂大理岩、建筑饰面石材、膨润土、瓷石（土）、黄铁矿以及宝石类和玉石类等10余种[1-4]。

通过综合分析区域地质背景和成矿环境关系，对比地层岩性柱，结合构造格局和构造旋回，划分岩浆岩类型、期次及演化特征，恢复区域变质等工作，总结区内非金属矿特征[5,6]。

一、平面上矿床（点）具成群成带分布特点

区内非金属矿产，由于成矿控制条件不同，在平面上成群、成带分布，北淮阳大理岩、萤石、明矾石、膨润土成矿带位于金寨—霍山复向斜内，主要受防虎山断裂和磨子潭—晓天深大断裂控制。矿带总的走向呈北西西向展布。矿带内发育有不同成因类型的矿床、点。

（1）次级控矿构造条件基本相同，不同成因类型的矿床、点可在同一空间成群成带出现。如总体上受北西西向龙门村—南港破碎带的控制，出现了同心寺霞石正长岩矿点，扫帚河、百花冲火山热液性明矾石矿床（点）群，下符桥、新街一带热液型萤石矿床（点）群，柴家冲火山热液—风华残余型瓷石（土）矿床等，呈带状分布。

（2）次级控矿条件不同，不同成因或相同类型矿床（点）则分别成群成带分布。①受金寨山前凹陷控制，出现了灰冲沉积变质型水泥大理岩与溶剂、耐火白云石大理岩，经接触变质作用叠加形成晶质石墨、红柱石矿点；②受佛子岭复向斜控制，则有诸佛庵—毛坦厂沉积变质型水泥大理岩与溶剂、耐火白云石大理岩及石英岩矿床成带分布；③受中生代火山岩凹陷盆地控制，出现了毛坦厂地区火山沉积—沉积型膨润土矿点群；④受中、新生代次级凹陷盆地控制，出现了大岗头沉积型膨润土矿床、三尖铺型砂矿床。

二、安徽北淮阳地区非金属矿划分为3个成矿区

（1）六安膨润土、型砂成矿区。

位于北部，西起石婆店，经苏家埠至舒城，呈北西西向带状展布。该成矿区以中、新生代沉积作用成矿为特征。

（2）毛坦厂明矾石、萤石膨润土成矿区。

*作者简介：方明（1982—），男，安徽淮北人，工程师，主要从事矿产资源勘查方面工作，E-mail：mingf@126.com，联系电话：18856437552。

呈北西西向展布，西起白大畈，经下符桥、毛坦厂、舒茶，转向西至单龙寺，呈向西分开的剪刀状。该成矿区以与燕山期岩浆侵入和火山喷发活动有密切联系的热液作用、火山作用成矿为特征。

区内分布地层主要为上侏罗系—下白垩系毛坦厂组、黑石渡组及下白垩系晓天组、白大畈组，岩性以中性火山岩及火山物质沉积为特征，部分地段分布有震旦系小溪河组片麻岩、片岩，轻微混合岩化。

岩浆岩发育，主要集中于成矿区东部河棚等地区，岩性有石英正长斑岩、花岗岩等，岩座、岩株规模。西部响洪甸地区发育碱性正长岩—霞石正长岩小侵入体。北西西向龙门冲—南港破碎带横贯全区，对区内矿产起重要控矿作用。

已知矿产主要有：中温热液交代型明矾石矿，中低温热液充填型萤石矿，火山沉积型或与之有成因联系的化学沉积型膨润土矿点，火山热液型瓷石（土）矿床，热液型彩石矿点；陶瓷用霞石正长岩，水泥用凝灰岩矿点等。

（3）仙人冲大理岩、石墨、硅石成矿区。

成矿区位于北淮阳成矿亚带南部，西起全军，经船板冲、诸佛庵至真龙地，呈北西西向带状展布。该区以沉积变质作用成矿为特征。

区内主要分布庐镇关群、佛子岭群及胡油坊组、杨小庄组地层，角闪岩相—绿片岩相变质程度，岩性以各种片岩为主，部分为片麻岩、千枚岩、大理岩等。

已知矿产有水泥大理岩小型矿床、溶剂耐火白云大理岩小型矿床、晶质石墨矿床，以及红柱石和磷灰岩等矿产。

参考文献

[1] 安徽省地质矿产勘查局313地质队，311地质队. 1986. 安徽省大别山成矿区非金属矿产成矿区划报告 [R].

[2] 邱军强. 2012. 安徽北淮阳成矿带特征及找矿远景 [J]. 矿床地质，s1：31-32.

[3] 安徽省地质调查院. 2013. 安徽省大别山宝玉石调查评价 [R].

[4] 朱及天，朱云辉. 1998. 浅论大别山区非金属矿开发 [J]. 安徽地质，8（4）：113-115.

[5] 杜建国，常丹燕，戴圣潜，等. 2001. 大别山区域成矿体系与成矿规律的初步研究 [J]. 安徽地质，11（2）：140-149.

[6] 陶维屏. 1985. 中国东部环太平洋带某些非金属矿的分布规律 [J]. 矿床地质，4（3）：61-69.

表面调控对硅藻蛋白石矿物载体性质改善的机制

刘　冬[1*]　宋雅然[1,2]　邓亮亮[1]　袁　鹏[1]

（1. 中国科学院广州地球化学研究所中科院矿物学与成矿重点实验室/
广东省矿物物理与材料研究开发重点实验室，广州 510640；
2. 中国科学院大学，北京 100049）

硅藻蛋白石是我国重要非金属矿产资源——硅藻土的主要组成矿物。其由硅藻生物的遗骸（即硅藻壳体）沉积而成，主要成分为无定形二氧化硅，具有蛋白石-A结构。硅藻蛋白石为硅藻壳体形貌的单体颗粒（我国主要矿床中硅藻蛋白石的颗粒直径通常为 3～40 μm），表面含有丰富的大孔（孔径为 0.04～0.8 μm）结构和硅羟基等活性基团。加之其具有较高的耐酸、耐高温性，硅藻蛋白石被认为是天然优良载体，已应用于多种复合材料的制备。

然而，必须指出的是，由于硅藻蛋白石的孔结构以大孔为主，其比表面较低（BET 比表面积通常＜60 m^2/g）；另一方面，作为主要吸附/负载位点的硅羟基，其密度也并不显著，且硅羟基的分布并不均匀。受这两方面性质制约，硅藻蛋白石作为载体应用时存在“短板”——负载量往往较低，且负载通常不均匀。

已有研究者通过对硅藻蛋白石表面进行改性等以提高其比表面积或活性位点[1-4]。例如，利用表面沸石化提高硅藻蛋白石的微孔性和比表面积；利用表面硅烷化（“嫁接”）等改善表面活性位点的数量或者增加相应官能团。然而，上述“改造”或流程复杂，需用到水热条件或者非水溶剂等；或使用造价昂贵的改性试剂，大大增加了产品的成本。因此，亟待寻找一种简单易行的方法，以解决目前硅藻蛋白石矿物作为载体所面临的上述“瓶颈”问题。

本研究发现，通过碱处理可对硅藻蛋白石表面进行微结构和性质调控。例如，使用弱碱（pH＝9.5 的氨水等）对硅藻蛋白石进行浸润等简单处理。一方面，可在较长时间（＞36 h）内使硅藻蛋白石表面形成微孔型“刻蚀坑”，并且随着处理时间的进一步增长，部分微孔间相互连通，形成介孔型“刻蚀坑”，进而形成具有微孔、介孔和大孔的多级孔道结构，其比表面积可增大至原来的几倍。另一方面，碱的溶蚀作用使得表面硅氧硅（Si－O－Si）键断裂，吸附溶液中的水分子后，其表面的硅羟基数量显著增加。

碱刻蚀导致了硅藻蛋白石表面微孔的出现、比表面积的增大以及硅羟基数量的增加，将显著改善其载体功能。例如，在制备可用于含磷废水处理的硅藻蛋白石—纳米氧化铝复合材料时，由于未经处理的硅藻蛋白石对氧化铝纳米颗粒的负载量少（最大负载量以 Al∶Si 质量百分比计仅约为 1∶6）且负载极不均匀，因此只能通过先负载锰氧化物再负载纳米氧化铝的方法提高铝的

* 通信作者：刘冬；E-mail：liudong@gig. ac. cn；手机号：13640293467。国家自然科学基金（41202024）和广东省自然科学基金杰出青年项目（2016A030306034）、科技专项计划（2017B020237003）和特支计划（609254605090）及中国科学院国际人才计划国际访问学者项目（Grant No. 2017VEA0009）等项目资助。

负载量（Al：Si 质量百分比>1：3）。然而，经碱预处理后的硅藻蛋白石对纳米氧化铝的负载量即可接近上述硅藻蛋白石—锰—铝氧化物中的铝负载量（图 1）。可见，通过碱预处理对硅藻蛋白石表面微结构和性质进行调控将显著提高硅藻蛋白石的负载能力。

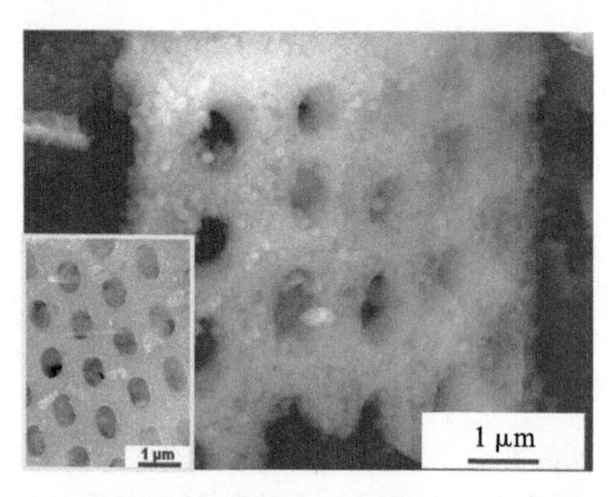

图 1　溶蚀后负载了纳米氧化铝颗粒的硅藻蛋白石

注：小图为浙江嵊州的直链藻起源硅藻蛋白石

基于硅藻蛋白石的矿物学性质，通过碱调控对硅藻蛋白石微结构和表面性质进行“改造”，可有效改善硅藻蛋白石的载体性能。

参考文献

［1］Yu W B，Deng L L，Yuan P，et al. 2015. Surface silylation of natural mesoporous/macroporous diatomite for adsorption of benzene［J］. Journal of Colloid and Interface Science，448：545 - 552.

［2］Yuan W W，Yuan P，Liu D，et al. 2016. A hierarchically porous diatomite/silicalite-1 composite for benzene adsorption/desorption fabricated via a facile pre-modification in situ synthesis route［J］. Chemical Engineering Journal，294：333 - 342.

［3］宋雅然，魏燕富，刘冬，等. 2017. 纳米氧化铝/氧化锰-硅藻土复合物对磷酸根阴离子的吸附［J］. 环境化学，36（10）：2265 - 2273.

［4］Zhang Y X，Huang M，Li F，et al. 2014. One-pot synthesis of hierarchical MnO_2-modified diatomites for electrochemical capacitor electrodes［J］. Journal of Power Sources，246：449 - 456.

不同氛围中高岭石/十八胺插层复合物的热解分析

刘庆贺　程宏飞*

（中国矿业大学（北京）地球科学与测绘工程学院，北京 100083）

高岭石插层复合物作为工业填料应用于聚合物复合材料中时，由于其纳米级颗粒带来的纳米效应、大比表面积及纳米矿物颗粒与聚合物基体之间强的界面作用使得高岭石/聚合物复合材料具有优于相同组分常规聚合物复合材料的力学性能和热力学性能，同时还具有原组分不具备的特殊性能。因此，对高岭石插层复合物的研究成为目前的热点之一。通过实验制备了高岭石/十八胺（Kaol/OA）插层复合物，使用 XRD、FTIR、TG-DSC 和 SEM 等表征手段对其进行表征。

实验采用河北张家口宣化地区的高岭石（Kaol），结晶度指数为 1.03，高岭石的含量约为 95%。二甲基亚砜（DMSO）、无水甲醇（Me）、十八胺（OA）均为分析纯；去离子水：实验室自制。

以高岭石/甲醇（Kaol/Me）为中间体经与十八胺（OA）进行插层反应后 Kaol 基底间距由 0.71 nm 逐步增加至 5.76 nm（图 1），说明 OA 分子成功插层进入高岭石层间。在 Kaol/OA 的 FTIR 谱（图 2）中位于 1 388 和 1 468 cm^{-1}处的振动峰被认为是由 OA 分子中的$-CH_3$与$-CH_2$的 C—H 键振动所导致。而在 2 850 和 2 918 cm^{-1}处新出现的两个的振动峰则是由 OA 分子中$-(CH_2)n-$的对称与非对称伸缩振动所导致，在 3 334 cm^{-1}处的峰是由 OA 分子中 N—H 键的伸缩振动导致。这些振动峰的出现均证明 OA 分子已成功插层进入高岭石层间。

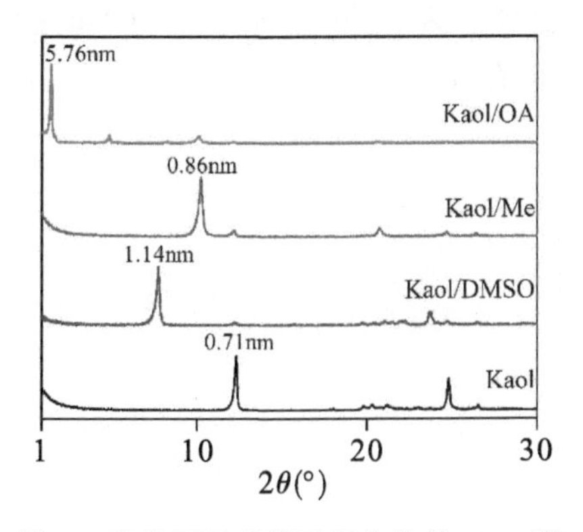

图 1　高岭石及其插层复合物的 XRD 谱

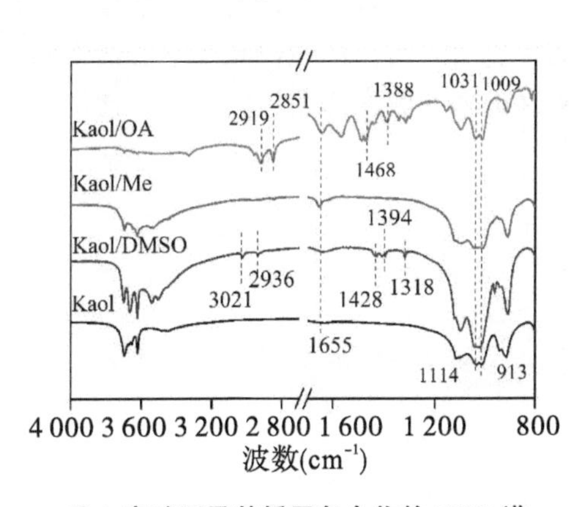

图 2 高岭石及其插层复合物的 FTIR 谱

将 Kaol/OA 插层复合物分别在氧气（Kaol/OA-O_2）和氮气（Kaol/OA-N_2）氛围下进行热分析，结果表明 Kaol/OA-N_2在热解过程中插层剂的脱嵌挥发可分为两个阶段，且其 DSC 曲线对应

* 通信作者：程宏飞；E-mail：h. cheng@cumtb. edu. cn；手机号：13426326160。国家自然科学基金（41602171）。

温度上只存在吸热峰。Kaol/OA－O_2的 TG 曲线上在 200～450 ℃插层剂的脱嵌则可分为 3 个台阶，与之对应的 DSC 曲线上在该温度区间内则出现了 3 个明显的放热峰。在 Kaol/OA－N_2的热解过程只存在吸热峰说明 OA 分子在氮气氛围下仅能通过吸热来完成脱嵌挥发，而 Kaol/OA－O_2在热解过程中出现了明显的放热峰，说明 Kaol 层间的 OA 分子在加热时与氧气发生了剧烈反应导致放热。

高岭石晶层由硅氧四面体和铝氧八面体共用氧原子而形成，结构上原子的错位导致其单层不能形成稳定的二维形态，因此在其片层剥离后由于结构的不稳定性而形成纳米卷（图 3）。纳米卷主要以两种形态存在，一种是晶层边缘发生卷曲，位于上层的晶层卷曲较为完整，向下层其晶层卷曲逐渐变得困难，仅产生半卷甚至不卷曲（图 3a）。另一种是单独剥离成卷，在连续的插层过程中，高岭石上部的晶层出现剥离，形成薄层，因其单层的不稳定性而形成完整的纳米卷（图 3b）。

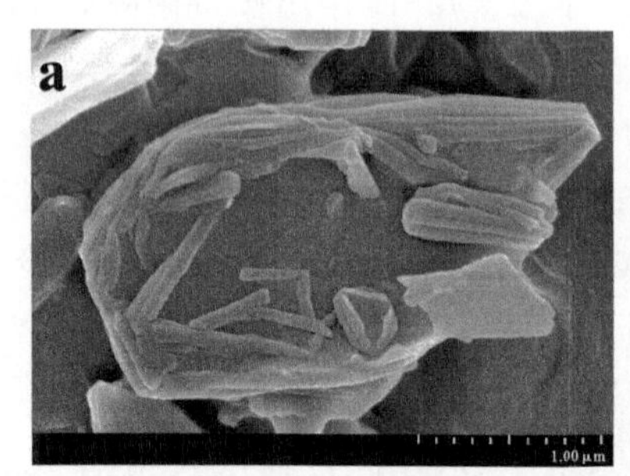

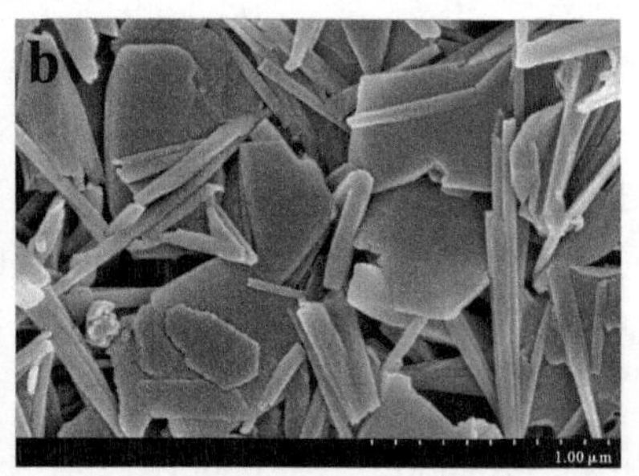

图 3　高岭石/十八胺的 SEM 照片和形貌示意

基于实验分析可知，Kaol/OA 插层复合物在氧气氛围下的热解反应之所以能够产生放热峰是由多种因素共同作用而产生的。一种是在插层的过程中随着碳链长度的增加，插层剂在 Kaol 层间逐渐变为胶结的块状存在，增大了 Kaol 层间插层剂的含量，且这种存在状态导致在插层完成后的洗涤过程中很难将多余的插层剂除去，在洗涤完成后仍会有部分插层剂因为胶结的原因而附着在高岭石的表面。此外，高产率的纳米卷也是一个重要因素，纳米卷的形成促使更多的插层剂被包裹在纳米卷的内部和层间，使得插层剂在 Kaol 层间大量存在。

The Challenge of Clay Minerals Characterization

Sabine Petit

(Institut de Chimie des Milieux et Matériaux de Poitiers,

UMR-CNRS 7285, Université de Poitiers, 86073 Poitiers, France)

Clay minerals are widespread at the Earth surface. They are widely used as mined state, as low cost impure materials, as well as refined, high purity materials for highest technological achievements. Indeed, clay minerals show a great diversity in compositions and properties. They are generally characterized by a fine size polydisperse distribution and a wide chemical and structural heterogeneity that makes them often difficult to characterize unambiguously, although a thorough characterization is generally mandatory for using them as advanced geomaterials or sources of advanced materials.

Spectroscopies offer powerful toolboxes to attain the short range orderand crystal-chemistry of clay minerals and to constrain their structural formulae. Among the others, IR spectroscopy is a very informative technique to characterize clay minerals and their crystal-chemistry as well as to study their surface properties via organic molecules interactions. It is also an appropriate technique to study poorly crystallized clay minerals, some of them sometimes behaving as XRD amorphous. Although the relationship between clay structure and IR spectrum is complex and has been only partially rationalized up to now, it is possible to attain information with this spectroscopy that could not be reached easily by other techniques. The improvement in spectral band attributions especially in the near infrared range contributes to develop the infrared analyses in field geology and remote sensing.

This lecture aims toillustrate the benefit to use IR spectroscopy and to attain crystal-chemistry of clay minerals for some industrial applications through chosen case studies.

World Class Industrial Minerals Deposits of Greece

George E. Christidis[1*] Ioannis Marantos[2]
(1. Technical University of Crete, School of Mineral Resources Engineering, Chania 73100, Greece;
2. Institute of Geological and Mineral Exploration, Spyrou Loui 1, Acharnes 13677, Greece)

Greece has been endowed with numerous deposits of industrial minerals and rocks such as bentonite, perlite, magnesite and Mg-compounds, pumice, pozzolan, gypsum and anhydrite, palygorskite, amphibolite, dunite, calcium carbonates, huntite, structural clays, emery, aggregates and construction stone, zeolites[1]. The main industrial mineral deposits are distributed in three main geological units, like the External Hellenides (EH), the Internal Hellenides (IH) and the South Aegean Volcanic Arc (SAVA). Some of these industrial minerals are considered as world class deposits due to their large reserves of very high quality. These deposits include the bentonite and perlite deposits of Milos Island (SAVA), and the attapulgite deposit of W. Macedonia (IH). Also very high quality deposits of $CaCO_3$ fillers (chalk) occur in Kefallonia Island (EH) (Fig. 1). The description of these deposits is the purpose of this presentation.

The Pleistocene stratabound bentonite deposits of Milos Island formed by low-T submarine hydrothermal alteration of pyroclastic flows and lavas of andesitic to rhyolitic composition. Cooling of the volcanic rocks coupled by the high heat flow sustained the hydrothermal system. Bentonites have been affected by subsequent illitization and S-alteration associated with barite emplacement and high sulfidation epithermal activity respectively. The deposits are composite consisting of several horizons. The reserves exceed 80 million tons, with average smectite content $>75\%$ and the main smectite is Ca-montmorillonite[2]. Milos is a major world bentonite producer (>1 mt/y) [1]. Major markets of the bentonite are the foundry and drilling industries, civil engineering, animal litters etc.

The Middle Pleistocene Miloan perlite deposits were formed by reaction of meteoric water with cooled rhyolitic lavas (SAVA). The perlites are not characterized by typical perlitic texture. The reserves exceed 60 million tons of rhyolitic glassy lavas consisting of $>95\%$ volcanic glass with 3.5%—4.2% H_2O, and microphenocrysts of quartz, oligoclase plagioclase, oxidized biotite and magnetite. Milos is the second world producer of perlite with production of $\sim$ 1 mt/y. The material is sold mainly to European countries and to North America in crude form. Expansion takes place at the receiving ports.

The Pleiocene palygorskite deposit in Ventzia basin, Grevena, W. Macedonia (IH), (Fig. 1)

* Corresponding author: George E. Christidis; E-mail: christid@mred. tuc. gr; Phone: +30—6980756616.

formed from conversion of Fe-rich smectitic lateritic soils which were transported to a lacustrine environment[3]. Alteration of smectite to Fe-rich palygorskite occurred mainly in the center of the lake，with detrital smectite remaining in the margins of the lake[3]. The overall reserves exceed 50 mt of palygorskite and Fe-rich smectite.

The Eocene-Oligocene filler/coating grade chalk of Kefallonia Island，W. Greece (EH) are unbedded massive slightly compacted ultra high purity limestones with 0.1—1 m thick intercalated chert beds. The limestones contain >99.5% calcite. The acid insoluble residue (~0.3%) contain mainly chromite，sphalerite，chalcopyrite and scarce pyrite. The calcite fillers have L* and b* values，96% and 1.7% respectively. Replacement of TiO_2 pigment by up to 50% calcite does not affect significantly the properties of paint colors，thereby decreasing production cost[4].

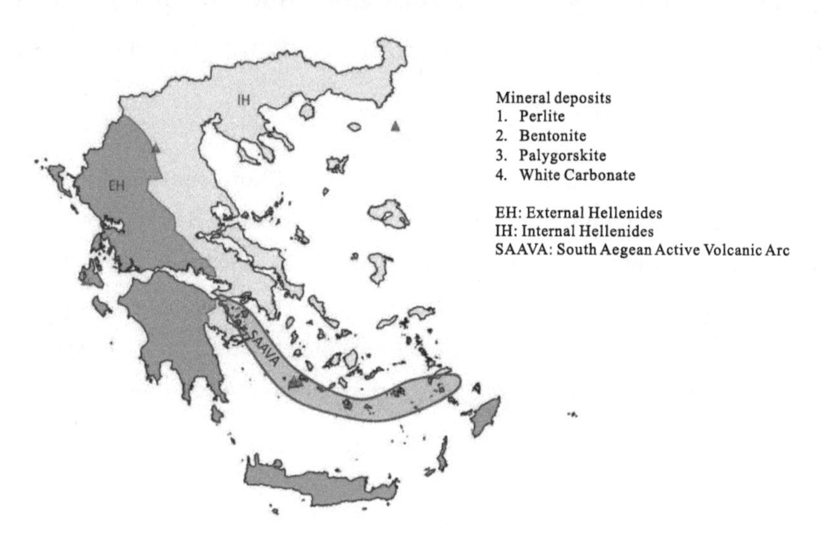

Fig. 1　Location of the world class industrial mineral deposits of Greece

References

[1] Tzeferis P. 2015. The mining/metallurgical industry in Greece. Commodity review for years 2013-2014. Mineral Resources Directorate，General Secretariat for Energy and Mineral Resources.

[2] Christidis G E，Huff W D. 2009. Geological aspects and genesis of bentonites [J]. Elements，5：93 - 98.

[3] Kastritis I D，Kacandes G H. 2003. The palygorskite and Mg-Fe-smectite clay deposits of Ventzia basin，W. Macedonia，Greece. in：Proceedings of the 7th SGA Meeting Millpress，Rotterdam，891 - 894 pp.

[4] Kalafati K，Christidis G E. 2007. Replacement of TiO_2 pigment by $CaCO_3$ from Kefalonia in emulsion paints [J]. Bulletin of Geological Society of Greece，40：759 - 768.

Magnesite Markets and Geological Factors in the Evaluation of Cryptocrystalline Magnesite from Proterozoic Marine Sedimentary Deposits, South Australia

John Keeling[1*] Ian Wilson[2]

(1. Geological Survey of South Australia, Adelaide 5000, Australia;

2. Industrial Minerals Consultant, Cornwall, PL30 5NW, United Kingdom)

World magnesite production in 2017 was estimated at 27 million tonnes (Mt) with China accounting for 67% with substantial production from Turkey (10%), Russia (5%), Brazil (4%) and Austria (3%); production was contributed also from Australia (2%), Slovakia (2%) Greece (1%), India (1%), Spain (1%) and North Korea (1%) (USGS 2018). Magnesite was used primarily in magnesia (MgO) production (world capacity ~11.9 Mt MgO-including sources from seawater, brines and from brucite) of which 80% was consumed by refractory markets and 20% by other markets, including magnesia cements, agricultural and environmental uses.

Australia' s 2017 production of 450, 000 tonnes was dominantly from Sibelco' s cryptocrystalline magnesite deposits at Kunwarara in Queensland used as feedstock for deadburned magnesia (DBM), caustic-calcined magnesia (CCM) and fused magnesia (FM) at the company' s plant at Parkhurst, Rockhampton. Australia has <5% of estimated world magnesite resources of ~13 billion tonnes but has over 50% of global cryptocrystalline magnesite resources that total ~8% of world magnesite resources[1]. The largest resources of cryptocrystalline magnesite are in South Australia (>500 Mt) and Queensland (129 Mt) [1,2]. South Australian production was only 3, 200 tonnes in 2017, largely from the Myrtle Springs mine, 26 km north-west of Leigh Creek Township.

Extensive cryptocrystalline magnesite resources in South Australia are peritidal facies of Neoproterozoic sedimentary rocks of Skillogalee Dolomite. Thecyclic deposits of thin magnesite-rich muds were desiccated and reworked as pebble conglomerate and clastic deposits, 0.1—5 m thick, interbedded with shallow marine dolomite. The magnesite mudflats exceeded 1 km width and extended for tens of kilometres along the Proterozoic shoreline. The sediments were lithified and folded into broad open folds, but were not recrystallized. Magnesite crystal size is typically 1—8 microns with some intergranular porosity. Magnesite content of individual beds varies between 80%—88% $MgCO_3$ with up to 20% contaminants consisting of very fine-grained dolomite (6%—14%), talc (1%—5%), quartz (0.1%—15%), with traces of albite and K-feldspar. Iron oxide content is low (<0.3% Fe_2O_3) but boron can exceed 200 mg/kg

* Corresponding author: John Keeling; E-mail: keelingl@bigpond. net. au; Mobile Phone No.: +61-401122015.

B，which reflects the marine origin.

The fine grained magnesite has a high surface area and high reactivity compared with ‘sparry’ macrocrystalline magnesite but is not easily beneficiated. Selectively mined magnesite is below quality specifications for refractory grade deadburned or fused magnesia. In the late 1990s，extensive investigations were made to assess the suitability for magnesium metal production. This included field mapping supported by airborne hyperspectral imagery and GPS-controlled ground survey of individual magnesite beds[3]，followed by diamond drilling and acid leach trials on bulk samples. Larges resources were confirmed and leach trials were largely successful，but increasing energy costs and lower than anticipated growth in demand for magnesium metal saw the project abandoned.

Calix Ltd acquired the Myrtle Springs mine in 2013 as feedstock for their Calix Flash Calcination（CFC）process，a modified kiln design that incorporates pure CO_2 capture. Indirect heating of magnesite at 780 ℃ creates a reactive magnesia product with high surface area（$>110\ m^2gm^{-1}$），which is easily converted to magnesium hydroxide liquid. This is used in sewer odour control and protection of sewer concrete infrastructure. The magnesia powder is being trialled also for markets in aquaculture，for conditioning pond water，and as fertiliser/pesticide in agriculture，for more sustainable and environmentally friendly farm practice.

References

［1］Wilson I. 2010. The world of magnesite［J］. Industrial Minerals，May 2010：50 - 67.

［2］Horn C M，Keeling J L，Olliver J G. 2017. Sedimentary magnesite deposits，Flinders Ranges，In：Phillips，G. N.（ed.）*Australian Ore Deposits*，The Australasian Institute of Mining and Metallurgy，Melbourne. pp. 671 - 672.

［3］Keeling J L，Mauger A M. 1998. New airborne HyMap data aids assessment of magnesite resources［J］. Department of Primary Industries and Resources，South Australia，MESA Journal，11：8 - 11.

Effect of Thermal Treatment on Structure and Surface Properties of Palygorskite

Ya Ting Cui[1] Jin Yao Cong[1] Yu Zheng[1] Wei Qing Wang[1,2*]
(1. School of Environment and Resource, Southwest University of Science and Technology, Mianyang 621010, China;
2. Key Laboratory of Solid Waste Treatment and Resource Recycle, Ministry of Education, Southwest University of Science and Technology, Mianyang 621010, China)

Palygorskite is a hydrated magnesium aluminum silicate with a particular crystal structure[1]. This provides it with a great specific surface area, good adsorbability and excellent cationic exchange performance. So, it is widely used as adsorbents, catalysts, ion-exchangers[2]. However, the properties and applications of activated palygorskite are better than palygorskite. Usually, suitable acid and thermal treatments can increase the catalytic and adsorbent activity of palygorskite[3].

In this study, the mineral sample was the palygorskite from Mingguang City, Anhui Province, China. It was ground using a jet mill to obtain palygorskite powder with the size of less than 0. 074 mm. Then the palygorskite powder was calcined at different temperatures (150 ℃, 250 ℃, 350 ℃, 450 ℃, 550 ℃, 650 ℃ and 750 ℃) for 3 h to prepare the thermal-activated palygorskite (TPx) samples. Then, the TPx samples were characterized by X-ray diffraction (XRD), thermogravimetric-differential thermal analysis (TG-DTA), Fourier transform infrared spectroscopy (FTIR), Brunauer-Emmett-Teller (BET) specific surface area measurements, X-ray photoelectron spectroscopy (XPS) and potentiometric titration. The objective of this study is to investigate the effect of thermal activation temperature on structure and surface properties of palygorskite.

The XRD and TG-DTA results showed that the crystal structure of TP250 had no obvious change, and the adsorbed water and part of coordinated water were lost. The FT-IR results indicated that TP250 had the most complete Si-O bands. The BET specific surface area measurement results showed that TP250 had the biggest specific surface area. The high-resolution XPS spectra indicated that thermal activation could change the chemical environment of element on palygorskite surface. As shown in Fig. 1, after thermal activation at 250 and 350 ℃, the binding energies of Mg1s were shifted by +0.29 and +0.39 eV,

* Corresponding author: Wei Qing Wang; E-mail: wangweiqing @ swust. edu. cn; Mobile Phone No.: ++86-8162419569. This study was financially supported by Postgraduate Innovation Fund Project by Southwest University of Science and Technology (18ycx052).

respectively. The evolution of Z versus pH for the samples (Fig. 2) showed that deprotonation reaction was predominant within the entire pH range. When the palygorskite was moderately treated, its surface showed a different behaviour. When the temperature reached 250 ℃, thermal activation made the surface of palygorskite produce new surface points with different acid alkaline properties.

During the thermal activation process, the chemical environment of the surface elements of the palygorskite changed obviously with the increasing temperature. When the temperature reached 250 ℃, thermal activation made the surface of palygorskite produce new surface points with different acid alkaline properties and the specific surface area of palygorskite maximize.

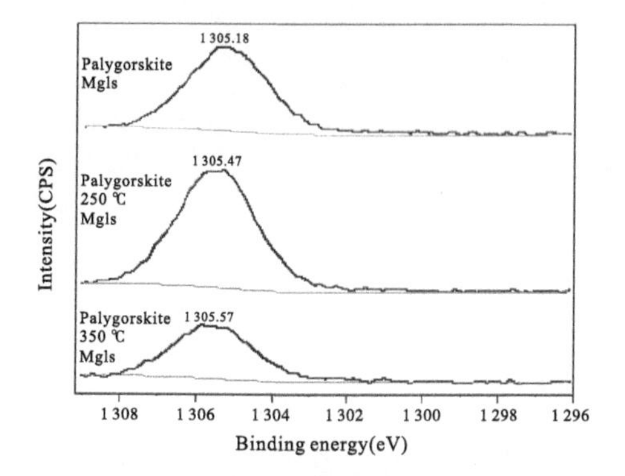

Fig. 1　The Mg1s high-resolution XPS spectra of P, TP250 and TP350

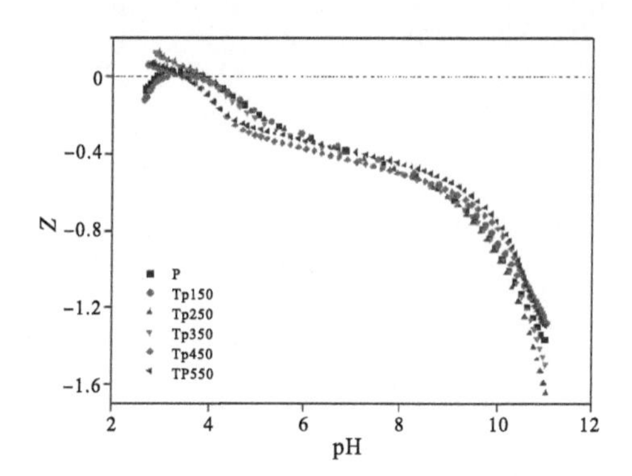

Fig. 2　Z versus pH curves of different suspension samples in 0.01 mol/L NaCl

References

[1] Bradley W F. 1940. The structure scheme of attapulgite [J]. American Mineralogist, 25: 405 - 410.

[2] Chen T, Liu H, Li J, et al. 2011. Effect of thermal treatment on adsorption-desorption of ammonia and sulfur dioxide on palygorskite: Change of surface acid-alkali properties [J]. Chemical Engineering Journal, 166: 1017 - 1021.

[3] Bamios M S, González L V F, Rodríguez M A V, et al. 1995. Acid activation of a palygorskite with HCl: Development of physico-chemical, textural and surface properties [J]. Applied Clay Science, 10: 247 - 258.

第三章

非金属矿—聚合物复合功能材料

黑液—蒙脱土复合物对氯醚橡胶机械和热性能的影响

于智鹏[1]　谭雅婷[2]　罗琼林[2]　王　曦[2]　苏胜培[2*]
（1. 湖南第一师范学院，长沙 410081；
2. 湖南师范大学化学化工学院资源精细化与先进材料
湖南省高校重点实验室，长沙 410081）

在双辊磨机上通过机械混合制备填充有黑液—蒙脱土复合物（BL－MMT）氯醚橡胶（ECO）复合材料。XRD 和 TEM 数据均表明，填料颗粒很好地分散在 ECO/BL－MMT 复合材料中。在添加 50％的 BL－MMT 下，橡胶复合材料的拉伸强度、断裂伸长率和 100％模量分别为 14.0 MPa、457％和 3.9 MPa。在空气循环烘箱中 100 ℃下热氧化老化 72 h 后，拉伸强度的保持率为 99％。实验结果表明，ECO/BL－MMT 复合材料具有良好的机械性能和热氧化老化性能。图 1 所示低放大倍率和高放大倍率的 TEM 图像。可以看出，黏土在 ECO/BL－MMT 复合材料中具有良好的分散性。

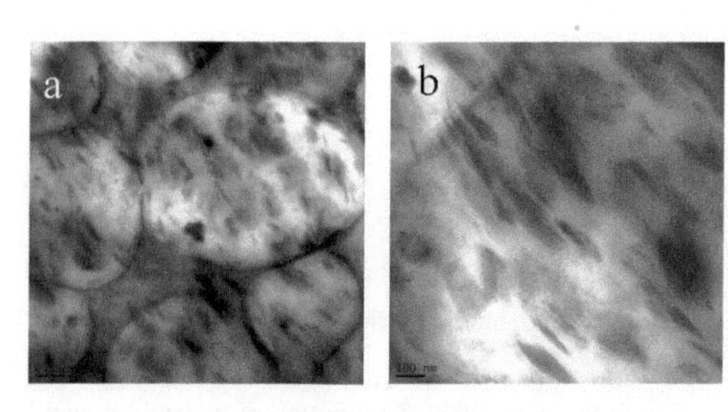

图 1　ECO/BL－MMT－30 复合物的 TEM 图像

注：a. 低倍；b. 高倍

图 2 所示，对于在 100 ℃空气中老化 72 h 的样品，从 ECO/BL－MMT 获得的拉伸强度保持率高于 ECO/N330。ECO/BL－MMT 复合材料在 10、30、50、70 和 90 phr 的拉伸强度保持率分别为 76％、106％、99％、102％和 108％[1-6]。

表 1 列出了 ECO/BL－MMT 复合材料的拉伸强度、断裂伸长率、100％模量和邵氏 A 硬度等力学性能数据。

* 通信作者：苏胜培；E-mail：sushengpei@yahoo. com；手机号：18670763879。

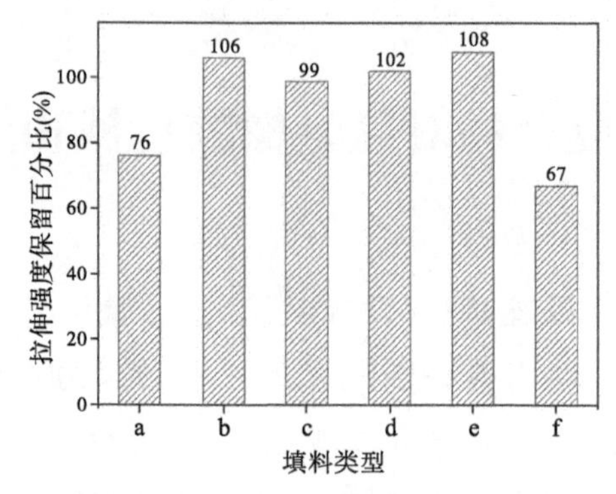

图 2 由 6 种物质获得的 ECO 复合材料的拉伸强度保留百分比

注：a. ECO/BL－MMT－10；b. ECO/BL－MMT－30；c. ECO/BL－MMT－50；d. ECO/BL－MMT－70；e. ECO/BL－MMT－90；f. ECO/N330－50

表 1 ECO 复合材料的力学性能

复合材料	拉伸强度（MPa）	断裂伸长率（%）	100%模量（MPa）	邵氏 A 硬度
ECO/BL－MMT－10	5.0	291	2.0	37
ECO/BL－MMT－30	9.1	307	3.0	50
ECO/BL－MMT－50	14.0	457	3.9	55
ECO/BL－MMT－70	13.5	335	5.6	67
ECO/BL－MMT－90	13.1	267	7.2	75

参考文献

[1] Gilman J W. 1999. Flammability and thermal stability studies of polymer layered-silicate（clay）nanocomposites [J]. Applied Clay Science，15：31－49.

[2] Chen R Y，Peng F B，Su S P. 2008. Synthesis and characterization of novel swelling tunable oligomeric poly（styrene-co-acrylamide）modified clays [J]. Journal ofApplied Polymer Science，108：2712－2717.

[3] Zang Y L，Xu W J，Liu G P，et al. 2009. Preparation of ultraviolet-cured bisphenol A epoxy diacrylate/montmorillonite nanocomposites with a bifunctional，reactive，organically modified montmorillonite as the only initiator via in situ polymerization [J]. Journal of Applied Polymer Science，111：813－818.

[4] Chen D，Zang Y L，Su S P. 2010. Effect of polymerically-modified clay structure on morphology and properties of UV-cured EA/clay nanocomposites [J]. Journal of Applied Polymer Science，116：1278－1283.

[5] Kosikova B，Gregorova A. 2005. Sulfur-free lignin as reinforcing component of styrene-butadiene rubber [J]. Journal of Applied Polymer Science，97：924－929.

[6] Cao Z L，Liao Z D. 2013. Preparation and properties of NBR composites filled with a novel black liquor-montmorillonite complex [J]. Journal of Applied Polymer Science，127：3725－3730.

矿物复合材料及其环境能源应用

张以河* 白李琦

（非金属矿物与固废资源材料化利用北京市重点实验室，
中国地质大学（北京）材料科学与工程学院，北京 100083）

介绍了以蒙脱石、云母、石墨及石墨烯、凹凸棒石、羟基磷灰石等矿物复合材料及其在生态环境材料、能源材料、生物材料等领域的应用。

通过对凹凸棒石复合材料的熔融共混制备，采用硅烷偶联剂改性，提高了 PBS 的力学性能和热性能，复合材料的弯曲性能和热稳定性也得到了显著的提高[1]，如图 1 所示。凹凸棒石添加到 PVDF - HFP 中，用过流延法制备复合薄膜，有效地提高了复合薄膜的介电性能；通过制备纳米管状二氧化硅低介电复合薄膜材料，添加到 3%时，介电常数降至 2.9[2]。通过原位聚合法制备了云母/PI 复合薄膜，该复合薄膜介电常数低、低温极化差且耐击穿性能优异。而蒙脱石的引入可有效降低 PI 介电常数、提高复合薄膜的力学性能，在低温时效果尤为明显。

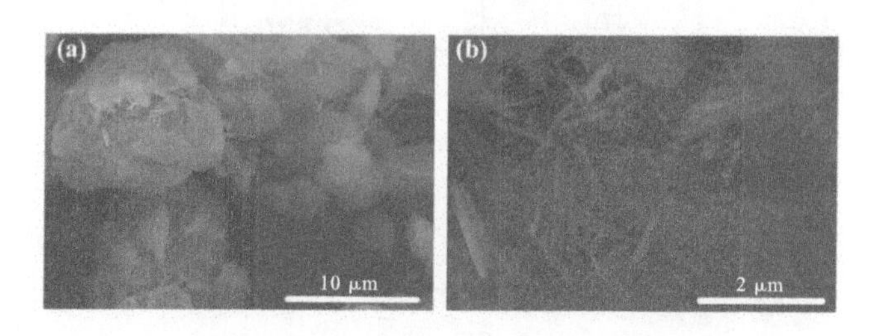

图 1　凹凸棒石改性前后的 SEM 照片[1]

通过改性羟基磷灰石而制备的羟基磷灰石/PBS 复合材料，具有良好的抗拉强度，可望用于生物医学领域。以基于天然石墨的石墨烯为基底，聚乙烯亚胺和聚丙烯酸为表面改性分子，通过层层自组装技术制备石墨烯多层膜复合材料，并以聚电解质—石墨烯复合结构为构筑基元制备具有核壳结构的复合材料，在物质检测和药物传递及释放等方面均具有的潜在应用价值[3]。将全氟叠氮苯甲酸改性石墨烯/PVDF - HFP 薄膜制备成发电储能一体化的功能薄膜（图 2）。通过手指弯曲使得薄膜产生电位差，并将其原位储存起来[4]。通过外电路控制释放时间，通过薄膜的串联，使得输出功率得以提升。制备的发电储能一体化功能复合薄膜材料可用于国防或地质野外作业能量转换和储存器件、传感器、记忆智能材料等方面。

通过石墨烯基全固态柔性能量转换存储薄膜与银纳米线复合结构（自供能—等离子共振耦合

*通信作者：张以河；E-mail：zyh@cugb. edu. com；手机号：13681112016。国家重点研发计划专项课题（2017YFC0703104）、国家自然科学基金（51572246、51772279）资助项目。

基底）来实现双重拉曼信号增强（图 3），不仅可以做成便携式可抛弃式的智能基底，也可以整合到手套、衣服上，做成具有拉曼信号增强作用的可穿戴设备，可以实时、现场获得拉曼信号[5]，在检测环境、水质、农药等方面具有潜在应用。

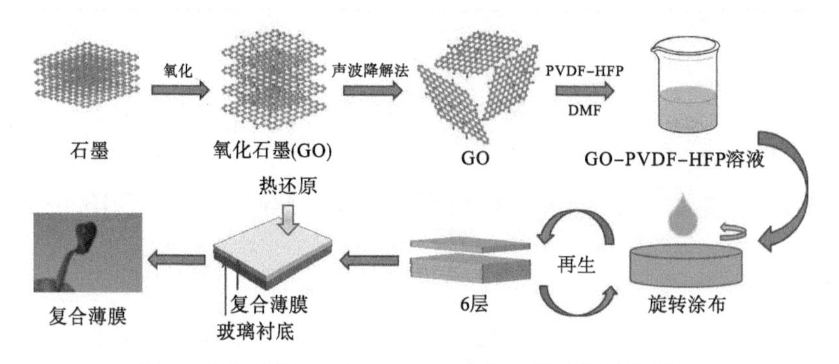

图 2　定向还原 GO/PVDF-HFP 复合薄膜材料的制备

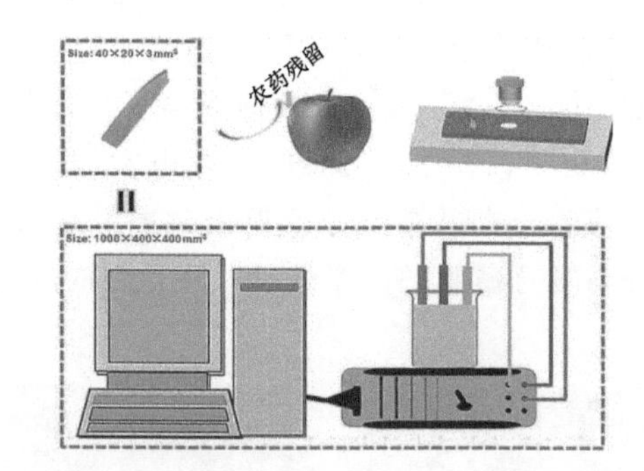

图 3　利用自组装衬底检测苹果表面农药残留的原理

参考文献

［1］Zhang Y，Yu C，Hu P，et al. 2016. Mechanical and thermal properties of palygorskite poly（butylene succinate）nanocomposite［J］. Appl Clay Sci.，119：96-102.

［2］Zhang Y H，Lu S G，Li Y Q，et al. 2005. Novel silica tube/polyimide composite films with variable low dielectric constant［J］. Adv Mater.，17（8）：1056-1059.

［3］Li X，Zhang Y，Wu Y，et al. 2015. Combined photothermal and surface-enhanced raman spectroscopy effect from spiky noble metal nanoparticles wrapped within graphene-polymer layers：Using layer-by-layer modified reduced graphene oxide as reactive precursors［J］. ACS Appl Mater Interfaces.，7（34）：19353-19361.

［4］Tong W，Zhang Y，Zhang Q，et al. 2015. An all-solid-state flexible piezoelectric high-k film functioning as both a generator and In situ storage unit［J］. Adv Funct Mater.，25（45）：7029-7037.

［5］Li H，Dai H，Zhang Y，et al. 2017. Surface-enhanced Raman spectra promoted by a finger press in an all-solid-state flexible energy conversion and storage film［J］. Angew Chemie Int Ed.，56（10）：2649-2654.

黏土基复合材料的制备及重金属离子吸附性能

张　丹[1]　王　兰[2]　王传义[1,2*]
（1. 陕西科技大学环境科学与工程学院，西安 710021；
2. 中国科学院新疆理化技术研究所，乌鲁木齐 830011）

重金属与人类社会的发展和进步息息相关，并已广泛应用在电子、机械制造、建筑、国防工业等领域，同时也造成了越来越严重的重金属污染。被重金属污染的废水和地表水是全世界的环境难题。吸附法因其简单、方便和高的去除率被认为是一种最常用去除重金属的方法。其中，黏土因其独特的优势被各国科学家广泛研究。

天然黏土矿物材料因具有离子交换能力、良好的化学稳定性和机械稳定性在重金属污染物去除方面有一定的优势，但其吸附容量较低、悬浮性和分散性等弊端在一定程度上也抑制了黏土的开发与利用。很多研究表明，天然黏土的重金属吸附容量随着改性而增加。因此，结合无机改性和有机改性技术，发展黏土复合重金属吸附材料的可控制备方法，并以环境中常见的重金属污染物Pb（Ⅱ）、Cd（Ⅱ）、Cu（Ⅱ）等离子为代表研究其环境功能与作用机制具有重要的理论意义和应用价值。

在前期工作中，我们开展了一系列黏土基环境功能材料的研究[1-6]。一个代表性的工作是蛭石为基础的铅选择性吸附剂，其制备方法如下：①取酸化蛭石（过 200 目筛）、甲苯、3－三乙氧基甲硅烷基丙基丙烯酸酯水浴搅拌回流、干燥，得到硅烷偶联剂改性蛭石（OVerm)；②将 OVerm 加入甲苯溶液，超声处理之后加入丙烯酰胺和偶氮二异丁腈，氮气氛围油浴搅拌、离心、干燥得聚丙烯酰胺/蛭石复合材料（PAM/OVerm）；③在 PAM/OVerm 中加入甲醛溶液，用 NaOH 调节 pH 值，水浴回流后再加入三乙烯四胺，搅拌、干燥获得表面有机改性修饰蛭石复合材料（g－PAM/OVerm）。

为了验证合成的 g－PAM/Overm 对 Pb（Ⅱ）的选择吸附性，设计了多组分重金属吸附实验（初始浓度均为 800 mg/L）。如图 1 所示，g－PAM/Overm 对 Pb（Ⅱ）的平衡吸附量相比 Cd（Ⅱ）、Cu（Ⅱ）和 Zn（Ⅱ）来说更高，说明 g－PAM/Overm 对 Pb（Ⅱ）离子具有明显的选择吸附能力，这是由于 Pb（Ⅱ）离子和－NH_2之间具有更强的键合力造成的。

通过结合酸改性和有机改性在蛭石表面接枝了大量对 Pb（Ⅱ）离子具有吸附作用的功能性－NH_2基团，这些改性明显改善了蛭石对 Pb（Ⅱ）离子的吸附能力。同时，制得的 g－PAM/Overm 相比 Cd（Ⅱ）、Cu（Ⅱ）和 Zn（Ⅱ）等离子，对 Pb（Ⅱ）离子具有明显的吸附选择性。说明该材料在去除水体中的 Pb（Ⅱ）方面具有很大的潜力，这为工业废水中重金属铅的去除提供了一条有效途径。

---

* 通信作者：王传义；E-mail：cywang@ms. xjb. ac. cn；手机号：13299131206。国家自然科学基金（U1403295）。

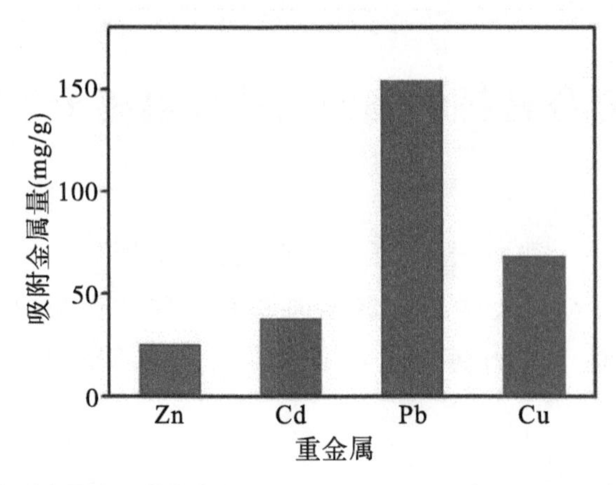

图 1 竞争重金属离子共存条件下 g - PAM/Overm 对 Pb（Ⅱ）的选择吸附性

注：吸附温度 T=30 ℃，吸附时间 t=120 min，pH=5.5

参考文献

[1] Mao X Y，Wang L，Gu S Q，et al. 2018. Synthesis of a three-dimensional network sodium alginate-poly（acrylic acid）/attapulgite hydrogel with good mechanic property and reusability for efficient adsorption of Cu^{2+} and Pb^{2+} [J]. Environ. Chem. Lett.，16（2）：653 - 658.

[2] Gu S Q，Wang L，Mao X Y，et al. 2018. Selective adsorption of Pb（Ⅱ）from aqueous solution by triethylenetetramine-grafted polyacrylamide/vermiculite [J]. Mater，11（4）：514.

[3] Zhu K C，Duan Y Y，Wang F，et al. 2017. Silane-modified halloysite/Fe_3O_4 nanocomposites：Simultaneous removal of Cr（Ⅵ）and Sb（Ⅴ）and positive effects of Cr（Ⅵ）on Sb（Ⅴ）adsorption [J]. Chem. Eng. J.，311：236 - 246.

[4] Zhu K C，Jia H Z，Wang F，et al. 2017. Efficient removal of Pb（Ⅱ）from aqueous solution by modified montmorillonite/carbon composite：Equilibrium，kinetics，and thermodynamics [J]. J. Chem. Eng. Data，62：333 - 340.

[5] Wang L，Wang X，Yin J，et al. 2016. Insights into the physicochemical characteristics from vermiculite to silica nanosheets [J]. Appl. Clay Sci.，132-133：17 - 23.

[6] Wang X H，Wang C Y. 2016. Chitosan-poly（vinyl alcohol）/attapulgite nanocomposites for copper（Ⅱ）ions removal：pH dependence and adsorption mechanisms [J]. Colloids Surf. A：Physicochem. Eng. Aspects，500：186 - 194.

聚偏氟乙烯/改性凹凸棒石纳米复合超滤膜

周守勇* 薛爱莲 李梅生 张 艳 赵宜江
（淮阴师范学院化学化工学院，江苏省环境功能材料工程实验室，
江苏省低维材料化学重点实验室，淮安 223300）

纳米复合超滤膜已成为膜材料研究领域的新热点，不同纳米材料的引入使得纳米复合膜在机械强度、热稳定性、抗污染能力、渗透性和选择性等方面得到了不同程度的提升。目前，用得较多的纳米材料是颗粒状的纳米 TiO_2、SiO_2、Al_2O_3和 ZrO_2等，这些纳米颗粒在膜制备和使用过程中会发生脱落，从而影响膜的性能和改性效果。相比之下，碳纳米管等一维纳米材料具有超强的力学性能、高的长宽比和高比表面，而且分散在高分子膜中的一维纳米材料，通过高分子链的螺旋缠绕可以有效提高其在膜材料中的稳定性。然而，碳纳米管等人工合成一维纳米材料制备成本高、纯度和产量低下、难以分散，大大限制了其在纳米复合膜中的规模化应用，导致纳米复合膜的发展目前还处于初级阶段[1,2]。

天然纳米凹凸棒石在形态、尺寸等外观特征上具有一维的纳米尺寸结构，与人工合成的一维结构纳米材料一致。但其性价比明显优于人工合成纳米纤维材料，可较好地解决人工纳米单元材料批量小、成本高等问题，这对于降低制膜成本以及商品化至关重要。此外，纳米凹凸棒石表面富含大量的羟基，亦为其进一步功能化修饰改性提供了便利，通过对纳米凹凸棒石进行适当的修饰改性，能够解决纳米凹凸棒石团聚及其与高分子膜材料相容性的问题，开发出具有实用化和产业化前景的新型膜材料及其制备技术[3,4]。

本研究利用化学接枝技术（图 1），合成可用于提高 PVDF 膜亲水性且性能稳定的改性凹凸棒石。采用浸没沉淀相转化法，以 PVDF/TEP/水为体系，制得的 PGS - g - PDMAEMA 粒子为共混改性剂，制备 PVDF/PGS - g - PDMAEMA 混合基质超滤膜，进一步研究其抗污染和易清洗性能。

利用化学接枝技术，在凹凸棒石表面用聚甲基丙烯酸、N,N -二甲氨基乙酯修饰。通过 FT-IR、XRD、TG、SEM、EDX、XPS 等一系列方法证明单体 PDMAEMA 成功接枝到凹凸棒石表面，并且没有改变凹凸棒石原有的晶型、形貌。PDMAEMA 的接枝量随着单体量的增加逐渐饱和，当单体量达到 6 g 时，接枝率和单体量为 4 g 的接枝率变化不大，呈过饱和状态。在进行 BSA 吸附实验时，PGS - g - PDMAEMA 对 BSA 的吸附量最小，仅为 5.03 mg/g。

利用浸没沉淀相转化法，以 PGS - g - PDMAEMA 粒子为共混改性剂制备 PVDF/PGS - g - PD-MAEMA 混合基质超滤膜。PGS - g - PDMAEMA 粒子的加入减少了膜表面的缺陷孔，使膜孔径

* 通信作者：周守勇；E-mail：z3517185@ 126. com；手机号：15996197498。国家自然科学基金（21476094）、江苏省自然科学基金（BK20171268）、淮安市科技支撑项目（HAG201609）、江苏省低维材料化学重点实验室开放课题（JSKC17005）、江苏省高校“青蓝工程”项目、江苏省“六大人才高峰”高层次人才项目。

分布更加均匀，平均孔径所占比例增加。添加了改性剂的混合膜其指状孔长度小于纯膜，皮层厚度小于纯膜。PGS－g－PDMAEMA 粒子在一定程度上起到致孔作用，使膜的孔隙率得到提高。改性后膜的机械强度、伸长率得到提高，膜的接触角下降，亲水性得到提升。其中，PVDF/P1 膜的表面粗糙度最小约为 14.46 nm，接触角最小为 75.81°，亲水性最佳。PVDF/P0 膜的纯水通量最低约为 123.28 L/（m^2 · h），随着改性剂 PGS－g－PDMAEMA 粒子的加入，纯水通量逐渐上升，PVDF/P7 膜的纯水通量最大，约为 271.23 L/（m^2 · h），是 PVDF/P0 膜的 2 倍。

对膜进行 BSA 过滤，PVDF/P7 膜过滤 BSA 时的通量衰减速度小于 PVDF/P0，在保证截留率均为 100％的情况下，PVDF/P7 膜的稳定渗透通量最大，约为 26.38 L/（m^2 · h），将近 PVDF/P0 膜的 2 倍。添加了 PGS－g－PDMAEMA 粒子的 PVDF 混合基质超滤膜的静态、动态吸附量小于纯 PVDF 膜，膜表面不易被蛋白吸附污染。并且，混合膜的通量恢复率均大于纯膜，PVDF/P1 的通量恢复率最大，达到 54.17％，具有最佳的易清洗性能。利用 Zeta 测试证明，膜的吸附主要受膜表面亲疏水性控制，表面电荷的影响不大。通过电镜、膜污染指数分析、膜污染阻力分析等方法，证明 PVDF/P0 膜表面的污染物堆积和膜孔堵塞现象严重，其中 PVDF/P1 膜受污染趋势最小，并且主要受可逆污染影响。

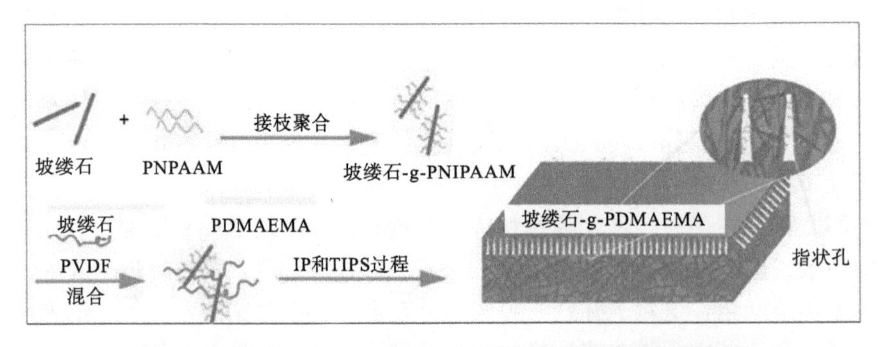

图 1　PVDF/PGS－g－PDMAEMA 混合基质超滤膜制备流程

参考文献

[1] 杨铁军. 2015. 产业专利分析报告（第 37 册）——高性能膜材料［M］. 北京：知识产权出版社.

[2] 王熙大，王志宁，高从堦. 2014. 纳米复合膜在膜分离领域的研究进展［J］. 应用化学，31（2）：123－132.

[3] Ji J，Zhou S，Lai C Y，et al. 2015. PVDF/palygorskite composite ultrafiltration membranes with enhanced abrasion resistance and flux［J］. Journal of Membrane Science，495：91－100.

[4] Cai J，Zhou S，Zhao Y，et al. 2016. Enhanced hydrophilicity of a thermo-responsive PVDF/palygorskite-g-PNIPAAM hybrid ultrafiltration membrane via surface segregation induced by temperature［J］. RSC Advances，6（67）：62186－62192.

十四碳烯琥珀酸酐表面改性电气石及功能聚合物合成

胡应模[*]　李梦灿
（中国地质大学材料科学与工程学院，北京 100083）

电气石为环状结构硅酸盐矿物，具有压电性、热释电性、远红外辐射和释放负离子性等独特性能，作为新型功能矿物材料受到人们高度重视，通过物理或化学方法与其他材料复合，可制得多种功能材料，被广泛应用于环保、电子、医药、化工、轻工、建材等领域，已成为一种高附加价值的新型工业矿物。为了开发电气石的高附加价值和更广的应用领域，近年来许多科技工作者对电气石的性能、改性及应用进行了深入探讨，为开发含电气石功能聚合物及其复合材料进行了大量工作[1-3]。

本文以十四碳烯琥珀酸酐为改性剂对电气石的表面改性进行了探讨，以改性产物对水的接触角以及在液体石蜡中的浊度为指标对十四碳烯琥珀酸酐改性电气石工艺条件进行了优化，结构分析表明，带有双键的有机链引入到了电气石表面，得到了表面有机化改性的可聚合十四碳烯琥珀酸电气石酯。电气石粉体的表面极性减小，团聚现象明显降低，分散性能得以改善，且其远红外辐射率及负离子释放量均具有明显提高。然后，将十四碳烯琥珀酸电气石酯与乙酸乙烯酯进行共聚反应，合成了含电气石的十四碳烯琥珀酸电气石酯—乙酸乙烯酯共聚物［p（TTDS－VA）］。通过 IR、XRD、SEM 等手段对其进行结构和形貌表征，实验结果表明，电气石粉体成功引入到共聚物中，具有优良的分散性和储存稳定性。且所得的十四碳烯琥珀酸电气石酯—乙酸乙烯酯共聚物成膜后具有良好的力学性能和优异的负离子释放量、远红外辐射性等性能。为含电气石的功能聚合物复合材料及其功能产品的开发提供了可行的基础数据。

可聚合十四碳烯琥珀酸电气石酯的红外光谱分析如图 1 所示。在改性电气石的 IR 图 1（b）中，在 1 802 cm^{-1}和 1 792 cm^{-1}处出现羰基的吸收峰，在 1 668 cm^{-1}处出现了碳碳双键吸收峰，电气石的特征吸收峰峰形基本没有发生变化，但均发生了几个波数的迁移，表明十四碳烯琥珀酸酐与电气石表面的羟基发生反应生成了酯基而引入了含双建的烷烃，得到了可聚合有机化电气石。

十四碳烯琥珀酸电气石酯—乙酸乙烯酯共聚物的负离子释放量、远红外辐射性性能如图 2 和图 3 所示。功能共聚物的负离子释放量随改性电气石用量的增加呈线性增加（图 2），其远红外辐射率均在 0.95 左右，随改性电气石用量的增加亦呈上升趋势。

* 通信作者：胡应模；E-mail：huyingmo@cugb. edu. cn；手机号：13911513729。国家自然科学基金（51372233）。

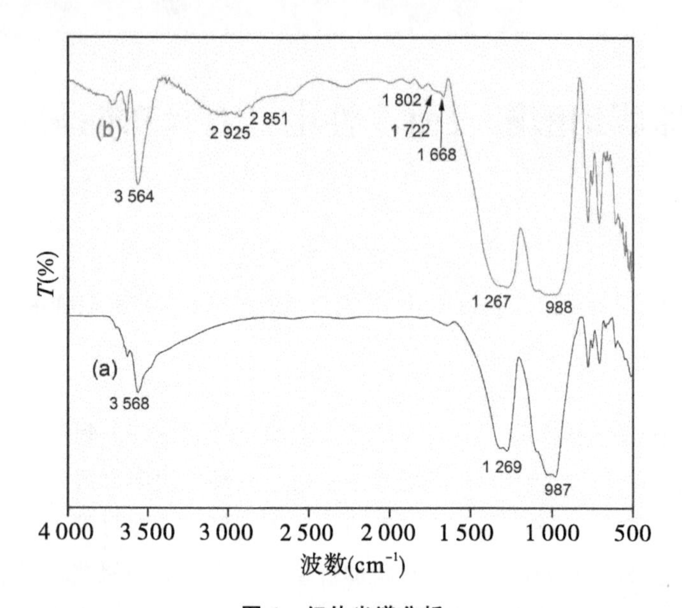

图1　红外光谱分析

注：(a) 未改性电气石；(b) 改性电气石

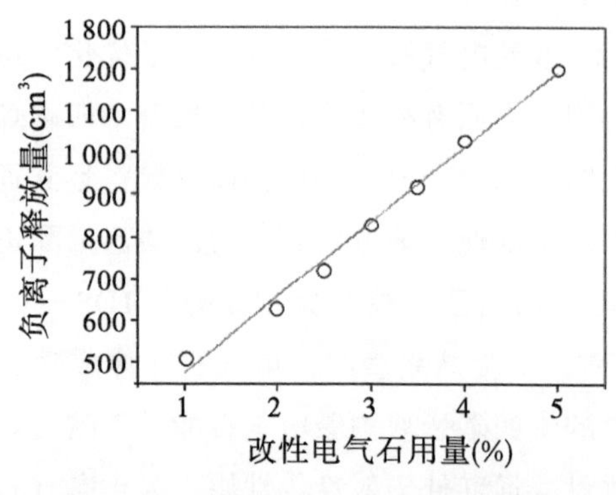

图2　p（TTDS－VA）的负离子释放量

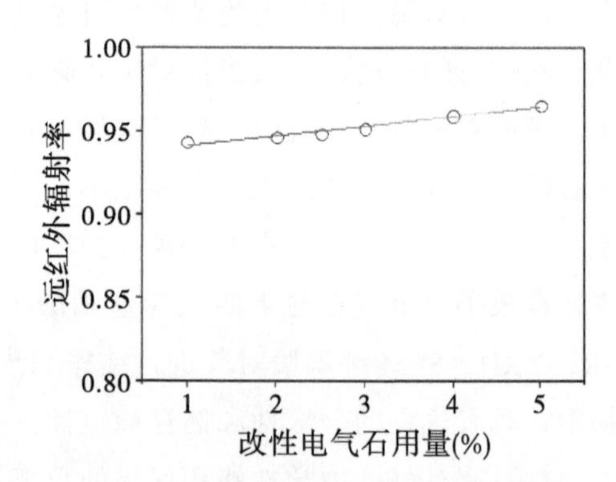

图3　p（TTDS－VA）的远红外辐射率

参考文献

[1] 胡应模，陈旭波，汤明茹. 2014. 电气石功能复合材料研究进展及前景展望［J］. 地学前缘，21（5）：331－337.

[2] Hu Y M，Yang X. 2012. The surface organic modification of tourmaline powder by span-60 and its composite［J］. Applied Surface Science，258：7540－7545.

[3] Hu Y M，Chen X B，Li Y H. 2016. The synthesis and characterization of poly（methyl methacrylate-tourmaline acrylate）［J］. Advances in Materials Science and Engineering，1－6.

海泡石基自组装涂层与软质聚氨酯泡沫阻燃性

潘　颖　赵红挺*
（杭州电子科技大学材料与环境工程学院，杭州 310016）

以聚乙烯亚胺、海藻酸钠和海泡石为带正负电荷的聚电解质和无机纳米粒子，利用层层自组装技术将海泡石基的涂层沉积到软质聚氨酯泡沫表面。通过锥形量热仪研究了组装层对软质聚氨酯泡沫的燃烧性能的影响。结果表明，组装 6 层海泡石基的涂层可以使聚氨酯泡沫的热释放速率峰值和总热释放分别降低 76％和 27％，显著降低了基体的火灾危险性。

海泡石是一种具有层链状结构的含水富镁硅酸盐黏土矿物，其化学式为 $Mg_8Si_{12}O_{30}(OH)_4(OH_2)_4 \cdot xH_2O$（x＝6～8）[1]。海泡石具有隔热、绝缘、抗腐蚀、抗辐射及热稳定等性能。软质聚氨酯泡沫作为一种三维结构的聚合物材料，被广泛应用于家具、建材、运输等领域。然而由于它的高度可燃性，且燃烧时伴有大量滴落以及烟气释放，为了符合阻燃规定，需要采取一些手段对聚氨酯泡沫进行阻燃改性[2]。

原料：软质聚氨酯泡沫，购自江苏绿源有限公司。海藻酸钠（alginate）、盐酸（HCl，37％）以及氢氧化钠购自国药集团化学试剂有限公司。聚乙烯亚胺（PEI）由上海阿拉丁生化科技股份有限公司提供。海泡石（sepiolite）购自洛南县腾发海泡石开发有限公司。

表 1　溶液浓度、样品所制备的层数以及样品增重

样品	聚乙烯亚胺（wt％）	海泡石（wt％）	海藻酸钠（wt％）	BL	增重（wt％）
PU－0	/	/	/	/	0
PU－1	0.5	/	0.3	3	2.3
PU－2	0.5	/	0.3	6	4.5
PU－3	0.5	0.5	0.3	3	8.9
PU－4	0.5	0.5	0.3	6	16.2
PU－5	0.5	1.0	0.3	3	16.9
PU－6	0.5	1.0	0.3	6	30.3

材料制备：首先将预处理过的聚氨酯泡沫在聚乙烯亚胺的溶液和海藻酸钠＋海泡石分散液中交替浸泡 2 min。当组装层达到需要的层数之后，将组装后的聚氨酯泡沫放入 70 ℃烘箱中干燥 12 h 并存放在干燥器中。所有的溶液浓度以及样品所制备的层数以及增重列于表 1。

* 通信作者：赵红挺；E-mail：info-iem@hdu. edu. cn；手机号：13656631588。国家自然科学基金（51573173）。

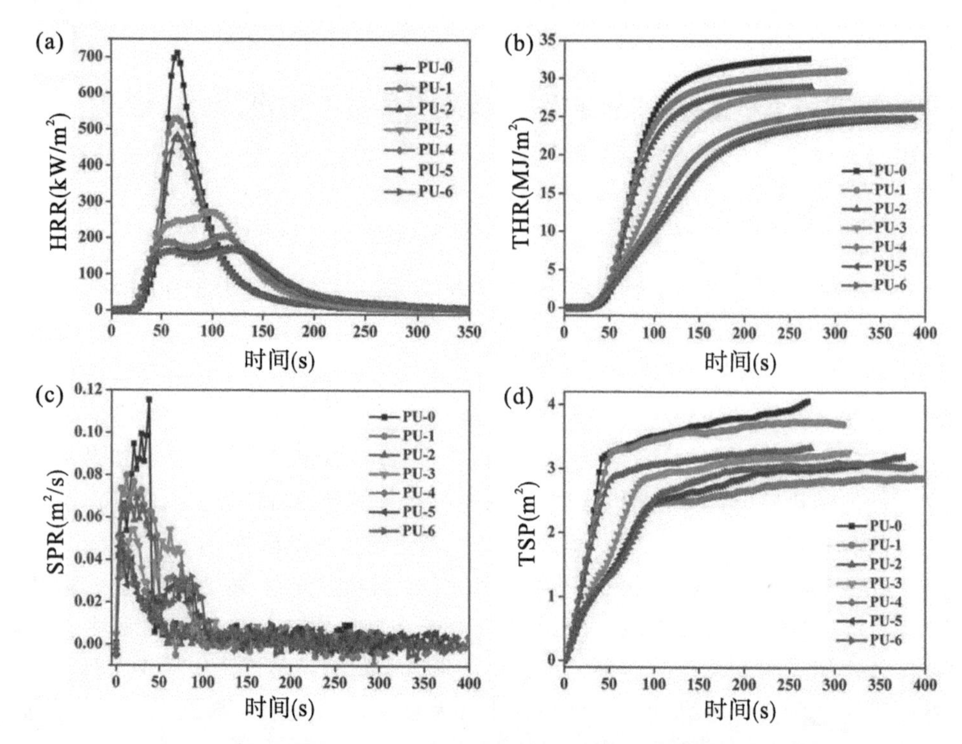

图 1　改性前后聚氨酯泡沫的热释放速率（HRR）、总热释放（THR）、烟气释放速率（SPR）以及总烟气释放量（TSP）曲线

图 1 给出了前后软质聚氨酯泡沫的 HRR、THR、SPR 和 TSP 曲线，可见仅由聚乙烯亚胺和海藻酸组成的涂层修饰聚氨酯泡沫之后各项参数均有一定程度的降低，如与 PU－0 相比，PU－2 的热释放速率峰值（PHRR）降低了 39.8%。当涂层中引入海泡石之后，海泡石基的涂层可以非常显著地降低基体树脂的 PHRR 以及烟气释放速率峰值（PSPR），如与 PU－0 相比，PU－6 的 PHRR 和 PSPR 分别降低了 76%和 58%。极大地减少了聚氨酯软泡的可燃性。

海泡石可通过层层自组装技术在软质聚氨酯泡沫表面成功组装涂层，且可明显降低基体树脂的火灾危险性。

参考文献

［1］Mora M，López M I，Carmona M Á，et al. 2010. Study of the thermal decomposition of a sepiolite by mid-and near-infrared spectroscopies［J］. Polyhedron，29：3046－3051.

［2］Boutin M，Lesage J，Ostiguy C，et al. 2004. Identification of the isocyanates generated during the thermal degradation of a polyurethane car paint［J］. Journal of Analytical & Applied Pyrolysis，71：791－802.

生物炭/凹凸棒石复合材料的制备及对 Cd（Ⅱ）的吸附

许琳玥[1,2]　何　跃[1,2*]　徐坷坷[1,2]　章　雷[1,2]
（1. 环境保护部南京环境科学研究所，南京 210042；
2. 南京国环环境研究院有限公司，南京 210042）

凹凸棒石黏土（Attapulgite）是以凹凸棒石为主要成分的一种多孔型链层状含水富镁铝硅酸盐类的黏土矿物，主要分布于我国江苏盱眙、安徽明光、甘肃临泽等地。凹凸棒石属于海泡石族，理论化学式为 $Mg_5Si_8O_{20}$（OH_2）$_4$ · H_2O，具有特殊的晶体结构和纤维状形貌特征，被广泛应用于环保领域[1]。

在应用过程中，由于凹凸棒石原矿含有大量类似蒙脱石、伊利石和碳酸盐等的杂质，影响凹凸棒石的使用性能，因此通常需要首先进行材料改性。常用改性方法包括热活化、酸活化、有机改性等。本研究以凹凸棒石热活化为基础，同时加入水稻秸秆炭化，制备获得生物炭/凹凸棒石复合材料（简称，BC/AT）。

将低品位凹凸棒石和水稻秸秆按质量比为 7∶3 的比例均匀混合，分别在 450、600 ℃下进行厌氧炭化活化，分别制备获得 2 种生物炭/凹凸棒石复合材料（BC/AT－450、BC/AT－600）。分别采用 SEM、FTIR、XRD、TG/DTA 对材料进行宏观形貌以及微观形态分析，并研究其对 Cd（Ⅱ）的吸附性能。SEM 表征结果如图 1 所示，材料的 FTIR 图和 XRD 图如图 2 所示。

材料表征实验表明：经厌氧炭化活化后，凹凸棒石能够均匀地包覆在生物炭表面，与凹凸棒石或生物炭相比，复合材料微孔数量明显增加，比表面积显著增加。AT 材料在 1 654 cm^{-1}处吸收峰为沸石水和表面吸附水的反对称伸缩振动峰[2]，而 BC/AT 两种复合材料在此处只有微弱的峰，表明随着温度的升高沸石水和表面吸附水逐渐脱去。加热过程中（≤600 ℃），凹凸棒石层链状结构未被破坏。

吸附实验表明：BC/AT 吸附平衡时间为 60 min。在污染溶液初始 pH 值为 7.0 时，BC/AT 对 Cd（Ⅱ）达到最大吸附量。使用 Langmuir、Freundlich 和 Redlich－Peterson 三种等温吸附模型对吸附过程进行拟合表明，Langmuir 模型拟合优度最高，BC/AT 对污染溶液中 Cd（Ⅱ）的吸附为单分子层化学吸附与不均匀层吸附并存。吸附动力学实验表明，BC/AT 复合材料对 Cd（Ⅱ）的吸附动力学为准二级反应动力学。

凹凸棒石或生物炭单独存在时，均能吸附 Cd 污染土壤中的 Cd（Ⅱ）[3,4]。实验室初步研究表明：BC/AT 材料对固化土壤中重金属镉存在一定作用，因此笔者下一步工作集中对 BC/AT 材料修复镉污染土壤方面开展深入研究。

---

* 通信作者：何跃；E-mail：heyue@nies. org；手机号：18905157095。中央级公益性科研院所基本科研业务专项（2017 年）、中科院盱眙凹土应用技术研发与产业化中心开放性课题资助项目（201412，201506）。

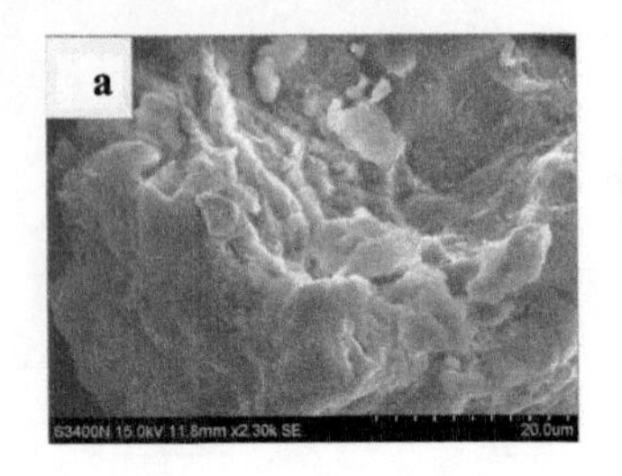

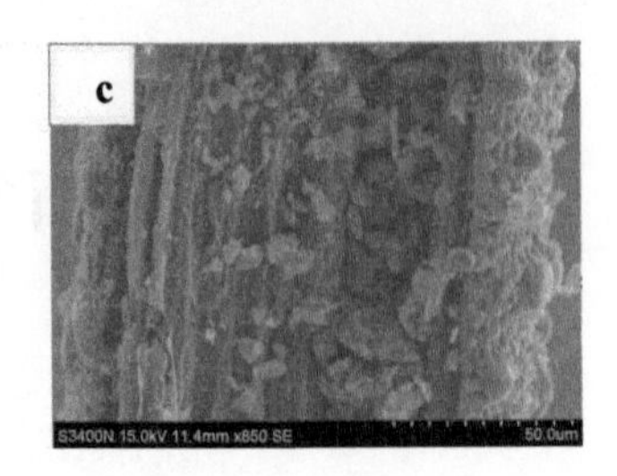

图 1　样品的 SEM 图

注：a. AT；b. BC/AT－450；c. BC/AT－600

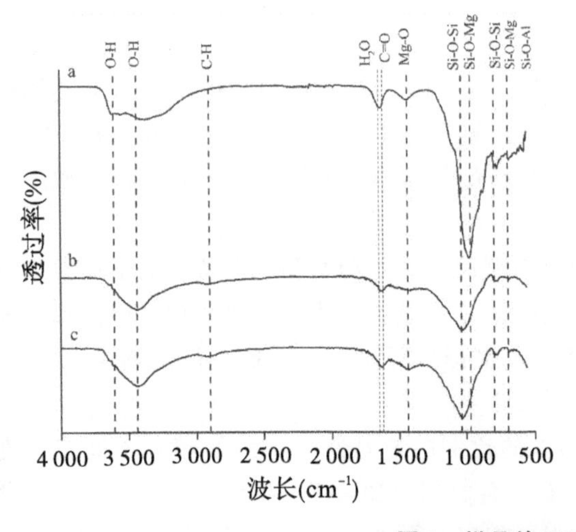

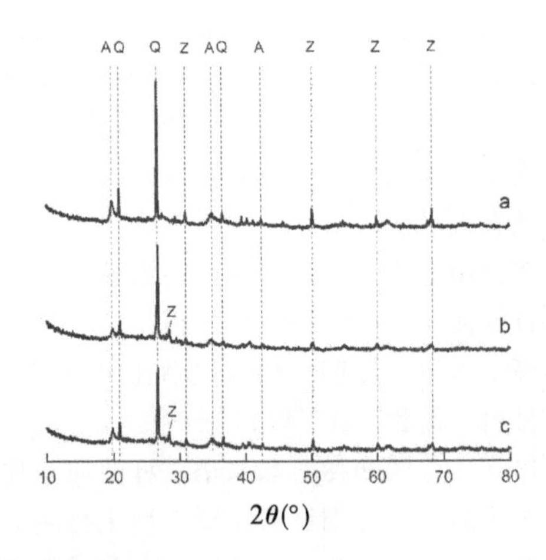

图 2　样品的 FTIR 图和 XRD 图

注：a. AT；b. BC/AT－450；c. BC/AT－600；A. 凹凸棒石；Q. 石英；Z. 沸石

参考文献

［1］谢晶晶，陈天虎，刘海波，等. 2018. 苏皖地区凹凸棒石黏土的特征和应用发展方向［J］. 硅酸盐学报，46：752－760.

［2］王爱勤. 2014 . 凹凸棒石棒晶束解离及其纳米功能复合材料［M］. 北京：科学出版社.

［3］廖启林，刘聪，朱伯万，等. 2014. 凹凸棒石调控 Cd 污染土壤的作用及其效果［J］. 中国地质，41：1693－1704.

［4］Zhang X W，Wang H L，He L Z，et al. 2013. Using biochar for remediation of soils contaminated with heavy metals and organic pollutants［J］. Environmental Science and Pollution Research，20：8472－8483.

重质碳酸钙吸油值的研究进展

李　敏[1]　许苗苗[1]　王建强[2]　陈建兵[1]　杨小红[1*]
（1. 池州学院非金属材料研究中心，池州 247000；
2. 池州金艺化工有限公司，池州 247000）

2017年，我国进口石材共1 466万吨，同比增长23.1%，国内对高端石材需求量增大；调查国内有规模的石材企业，其收入和利润双双回落，主要问题是遭遇前所未有的环保压力。

树脂基人造石材是发展最快的品种，重质碳酸钙是其最主要填料，一般占人造石材重量的75%，碳酸钙填充量越大，树脂用量越少，越环保。其关键在于降低碳酸钙填充料的吸油值。

吸油值也称树脂吸附量，表示填料对树脂吸收量的一种指数。大多数填料用吸油值来预测填料对树脂的需求量。填料吸油值大，树脂消耗量增加，人造石材环保差，同时成本增加[1-3]。

吸油值与晶形、颗粒比表面积、颗粒表面官能团、电性、pH值、润湿性等有关。吸油值通常0.5 g改性碳酸钙粉体，滴加邻苯二甲酸二辛酯，用调墨刀研压使之成团不散。吸油值计算方法如下：吸油值＝吸油量/样品质量。

本文综述了有关碳酸钙改性吸油值的研究，旨在从中吸取经验和扩展创新思维，提出有益的探索。

吴翠平等[4]采用聚乙二醇-200、一缩二乙二醇、三乙醇胺和氨基硅油-804等改性剂对重质碳酸钙进行干法改性，结果表明采用用量为1.00%氨基硅油-804，改性样品的吸油值达到0.115 mL/g。反应机理见图1。

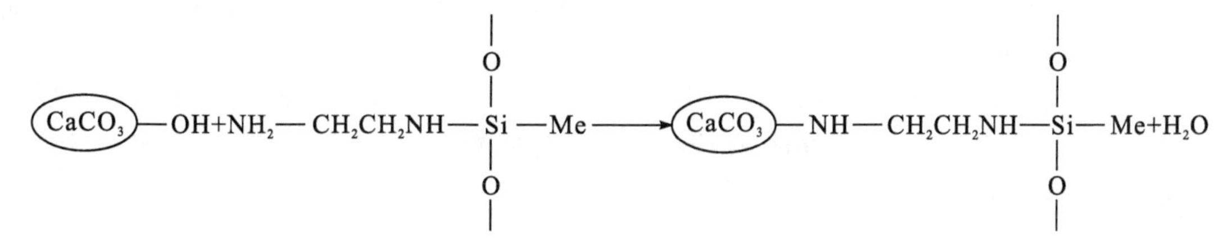

图1　反应机理

周国永等[5]用马来酸酐（MA）-丙烯酰胺（BA）-甲基丙烯酸正丁酯（BMA）共聚物湿法改性重质碳酸钙粉体，其最佳用量为（改性剂相对于重钙粉体的质量分数）为2%，改性温度为85 ℃，改性时间为120 min，搅拌速度为500 r/min，获得碳酸钙粉体吸油值为0.20 mL/g。

王友[6]以Span60为改性剂，乙醇为分散剂，采用球磨法对重质碳酸钙粉体进行表面改性，

---

*通信作者：杨小红；E-mail：yxh6110 @ yeah. net；手机号：18805665082。安徽省国际合作计划项目（1403062016）、安徽省教育厅重大自然科学项目（KJ201748）资助。

确定了最佳改性条件：球磨转速为 300 r/min，球磨时间为 0.5 h，球料比（球磨珠与碳酸钙的质量比）为 10∶1，改性剂用量为 1.5%（质量分数），获得碳酸钙吸油值为 0.20 mL/g。

王友等[7]以 2.0wt%硬脂酸和钛酸酯偶联剂为复合改性剂，期中 $m_{硬脂酸}$∶$m_{钛酸酯}$＝1∶3。无水乙醇为分散剂，通过优化湿球磨法工艺，对重质碳酸钙粉体进行改性，获得碳酸钙吸油值为 14.27 g/100 g。

刘立华采用 2%硬脂酸镁对碳酸钙进行表面改性处理，改性温度为 70 ℃，改性时间为 30 min，硬脂酸镁与碳酸钙之间发生了化学键合，获得碳酸钙吸油值为 10.91%。

陆宏志采用 3%钛酸酯偶联剂对纳米碳酸钙进行表面改性，改性时间为 1 h，改性温度为 80 ℃，钛酸酯偶主要以化学键的形式包覆在碳酸钙粉体表面，获得碳酸钙吸油值为 25.40 g/100 g。

周国永等采用 1.6%聚乙二醇湿法活化处理重质碳酸钙粉体，活化温度为 70 ℃，活化时间为 43 min，获得碳酸钙吸油值为 0.22 mL/g。

池州学院杨小红课题组根据碳酸钙表面官能团、pH 值等性质，选用硬脂酸等复配改性剂，采用球磨法对重质碳酸钙粉体进行表面改性，改性剂用量为 0.6wt%，通过选择合适改性工艺，获得碳酸钙粉体吸油值为 0.12 mL/g。

目前，碳酸钙改性主要是选择不同改性剂和改性工艺的研究，而且机理研究不够深入，需要大力加强基础研究。

参考文献

[1] 张苏. 2009. 复合改性剂对油墨用纳米碳酸钙改性研究 [D]. 沈阳：东北大学.
[2] 刘立华，宋云华，陈建铭，等. 2004. 硬脂酸镁改性纳米氢氧化镁效果研究 [J]. 北京化工大学学报，31（3）：31－34.
[3] 中国石材协会. 2018. 2017 年中国石材行业经济运行分析及 2018 年展望 [J]. 石材，4：3－8.
[4] 吴翠平，郭永昌，魏晨洁，等. 2016. 人造石材用重质碳酸钙填料的表面改性研究 [J]. 非金属矿，39（4）：21－23.
[5] 周国永，曾一文，李伦满，等. 2014. MA－BA－BMA 三元共聚物改性重钙粉体的研究 [J]. 无机盐工业，46（3）：26－30.
[6] 王友. 2016. Span60 表面改性重质碳酸钙粉体研究 [J]. 无机盐工业，48（7）：25－28.
[7] 王友，曾一文，覃康玉，等. 2016. 硬脂酸—钛酸酯偶联剂改性重质碳酸钙粉体研究 [J]. 无机盐工业，48（6）：38－40.

透明聚氨酯/氧化硅纳米纤维复合涂层的制备及性能

张军瑞[1*]　蒋国军[1]　黄田浩[1]　周春晖[2,3*]
（1. 浙江工业大学之江学院理学院，绍兴 312030；
2. 浙江工业大学化学工程学院，杭州 310014；
3. 青阳非金属矿研究院，青阳 242800）

纤蛇纹石俗称温石棉，是蛇纹石的一种，属于硅酸盐矿物的一种，其理想的分子结构式为 Mg_6（Si_4O_{10}）（$OH)_8$，具有耐热、抗拉伸、对可见光透明等特性，广泛应用于建筑、化工、冶金、机械等现代工业领域[1-4]。纤蛇纹石具有天然的一维纳米管状结构，具有较好的热稳定性、低导热率、高电阻率、较高的表面活性，同时由于天然产出的低成本也是其他成本昂贵的人工材料所不能比拟的。但纤蛇纹石具有生物毒性，长期接触纤蛇纹石会损害人体健康，因此对其进行合理的无害化改性并对其性能进行利用，发挥其低成本、独特性能等优势，对制备特殊功能的纳米材料具有重要意义。

本文将阴离子表面活性剂化学分散的纤蛇纹石通过盐酸酸浸除去 Mg—（OH）O 八面体及其他金属离子，制备了分散性好、纯度高、尺寸均匀的无定形氧化硅纳米纤维。通过共混的方法将其添加到聚氨酯基体中，制备了一系列不同氧化硅纳米纤维添加量的聚氨酯涂层，通过对其性能进行测定，研究了氧化硅纳米纤维的加入量对透明聚氨酯涂层性能的影响。

正硅酸乙酯（SiO_2含量为 28.5wt%）、无水乙醇（99.7%）、氨水（25%～28%）购于广州化学试剂厂，己二酸：化学纯，上海五联化工厂；1,4-环己烷二甲酸（1,4-CHDA）、1,4-环己烷二甲醇（1,4-CHDM）：韩国 SK 公司；1,6-己二醇、三羟甲基丙烷（TMP）：美国 Alfa Aesar 公司；异氰酸酯 Z4470（IPDI 固化剂，NCO 含量为 11.9%）德国拜耳公司；BC-98 聚酯合成催化剂：市售；BYK141：德国毕克公司；其他试剂均为分析纯。

FT-IR 采用 Bruker VECTOR-33 红外光谱仪检测，聚酯树脂使用 KBr 压片模式，薄膜使用衰减全反射模式。聚酯的分子量和分子量分布采用凝胶渗透色谱法（GPC），在 Perkin Elmer Series-200（300 mm×7.5 mm）仪器上测定，聚苯乙烯为校准曲线标样，四氢呋喃为流动相，流速为1.0 mL/min。

冲击强度、柔韧性、剪切强度、邵氏硬度分别按照 GB/T 1732—93、GB/T 1731—93、GB/T 10007—2008、GB/T 531—1999 标准进行检测。耐磨性能按照 GB/T 1768—79 标准进行检测。透光率在紫外—可见分光光度计（LAMBDA950，PerkinElmer，USA）上检测。波长范围为 200～

* 通信作者：张军瑞；E-mail：jrzhang0520@163. com；手机号：15088742571。周春晖，E-mail：chc. zhou@aliyun. com；电话：13588066098。浙江省青年自然科学基金（LQ18E030014）、国家自然科学基金（41672033）。

900 nm，并采用积分球模式消除膜厚不均造成的误差。广角 X 射线衍射（WAXD）在 Bruker－D8 ADVANCE 仪器上检测（室温，铜靶，LynxExe 阵列探测器，40 kV，40 mA），扫描步长 0.02°（2θ），扫描速度 0.1 s/step。氧化硅纳米纤维的聚集态结构在 Hitachi H－600 TEM 电镜（Hitachi Corporation，日本）上测定，样品用无水乙醇稀释，滴在醋酸纤维覆盖的铜网上，干燥后进行测定。

图 1 为添加不同质量的氧化硅纳米纤维/聚氨酯复合涂层在紫外—可见光区的透射光谱。可以看出，复合涂层的透光率随着氧化硅纳米纤维添加量的增加而降低，当聚氨酯中加入的氧化硅纳米纤维的质量达到或超过 1wt%时，复合涂层的透光率急剧下降，且表面发黑，这是由于氧化硅纳米纤维中含有一定的 Fe、Mg 等金属离子造成的。当聚氨酯中添加的氧化硅纳米纤维的量过多时，金属离子与聚氨酯基体之间较大的折射率差异使照射在聚氨酯表面上的光线分散，导致其透明度大幅下降。因此为使聚氨酯/氧化硅纳米纤维具有较好的透光率，应使氧化硅的加入量控制在 1wt%以下。

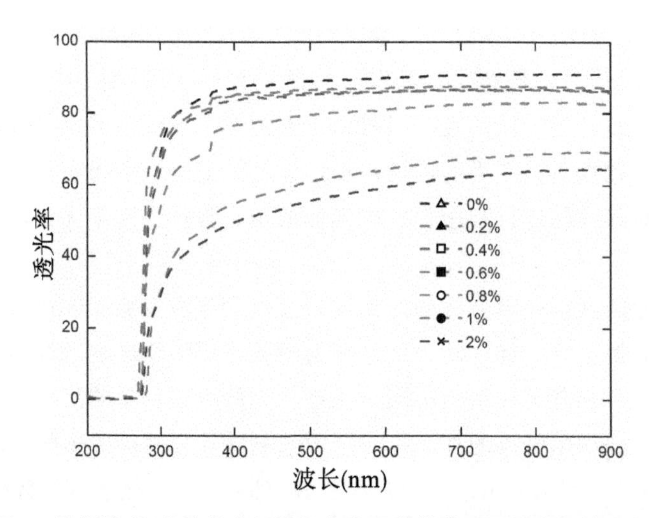

图 1　聚氨酯/氧化硅纳米纤维复合涂层在紫外—可见光区的透光率

表 1 为加入不同质量的氧化硅纳米纤维/聚氨酯复合涂层机械性能的测试结果。由表 1 可以看出，氧化硅纳米纤维的加入对聚氨酯涂层的机械性能有很大程度的影响。虽然聚氨酯涂层已具有较高的硬度和剪切强度，但氧化硅纳米纤维的加入对复合涂层的硬度和剪切强度的增强效果依然很明显，这是因为氧化硅纳米纤维具有较大的比表面积和多孔结构，对高分子材料具有较好的补强作用。而聚氨酯/氧化硅纳米纤维复合涂层的冲击强度则随着氧化硅纳米纤维的加入而下降，这是由于氧化硅的补强作用使涂层具有较高的硬度，从而使其韧性降低，因此使复合涂层的冲击强度呈现一定程度的下降。

图 2 为聚氨酯/氧化硅纳米纤维复合涂层的耐磨性能。由图 2 可知，复合涂层的耐磨性能随着氧化硅纳米纤维添加量的增加而增加，这是由于聚氨酯的硬度随着氧化硅纳米纤维添加量的增加而增加，硬度的增加使涂层表面具有较好的耐磨性能，因此通过在聚氨酯涂层中添加氧化硅纳米纤维可以提高涂层的耐磨性能。

表 1　氧化硅纳米纤维/聚氨酯复合涂层的机械性能

样品	抗冲击性（kg·cm）	弹性（mm）	邵氏 A 硬度	剪比强度（MPa）
PU	90	1	65.2	49.1
0.2%	90	1	65.6	50.3
0.4%	85	1	65.9	51.2
0.6%	80	1	66.8	53.5
0.8%	70	1	67.6	55.5
1%	65	1	68.2	56.3
2%	60	1	72.5	60.2

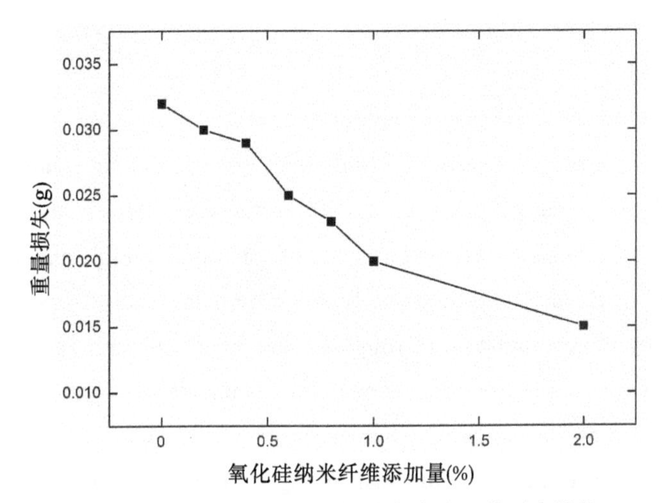

图 2　聚氨酯/氧化硅纳米纤维复合涂层的耐磨性能

参考文献

[1] Iijima S. 1991. Helical mocritubules of graphite carbon [J]. Nature，354：56－58.

[2] 杨艳霞. 2007. 纤蛇纹石酸浸及其制备氧化硅纳米线的研究 [D]. 长沙：中南大学.

[3] Jiang X C，Herricks T，Xia Y N. 2002. Cuo nanowires can be synthesized by heating copper substrates in air [J]. Nano Letters，2 (12)：1333－1338.

[4] Chen C C，Yeh C C，Chen C H，et al. 2001. Catalytic growth and characterization of gallium nitride nanowires [J]. Journal of the American Chemical Society，123 (12)：2791－2798.

Superhydrophobic and Superamphiphobic Coatings Based on Palygorskite

Jun Ping Zhang* Bu Cheng Li
(Key Laboratory of Clay Mineral Applied Research of Gansu Province, Lanzhou Institute of Chemical Physics, Chinese Academy of Sciences, Lanzhou 730000, China)

Biomimetic superhydrophobic and superamphiphobic coatings are receiving extensive attention because of their unique self-cleaning properties[1]. Superhydrophobic and superamphiphobic coatings have wide potential applications in many fields, e. g., oil/water separation, anticorrosion, water collection, and directional water transport, *etc*.. Thousands of superhydrophobic and superamphiphobic coatings have been prepared in the past two decades. Superhydrophobic and superamphiphobic coatings are in most cases prepared by the combination of rough surface microstructure and materials of low surface energy. Most of the microstructures are based on synthetic nanomaterials, which are expensive and complicated. Also, preparation of the synthetic nanomaterials is often pollutive.

Moreover, the mechanical durability of most of superhydrophobic and superamphiphobic coatings is low. The micro-/nanostructures of superhydrophobic and superamphiphobic coatings are inherently weak to mechanical damage. The loss of microstructure increases the solid - liquid contact area, causing loss of the self-cleaning property.

Here we report our recent progress about superhydrophobic and superamphiphobic coatings based on palygorskite, the natural nanomaterial (Fig. 1)[2-4]. The coatings are prepared by modification of palygorskite with silanes followed by spray-coating onto substrates. The influences of clay minerals, activation of clay minerals and their contents on surface microstructure, wettability and stability of the coatings are discussed. The coatings showed excellent superhydrophobicity and superamphiphobicity as well as high stability. We also prepared self-cleaning Maya Blue-like pigments by the combination of Maya Blue-like pigment and the superhydrophobic and superamphiphobic coatings (Fig. 2). Stability of the pigments was evidently improved by the superhydrophobic and superamphiphobic coatings by reducing the contact area and contact time between the pigments and the corrosive liquids.

---

* Corresponding author: Jun Ping Zhang; E-mail: jpzhang@ licp. cas. cn; Mobile phone: 18919857738。Supported by the hundred Talents Program of the Chinese Academy of Sciences.

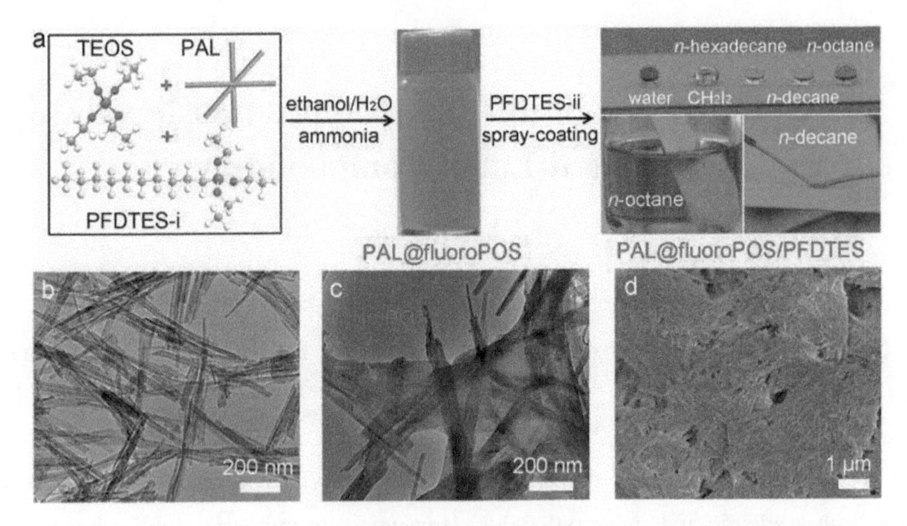

Fig. 1 Superamphiphobic coatings with low sliding angles based on palygorskite

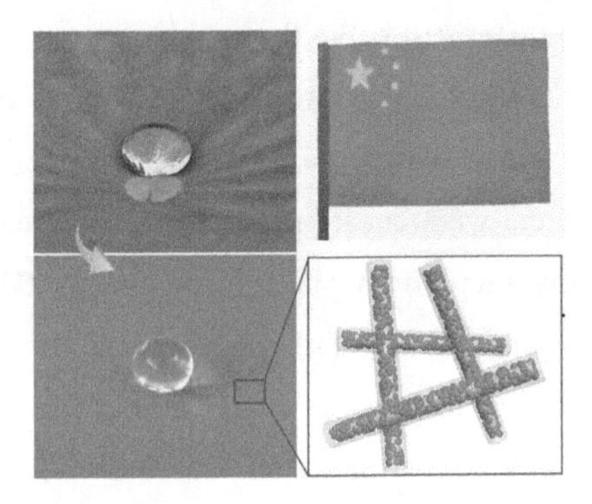

Fig. 2 Self-cleaning Maya Blue-like pigments with different color

References

[1] Chu Z, Seeger S. 2014. Superamphiphobic surfaces [J]. Chemical Society Reviews, 43: 2784 - 2798.

[2] Li B C, Zhang J P. 2016. Durable and self-healing superamphiphobic coatings repellent even to hot liquids [J]. Chemical Communications, 52: 2744 - 2747.

[3] Dong J, Wang Q, Zhang Y, et al. 2017. Colorful superamphiphobic coatings with low sliding angles and high durability based on natural nanorods [J]. ACS Applied Materials & Interfaces, 9: 1941 - 1952.

[4] Li B C, Zhang J P, Wu L, et al. 2013. Durable superhydrophobic surfaces prepared by spray coating of polymerized organosilane/attapulgite nanocomposites [J]. ChemPlusChem, 78: 1503 - 1509.

Exfoliated Kaolinite/SBR Nanocomposites by Combined Latex Compounding

Yong Jie Yang Qin Fu Liu* Zhi Chuan Qiao Ke Nan Zhang
(China University of Mining and Technology, Beijing 100083, China)

Organic polymer/inorganic clay composite has become a new star in recent years, it attracted much attention from both academic and industrial fields. However, its industrial application was limited because of the poor compatibility and unsatisfactory dispersion of clay minerals. By using in *situ* modification and latex compounding method can well solve the problem.

The exfoliated surface of the kaolinite is electro-negative with a large number of Al—O and Si—O broken bonds existed on the broadsides, which provides active sites formodification[1]. Therefore, two kinds of nitrogen-containing modifiers were chose in order to gain interfacial interaction with kaolinite. One is organic cation cetyltrimethylammonium chloride named as C16, the other is triaminosilane-KH892. The preparation procedure divided into three steps (Fig. 1), ①in *situ* modified kaolinite is stirred vigorously with ethanol, and then centrifuged after three times washing through deionized water. ②the re-dispersed kaolinite slurry is mixed with styrene-butadiene latex. ③kaolinite/SBR nanocomposites are finally made after flocculated by adding dropwise with dilute sulfuric acid.

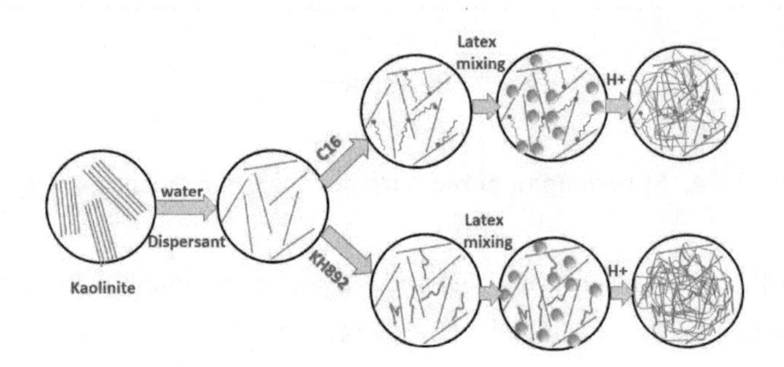

Fig. 1 Schematic diagram of blending process of kaolinite with styrene-butadiene latex

The coagulated compounds were mixed with rubber ingredients on a two-roll mill for 10 min. Then, the composites were vulcanizated at 150 ℃ for the optimum cure time (T90). The tensile properties tests were conducted on an electronic tensile testing machine at 23 ℃.

* Corresponding author: Qin Fu Liu; E-mail: lqf@cumtb. edu. cn; Mobile Phone No.: ++86-13911683809.

Table 1 Mechanical properties of SBR/Kaolin nanocomposites

Nanocomposite	Tensile strength (MPa)	Elongation at break (%)	Stress at 100% strain (MPa)	Stress at 300% strain (MPa)
SBR/Kaolin	7.73	900.15	1.00	1.34
SBR/Kaolin - g - C16	11.12	682.67	1.35	1.97
SBR/Kaolin - g - KH892	15.54	638.4	1.60	5.60
SBR/Kaolin - g - KH892/Si69	17.28	516.30	1.77	8.75

Table 1 lists the mechanical properties of unmodified, single and synergistic modified nanocompound. For C16 and KH892 nanocomposite, the tensile strength is improved to 11.12 MPa and 15.54 MPa, respectively. The KH892 and Si69 synergistic modified compound have the best mechanical properties, which tensile strength increased by 124% compared to unmodified composite.

The TEM images (Fig. 2) exhibits that latex blending technique contributes to the dispersion of kaolinite in rubber. The kaolinite layer is exfoliated sufficiently and dispersed uniformly into the rubber matrix without re-agglomeration.

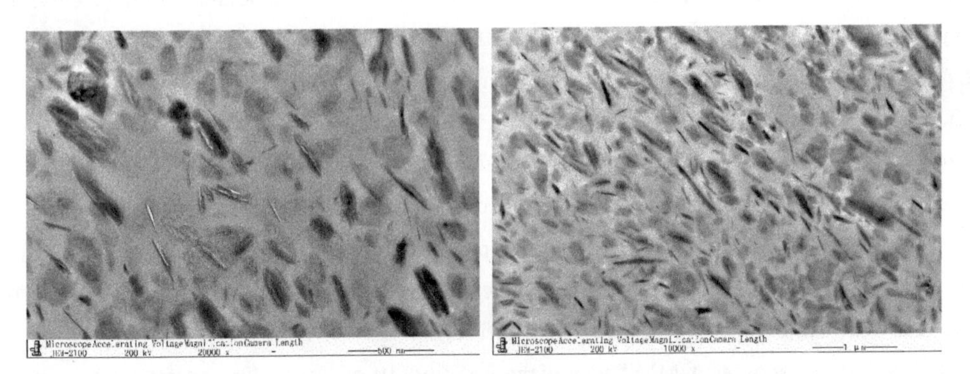

Fig. 2 TEM images of SBR/Kaolin nanocomposites

The resultsilluminate that not only C16 but also KH892 plays a significant role in the SBR/Kaolin nanocomposites. Furthermore, both types of modifiers can be grafted on kaolinite according to the infrared and thermogravimetric data. As a consequence, the kaolinite/SBR nanocomposites possess highly uniformly dispersed kaolinite sheets that ultimately achieve covalent interfacial interactions and exhibit high performance.

References

[1] Young R A, Hewat A W. 1988. Verification of the triclinic crystal structure of kaolinite [J]. Clays Clay Miner, 36 (3): 225.

Eco-friendly Pickering Medium Internal Phase Emulsions for Formation of Macroporous Adsorbent

Yong Feng Zhu　Feng Wang　Wen Bo Wang*　Ai Qin Wang*
(Key Laboratory of Clay Mineral Applied Research of Gansu Province, Center of Eco-material and Green Chemistry, Lanzhou Institute of Chemical Physics, Chinese Academy of Sciences, Lanzhou 730000, P. R. China)

Pickering emulsion has often been applied in pharmaceutical, food, cosmetic, microreactor, and also been used as template to create porous polymer monoliths for various practical applications due to the its unique properties[1-3]. While, the dispersion phase of O/W Pickering HIPEs mostly focused on organic solvent, such as paraffin, hexadecane, toluene, hexane, P-xylene and so on[4]. Obviously, the large quantity of organic solvent not only increased the cost, but also harmful to human health and environment. Thus, replaced poisonous organic solvent with low-cost and eco-friendly plant oil is very meaningful.

Herein, we report a novel O/W Pickering emulsion stabilized by Montmorillonite (MMT) clay in combination with a trace amount of hydrophilic surfactant Tween-20. The non-toxic, low-cost, eco-friendly and food-grade flaxseed oil was used as dispersion phase with less than 50% internal phase volume ratio to create the emulsion. The prepared macroporous material was employed to adsorption rare earth metal Ce (Ⅲ) and Gd (Ⅲ) in aqueous solution.

Briefly, a certain amount of MMT was dispersed into 10 mL of water containing an appropriate amount of Tween-20 and polymerizable monomer under stirring condition. Then, 10 mL flaxseed oil was added into the mixture and emulsified with a GJD-B12K homogenizer at 11 000 r/min for 5 min to form the Pickering MIPEs. After that, obtained Pickering-MIPEs transferred into centrifuge tubes, sealed and immersed into in a 65 ℃ water bath for 24 h to polymerize. The prepared monolithic polymers were cut into slices and washed with acetone for 12 h then dried in a vacuum oven at 60 ℃.

The prepared macroporous monolithsshown fast adsorption rate for Ce (Ⅲ) and Gd (Ⅲ) and the adsorption equilibrium can be reached within 30 min and 25 min for Ce (Ⅲ) and Gd (Ⅲ), respectively. The fast adsorption rate is derived from the internal porous structure and the sufficient adsorption groups. The macro-pores allow fast and efficient mass transport and also act as an impounding reservoir to provide sufficient contact between active groups and adsorbates.

* Correspondence authors. Wen Bo Wang, E-mail: boywenbo@126.com; Ai Qin Wang, E-mail: aqwang@licp.cas.cn; Tel.: +86 931 4968118; Fax: +86 931 4968019.

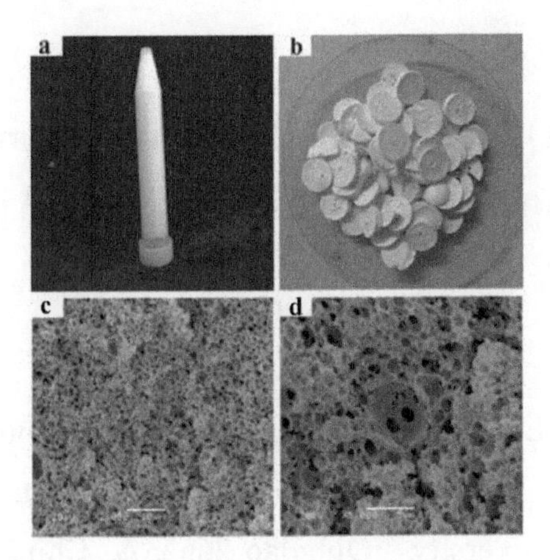

Fig. 1 Digital photos of the Pickering MIPEs (a), polymeric monolith (b) and the SEM of porous monolith (c, d)

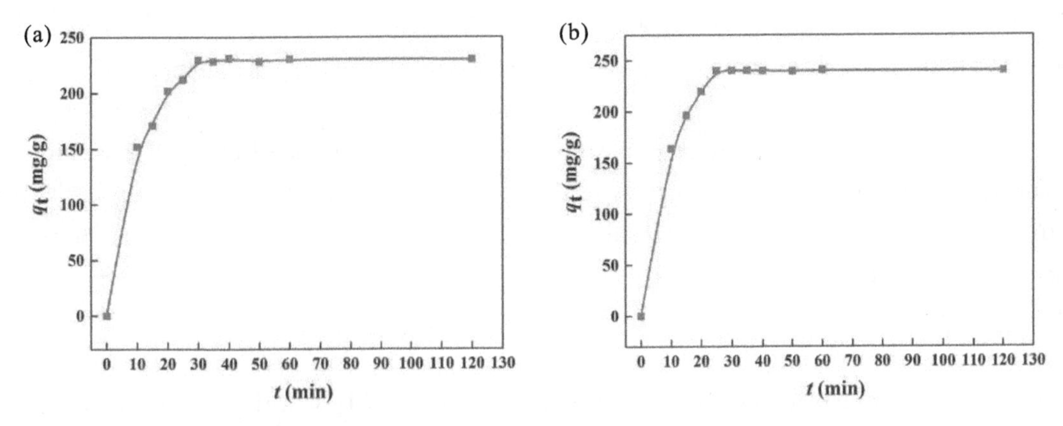

Fig. 2 Adsorption kinetic curves of the porous monolith for Ce (Ⅲ) (a) and Gd (Ⅲ) (b)

References

[1] Kalashnikova I, Bizot H, Cathala B, et al. 2011. New Pickering Emulsions Stabilized by Bacterial Cellulose Nanocrystals [J]. Langmuir, 27: 7471－7479.

[2] Ye A, Zhu X, Singh H. 2013. Oil-in-water emulsion system stabilized by protein-coated nanoemulsion droplets [J]. Langmuir, 29: 14403－14410.

[3] Dong J, Worthen A J, Foster L M, et al. 2014. Modified montmorillonite clay microparticles for stable oil-in-seawater emulsions [J]. ACS Applied Materials & Interfaces, 6: 11502－11513.

[4] Zhu Y, Zheng Y, Wang F, et al. 2016. Monolithic supermacroporous hydrogel prepared from high internal phase emulsions (HIPEs) for fast removal of Cu^{2+} and Pb^{2+} [J]. Chemical Engineering Journal, 284: 422－430.

Palygorskite/TiO_2 Incorporated Thin Film Nanocomposite Membranes for Reverse Osmosis Application

Tian Zhang Zhi Ning Wang*

(Shandong Key Laboratory of Water Pollution Control and Resource Reuse, School of Environmental Science and Engineering, Shandong University, Qingdao 266237, China)

Palygorskite (Pal) is a kind of low-cost and environment-friendly natural nanoclay with rod-like crystal structure, nano-tunnels extending along c axis and excellent hydrophilicity. TiO_2 nanoparticles, especially anatase phase, have prominent photocatalytic bactericidal and organic pollutant decomposition activities[1,2]. In this work, Pal and Pal/TiO_2 nanocomposite were successfully embedded in the polyamide (PA) selective layer of the reverse osmosis (RO) membranes via interfacial polymerization. The nano-tunnel structure of Pal possesses a cross-sectional area of 0.37 nm×0.63 nm, which facilitates the selective transport of water molecules through the PA layers. The water flux of Pal incorporated TFN membrane increases to approximately 40 $L \cdot m^{-2} \cdot h^{-1}$, which is 1.6-fold higher than the reference TFC membrane. Meanwhile, the NaCl rejection is maintained at approximately 98%. Although the Pal/TiO_2 incorporated TFN membrane exhibited slightly lower flux (1.4-fold higher than TFC), the embedded Pal/TiO_2 contributed to the antifouling and photocatalytic bactericidal capacities, which are greatly desired in the membrane desalination and water reclamation processes (Fig. 1 and Fig. 2).

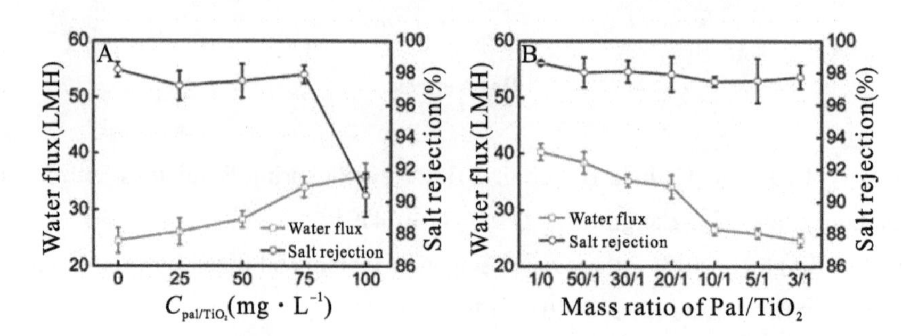

Fig. 1 Water flux and NaCl rejection of the prepared TFN－Pal/TiO_2 membranes with different Pal/TiO_2 concentration (Pal/TiO_2＝20/1) (A) and with different Pal/TiO_2 mass ratio (Pal/TiO_2 concentration ＝ 75 mg/L) (B)

---

* Corresponding author: Zhi Ning Wang; E-mail: wangzhn@sdu. edu. cn; Mobile Phone No.: ＋86-13573895003. NSFC (21476219) and the key research and development plan (public interest category) of Shandong Province (2017GSF17102).

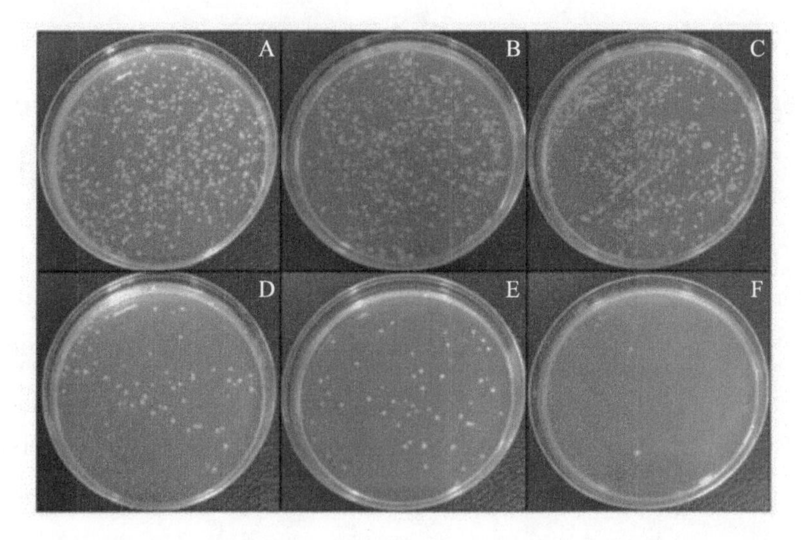

Fig. 2　Antibacterial properties of the membranes shown on total agar plate counts

Note: Representative photographs of the recultivated *E. coli* colonies on agar after contacting with TFC (A), TFN75 - Pal (B) and TFN75 - Pal/TiO_2 (20/1) (C) in dark and after contacting with TFC (D), TFN75 - Pal (E) and TFN75 - Pal/TiO_2 (20/1) (F) under UV illumination

References

[1] Werber J R, Osuji C O, Elimelech M. 2016. Materials for next-generation desalination and water purification membranes [J]. Nature Reviews Materials, 1: 16018.

[2] Gao X, Li Y, Yang X, et al. 2017. Highly permeable and antifouling reverse osmosis membranes with acidified graphitic carbon nitride nanosheets as nanofillers [J]. Journal of Materials Chemistry A, 5: 19875 - 19883.

Preparation and Electrorheological Performance of ATP/TiO_2/PANI Based ER Fluid

Ling Wang[1,2] Ting Zhou[1] Chen Chen Huang[1] Feng Hua Liu[1*] Gao Jie Xu[1*]

(1. Ningbo Institute of Materials Technology & Engineering, Chinese Academy of Science, Ningbo 315210, China;

2. School of Materials Science and Engineering, Jiangxi University of Science and Technology, Ganzhou 341000, China)

Electrorheological (ER) fluid is a smart material, which is made of dielectric particles dispersing in an insulating liquid. Under an applied electric field, the dispersed dielectric particles will be polarized and attracted to each other to form chain or column structures. These chains and columns enable ER fluid to suddenly increase its viscosity and even change from a liquid-like state to a solid-like state. The viscosity change or liquid-solid state transition of ER fluid is rapid and reversible as the change of applied electric field. Because of its controllable viscosity and short response time, ER fluid has attracted much interest in various areas, such as clutches, damping devices, display, human muscle simulator, and so on[1].

Theactivity of ER fluids is closely related to the polarization ability of dielectric particles, and the shape of dielectric particles is an important factor. The theoretical and experimental investigations of the effect of particle shape on the ER performance and sedimentation stability indicate that the one-dimension nanomaterials, such as nanowire and nanotubes, can improve the suspended stability of ER fluids and even enhance ER effect, due to the characteristics of nanosize and anisotropic morphology. Titanium oxide (TiO_2) and polyaniline (PANI) are considered to be the most promising two kinds of electrorheological materials. However, the one dimension structure of TiO_2 and PANI, especially the one-dimensional composite structure of TiO_2/PANI, is very difficult to synthesize, and the cost of preparation is very high[2-4].

In this paper, the natural one-dimensional nanometer material attapulgite (ATP) with low density, suitable size and stable structure was used as template to prepare a new kind ofone-dimension ATP/TiO_2/PANI nanocomposite via the versatile kinetics-controlled coating and in-situ polymerization. The results of SEM and TEM analysis indicate that the controllable preparation of one-dimensional ATP/TiO_2/PANI core shell particles can be achieved by means of kinetics-controlled coating and in-situ polymerization

* Correspondence authors. Feng Hua Liu, E-mail: lfh@nimte.ac.cn; Gao Jie Xu, E-mail: xugj@nimte.ac.cn; Tel.: +86 574 86685163; Fax: +86 574 86685163.

(Fig. 1). Compared to conventional TiO_2 or PANI ER fluid, the ATP/TiO_2/PANI ER fluid exhibited distinctly improved electrorheological activity and suspended stability (Fig. 2).

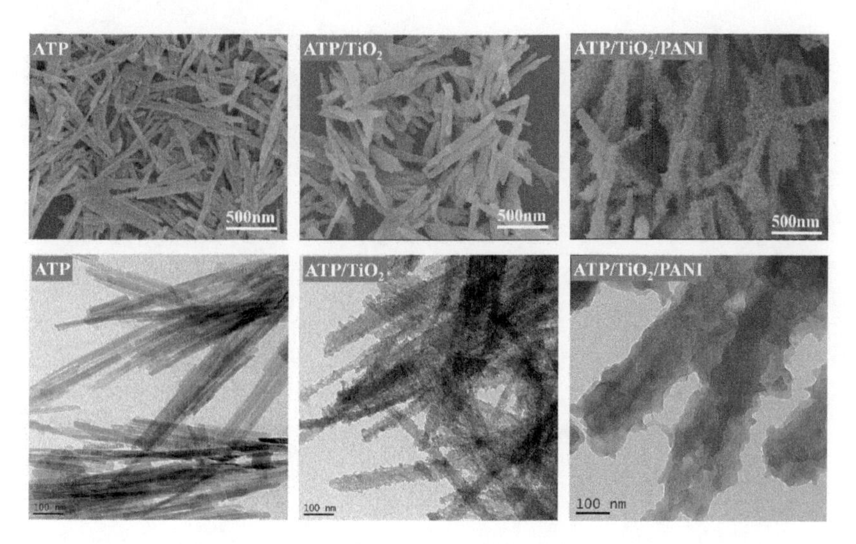

Fig. 1　The SEM and TEM of ATP, ATP/TiO_2 and ATP/TiO_2/PANI

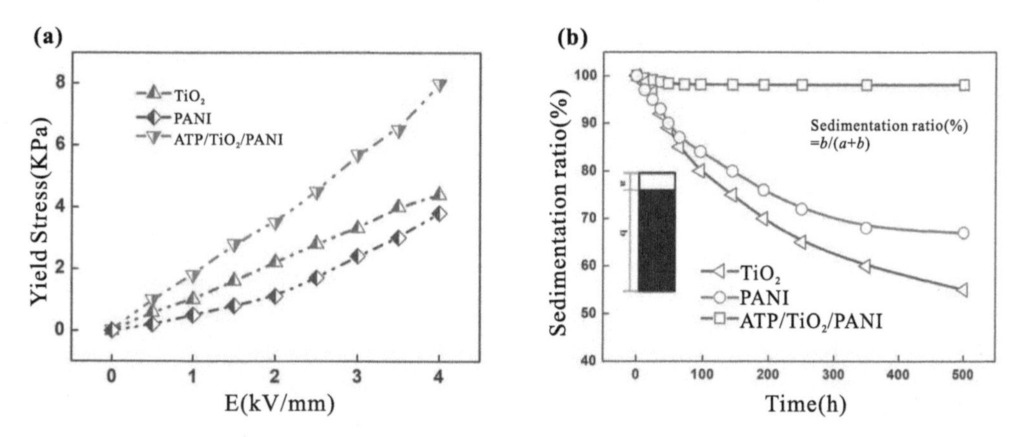

Fig. 2　The yield stress and suspended stability of TiO_2, PANI, ATP/TiO_2/PANI based fluids

References

[1] Halsey T C. 1992. Electrorheological fluids [J]. Science, 258: 761 - 766.

[2] He K, Wen Q, Wang C, et al. 2017. The preparation and electrorheological behavior of bowl-like titanium oxide nanoparticles [J]. Soft Matter, 13: 7677 - 7688.

[3] Chen D, Caruso R A. 2013. Recent progress in the synthesis of spherical titania nanostructures and their applications [J]. Advanced Functional Materials, 23: 1356 - 1374.

[4] Wang W B, Wang A Q. 2016. Recent progress in dispersion of palygorskite crystal bundles for nanocomposites [J]. Applied Clay Science, 119: 18 - 30.

[illegible]

[illegible]

[illegible]

References

[1] [illegible]

[2] [illegible]

[3] [illegible]

[4] [illegible]

第四章

非金属矿光、电、磁、声、热材料

利用废弃粉矿制备隔热材料

王　刚* 韩建燊　张　琪　袁　波　李红霞
（中钢集团洛阳耐火材料研究院有限公司，
先进耐火材料国家重点实验室，洛阳 471039）

废弃粉矿来源于矿物在开采或加工过程中产生的废弃物或未能利用到的尾矿等。目前，我国大量废弃粉矿堆砌如山，严重污染环境，同时易造成山体滑坡等危害。废弃粉矿得不到有效的利用，造成了极大的资源浪费。例如，河南西峡大量的镁橄榄石细粉堆积如山，无法仓储，工厂不得不将细粉用于填沟，造成了资源的浪费，湿法生产线的细粉除在沉淀池沉淀外，部分已随河水漂流，严重污染环境[1]，同样的问题也存在于其他地区[2,3]。再例如，火力发电厂煤燃烧烟气中收集的细灰，采煤和洗煤过程中排出的废弃物煤矸石、粉煤灰等，其耐火度较低，总体利用价值不高，甚至被用作路基填埋[4]。因此，废弃粉矿亟待开发再利用，使其变废为宝。

以废弃矿粉制备轻质保温材料是顺应时代对节能环保要求的一种良好途径。由于不同成分的废弃粉矿耐火度不同。因此，需要采用不同方法制备隔热材料。对于耐火度高的一类高温粉矿，采用发泡法可制备出微孔、高强、隔热效果好的隔热材料，该方法只需加入极少量有机或无机发泡剂发泡，依靠空气造孔，烧成过程污染物排放量极少，符合节能环保、绿色制造的发展宗旨。对于耐火度较低的一类低温粉矿，添加硼酸、氧化钠等低熔点物质继续降低其熔点，添加发泡剂可制备质轻、隔热及隔音性能好的泡沫玻璃。

高温粉矿以镁橄榄石粉矿为例。将分散剂六偏磷酸钠、稳泡剂 CMC、结合剂硅溶胶、菱镁矿细粉依次加入水中，加入发泡剂搅拌发泡，将镁橄榄石废弃粉矿和镁砂干混均匀后加入泡沫中，混合均匀后倒入模具，静置凝胶固化成型；烘干后置于电炉中高温烧成。图 1 为 1 500 ℃下烧成试样的 SEM 图。由图 1 可知，试样中气孔均匀，孔壁晶粒发育良好，结合紧密。新开发的轻质镁橄榄石隔热材料使用温度可达 1 500 ℃，导热系数最低可达 0.15 W/M・K（1 000 ℃），几乎与莫来石纤维制品相当，具有很好的应用前景。

低温粉矿以粉煤灰为例。以粉煤灰粉矿为主要原料，加入玻璃粉、碳酸钠、发泡剂等混合磨细，制成坯体；将坯体加热到 900 ℃，由于发泡剂生成大量气体，在坯体内形成细小的封闭气孔，可制备出性能优良的泡沫玻璃。图 2 为粉煤灰制备泡沫玻璃试样的 SEM 图。从图 2 可以看出，试样中孔径分布均匀。试样的导热系数可达 0.08 W/M・K（1 000 ℃）。

根据废弃粉矿耐火度的不同，采用合适的工艺制备不同轻质隔热材料，能更有效地利用废弃资源优势，更有针对性地开发产品。这一思路为废弃粉矿的资源化利用开辟了新途径。

* 通信作者：王刚；E-mail：wangg@ lirrc. com；手机号：13837975338。国家自然科学基金（NSFC-No. 51672256）、中国国际合作项目(ISTCP-No. 2014DFA50240）资助项目。

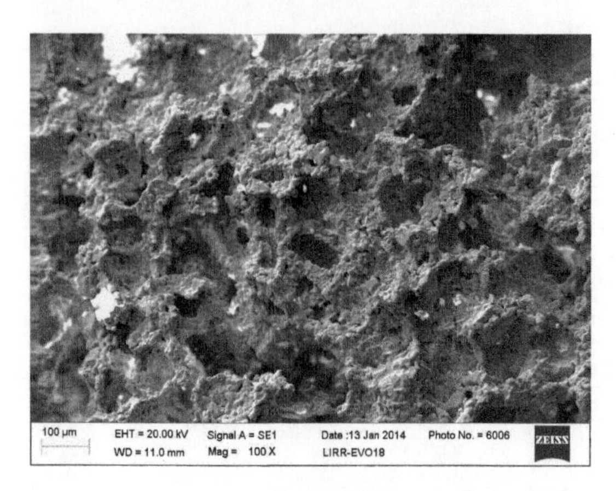

图1　1 500 ℃烧成橄榄石轻质试样的SEM图

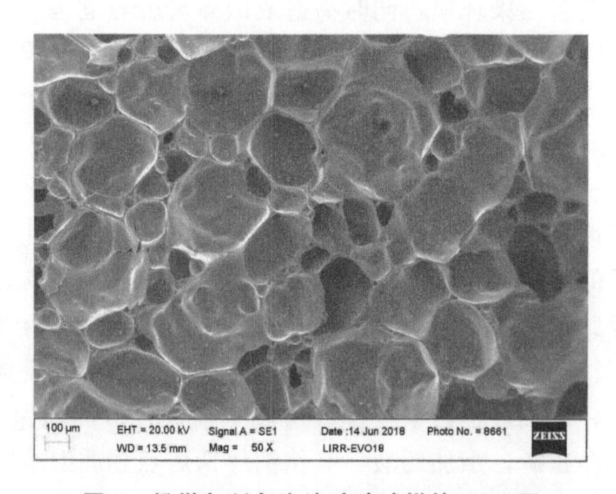

图2　粉煤灰制备泡沫玻璃试样的SEM图

采用发泡法制备隔热材料，其工艺条件可控，以废弃粉矿为原料，其产品附加值高，具有一定的市场价值。

参考文献

[1] 靳亲国. 2003. 河南橄榄石的生产加工及开发利用 [J]. 耐火材料，20：116－118.

[2] 徐云朋，张双明. 1994. 湖北宜昌镁橄榄石矿的开发利用途径研究 [J]. 湖北地质，8（1）：79－84.

[3] Cheng T W，Ding Y C，Chiu J P. 2002. A study of synthetic forsterite refractory materials using waste serpentine cutting [J]. Minerals Engineering，13：271－275.

[4] 方荣利，刘敏，周元林. 2003. 利用粉煤灰研制泡沫玻璃 [J]. 保温材料与建筑节能，6：38－40.

再生硅藻土负载二氧化钛室温光催化氧化甲醛

孙怀虎[1,2]　黄　浅[1,2]　彭　鹏[1,2]　王海波[3]　张世英[1]　张向超[1*]
（1. 长沙学院环境光催化应用技术湖南省重点实验室，长沙 410022；
2. 长沙学院生物与环境工程学院，长沙 410022；
3. 华东交通大学理工学院，南昌 330100）

硅藻土是由古生物硅藻尸骸经过许多年沉积而形成的天然硅酸盐矿物材料，具有质轻、松散、多孔等特性，独特的硅藻壳体结构、强吸附性、大比表面积、高孔隙度、耐高温等优良性质决定了它是一种得天独厚的载体材料，在食品、药品、化工、环保等领域具有广泛的用途[1]。特别是作为生产饮料（啤酒、葡萄酒、果汁等）助滤剂，因表面吸附着啤酒中的有机物质，助滤作用后只能废弃。啤酒过滤过程中硅藻土的消耗量为 1～2 g/L。在欧洲，硅藻土耗费量约是 1.7 g/L。我国的啤酒工业随着改革开放的高速步伐与人民生活水平的稳步提高而迅猛发展，2016 年我国国产啤酒产量达 4 506 万千升，约产生 9 万吨的硅藻土固体废弃物。目前大多采用填埋法处理废弃硅藻土，但这样占用大量土地，另一方面由于填埋不当等因素会使其暴露于空气中或渗入地下水，从而对环境造成更大的污染和破坏[2]。当前，国内外对废弃硅藻土的利用主要是简单处理，用作土壤改良剂或建筑材料的原料，产品附加值不高。因此，废弃硅藻土吸附特性等功能化再生，不仅能实现固体废弃物的综合利用，更重要的是充分利用其特性，最大限度发挥或挖掘其天然禀赋的材料或制品，通过环境净化材料的协同作用，将在降解甲醛等有机挥发物等领域具有广阔的应用前景[3]。

以废弃硅藻土为研究对象，通过加碱和焙烧等方法对废弃硅藻土进行表面改性，利用热分析（TG－DTA）、X 射线衍射（XRD）、扫描电镜（SEM）和 N_2吸附—脱附等测试分析手段，研究加碱和焙烧改性工艺对废弃硅藻土微观结构的影响，将焙烧处理的功能再生硅藻土复合二氧化钛（商用 P25），以气相甲醛为目标降解物，探讨二氧化钛掺入量对其光催化性能的影响。

实验试剂包括废弃硅藻土、二氧化钛（商品级 P25）、硫酸（98%）和氢氧化钠（固体）。实验仪器包括 X 射线衍射分析仪、扫描电镜、红外声气体检测仪等。

实验过程包括加碱处理、焙烧处理、复合二氧化钛。

利用 X 射线衍射对加碱处理和焙烧 650 ℃后复合 TiO_2的样品进行了分析，如图 1 所示。由图 1（a）可知，除了 SiO_2的（101）、（102）、（111）和（200）的特征峰，还存在一定量的杂质峰；而加碱处理的硅藻土样品仅有 SiO_2的特征峰，表明废弃硅藻土经加碱处理能有效去除所吸附的有机物。图 1（b）为焙烧 650 ℃后复合不同 TiO_2含量样品的 XRD 图。

图 2 是废弃硅藻土功能化再生样品的光催化降解甲醛性能图。废弃硅藻土 650 ℃焙烧 2.5 h

* 通信作者：张向超；E-mail：xczhang@ccsu. edu. cn；手机号：13874998346。湖南省“环境与能源光催化”2011 协同创新中心（湘教通［2016］429 号）、长沙学院青年英才计划。

复合 5wt% TiO_2样品，由于废弃硅藻土再生的吸附特性与 TiO_2的光催化降解协同作用，对甲醛的降解率高达 95.59%。

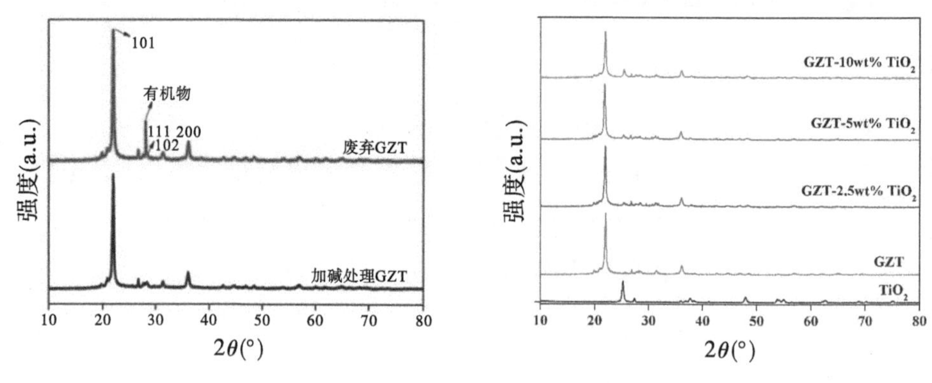

图 1　废弃硅藻土功能化样品的 X 射线衍射图

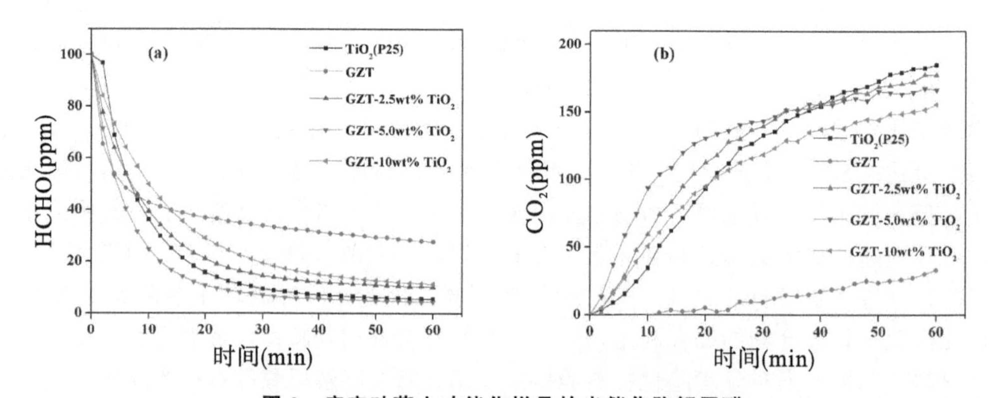

图 2　废弃硅藻土功能化样品的光催化降解甲醛

注：a. 甲醛浓度随时间的变化；b. CO_2浓度随时间的变化

加碱和焙烧处理能有效去除废弃硅藻土吸附的有机污染物，再生硅藻土负载 P25 对甲醛的降解率达 95.59%，研究结果为废弃硅藻土的功能再生奠定了实验基础，将在降解甲醛等净化室内空气领域具有广阔的应用前景。

参考文献

[1] 郑水林. 2007. 非金属矿物材料［M］. 北京：化学工业出版社.

[2] Wang B，Zhang G，Leng X，et al. 2015. Characterization and improved solar light activity of vanadium doped TiO_2/diatomite hybrid catalysts［J］. Journal of Hazardous Materials，284：212 - 220.

[3] Padmanabhan S，Pal S，Haq E. 2014. Nanocrystalline TiO_2- diatomite composite catalysts：Effect ofcrystallization on the photocatalytic degradation ofrhodamine B［J］. Applied Catalysis A：General，485：157 - 162.

凹凸棒石黏土无溶剂绿色合成 ZSM－5 沸石

蒋金龙* 张鹏宇　刘永魁

（淮阴工学院化学工程学院，江苏省凹土资源利用重点实验室，淮安 223003）

沸石具有规则的孔道、较大的比表面积、水热稳定性和优良的吸附性能，被广泛应用于催化、吸附和分离等领域。传统沸石的制备是在大量溶剂存在下进行的，导致大量废液的产生、硅铝组分的流失以及沸石生产效率的低下。为此，一种通过碾磨、加热进行的简便、无溶剂合成沸石的方法被开发出来制备各种沸石[1]。这种无溶剂合成沸石的方法避免了溶剂的使用，减少了环境污染，提高了合成效率，并且所得沸石的性能也优于水热法合成的沸石。采用的硅源主要有煅烧氧化硅、硅酸钠、固体硅胶、水合氧化硅等合成原料，与黏土矿物相比，其价格较高，不利于沸石成本的降低。

黏土矿物含有大量的硅和铝，在经过处理后可以用来合成各种沸石以降低其生产成本。硅藻土与拟薄水铝石混合碾磨后通过无溶剂的方法可以制备方钠石，在模板剂作用下也可以无溶剂合成 ZSM－5 沸石[2,3]。但是，硅藻土的二氧化硅活性大，不需要进行处理即可转化为沸石。而黏土的种类众多，结构和组成也不相同，因此以黏土为原料无溶剂制备沸石还存在挑战。本文以凹凸棒石黏土（简称凹土）为硅源，无溶剂制备 ZSM－5 沸石，考察了沸石合成影响因素，对其催化性能也进行了考察。

典型沸石合成如下：将凹土用 3 mol/L HCl 在 80 ℃浸泡处理 48 h，抽滤、水洗至中性，得到酸化凹土，然后将其与硅酸钠、四丙基溴化铵、氯化铵混合碾磨，180 ℃晶化处理一定时间，产品冷却、水洗多次后烘干，550 ℃煅烧去除模板剂，得到 ZSM－5 沸石，随后与 NH_4Cl 溶液交换 2 次煅烧制备 H 型 ZSM－5 沸石。

沸石利用 XRD、SEM、固体核磁、N_2吸附脱附进行表征，以环已酮与甲醇、乙二醇的缩酮反应为探针反应考察其催化性能并与水热法正硅酸四乙酯为硅源合成的 ZSM－5 沸石性能相比较。

考察了沸石合成过程中铝酸钠添加量的影响，从图 1 的 XRD 图谱可知，凹土在酸化后其特征衍射峰消失，在没有铝酸钠添加情况下，可以在模板剂和氯化铵矿化剂存在下与硅酸钠合成 ZSM－5，随后添加铝酸钠，均可以合成 ZSM－5，但是沸石特征峰随铝酸钠增加下降，同时还出现了明显的方沸石特征衍射峰。凹土在低倍图像下呈现无规则固体粒子形态，酸化对其形貌没有影响，但是形成 ZSM－5 沸石后，转变为球形，有孪晶现象出现，这是典型的 ZSM－5 沸石形貌，随着铝酸钠添加量的增多，沸石呈现聚合晶体形貌并且还有纤维状棒晶出现，这种纤维状棒晶为方沸石。

图 2 为合成 ZSM－5 沸石的 Al 的固体核磁谱和 N_2－吸附脱附曲线。化学位移为 55.39 处出现的峰表明 Al 元素进入沸石骨架，而化学位移为 0 处的峰表明有部分 Al 在沸石骨架外。从其

* 通信作者：蒋金龙；E-mail：jiangjinlong75@163. com；手机号：13915100532。国家自然科学基金（51574130）、江苏省青蓝工程资助项目。

N_2-吸附脱附曲线可知，沸石在 P/P_0为 0.4～1.0 高压范围出现明显回滞环，说明沸石中存在介孔。

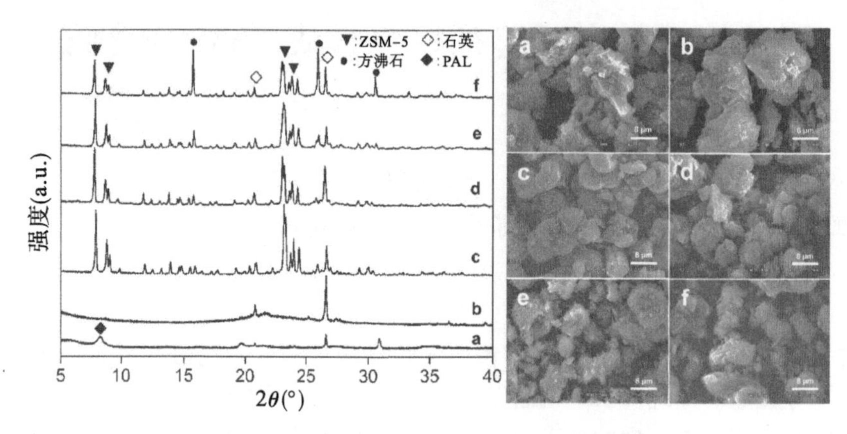

图 1　PAL (a)、APAL (b) 和添加 0 g (c)、0.02 g (d)、0.08 g (e)、0.12 g (f) $NaAlO_2$合成的 ZSM－5 沸石的 XRD 图谱和 SEM 照片

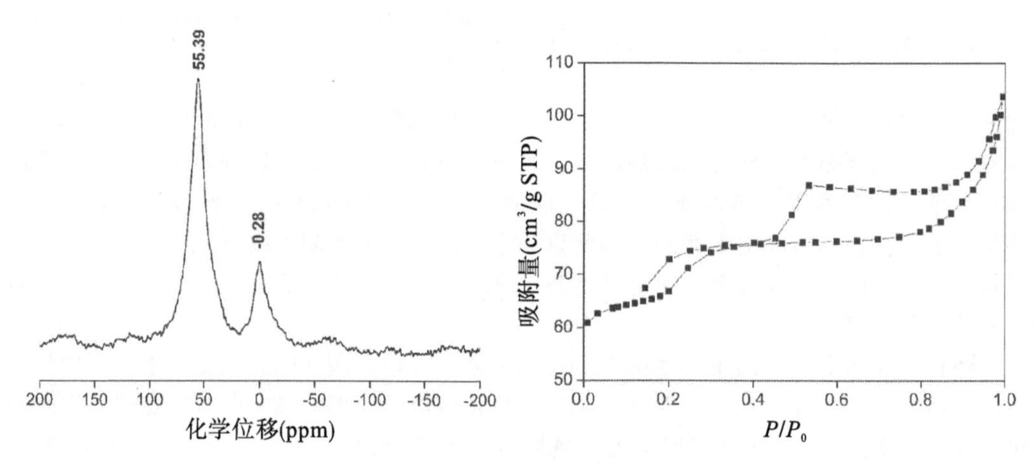

图 2　ZSM－5 沸石的固体核磁和 N_2-吸附脱附曲线

其催化性能表明，合成沸石环已酮转化率为 66%（甲醇）和 98%（乙二醇），而水热法合成的 ZSM－5 沸石环已酮转化率为 90%（甲醇）和 99%（乙二醇）。

上述实验表明，以凹土替代煅烧氧化硅等合成原料无溶剂合成 ZSM－5 沸石是可行的，对其他类型黏土无溶剂绿色制备沸石具有较大的借鉴意义。

参考文献

[1] Ren L M，Wu Q M，Yang C G，et al. 2012. Solvent-free synthesis of zeolites from solid raw materials [J]. J. Am. Chem. Soc.，134：15173－15176.

[2] Zeng S J，Wang R W，Zhang Z T，et al. 2016. Solventless green synthesis of sodalite zeolite using diatomite as silica source by a microwave heating technique [J]. Inorganic Chem Commun，70：168－171.

[3] 罗永明，万耿平，周元. 一种多级孔 ZSM-5 分子筛的制备方法. CN201610527651. 5.

蒙脱石负载型铁复合材料及微波降解罗丹明 6G

饶文秀　吕国诚*　梅乐夫　王丹宇　马帅飞　廖立兵*
（中国地质大学（北京），材料科学与工程学院，北京 100083）

印染废水是加工棉、麻、化学纤维及其混纺产品、丝绸为主的印染及丝绸厂等排出的废水，是中国和其他地方水污染的主要来源。每年，大量的含有大量有机污染物和复杂成分的废水未经处理就被排放，这将极大地影响人类的用水安全。此外，在纺织工业中生产有机染料期间产生大量污染物，如果未经处理就排出，将会造成严重的环境污染。印染废水因其成分复杂、难降解物浓度高、色度和化学需氧量（COD）含量高等特点，难以被去除。高浓度印染废水中的染料吸收光线，降低水体透明度，影响水生生物微生物生长，不利于水体自净，也会造成视觉污染，影响人类健康[1]。因此，在染料生产和印染行业中迫切需要寻找有效且经济的染料废水处理方法。

纳米零价铁作为一种新型的纳米材料，具有较大的比表面积和较高的反应活性，能有效去除水体中的重金属及有机污染物。然而普通的纳米零价铁具有稳定性差、易团聚、易氧化等缺点，从而降低其反应活性，限制了它在工业生产中的实际应用[2,3]。蒙脱石是一种 2∶1 型的硅酸盐黏土矿物材料，由两层硅氧四面体片和中间的铝氧八面体片组成，由于其独特的层状结构，蒙脱石可以吸附各种有机污染物。越来越多的复合材料被用作催化剂，我们期望通过将纳米零价铁负载到蒙脱石薄片上形成复合材料（零价铁/蒙脱石）来保护纳米零价铁以避免其氧化。

微波诱导氧化技术是一种通过载体上的催化剂位点吸收微波，使位点温度迅速上升，达到使有机物在高温下解离或加速反应的效果的一种新型的高效有机污染物处理技术，已成为国内外研究的热点，具有用时短、效率高、选择性加热、无二次污染等优点[4]。由于微波的强烈穿透作用，它可以直接加热反应物分子，并改变系统的热力学功能。在这些相互作用过程中，降低了反应的活化能和分子的化学键强度，而 Fe^{2+} 和 · OH 的生成速率加快，导致降解反应速率加快。

本文采用无机材料负载法，将零价铁负载在蒙脱石上，制备出零价铁/蒙脱石复合材料（图 1），将其用作在微波辐射下去除罗丹明 6G 的非均相芬顿催化剂。目前，纳米零价铁技术被广泛应用于罗丹明 6G 的降解，因此研究优化的纳米零价铁对罗丹明 6G 污染的修复效果具有实际意义。

对零价铁/蒙脱石进行物相、成分、结构分析，发现负载在蒙脱石上的零价铁可以有效缓解单纯纳米零价铁的团聚，增加其抗氧化能力。负载在蒙脱石上的纳米零价铁尺寸多为 10 nm。零价铁/蒙脱石在微波辐射下既会产生介电损耗又会产生磁损耗，反应以介电损耗为主。对罗丹明 6G 初始浓度、微波功率、pH 值等各种参数对降解的影响进行分析，发现零价铁/蒙脱石对罗丹明 6G 的去除能力很好，在 15 min 内去除量达到 500 mg/g，其中羟基自由基对降解起主要作用，

---

* 通信作者：吕国诚，E-mail：guochenglv@cugb. edu. cn；廖立兵，E-mail：lbliao@cugb. edu. cn。国家重点科技攻关计划（2017YFB0310704）、国家自然科学基金青年基金（51604248）资助。

降解速率达到 0.436 5 min^{-1}，明显高于以前报道过的去除罗丹明 6G 的方法。

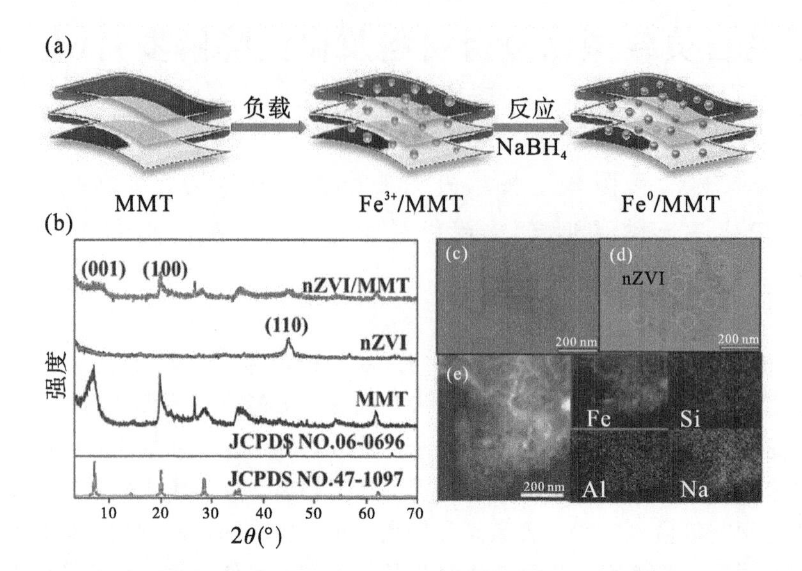

图 1　蒙脱石负载型纳米零价铁复合材料的合成过程（a），MMT、nZVI、nZVI/MMT 的 XRD 图（b），MMT 的透射电镜图（c），nZVI/MMT 的透射电镜图（d）以及 nZVI/MMT 的 EDS 元素映射图像（e）

参考文献

［1］Robinson T，Mcmullan G，Marchant R，et al. 2001. Remediation of dyes in textile effluent：a critical review on current treatment technologies with a proposed alternative［J］. Bioresource Technology，77：247－255.

［2］Deng J M，Dong H R，Zhang C，et al. 2018. Nanoscale zero-valent iron/biochar composite as an activator for Fenton-like removal of sulfamethazine［J］. Separation and Purification Technology，202：130－137.

［3］Huang P P，Ye Z F，Xie W M，et al. 2013. Rapid magnetic removal of aqueous heavy metals and their relevant mechanisms using nanoscale zero valent iron（nZVI）particles［J］. Water Research，47：4050－4058.

［4］Jiang B H，Zhao Y，Jin Y，et al. 2011. Study on coupled oxidation and microwave process in treating urban landfill leachate by Fenton and Fenton-like reaction［J］. Advanced Materials Research，393-395：1443－1446.

$BiVO_4$/HNTs复合材料的制备与光催化性能

孙 青[1,2] 秦 丰[1] 张 俭[1] 马俊凯[1] 盛嘉伟[1,2*]

（1. 浙江工业大学材料科学与工程学院，杭州 310014；

2. 浙江工业大学温州科学技术研究院，温州 325011）

窄禁带半导体 $BiVO_4$具有可见光下降解有机污染物的能力，近年来广受关注[1]，然而纯 $BiVO_4$存在吸附能力差和分离回收困难的不足，限制了它的应用。埃洛石纳米管（HNTs）是一种纳米管状黏土矿物，因其具备较大的比表面积、良好的吸附性能，HNTs 作载体材料多用于负载传统紫外光响应光催化剂[2,3]，相较而言，HNTs 负载 $BiVO_4$等铋基可见光催化材料的研究偏少。

本研究采用水解沉淀法制备了 $BiVO_4$/HNTs 复合光催化材料，并采用 XRD、TEM 等手段对 400 ℃下保温 2 h 得到的复合材料（BH－400 ℃）的结构进行表征，采用 1 000 W 氙灯模拟太阳光照射，研究了复合材料对 25 mg/L 亚甲基蓝（MB）溶液的光催化降解性能。以与样品中 $BiVO_4$等质量的纳米 TiO_2 Degussa P25 为参比光催化剂。

图 1 为原矿 HNTs（a）及 BH－400 ℃（b）样品的 XRD 图谱。由图 1 可知，相比原矿 HNTs，样品 BH－400 ℃中 HNTs 的衍射峰逐渐减弱，这是可能由于 HNTs 在高温煅烧时脱去结合水影响了其结构的完整性，BH－400 ℃中 $BiVO_4$为单斜相和四方相混晶结构，使其较优异的光催化性能[4]。

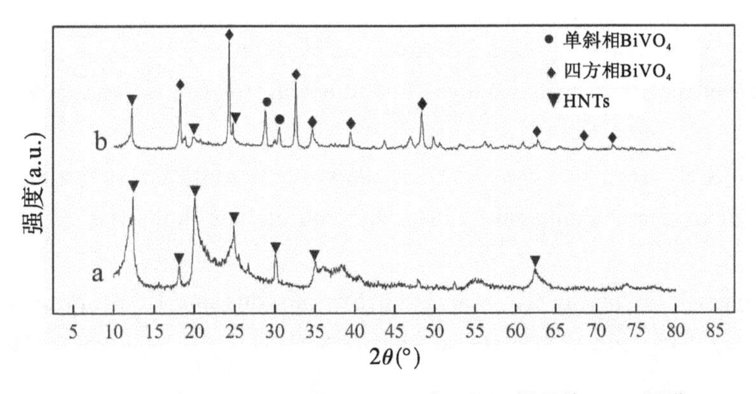

图 1 原矿 HNTs（a）及 BH－400 ℃（b）样品的 XRD 图谱

图 2 为 BH－400 ℃样品的 HRTEM 及不同样品对 MB 的光催化降解效果。由图 2a 可见，HNTs 上分布着尺寸为 5～15 nm 的 $BiVO_4$颗粒，$BiVO_4$颗粒未见有明显团聚；由图 2b 可见，由于

---

* 通信作者：盛嘉伟；E-mail：jw-sheng@ zjut. edu. cn。国家自然科学基金（51604242）；温州市重大科技专项（ZG2017029）；浙江省大学生科技创新活动计划（新苗人才计划）项目（2018R403042）。

载体 HNTs 的吸附作用，$BiVO_4$/HNTs 对 MB 的吸附去除率明显高于 Degussa P25，当光照时间为 3 h 时，样品 BH-400 ℃对 MB 的去除率接近 100%，光催化降解效果优于 Degussa P25 和空白对比。

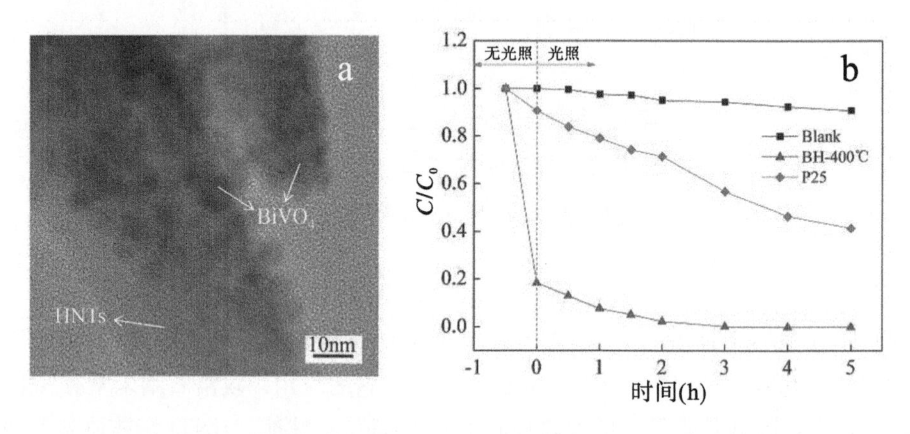

图 2　BH-400 ℃样品的 HRTEM 及不同样品对 MB 的光催化降解效果

采用水解沉淀法制备的 $BiVO_4$/HNTs 复合光催化材料中，$BiVO_4$ 以 5～15 nm 颗粒依附于 HNTs 上，$BiVO_4$为四方相和单斜相混晶结构，光照时间为 3 h 时对 25 mg/L MB 溶液的去除率接近 100%。

参考文献

[1] Malathia A，Madhavana J，Ashokkumar M，et al. 2018. A review on $BiVO_4$ photocatalyst：Activity enhancement methods for solar photocatalytic applications [J]. Applied Catalysis A：General，555：47-74.

[2] Jiang L，Huang Y P，Liu T X. 2015. Enhanced visible-light photocatalytic performance of electrospun carbon-doped TiO_2/halloysite nanotube hybrid nanofibers [J]. Journal of Colloid and Interface Science，439 (439)：62-68.

[3] Peng H X，Liu X H，Tang W，et al. 2017. Facile synthesis and characterization of ZnO nanoparticles grown on halloysite nanotubes for enhanced photocatalytic properties [J]. Scientific Reports，7 (1)：2250.

[4] Saison T，Chemin N，Chanéac C，et al. 2015. New insights into $BiVO_4$ properties as visible light photocatalyst [J]. The Journal of Physical Chemistry C，119 (23)：12967-12977.

埃洛石/碳钴复合材料用于宽频带微波吸收

刘天豪[1,2]　张　毅[1,2]　杨华明[1,2]　欧阳静[1,2*]
（1. 中南大学资源加工与生物工程学院无机材料系，长沙 410083；
2. 矿物材料及其应用湖南省重点实验室，长沙 410083）

近年来，辐射电磁波已经成为一个严重的环境问题，不仅影响电子设备的运行，而且影响人类的健康和军事武器的应用。因此，作用于 2～18 GHz 频率的高性能微波吸收材料引起了研究者们的广泛关注[1]。损耗能力和阻抗匹配是影响其性能的最关键的两个因素。最近，还原氧化石墨烯基复合材料[2]、核壳[3]或蛋壳复合材料[1]等是该领域的研究热点，但是昂贵的原材料、复杂的工艺过程等问题阻碍着它们的实际应用。埃洛石是一种天然的中空管状硅酸盐黏土矿物，其化学式为 $Al_2Si_2O_5(OH)_4 \cdot nH_2O$，产量丰富[4]。$SiO_2$作为埃洛石的主要成分之一，可以用来调节吸波材料的介电常数[5]，从而优化其阻抗匹配。据我们所知，很少有报道将埃洛石用于微波吸收。同时，埃洛石独特的中空管状形貌可能更有利于“捕获”电磁波。

图 1　埃洛石碳钴的 SEM 图

本文以埃洛石为基体，聚乙二醇为碳源，六水合硝酸钴为钴源，使用一步热解法制备了埃洛石碳钴复合物，之后通过扫描电镜（SEM）来表征它的微观形貌，使用矢量网络分析仪（VNA）测试其在 2～18 GHz 范围内的电磁参数（复介电常数的实部和虚部、复磁导率的实部和虚部），

* 通信作者：欧阳静；E-mail：jingouyang@csu. edu. cn；手机号：15116338598。国家自然科学基金（51774331）、湖南省自然科学基金（2017JJ2313）资助项目。

然后利用吸波公式计算其在 2～18 GHz 波段内的反射损耗。如图 1 所示，埃洛石碳钴复合物整体是中空管状形貌，由埃洛石纳米管和碳纳米管所组成，其中碳纳米管表面有部分突起，表明钴纳米粒子被包在了碳纳米管的管内。

图 2 为埃洛石碳钴的反射损耗图，其表现出优异的微波吸收性能，在 2.72 mm 频宽可达 6.5 GHz，7.46 mm 时在 4.24 GHz 处反射损耗最小为－40.43 dB。埃洛石碳钴复合物杰出的微波吸收性能可归因于其独特的中空管状结构以及多组分之间的界面极化。

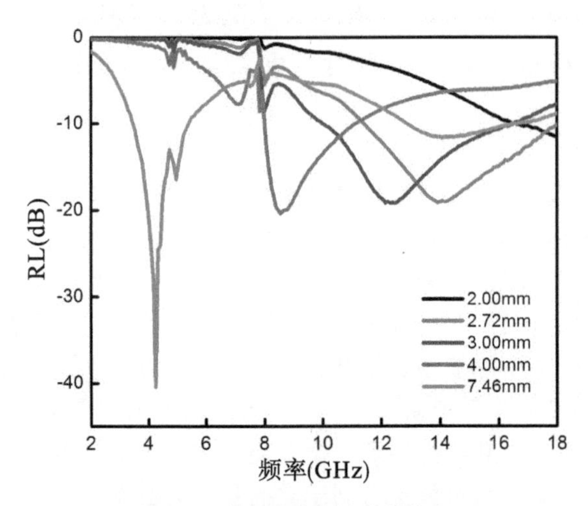

图 2　埃洛石碳钴的反射损耗

本研究通过一步热解法制备了埃洛石碳钴复合物，原料产量丰富，制备工艺简单，为低成本制备高效微波吸收剂提供了新思路。

参考文献

[1] Zhao B，Guo X，Zhao W，et al. 2016. Yolk-shell Ni@ SnO_2 composites with a designable interspace to improve electromagnetic wave absorption properties [J]. Acs Applied Materials & Interfaces，8 (42)：28917 - 28925.

[2] Han M，Yin X，Kong L，et al. 2014. Graphene-wrapped ZnO hollow spheres with enhanced electromagnetic wave absorption properties [J]. Journal of Materials Chemistry A，2 (39)：16403 - 16409.

[3] Liu Q H，Cao Q，Bi H，et al. 2015. Coni@ sio2 @ tio2 and coni@ air@ tio2 microspheres with strong wideband microwave absorption [J]. Advanced Materials，28 (3)：486 - 490.

[4] 马智，王金叶，高祥，等. 2012. 埃洛石纳米管的应用研究现状 [J]. 化学进展，24 (2)：275 - 283.

[5] Wen B，Cao M S，Hou Z L，et al. 2013. Temperature dependent microwave attenuation behavior for carbon-nanotube/silica composites [J]. Carbon，65 (12)：124 - 139.

$BiOCl/TiO_2$/硅藻土材料制备及可见光催化降解有机染料

敖敏琳[1,2]　唐学昆[1,2]　李自顺[1,2]　彭　倩[1,2]　刘　琨[1,2*]
（1. 中南大学资源加工与生物工程学院，长沙 410083；
2. 中南大学矿物材料及其应用湖南省重点实验室，长沙 410083）

近几十年来，由于印染、纺织、皮革加工等工业的不断发展，工业废水排放量不断增大，该类废水的治理已成为当前我国环境保护工作的重中之重[1]。在众多的有机废水治理技术中，光催化氧化技术展现出广阔的应用前景，其仅仅在紫外光或可见光光照催化剂的条件下就能够将有机污染物完全矿化，具有成本低、应用安全、效率高、不产生二次污染的特点。光催化的核心之一是光催化剂。纳米二氧化钛作为很有发展前途的光催化剂之一，由于其化学稳定性、催化活性高、无毒性、环境友好性和廉价易得等优异性能，被认为是在环境保护领域中很具应用前景的光催化材料之一。然而，纳米二氧化钛颗粒具有极高的表面能，在制备或使用的过程中极易团聚，使其光催化活性大大降低[2]。此外，二氧化钛仅仅在紫外光下具有催化活性，无法利用自然界中广泛存在的可见光，这些缺点都极大地限制了纳米二氧化钛在实际环境中的应用。

针对存在的这些问题，本文以硅藻土作为载体，采用改性的溶胶—凝胶法均匀负载二氧化钛（TiO_2）纳米颗粒层，解决 TiO_2 容易团聚的问题，得到在硅藻土载体上分散性良好的 TiO_2 纳米颗粒层。再通过沉淀—煅烧法在其表面负载均匀的氯氧铋（BiOCl）纳米片层，形成了 $BiOCl/TiO_2$/硅藻土复合光催化材料，实现其在可见光条件下的催化性能。通过 X 射线衍射（XRD）、扫描电子显微镜（SEM）、透射电子显微镜（TEM）、紫外—可见漫反射光谱（UV-Vis）、X 射线光电子能谱分析（XPS）等测试方法对样品的物理化学性质进行了表征。结果表明，高度分散的 TiO_2 纳米颗粒与 BiOCl 纳米片依次覆盖于硅藻土载体的表面，形成双层结构。此外，在 $BiOCl/TiO_2$/硅藻土复合光催化材料中，TiO_2 与 BiOCl 的紧密结合可以有效地促进光生电子—空穴对的分离，大大提高了复合材料对光子的利用率[3]。以罗丹明 B 溶液为目标降解反应物，以可见光（$\lambda >$ 400 nm）作为光源，对所制备复合材料的光催化性能进行表征。实验结果表明，$BiOCl/TiO_2$/硅藻土复合材料在可见光下的催化性能要远远高于单一的 TiO_2 与 BiOCl（图 1）。当 BiOCl 与二氧化钛摩尔比为 1∶1 时，焙烧温度为 400 ℃时所制备的 $BiOCl/TiO_2$/硅藻土复合材料具有最佳的催化效果。通过研究复合材料的光催化反应机理分析表明，$BiOCl/TiO_2$/硅藻土复合材料具有较强可见光催化活性的原因是光生电子的跃迁作用，光生电子从 BiOCl 的导带向 TiO_2 表面迁移，从而降低了光生电子—空穴对的复合率（图 2）。由于 $BiOCl/TiO_2$/硅藻土复合材料在可见光下具有优异的催化性能，使其在太阳能驱动下的污水环境处理方面具有广阔的应用价值。

* 通信作者：刘琨；E-mail：kliu@csu.edu.cn；手机号：15873150374。国家自然科学基金（51774330）、中南大学研究生自主探索创新项目（2018zzts778）资助项目。

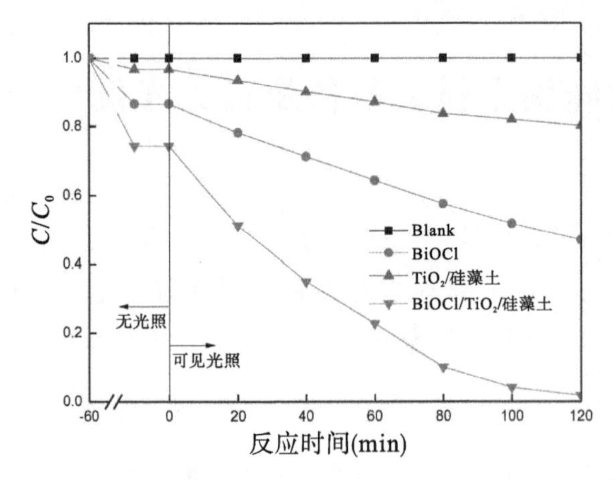

图 1　所制备样品在可见光（λ>400 nm）照下的光催化性能

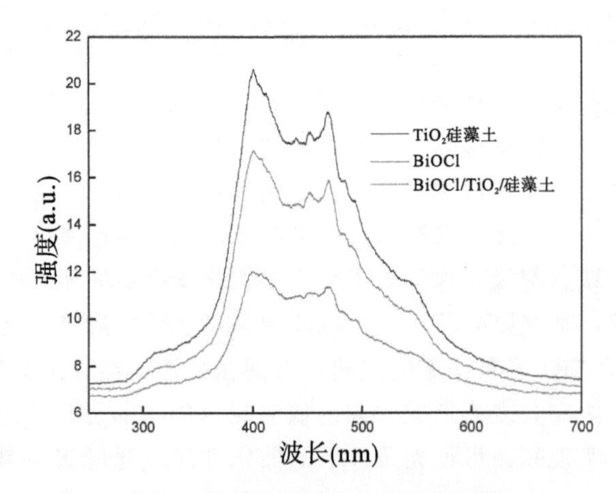

图 2　所制备样品的光致发光光谱

参考文献

［1］Wang B，Zhang G，Leng X，et al. 2015. Characterization and improved solar light activity of vanadium doped tio2/diatomite hybrid catalysts［J］. Journal of Hazardous Materials，285：212－220.

［2］Tang X，Feng Q，Liu K，et al. 2016. Synthesis and characterization of a novel nanofibrous TiO_2/SiO_2，composite with enhanced photocatalytic activity［J］. Materials Letters，183：175－178.

［3］Zhang G，Sun Z，Hu X，et al. 2017. Synthesis of BiOCl/TiO_2-zeolite composite with enhanced visible light photoactivity［J］. Journal of the Taiwan Institute of Chemical Engineers，81：435－444.

矿物- TiO_2复合乳浊剂在陶瓷洁具中的应用

敖卫华[1] 夏文华[2] 潘 伟[3] 高亮长[4] 常 亮[1] 丁 浩[1*]
（1. 中国地质大学（北京）材料科学与工程学院，北京 100083；
2. 唐山中陶卫浴制造有限公司，唐山 063611；
3. 辽宁苏泊尔陶瓷工业有限公司，沈阳 110400；
4. 许昌市恒大陶瓷有限公司，许昌 461694）

乳浊剂是建筑卫生陶瓷行业用来遮盖坯体以增加白度的添加剂。目前国内外主要使用硅酸锆（$ZrSiO_4$）作为乳浊剂，它具有性质稳定、适应性好的特点，但在长期使用过程中仍存在诸如产品价格昂贵及存在放射性的问题[1,2]。二氧化钛（TiO_2）具有优异的乳浊性能[3,4]，但相比 $ZrSiO_4$ 而言，使用 TiO_2对陶瓷烧成温度、气氛等条件的适应性收窄，并往往导致釉面呈现黄色调而影响外观，一直未得到广泛应用。

本文所述的矿物- TiO_2复合乳浊剂具备特定的有序结构和颗粒间界面结合作用，可望对釉料高温生成榍石相的反应产生诱导[5,6]，从而提升釉面品质。另外，其不含放射性物质，故用此可制得无放射性危害的安全陶瓷卫生洁具。

色度检测采用 SP 系列分光光度计，测试粉体压片和陶瓷釉面的明度（L* ）、红绿值（a* ）和蓝黄值（b* ）。陶瓷釉层物结构及物相特征通过扫描电镜（日本日立电子显微镜公司 S－3500N 型）、（日本理学 D/MAX. RC 型 X 射线衍射仪）来观察和测试。

将矿物- TiO_2复合乳浊剂按试验所在生产企业基础釉配方配制成浆料，均匀施釉，在工厂隧道窑中采用氧化气氛高温烧制，烧制周期为 13～17 h。成品件釉层均匀，乳浊效果好，釉面细腻、光泽强。表 1 为河北唐山工厂烧制的成品马桶釉面不同位置的色度值。结果显示，加入矿物- TiO_2复合乳浊剂卫士洁具的釉面达到和优于使用 $ZrSiO_4$的釉面。

按照国标《室内装饰装修材料—建筑材料放射性核素限量》GB 6566—2010 对 2 种卫浴产品进行了放射性核素的检测。结果显示，使用矿物- TiO_2复合乳浊剂釉层物质的 I_{Ra}和 I_Y值极低，仅分别为 0.05 和 0.24，表明无放射性辐射，而使用 $ZrSiO_4$釉层 I_{Ra}和 I_Y分别高达 1.37 和 1.18，按标准规定仅可用于建筑物外部。从图 1（a）看出，晶相包括榍石和石英，高温乳浊相为榍石，与文献的结论一致[7]。从图 1（b）可见，釉层中乳浊颗粒尺度均匀，反映了榍石在釉面玻璃相中均匀分布。

使用矿物- TiO_2作为陶瓷乳浊剂，其产品的釉面效果与 $ZrSiO_4$乳浊剂釉面相当，可制备出无放射性、对健康和环境无危害的卫生洁具，具备替换硅酸锆乳浊剂的潜力[8]。

*通信作者：丁浩；E-mail：dinghao @ cugb. edu. cn；手机号：13501059304。中央高校基本科研业务费（2652015099）资助项目。

表1　卫生洁具釉面的CIE色度

测试位置		加入矿物-TiO_2复合乳浊剂			加入$ZrSiO_4$		
		L*	a*	b*	L*	a*	b*
外部	左侧	90.15	−0.83	2.64	89.75	−0.43	2.49
	右侧	89.34	−0.79	2.87	89.47	−0.54	2.33
	水箱正面	89.58	−0.8	2.84	89.70	−0.43	2.27
内部	位置1	88.82	−0.68	2.79	86.53	−0.47	2.15
	位置2	87.81	−0.71	2.56	89.94	−0.50	2.43

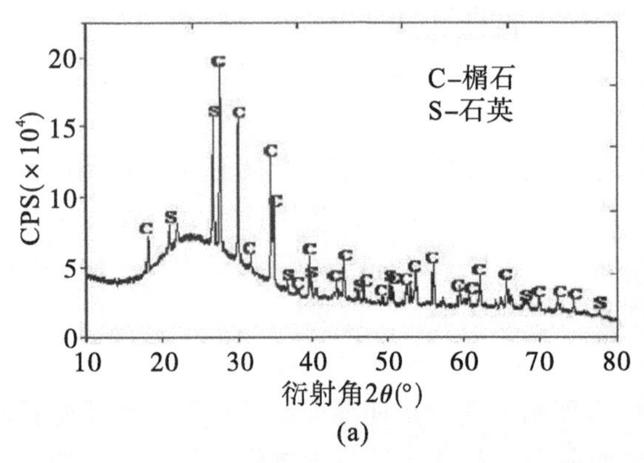

(a)

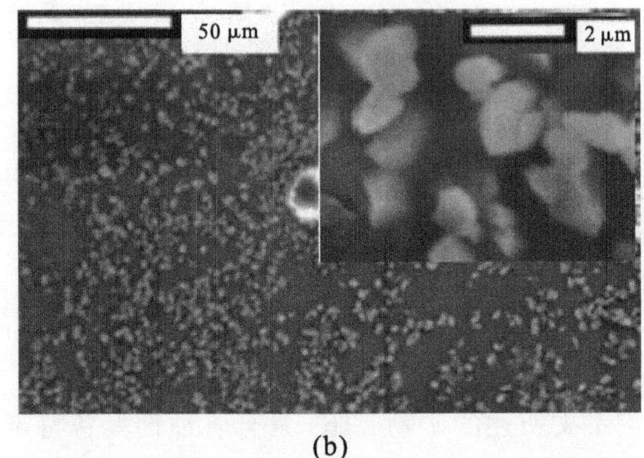

(b)

图1　加入矿物-TiO_2复合乳浊剂釉层的XRD（a）和SEM（b）

参考文献

[1] 关振铎，张中太，焦金生. 2011. 无机材料物理性能（第2版）[M]. 北京：清华大学出版社.

[2] 张朝春，吴政宙，张洪元. 2002. 进口锆英砂的放射性对人体的危害 [J]. 中国国境卫生检疫杂志，25（4）：226-227.

[3] 盛忠旗，黄平. 2005. 钛白乳浊釉在建筑陶瓷上的应用 [J]. 佛山陶瓷，（7）：14-16.

[4] Gao Q，Wu X M，Fan Y M. 2012. The effect of iron ions on the anatase-rutile phase transformation of titania（TiO_2）in mica-titania pigments [J]. Dyes and Pigments，95：96-101.

[5] Pekka E，Mikko R，Markku L. 1993. The effect of calcinations on the surface composition and structure of titanium dioxide coated mica particles [J]. Journal of Solid State Chemistry，103：160-169.

[6] 丁浩，林海，邓雁希，等. 2016. 矿物-TiO_2微纳米颗粒复合与功能化 [M]. 北京：清华大学出版社.

[7] 王春玲，罗宏杰，封鉴秋，等. 1997. 中高温钛乳浊釉显微结构的研究 [J]. 硅酸盐通报，16（4）：10-13.

[8] 杨萍，杨中喜，周广军. 2000. 钛釉乳浊机理的研究 [J]. 佛山陶瓷，40：7-9.

水滑石光催化氧化协同降解亚甲基蓝

潘国祥* 徐敏虹　伍　涛　王永亚　郭玉华

（湖州师范学院材料工程系，湖州 313000）

染料废水是一种浓度高、生物降解性差、色度高的难处理工业废水。随着印染行业的快速发展，日益增多的排放导致了水环境质量的急剧下降。近年来，利用光催化降解技术来控制环境污染问题是一种极具应用前景的技术，寻找高效光催化剂对研究此技术推广至关重要。类水滑石是一类阴离子型层状黏土材料，由于 LDHs 具有阴离子可交换性、可调变性的孔径以及热稳定性，使其在催化、吸附、医药等方面展示了广阔的应用前景[1,2]。

本文采用共沉淀法制备 M/Cr 水滑石。在可见光照射下，以亚甲基蓝溶液为模拟废水，考察了反应条件对催化降解亚甲基蓝溶液的影响，并探讨了光催化反应机理。

采用共沉淀法制备 M_3Cr-CO_3-LDHs（M＝Co，Mg，Ni，Zn，Cu）。称取 0.5 g M_3Cr-CO_3-LDHs 进行光催化实验。将该水滑石置于 25 mL 5 mg/L 亚甲基蓝溶液中，加入 0.5 mL H_2O_2，在可见光照射下考察催化时间、催化剂用量、不同催化剂种类等因素对亚甲基蓝催化降解效果的影响。亚甲基蓝浓度使用可见分光光度计进行测定。通过催化降解率评价该材料的催化能力。

XRD 测试表明，5 种含 Cr 水滑石都具有水滑石典型的层状结构。并且含 Cr 水滑石的衍射峰较强并且尖锐，表明所合成的类水滑石结晶度较好。FT－IR 测试表明，样品合成的含 Cr 水滑石具有水滑石类化合物的典型结构特征。TG－DTA 曲线可知，5 种含 Cr 水滑石的热分解主要包括 2 个明显的吸热过程：在 50～200 ℃的吸热峰，主要是由于表面吸附水及层间水分子的脱除引起的；265～450 ℃的吸热峰，表明含 Cr 水滑石层板上的 OH^- 和层间的 CO_3^{2-} 分解生成 H_2O 和 CO_2。5 种含 Cr 水滑石的热稳定性如下：$Cu_3Cr-CO_3-LDHs > Mg_3Cr-CO_3-LDHs > Ni_3Cr-CO_3-LDHs > Zn_3Cr-CO_3-LDHs \approx Co_3Cr-CO_3-LDHs$。紫外可见漫反射光谱分析表明，$Zn_3Cr-CO_3-LDHs$ 水滑石的吸收区域最宽，Ni_3Cr-CO_3-LDHs 的次之，Mg_3Cr-CO_3-LDHs 和 Co_3Cr-CO_3-LDHs 的再次之，而 Cu_3Cr-CO_3-LDHs 的最窄。

用不同的金属离子取代的 $MCr-CO_3-LDH$ 分别对 MB 进行光催化氧化降解性能评价，如图 1 所示。结果表明，$CuCr-CO_3-LDHs$ 的降解率在 60 min 后接近 100%，$MgCr-CO_3-LDHs$、$CoCr-CO_3-LDHs$ 和 $ZnCr-CO_3-LDHs$ 的降解率大于 90%。但在相同条件下，$NiCr-CO_3-LDHs$ 的降解效率较差，MB 降解率为 76.43%。

* 通信作者：潘国祥；E-mail：pgxzjut@163. com；手机号：13665714786。浙江省自然科学基金（LY17E040001）资助项目。

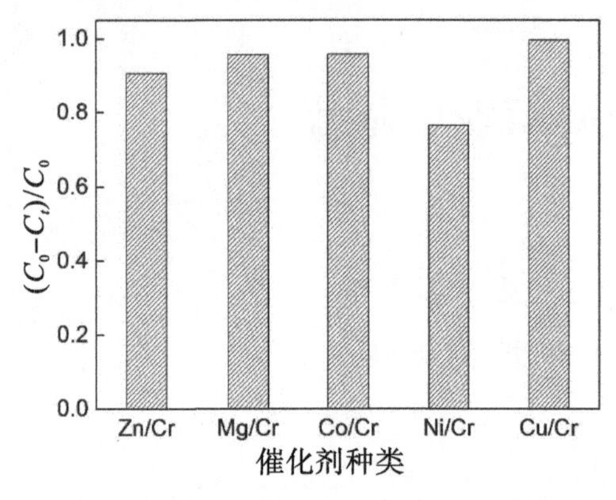

图 1　催化剂种类对光催化降解性能的影响

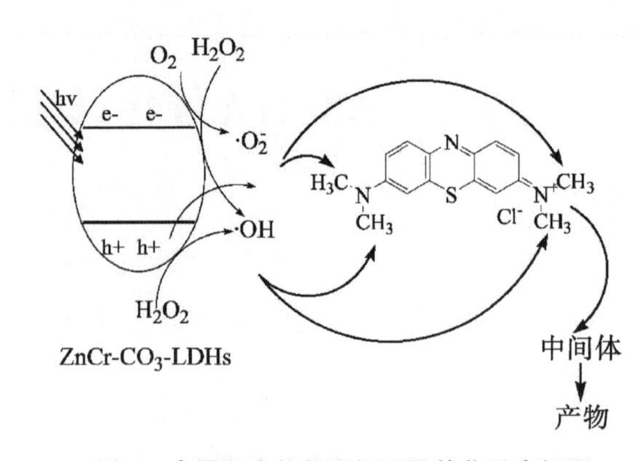

图 2　水滑石光催化降解亚甲基蓝反应机理

当 ZnCr－CO_3－LDHs 样品连续重复使用 3 次时，MB 溶液的降解率均大于 90%。XRD 测试催化反应前后的催化剂结构，使用 3 次均未发生变化，表明 ZnCr－CO_3－LDHs 在光催化反应中是稳定的，未发生光腐蚀。

紫外—可见光谱仪对 MB 降解的中间产物进行了测试，随着反应时间的延长，MB 溶液的最大吸收峰（664 nm）逐渐减小，并发生偏移，这可能是由于光催化降解 MB 时发生了去甲基化反应。在紫外区域（372 nm）出现新的吸收峰，表明在 MB 的降解过程中出现了新的物质。

加入异丙醇对亚甲基蓝的光降解有轻微的抑制作用，表明 MB 降解主要是由 · OH 自由基参与反应。但当加入 EDTA 和 p－苯醌时，反应加快，以上表明反应活性物种含有 · O_2^- 、 · OH 和 h^+，催化反应机理如图 2 所示。

参考文献

［1］Da S E，Prevot V，Forano C. 2014. Heterogeneous photocatalytic degradation of pesticides using decatungstate intercalated macroporous layered double hydroxides ［J］. Environmental Science & Pollution Research International，21：11218－11227.

［2］Wu S Z，Li N，Zhang W D. 2014. Attachment of ZnO nanoparticles onto layered double hydroxides microspheres for high performance photocatalysis ［J］. Journal of Porous Materials，21（2）：157－164.

海泡石黏土修饰电极材料的制备及应用

闫　鹏　唐爱东*
（中南大学化学化工学院，长沙 410083）

摘要：本文制备了一种以氮掺杂碳包覆海泡石为载体复合 Cu_2O 纳米颗粒修饰玻碳电极（ASEP/C－N/Cu_2O/GCE），用于电化学检测 H_2O_2。结果发现，该电极的电化学活性有效面积为 0.422 cm^2，反应速率常数为 8.68×10^7 $cm^3\cdot mol^{-1}\cdot s^{-1}$，电化学阻抗最小（165.20 Ω），检测$H_2O_2$的线性范围为 5 μM ～ 0.83 mM，灵敏度为 1 905.3 $\mu A\cdot mM^{-1}\cdot cm^{-2}$，检测限为 0.32 μM（S/N=3）。其优异的电化学性能是由于 C－N@ASEP 提高了材料的导电性，同时 C-N@ASEP 与 Cu_2O 之间的协同效应也促进了电子传递，而且海泡石的存在有助于提高材料的导电性。因此，所制备的修饰电极可应用于快速、灵敏、准确地检测液相中 H_2O_2的含量。

前言：海泡石是一种含水的镁硅酸盐矿物，具有层状和链状纤维的过渡型结构特点，它具有热稳定性好、储量丰富、价格低廉、环境友好等优势。然而，天然的海泡石由于其本身的结构缺陷，导电性较差，因此需要对海泡石原矿作进一步的处理来改善黏土的结构及导电性，以便于在电化学中的应用。本文通过氮掺杂、碳包覆和负载 Cu_2O 纳米材料等多种方式改善矿物的导电性，制备的黏土修饰电极具有较好的电化学性能。

电极材料制备方法：首先对海泡石进行酸化改性，再与壳聚糖溶液充分搅拌，在一定温度下将混合溶液蒸干，所得前驱物于惰性气氛下（N_2）煅烧得到氮掺杂碳包覆酸改性海泡石，在表面活性剂（PVP）与碱性条件（NaOH 溶液）下，加入铜源（Cu（NO_3）$_2$ · $3H_2O$）及还原剂（$N_2H_4\cdot H_2O$），反应所得物即为氮掺杂碳包覆海泡石负载 Cu_2O 复合物（ASEP/C－N/Cu_2O）。

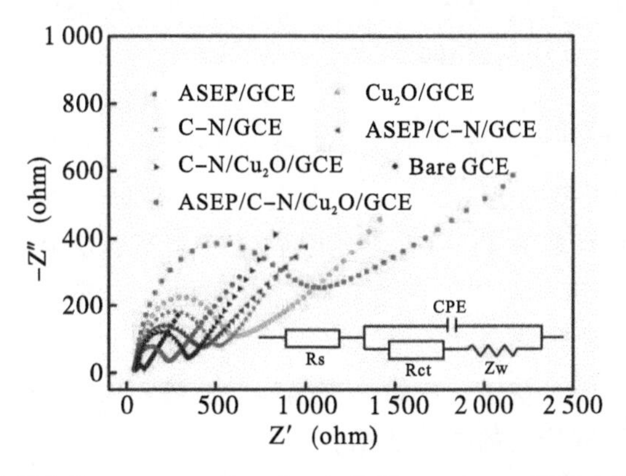

图 1　不同电极在含 5 mM K_3［Fe（CN）$_6$］的 0.1M KCl 溶液中的奈奎斯特图

*通信作者：唐爱东；E-mail：adtang@csu. edu. cn；手机号：13657354762。国家自然科学基金（51674293）资助项目。

不同电极的阻抗如图 1 所示，ASEP/GCE 的阻抗最大说明海泡石的导电性很差。相比之下，C－N 包覆海泡石和海泡石负载 Cu_2O 均能降低电极的阻抗，提高其导电性能。由图 1 可知，除了裸玻碳电极以外，ASEP/C－N/Cu_2O/GCE 的阻抗最小，仅 165.20 Ω。研究海泡石对电极阻抗的影响发现，ASEP/C－N/GC 的阻抗小于 C－N/GCE 的；ASEP/C－N/Cu_2O/GCE 的阻抗小于 C－N/Cu_2O/GCE 的，说明海泡石不仅起着载体的作用，同时由于海泡石的亲水性能使制备的材料均匀分散，从而降低了电极的阻抗。

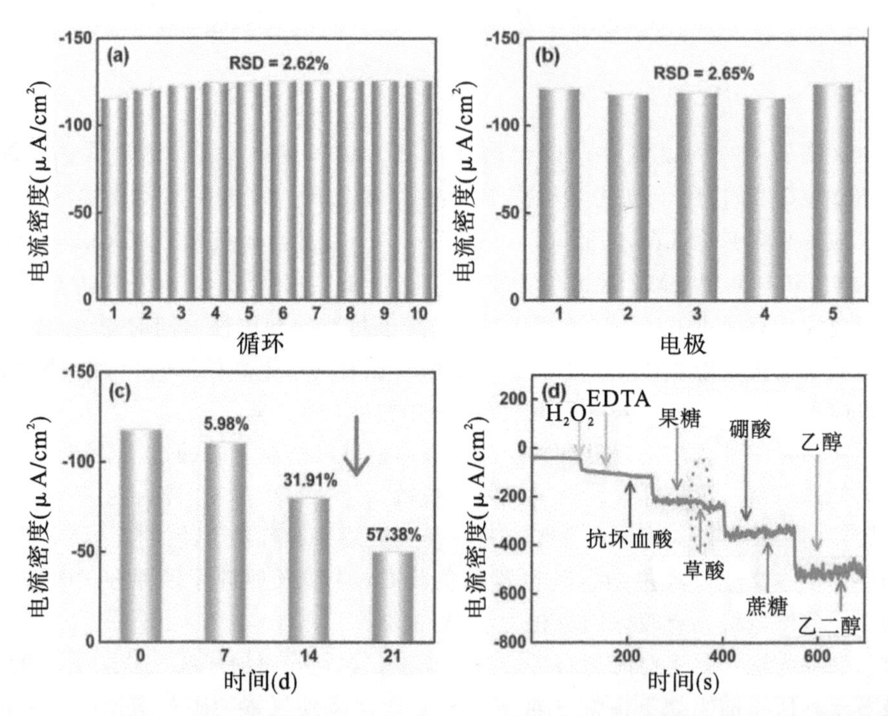

图 2　(a) ASEP/C－N/Cu_2O/GCE 电极在 0.2 mM 的 H_2O_2 溶液中进行循环伏安测试 10 次所得到还原峰电流的柱形图；(b) 5 个 ASEP/C－N/Cu_2O/GCE 电极在同一条件下测定 H_2O_2 得到还原峰电流的柱形图；(c) ASEP/C－N/Cu_2O/GCE 电极室温下空气中存放 7、14、21 d 后测定 H_2O_2 得到还原峰电流的柱形图；(d) 干扰物质（EDTA、抗坏血酸、果糖、草酸、硼酸、蔗糖、乙醇和乙二醇）在 ASEP/C－N/Cu_2O/GCE 电极上的安培计时电流响应，干扰物质浓度为 H_2O_2 的 100 倍

图 2（a）表明，10 次重复检测的相对标准偏差为 2.62％，说明电极重复性好。图 2（b）表明，5 个电极检测的相对标准偏差为 2.65％，说明电极重现性好。图 2（c）表明，电极放置在室温下空气中不稳定，因为活性成分 Cu_2O 在空气中放置过久容易发生氧化，因此需保存在低温和干燥的环境中。图 2（d）表明，草酸能产生较低的响应电流，若能消除草酸的干扰，ASEP/C－N/Cu_2O/GCE 将可用于 H_2O_2 的测定。

结论：所制备的修饰电极可快速、灵敏、准确地检测液相中 H_2O_2 的含量，为海泡石基 H_2O_2 传感器的设计与应用提供了有价值的参考依据。

碳/埃洛石/钡铁氧体复合材料的制备及电磁性能

穆大伟[1,2]　欧阳静[1,2*]　杨华明[1,2]
（1. 中南大学资源加工与生物工程学院，长沙 410083；
2. 矿物材料及其应用湖南省重点实验室，长沙 410083）

现代手机以及各种通讯设备已经成为人们生活中不可或缺的一部分，而现代通信和电子技术的飞速发展也使得电磁波污染成为继水、空气、噪声污染后的第四大污染，危害着人们的生活和健康。因此，微波吸收材料也成为材料领域研究的一个热点。但单一的微波吸收材料不能够同时满足材料对“薄、轻、宽、强”的要求，研究高效的复合吸波材料也是迫在眉睫。钡铁氧体由于其具有高的电阻率、单轴磁晶各向异性、较高的矫顽力、较强的磁滞损耗、优异的抗氧化性和耐腐蚀性和低成本等优点被大量用微波吸收材料。埃洛石是一种天然的硅酸盐矿物，在自然界中通常以纳米管的形态存在，由于具有高比表面积、无毒等优点而在催化和药物载体等方面被大量应用。

碳纳米管是一种常用的微波吸收材料，而非晶态碳纳米管的管壁结构具有短程有序和长程无序的结构特征[1,2]。由于具有良好的导电性，使得非晶态碳纳米管具有明显的介电损耗。非晶态碳纳米管的阻抗接近自由空间的阻抗能够更好地与其他材料进行匹配。本研究以埃洛石矿物为硬模板[3]，以聚乙烯醇为原料制备碳/埃洛石复合材料，与制备的钡铁氧体进行球磨水热制备碳/埃洛石/钡铁氧体三元复合材料，提高了埃洛石导电性的同时，也降低了材料的重量，是一种优异的微波吸收材料。

埃洛石（HNTs）来源于山西省临汾市；硝酸钡购自西陇化工股份有限公司；九水硝酸铁、一水柠檬酸购自国药集团化学试剂有限公司；聚乙烯醇（PVA）购自国药集团化学试剂有限公司。取一定量的埃洛石于 500 ℃下煅烧 2 h，得到煅烧埃洛石；称取一定量的 PVA 于 90 ℃溶解配制成（5%wt）的水溶液，加入煅烧的埃洛石（PVA∶HNTs＝2∶3）搅拌于 80 ℃蒸干，将混合物 50 ℃烘干后在氮气气氛下 600 ℃煅烧 3 h，得到 CHNTs。按照摩尔比 1∶11.8∶12.8 称取一定量的硝酸钡、九水硝酸铁、一水柠檬酸溶解于 30 mL 超纯水中，加入若干氨水调节溶液 pH＝7 后于 80 ℃下蒸干，将得到的胶体溶胶在 120 ℃下干燥，在 950 ℃下煅烧 2 h 得到钡铁氧体。将 CHNTs 与钡铁氧体混合后研磨 2 h，加入一定量去离子水超声 30 min 后蒸干并烘干。

图 1 为 CHNTs 与 CHNTs/$BaFe_{12}O_{19}$复合材料的 TEM 图。由图 1（a）中的插图可以看出，HNTs 以一维中空管状结构为主，尺寸相对均匀，比较容易团聚，且可以明显看到管子是两端开口的，可以粗略推算出其长度为 200～1 000 nm，内径为 15～25 nm。图 1（a）可以看出产物呈现

* 通信作者：欧阳静；E-mail：jingouyang@csu. edu. cn；手机号：15116338598。国家自然科学基金（51774331）、湖南省自然科学基金（2017JJ0351）资助项目。

出管状结构，并且在管外有一些杂质存在，从图中可以看出所制备的碳纳米管的形状与 HNTs 非常类似，直径在 20 nm 左右，长度为几百纳米不等。这说明所制备的碳材料很好地利用了 HNTs 这一模板，得到碳纳米材料。从图 1（b）可以看出，有大量的钡铁氧体纳米片沉积在 CHNTs 表面。

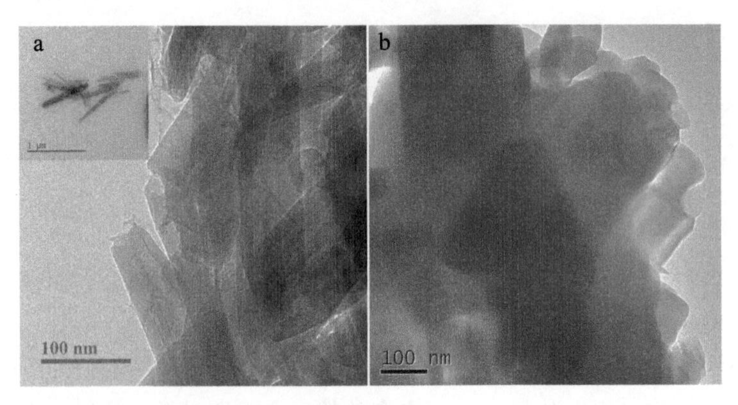

图 1 CHNTs (a) 和 CHNTs/$BaFe_{12}O_{19}$ (b) 的 TEM

参考文献

［1］Zhao T，Ji X，Jin W，et al. 2017. Electromagnetic wave absorbing properties of aligned amorphous carbon nanotube/$BaFe_{12}O_{19}$，nanorod composite［J］. Journal of Alloys & Compounds，703：424 - 430.

［2］Zhao T K，Ji X L，Jin W B，et al. 2017. Direct in situ synthesis of a 3D interlinked amorphous carbon nanotube/graphene/$BaFe_{12}O_{19}$ composite and its electromagnetic wave absorbing properties［J］. RSC Advances，7：15903 - 15910.

［3］Cheng Z L，Liu Y Y，Liu Z. 2016. Novel template preparation of carbon nanotubes with natural HNTs employing selective PVA modification［J］. Surface & Coatings Technology，307：633 - 638.

喷雾干燥法制备埃洛石微球及颗粒分级

贺子龙[1,2] 欧阳静[1,2*]
（1. 中南大学资源加工与生物工程学院，长沙 410083；
2. 中南大学矿物材料及其应用湖南省重点实验室，长沙 410083）

喷雾干燥是目前较为常用的一种干燥形式，广泛用于食品、化工、生物制品等领域，喷雾干燥制球是将一定浓度的溶液或者悬浮液经喷雾干燥固化后制成一定形状和强度的细小颗粒的一种方法。一般来讲，喷雾干燥制得的为实心球形颗粒，少数情况下会出现颗粒表面呈多孔或酒窝等形貌，另外，空心球、破损的球壳和碎片也会出现，这些不同形貌和结构的颗粒在工业生产和生活中具有不同的作用[1-4]。

埃洛石是一种天然纳米管状黏土矿物，具有球状、片状和管状结构，通常情况下是管状结构，分子式可以表示为 $Al_2Si_2O_5(OH)_4 \cdot nH_2O$，n＝0 或 2，分别代表脱水和水化状态，管内径一般为 15～100 nm，管长一般为 500～1 000 nm，埃洛石作为一种硅酸盐矿物，其资源蕴含丰富，分布较广且开采相对简单，由于其化学稳定性、无毒性、耐高温、环境友好性和抗氧化性等优点，近年来被充分挖掘，应用于化工、生物等领域。

埃洛石来源于安徽铜陵；聚乙二醇 2000 购自国药集团化学试剂有限公司；六偏磷酸钠购自西陇科学股份有限公司。仪器：实验室用喷雾干燥机为雅程 Yc－500。

以埃洛石矿为原料，经破碎，研磨，水洗和酸浸除杂，过 200 目筛后，加入一定量的六偏磷酸钠或者聚乙二醇 2000 作为分散剂，以水为溶剂，配制成一定固含量的悬浮液，磁力搅拌一定时间后，通过喷雾干燥机喷雾造球。通过控制固含量、分散剂的加入种类和加入量、喷雾干燥机进口温度和出口温度以及进料速度、喷雾压力，制备出球形颗粒较好的埃洛石微球，由于制备的微球粒径分布较广，通常为 1～50 μm，会限制一些方面的应用，又进一步将喷雾干燥制得的埃洛石微球经 400～600 ℃高温处理成陶瓷微球，将陶瓷微球分散在水中，通过机械搅拌、超声分散、重力沉降等步骤制得粒径在 1～10 μm 的陶瓷微球。

埃洛石矿喷雾干燥制球过程中，加入分散剂是必须的，聚乙二醇 2000 和六偏磷酸钠均可以作为分散剂加入，由于埃洛石本身为黏土矿物，有一定黏结力，所以不需要加入黏结剂就可以成球，固含量在 10%～50%都可制得球形颗粒，随固含量的降低，喷雾温度也需要升高，进料速度和喷雾压力会同时影响到埃洛石球形的粒径大小和粒径分布范围，喷雾压力越大，平均粒径越小，进料速度越大，平均粒径越大，进料温度在 150～200 ℃，出口温度大于 70 ℃才可以正常干燥。制得的埃洛石微球经 500 ℃高温处理 2 h 后得埃洛石陶瓷微球，超声加机械搅拌后重力沉降 30 min，取上清液离心得 1～10 μm 的陶瓷微球。

---

*通信作者：欧阳静；E-mail：lhitxu@126.com；手机号：15116338598。国家自然科学基金（51774331，51304242，51374250）、湖南省自然科学基金（2017JJ0351）资助项目。

本文利用喷雾干燥技术制作埃洛石微球，并通过高温处理成陶瓷微球，通过重力沉降实现分级，后期希望通过制得的微球与磁性材料复合，通过球形颗粒的特殊形貌，实现对磁性材料吸波性能的提升。

参考文献

[1] Sun R，Lu Y，Chen K. 2009. Preparation and characterization of hollow hydroxyapatite microspheres by spray drying method [J]. Materials Science & Engineering C，29 (4)：1088 - 1092.

[2] Bastan F E，Erdogan G，Moskalewicz T，et al. 2017. Spray drying of hydroxyapatite powders：The effect of spray drying parameters and heat treatment on the particle size and morphology [J]. Journal of Alloys & Compounds，724.

[3] Wang P，Lv A，Hu J，et al. 2008. In situ synthesis of SAPO-34 grown onto fully calcined kaolin microspheres and its catalytic properties for the MTO reaction [J]. Industrial & Engineering Chemistry Research，50 (17)：12741 - 12749.

[4] Cotabarren I M，Bertín D，Razuc M，et al. 2018. Modelling of the spray drying process for particle design [J]. Chemical Engineering Research and Design，132：1091 - 1104.

天然黏土功能设计新型矿物材料的微观机制

傅梁杰[1,2*]　燕昭利[1,2]　杨华明[1,2]　胡岳华[1,2]
（1. 中南大学资源加工与生物工程学院，长沙 410083；
2. 矿物材料及其应用湖南省重点实验室，长沙 410083）

具有特殊形貌的天然矿物，作为一种廉价的基体材料，在催化、吸附和生物医学等很多领域具有潜在价值。对这类矿物进行结构改性，可以进一步提高其应用价值，然而，实现可控的微观结构的调整是这类研究的难点。

一般来说，矿物基体结构调控制备高性能材料的可以分为表面改性、结构改型和功能组装。近年来，结合实验表征与理论计算，天然黏土矿物结构特征对矿物功能材料应用性能的影响获得了微观结构上的阐释。其中，蒙脱石、高岭石、埃洛石等矿物基体或结构改型后的基体与其他功能组分进行界面复合被用于催化、储热、重金属吸附和功能负载领域。这些基于天然矿物结构的纳米材料充分利用了其矿物结构特性，展现了矿物基复合材料的潜在应用[1,2]。

图 1 以天然矿物为基体，探索了结合酸浸和改进的赝结构重整方法来功能设计形貌结构可控的介孔纳米管，并阐释其中原子尺度结构演变的机理。

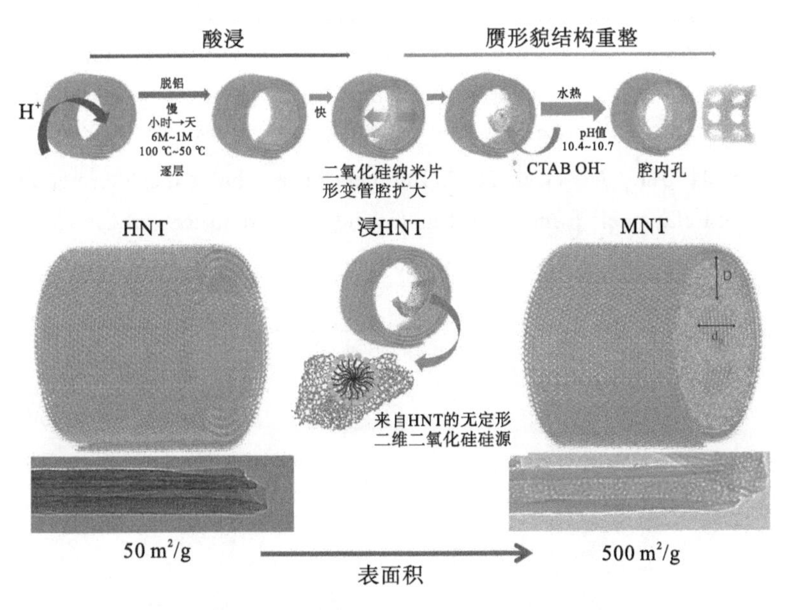

图 1　埃洛石微观结构调控功能设计管状纳米容器

---

* 通信作者：傅梁杰；E-mail：franch@csu. edu. cn；手机号：15673159391。中南大学创新驱动项目（2018CX018）资助项目。

图2以高岭石为原料，通过煅烧、酸浸和表面无机改性，得到一种表面具有大量高度分散并稳定存在的钛羟基基团的纳米矿物基杂化材料。结合理论计算，发现表面改性后的黏土矿物基体可以在稳定功能官能团和纳米结构上起到重要作用，是可以用于重金属吸附及环境恢复的新型矿物材料。

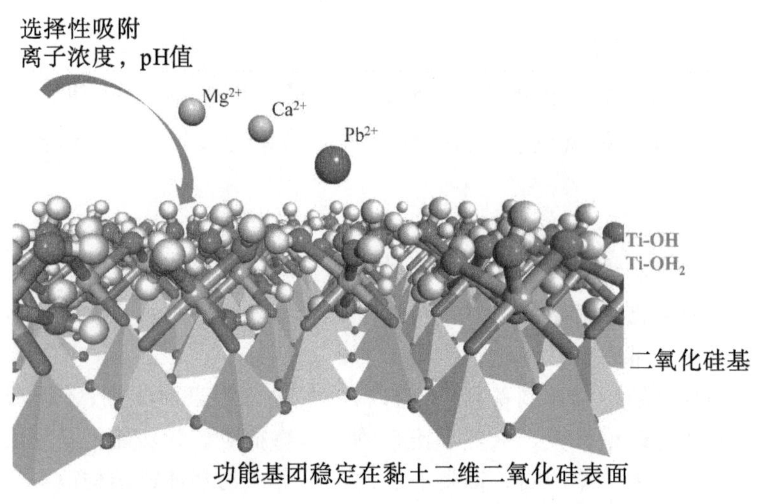

图2　高岭石功能设计杂化二维纳米吸附材料

参考文献

[1] Yan Z L，Fu L J，Yang H M. 2018. Functionalized 2D clay derivative：Hybrid nanosheets with unique lead sorption behaviors and interface structure [J]. Advanced Materials Interfaces，5（4）：1700934.

[2] Fu L J，Yang H M，Tang A，et al. 2017. Engineering a tubular mesoporous silica nanocontainer with well-preserved clay shell from natural halloysite [J]. Nano Research，10（8）：2782－2799.

g - C_3N_4/累托石复合材料及可见光光催化性能

张祥伟 孙志明* 董雄波
（中国矿业大学（北京）化学与环境工程学院，北京 100083）

传统的污水处理方式，如吸附、超滤、生物处理等方式往往成本较高，易造成二次污染。半导体光催化技术因其高效、绿色环保、低成本等优势，在污水处理领域具有良好的应用前景[1,2]。近年来，石墨相氮化碳（g - C_3N_4）因其禁带宽度较小（2.7 eV）、吸收光谱范围较宽等优势成为光催化领域的研究热点。但传统的高温聚合工艺所制备的材料易团聚，纳米片堆积造成反应活性位点减少及电子—空穴对复合速率较高，极大地限制了其在水处理领域的实际应用[3]。以来源丰富、成本低廉的天然矿物为光催化剂载体进行复合改性，可以显著提高光催化剂的分散性及对污染物的吸附捕捉性能，进而有效提高材料的光催化性能[4]。本文选用成本低廉、来源广泛的累托石作为 g - C_3N_4载体来制备 g - C_3N_4/累托石复合光催化材料。

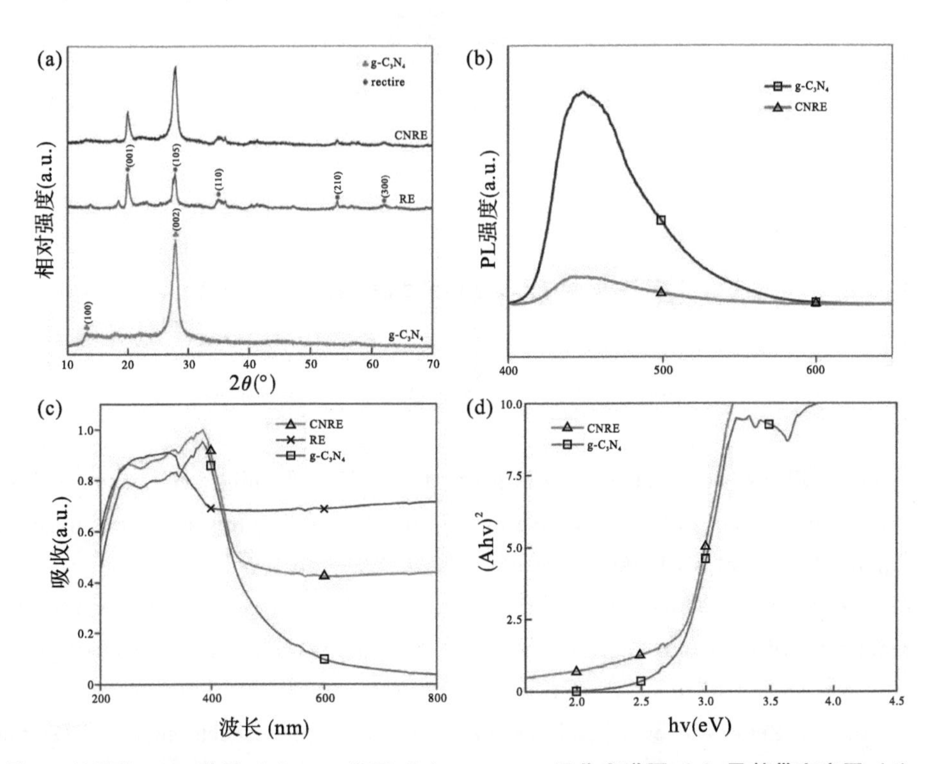

图 1 材料的 XRD 谱图（a）、PL 谱图（b）、UV - Vis 吸收光谱图（c）及禁带宽度图（d）

* 通信作者：孙志明；E-mail：zhimingsun@ cumtb. edu. cn；手机号：13466774499。国家自然科学基金（51504263）资助项目。

累托石、g－C_3N_4及 g－C_3N_4/累托石复合材料（CNRE）的 XRD 谱图见图 1（a）。复合材料中同时出现了累托石和 g－C_3N_4的特征衍射峰，且二者特征峰均未发生显著偏移，表明累托石和 g－C_3N_4实现了有效复合。图 1（b）为 g－C_3N_4和 g－C_3N_4/累托石复合材料的 PL 谱图。由图 1（b）可知，复合材料的 PL 谱相对强度低于 g－C_3N_4，表明复合材料中 g－C_3N_4的电子—空穴对分离率有所提高。图 1（c、d）为不同产品 UV－Vis 谱图和禁带宽度图。由图 1（c）可知，g－C_3N_4/累托石复合材料相对于纯 g－C_3N_4，在整个波长范围内都有较强的吸收能力。图 1（d）为g－C_3N_4、g－C_3N_4/累托石复合材料的禁带宽度图，由禁带宽度公式计算得到 g－C_3N_4禁带宽度为 2.78 eV，而复合材料禁带宽度降至2.73 eV。累托石、g－C_3N_4、g－C_3N_4/累托石复合材料在可见光下对罗丹明 B 的去除效果如图 2（a）所示。暗吸附 30 min，达到吸附—解吸平衡后，光反应6 h，最终 g－C_3N_4对罗丹明 B 的降解率为 55%，而 g－C_3N_4/累托石复合材料在相同条件下降解率可达 78%。这主要是因为复合材料具有更强的吸附能力和更多的反应活性位点。图 2（b）为 g－C_3N_4和复合材料降解罗丹明 B 的准一级动力学曲线，结果表明复合材料的反应速率是纯 g－C_3N_4的 1.51 倍。累托石与 g－C_3N_4的有效复合降低了 g－C_3N_4纳米片的团聚程度，增加了反应活性位点，降低了电子—空穴对的复合速率，提高了 g－C_3N_4的量子利用效率。

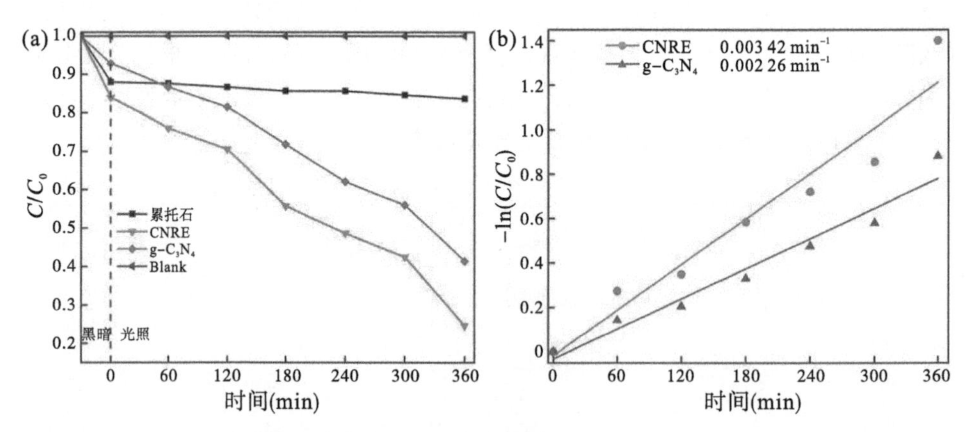

图 2　可见光下不同材料对罗丹明 B 的降解曲线（a）及光反应动力学曲线（b）

参考文献

［1］Li C，Sun Z，Zhang W，et al. Highly efficient g-C_3N_4/TiO_2/kaolinite composite with novel three-dimensional structure and enhanced visible light responding ability towards ciprofloxacin and S. aureus ［J］. Applied Catalysis B Environmental，220：272－282.

［2］Maeda K，Wang X，Nishihara Y. Photocatalytic activities of graphitic carbon nitride powder for water reduction and oxidation under visible light ［J］. Journal of Physical Chemistry C，113：4940－4947.

［3］Li Y，Zhou Y，Zeng W，et al. Acid-exfoliated g－C_3N_4 nanosheets coated silver nanoparticles with tunable loading：an efficient catalyst for visible light photocatalytic reaction ［J］. Chemistryselect，2：9947－9952.

［4］Zhang G，Sun Z，Duan Y，et al. Synthesis of nano－TiO_2/diatomite composite and its photocatalytic degradation of gaseous formaldehyde ［J］. Applied Surface Science，412：105－112.

离子液稀土 Ce 掺杂 ZnO 微纳米材料光催化性能

罗锡平[1,2*]　杨胜祥[1,2]

（1. 浙江省林业生物质化学利用重点实验室，杭州 311300；

2. 浙江农林大学理学院，杭州 311300）

离子液从绿色化学溶剂和催化领域迅速扩展到功能材料、光热材料和光电材料等领域，并极大地影响这些领域的发展。在半导体光催化剂中掺杂适当的金属离子能够有效地调控其电子能态结构，并改变其表面状态，进而起到提高光催化性能的作用。金属或非金属掺杂对 ZnO 的晶体结构、能带结构、表面微结构和紫外、可见光催化活性产生重大影响[1-3]。

本文选择稀土元素 Ce 对不同形貌的 ZnO 进行掺杂，利用粉末 X 射线衍射（XRD）、扫描电镜（SEM）、紫外—可见（UV－Vis）光谱等技术对制备光催化剂进行表征，探讨其对亚甲基蓝的光催化降解性能。2－甲基咪唑和锌离子构筑的 ZIF－8 金属有机框架合成微纳米材料（ZnO－A）：先将溶剂热法制得的 ZIF－8 样品放入马弗炉中，然后通入流速为 V＝80 cc/min 的氮气，以升温速率为3 ℃/min将样品从常温加热到 5 000 ℃并保温 1 h，自然冷却至室温，将得到的黑色粉末再次放入马弗炉中，并通入流速为 V＝80 cc/min 的氧气，以升温速率为 30 ℃/min 将样品从常温加热到3 500 ℃并保温 1 h，自然冷却至室温，所得黑色粉末命名 ZnO－A。

ZnO－B 微纳米材料合成：利用直接沉淀法制备 ZnO 纳米丛（nanobushes，ZNB），在搅拌条件下将 50 mL 0.1 mol/L ZnCl 溶液缓慢滴加到 110 mL 0.1 mol/L $NaHCO_3$溶液中，在室温(20 ℃)下继续搅拌 10 h 后抽滤，沉淀经水洗和干燥，得到的前驱体再置于马弗炉中 200 ℃的空气氛围下煅烧 3 h，即得 ZNB，标记为 ZnO－B。稀土 Ce 掺杂的 ZnO 微纳米材料的制备：选择催化降解性能较好的 ZnO－B 微纳米材料，进行稀土 Ce 掺杂。100 mL 5.0×10^{-2} mol/L 的硝酸锌溶液中加入 44 mg 六水硝酸铈，在搅拌条件下滴加一定量浓氨水，继续搅拌数小时，抽滤，洗涤，干燥，将得到的前驱体置于马弗炉中 500 ℃的空气氛围下煅烧 2 h，即得 Ce 掺杂 ZnO 微纳米材料。

HNTs 负载掺氮纳米氧化锌的制备及其光催化性能研究。称取 1 g 预处理过的 HNTs 粉末放入 120 ℃恒温箱中恒温，然后在恒温水浴条件下配制 25 ℃下硝酸锌（六水）的饱和水溶液，并不断搅拌条件下加入 1 g 尿素配成混合溶液。将混合溶液逐滴滴加在刚取出的热的 HNTs 上，使其充分吸收，以 HNTs 刚好湿润为佳。将充分吸收了硝酸锌溶液的 HNTs 放入 150 ℃恒温箱中恒温反应一段时间，并重复此操作 2～3 次，最后将其放入马弗炉分别在 400、500 ℃下锻烧 3 h。吸附柱内填充一定质量和高度的 HNTs 复合颗粒，亚甲基蓝溶液通过恒流蠕动泵，以均匀的速度从吸附柱的上端流入，下端连接到自动收集器，每隔一定时间自动取样，同样采用分光光度计检测。在暗室中用 0.02 mg/mL 亚甲基蓝溶液浸泡 30 min 达到吸附/脱附平衡后，置于紫外灯下，每

* 通信作者：罗锡平；E-mail：luoxiping126@126. com；电话：0571-63740810，手机号：15968882838。国家 948 项目（2014-4-29）资助。

隔 15 min 取 3 mL 样品。

2 种 ZnO 材料的光催化降解性能（表 1）显示：在其他条件不变的情况下，ZnO－B 对亚甲基蓝的光催化降解率高于 ZnO－A。

表 1　两种 ZnO 吸光度与降解率

时间（min）	ZnO－A		ZnO－B	
	吸光值 A	降解率 η（%）	吸光值 B	降解率 η（%）
0	$A_0=1.948$	0	$B_0=1.752$	0
15	$A_1=1.558$	20	$B_1=1.293$	25
30	$A_2=1.383$	29	$B_2=1.173$	32
45	$A_3=1.266$	35	$B_3=1.103$	37
60	$A_4=1.110$	43	$B_4=0.963$	45
75	$A_5=0.974$	50	$B_5=0.605$	54

掺杂稀土 ZnO 对亚甲基蓝降解性能影响（图 1）显示，掺杂铈的 ZnO 比未掺杂铈的 ZnO 对亚甲基蓝的光催化降解性能好。因此，稀土 Ce 掺杂的 ZnO－B 作为催化剂大幅度提高了对亚甲基蓝的光催化降解性。HNTs 负载掺氮纳米氧化锌光催化性能（图 2）显示，负载的 ZnO 相对于未负载的 ZnO 对亚甲基蓝的降解率更高，催化降解效果更好。

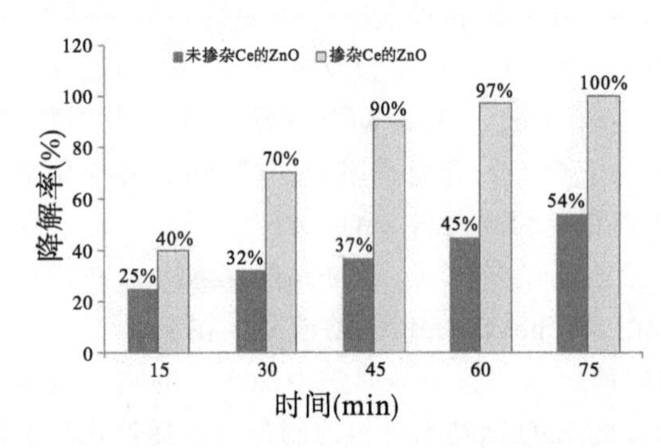

图 1　未掺杂铈 ZnO 与掺杂铈 ZnO 的降解性能比较

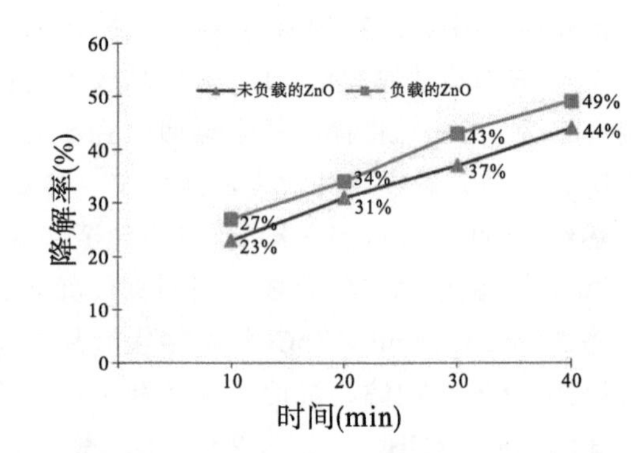

图 2　未负载 ZnO 与负载 ZnO 的降解率

在半导体光催化剂中掺杂适当的金属离子能够有效地调控其电子能态结构，提高光催化性能。离子液为新型材料研制的绿色溶剂，为蒙脱土改性处理，研制新型功能化材料提供了依据。

参考文献

［1］Liu W，Zhao Y J，Zeng C F，et al. 2017. Microfluidic preparation of yolk/shell ZIF-8/alginate hybrid microcapsules from pickering emulsion［J］. Chemical Engineering Journal（Amsterdam，Netherlands），307：408－417.

［2］Hai Y J，Liu J J，Cheng K，et al. 2013. Enhanced visible light photocatalysis of Bi_2O_3 upon fluorination［J］. J. Phys. Chem. C. 117：20029－20036.

［3］Zhang H，Guo L H，Zhao L X，et al. 2015. Switching oxygen reduction pathway by exfoliating graphitic carbon nitride for enhanced photocatalytic phenol degradation［J］. J. Phys. Chem. Lett. 6：958－963.

Comparative Study on CdS/clay Mineral Nanocomposites for Photocatalytic Degradation of Congo Red

Xiao Wen Wang[1,2,3]　Bin Mu[1,3*]　Yu Ru Kang[1,3]　Ai Qin Wang[1,3*]

(1. Key Laboratory of Clay Mineral Applied Research of Gansu Province, Center of Eco-Materials and Green Chemistry, Lanzhou Institute of Chemical Physics, Chinese Academy of Sciences, Lanzhou 730000, P. R. China;
2. University of Chinese Academy of Sciences, Beijing 100049, P. R. China;
3. Center of Xuyi Palygorskite Applied Technology, Lanzhou Institute of Chemical Physics, Chinese Academy of Sciences, Xuyi 211700, P. R. China)

Due to natural micro/nano-sized structures, high specific surface area and superior physico-chemical properties, clay minerals have attracted increasingly attention for synthesis of CdS/clay mineral nanocomposites to improve the aggregation of CdS nanoparticles and the recombination of photoexcited charge carriers of CdS[1]. However, the relevant studies was mainly focused on the single clay mineral to construct CdS/clay mineral photocatalysts and investigate the photocatalytic degradation of organic dyes, and it is scarcely reported the comparative study on the photocatalytic properties of CdS/clay mineral nanocomposites prepared using different clay minerals. In this work, CdS/clay mineral nanocomposites were prepared for photocatalytic degradation of Congo red (CR) using different clay minerals of montmorillonite (MMT) and Kaolin (Kal), halloysite nanotubes (HNTs) and sepiolite (Sep), and the photocatalytic properties of photocatalysts derived from different clay minerals were comparatively studied. HNTs, Kal, Sep and MMT were procured from Zhengzhou Las Vegas sun ceramics Co., Ltd (Henan, China), Long Yan Kaolin Clay Co., Ltd (Fujian, China), Yixian Dazhi thermal insulation materials Co., Ltd (Hebei, China), and Inner Mongolia Ningcheng Chemical Co., Ltd (China), respectively. Thiourea (NH_2CSNH_2), Cadmium chloride hydrate ($CdCl_2 \cdot 2.5H_2O$) and CR were purchased from Tianjin Kemiou Chemical Reagent Co., Ltd and Shanghai Sanyou Reagent Factory, Tianjin Kemiou Chemical Reagent Co., Ltd (Tianjin, China), respectively. CdS/clay mineral nanocomposites were prepared via an effective and simple hydrothermal decomposition of $[Cd(NH_2CSNH_2)_2]Cl_2$ in the presence of clay minerals[2]. Photocatalytic activities from an aqueous solution using the as-prepared CdS/clay mineral photocatalysts were systematically evaluated by photocatalytic degradation of CR

* Corresponding authors. E-mail addresses: mubin@licp.cas.cn (B. Mu) and aqwang@licp.cas.cn (A. Wang); Tel: +86 931 4868118. The authors are grateful for financial support of the Major Basis Research Projects of Gansu Provincial Science and Technology Department, China (18JR5RA001).

under visible light irradiation. As shown in Fig. 1, the decomposition percentage increased gradually with increase in the visible light irradiation time for different CdS/clay mineral samples. However, the distinct differences were observed after incorporating of different clay minerals. The degradation ratio of CR reached 94.3%, 82.2%, 93.5% and 85.4% for HNTs/CdS, Kal/CdS, Sep/CdS and MMT/CdS samples at 80 min, respectively (Fig. 1a), and the orders of k_{app} values were HNTs/CdS (0.033 70 min^{-1}) >Sep/CdS (0.028 80 min^{-1}) >APT/CdS (0.027 50 min^{-1}) >MMT/CdS (0.016 44 min^{-1}) >Kal/CdS (0.013 85 min^{-1}) (Fig. 1b and Fig. 1c), which were in accordance with photocatalytic activities of the samples. Furthermore, the fading of liquid supernatants was obvious when the irradiation time was up to 30 min (Fig. 1d and insert). It indicated that photocatalytic degradation activities for CR of CdS/clay mineral nanocomposites prepared with two-dimensional clays were superior to that of one-dimensional clay minerals.

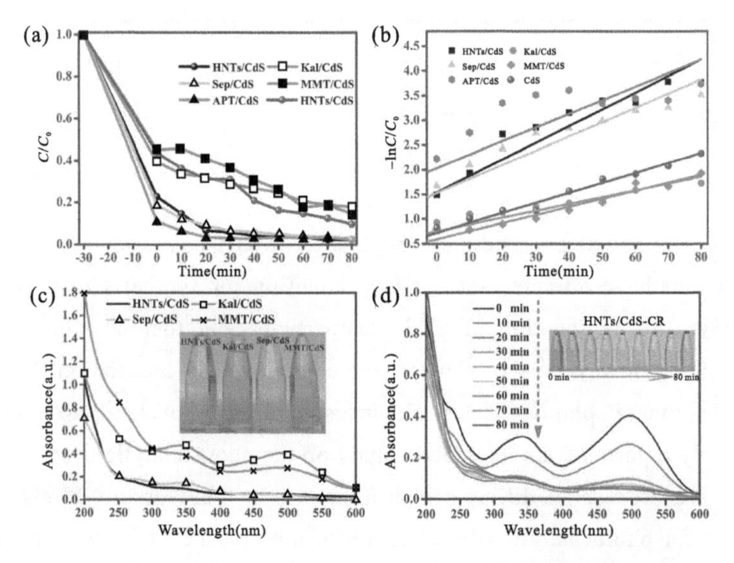

Fig. 1 Photocatalytic activity of clay mineral/CdS nanocomposites

Note: (a) Plot of (C/C_o) vs. t; (b) Plot of $-\ln$ (C/C_o) vs. t; (c) and (d) Absorption spectra of CR after incorporation of clay mineral/CdS samples under visible light irradiation for 80 min and different time, respectively

References

[1] Vinokurov V A, Stavitskaya A V, Ivanov E V, et al. 2017. Halloysite nanoclay based CdS formulations with high catalytic activity in hydrogen evolution reaction under visible light irradiation [J]. ACS Sustainable Chemistry & Engineering, 5: 11316 - 11323.

[2] Wang X, Mu B, An X, et al. 2018. Insights into the relationship between the color and photocatalytic property of attapulgite/CdS nanocomposites [J]. Applied Surface Science, 439: 202 - 212.

Surface Modified Sepiolite Nanofibers as A Novel Lubricating Oil Additive

Fei Wang[1,2*] Ting Ting Zhang[1,2] Jin Sheng Liang[1,2] Bai Zeng Fang[3]
Hui Min Liu[1,2] Pei Zhang Gao[1,2]

(1. Key Laboratory of Special Functional Materials for Ecological Environment and Information, Hebei University of Technology, Ministry of Education, Tianjin 300130, China;
2. Institute of Power Source and Ecomaterials Science, Hebei University of Technology, Tianjin 300130, China;
3. Department of Chemical & Biological Engineering, University of British Columbia, 2360 East Mall, Vancouver, B. C. V6T 1Z3, Canada)

Formechanical equipments, small size, light weight, high power, high efficiency, high reliability and environmentally friendly features are highly desired, and the demands and requirements for lubricating oils and additives are increasing yearly[1,2]. Nowadays, traditional lubricating oils still dominate in the market, inspite of many limitations under the conditions of high load, high speed, long-term and harsh operation. Lubricating oils with excellent anti-friction and anti-wear properties can significantly reduce the friction of machinery, and thus leading to the effective improvement of its reliability, service life and energy consumption reduction. With the increasing awareness of sustainable development in the environment, environmental pollution caused by lubricating oils has drawn wide attention. Traditional lubricating oils and additives have poor biodegradability and toxicity accumulation, leading to the serious pollution of the land, rivers and lakes[3].

In this work, sepiolite mineral materials were adopted as a lubricating oil additive, and the effects of the surface properties and the added amount of the sepiolite nanofibers on the performance of lubricating oils were investigated systematically. It was noted that the addition of an appropriate amount of the sepiolite nanofibers could significantly enhance the performance of the lubricating oils.

The effects of the sepiolite addition amount on the lubricating oil sample viscosity at 40 ℃ and 100 ℃ are illustrated in Fig. 1 (A) and (B), respectively. It is evident that the oil samples containing the different amounts of sepiolite nanofibers have the same change tendency of viscosity at 40 ℃ and

* Corresponding author: Fei Wang; E-mail: wangfei@hebut. edu. cn. This work was financially supported by Excellent Young Scientist Foundation of Hebei Province, China (No. E2018202241).

100 ℃, and the viscosity increased with the increase in the addition amount of the sepiolite nanofibers. Specifically, the viscosity of the lubricating oil with 2% sepiolite addition was much higher than that of the control sample, which is favorable for the lubricating oil utilization under normal operating conditions. In the Fig. 2, the copper corrosion levels of sample a, b and e are 2c as the moderate discoloration, while the level of the lubricating oil containing 1% and 1.5% modified sepiolite nanofibers (sample c and d) is 1b and 1a as the mild discoloration. Specifically, the sample d looks like a newly polished copper, which is mainly attributed to the best improvement in the dispersion and stability in the lubricating oil by the addition amount optimization for the modified sepiolite nanofibers.

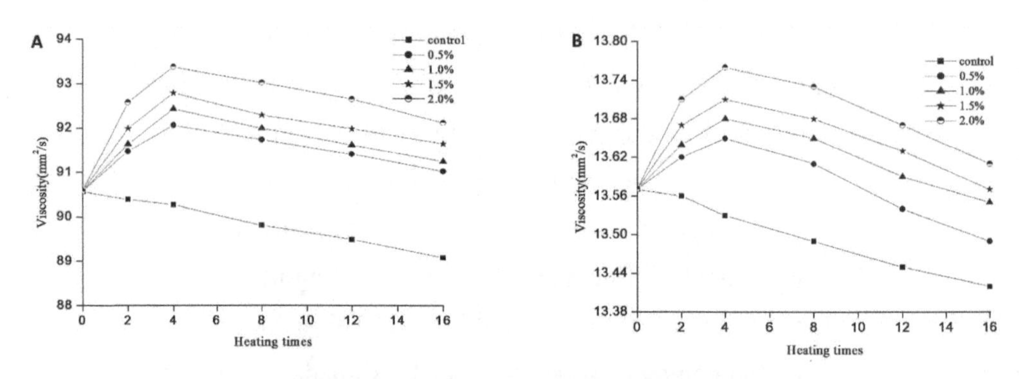

Fig. 1 Effects of the sepiolite addition amount on the lubricating oil sample viscosity at 40 ℃ (A) and 100 ℃ (B)

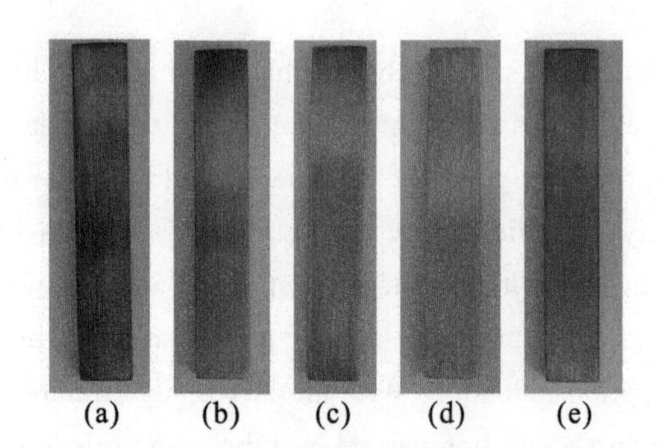

Fig. 2 Effect of the sepiolite addition amount on the oil sample copper corrosion

Note: a. Control; b. 0.5%; c. 1%; d. 1.5%; e. 2%

References

[1] Zhu X, Zhong C, Zhe J. 2017. Lubricating oil conditioning sensors for online machine health monitoring: A review [J]. Tribol. Int., 109: 473 - 484.

[2] Dai W, Kheireddin B, Gao H, et al. 2016. Roles of nanoparticles in oil lubrication [J]. Tribol. Int., 102: 88 - 98.

[3] Zubaidi I A, Tamimi A A. 2018. Soil remediation from waste lubricating oil [J]. Environ. Technol. Inno., 9: 151 - 159.

Removal of Cd^{2+} from Aqueous Solution Using Sepiolite Mineral Nanofibers

Fei Wang[1,2*]　Mao Mao Zhu[1,2]　Jin Sheng Liang[1,2]　Bai Zeng Fang[3]
Zeng Yao Shang[1,2]　Li Cui[1,2]

(1. Key Laboratory of Special Functional Materials for Ecological Environment and information, Hebei University of Technology, Ministry of Education, Tianjin 300130, China;
2. Institute of Power Source and Ecomaterials Science, Hebei University of Technology, Tianjin 300130, China;
3. Department of Chemical & Biological Engineering, University of British Columbia, 2360 East Mall, Vancouver, B. C. V6T 1Z3, Canada)

Nowadays, industrial wastewater treatment has become one of the most important solutions to solve environmental problems, especially in the field of waste water containing heavy metal ions. Adsorption method as a simple and effective way has been widely used in the removal of heavy metal ions, and activated carbon materials often used as adsorbent has disadvantage of high cost and environmental pollution[1]. Sepiolite is a type of fibrous silicate clay mineral rich of magnesium with a unit cell formula of $Si_{12}O_{30}Mg_8(OH)_4(H_2O)_4 \cdot 8H_2O$ and fine microporous channels of dimensions of 0.37 nm×1.06 nm running parallel to the direction of fibers. Benefitting from its large reservation, low cost and reusable attribute, sepiolite has been intensively utilized in the field of environmental improvement[2,3].

In this work, sepiolite mineral nanofibers were prepared by microwave chemical method under different microwave power, and the Cd^{2+} removal performance of sepiolite nanofibers were also studied systematically.

Fig. 1 shows the effect of sepiolite mineral nanofibers on the adsorption capacity of cadmium ions in the dosage of MH-SEP range of 2—8 g/L. In Fig. 1, Cd^{2+} adsorption rate increases obviously in the initial stage along with the increase of sepiolite mineral nanofibers dosage, and the adsorption tends to balance when the addition amount exceeds 5 g/L. The adsorption rate of Cd^{2+} is increased from 59.8% to 70.8%, and the equilibrium adsorption capacity is decreased by 29.9 mg/g.

The fitting parameters are shown in Table 1. In Table 1, the adsorption capacity $q_{e(cal.2)}$ of the pseu-

* Corresponding author: Fei Wang; E-mail: wangfei@hebut. edu. cn. This work was financially supported by Excellent Young Scientist Foundation of Hebei Province, China (No. E2018202241) and National Natural Science Foundation of China (Grant No. 51404085).

do-second-order equation increases gradually along with the temperature increase, and the pseudo-second rate constant increases first and then decreases, which indicates that the increase of temperature is propitious to overcome the adsorption mass transfer resistance, leading to the migration acceleration of cadmium ions on the surface and pore channels of sepiolite mineral nanofibers.

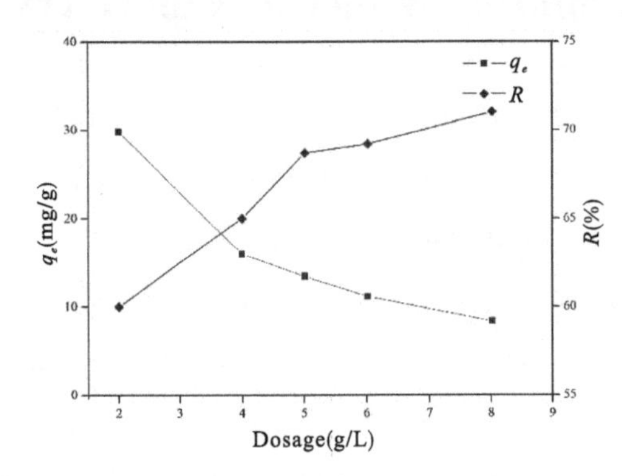

Fig. 1 Effect of dosage of sepiolite mineral nanofibers on the adsorption of Cd^{2+}

Table 1 Pseudo-first-order, pseudo-second-order kinetics and intraparticle diffusion fitting parameters at different temperature (pH=5.0, m/V=5 g/L, C_0=100 mg/L)

Kinetics	Parameter	298 K	308 K	318 K
Pseudo-first-order	$a_{e(exn.)}$ (mg/g)	13.72	15.18	17.39
	$a_{e(cal.1)}$ (mg/g)	4.22	4.72	4.56
	k_1 (min^{-1})	0.056	0.063	0.051
	R^2	0.879 9	0.911 2	0.818 7
Pseudo-second-order	$a_{e(cal.2)}$ (mg/g)	13.89	15.30	17.52
	k_2 (g/mg/min)	0.058	0.067	0.058
	R^2	0.999 3	0.999 4	0.999 3
Intraparticle diffusion	k_1 ($mg/g/min^{1/2}$)	0.852	0.778	0.900
	R^2	0.877 1	0.872 7	0.871 5
	k_2 ($mg/g/min^{1/2}$)	0.120	0.106	0.122
	R^2	0.893 5	0.894 3	0.896 3

References

[1] Lelifajri R, Nurfatimah R. 2018. Preparation of polyethylene glycol diglycidyl ether (PEDGE) crosslinked chitosan/activated carbon composite film for Cd^{2+} removal [J]. Carbohydrate Polymers, 199: 499 - 505.

[2] Li Y, Wang M, Sun D, et al. 2018. Effective removal of emulsified oil from oily wastewater using surfactant-modified sepiolite [J]. Applied Clay Science, 157: 227 - 236.

[3] Al-Ani A, Gertisser R, Zholobenko V. 2018. Structural features and stability of Spanish sepiolite as a potential catalyst [J]. Applied Clay Science, 162: 297 - 304.

From Natural Minerals to Low-dimensional Functional Nanomaterials: The Case of Molybdenite and Flake Graphite

De Liang Chen* Jia Heng Li Hui Na Dong Kai Wang Zhao Wu Wang
(School of Materials Science and Engineering, Zhengzhou University,
Zhengzhou 450001, People' s Republic of China)

Few-layered MoS_2 (FL-MoS_2) and graphene have attracted more and more wide attention due to their unique physical properties and wide application in electronics, energy storage, sensor and catalysis. The available preparation methods can be categorized into two types: 'bottom-up' and 'top-down'. The 'bottom-up' method is the common chemical synthetic process, but sometimes complex. The 'top-down' process can prepare 2D nanocrystals through the exfoliation of layered solids with an advantage of low cost. This presentation reports a synthesis of FL-MoS_2 nanoplates and graphene via the synergistic effect of multi forces (ultrasonication-milling (U&M)) by exfoliating natural layered solids (e. g., molybdenite and flake graphite). The properties of the as-obtained FL-MoS_2 nanoplates and graphene in hydrogen evolution reaction (HER) is also investigated. This direct exfoliation of natural layered solids is an attractive method for large-scale preparation of 2D nanoplates[1].

Using NMP asthe exfoliation solvent, the yield of the FL-MoS_2 nanoplates reaches 21.6%, under the following situations: the initial molybdenite concentration=45 g/L, ultrasonic power=280 W, rotation speed of sand mill=2 250 r/min, exfoliation time=6 h and the molar PVP-unit-to-MoS_2 ratio=0.5. The thickness of the FL-MoS_2 nanoplates is about 2—3 nm according to AFM data, and its surface specific area is up to 923 m^2/g. Typical results are shown in Figs. 1—3. In addition, graphene was also prepared by exfoliating natural flake graphite via the U&M approach using ethanol /water mixture (mass ratio=1 : 1) as the exfoliation media. Using tween-80 as the surfactant, and the initial concentration of graphite is 15 g/L, the exfoliation yield of the graphene with a thickness of 5—7 nm reaches 7.5%.

* Corresponding author: De Liang Chen; E-mail: dlchennano@qq. com; Mobile Phone No.: +86-13592545475. Acknowledge: National Natural Science Foundation of China (NSFC No. 51172211, 51574205), Program for Science and Technology Innovation Talents in Universities of Henan Province (14HASTIT011), and Plan for Scientific Innovation Talent of Henan Province (154100510003).

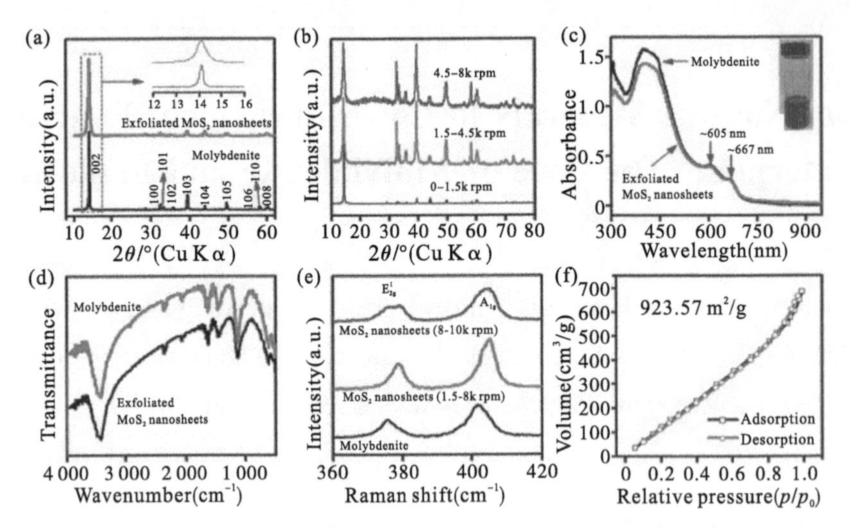

Fig. 1 Composition characterization of MoS_2 nanoplates exfoliated from molybdenite via the coupled ultrasonication-milling approach

Note: a. XRD patterns; b. XRD patterns; c. UV-vis spectra before and after standing for 42 days; d. FT-IR spectra; e. Raman spectra; f. N_2 adsorption-desorption isotherms

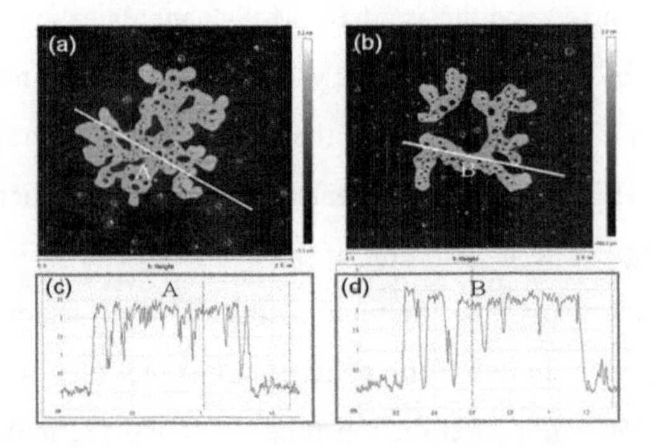

Fig. 2 Typical atomic force images of MoS_2 nanoplates exfoliated from molybdenite via the coupled ultrasonication-milling approach

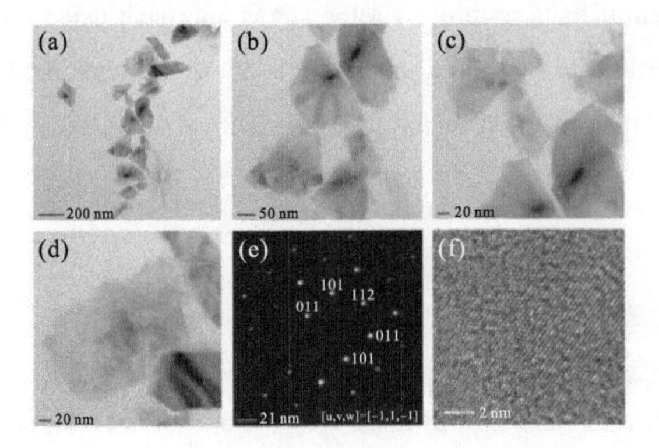

Fig. 3 Typical TEM observations of MoS_2 nanoplates exfoliated from molybdenite via the coupled ultrasonication-milling approach

References

［1］Dong H，Chen D，Wang K，et al. 2016. High-yield preparation and electrochemical properties of few-layer MoS_2 nanosheets by exfoliating natural molybdenite powders directly via a coupled ultra-sonication-milling process［J］. Nanoscale Research Letter，11（1）：409.

第五章

非金属矿和新型化工、建材、能源

NaY 沸石/珍珠岩复合材料的合成和性质

申宝剑[1*] 王闻年[1] 郭巧霞[2]

（1. 中国石油大学（北京）重质油国家重点实验室，中国石油天然气集团公司催化重点实验室，化学工程学院，北京 102249；

2. 中国石油大学（北京）重质油国家重点实验室，理学院，北京 102249）

膨胀珍珠岩中含有约 70wt.%的氧化硅、15wt.%的氧化铝，其余为钠、钾等金属阳离子。它是在火山喷发过程中瞬间高温膨胀形成的无定形矿物，硅铝物种活性较高，是制备沸石材料的潜在矿物原料。本文利用膨胀珍珠岩片状结构，通过水热合成法制备了具有单层分布的小晶粒 Y 沸石/珍珠岩复合材料，对其性质进行了表征研究，并以其改性产物为活性组分制备了催化裂化催化剂，研究了其催化性能。

NaY 沸石/珍珠岩复合材料的合成：首先制备合成 NaY 的导向剂，之后依次加入导向剂、水玻璃、膨胀珍珠岩，混合搅拌均匀后，以 NaOH 调节体系碱度。成胶完成后，装入晶化釜中进行水热晶化一定时间。晶化完成后，过滤、洗涤、干燥得到钠型产品。样品命名为 NaY/perlite。

NaY/perlite 改性：按照 NaY 沸石改性常规的“两交两水热”方法进行。所得样品命名为 USY/perlite。作为对比，将普通 NaY 沸石按照相同的处理方法制备，并将之命名为 USY。

NaY/珍珠岩复合材料的合成中研究了一系列合成化学条件（如碱度、硅铝比、晶化温度等）对复合材料晶化的影响，对合成过程进行了跟踪研究。不同晶化时间所得样品的相对结晶度见图 1。该 NaY 合成体系的晶化以典型的 S 型曲线进行，晶化诱导期在 2 h 左右，比常规 NaY 的晶化诱导期明显缩短；快速生长期在 2～10 h，明显比常规 NaY 的快速生长期长；最后 12～24 h 样品的相对结晶度增加非常缓慢，晶化 20 h 后，样品的相对结晶度不再增加，保持在 70%左右。XRD 表征结果表明，所得样品主要为纯相的 NaY 沸石，因为一部分珍珠岩没有转化成 NaY，所以产品中 NaY 相对结晶度不高。这也是本工作的特点，因为我们留下一定量的未转化（晶化）的珍珠岩作为基体，来保持复合材料的特殊形貌。

NaY/perlite 复合材料的 SEM 电镜图片列于图 2，复合材料中 NaY 沸石的晶粒以单层的形式生长在膨胀珍珠岩表面，大小在 200～300 nm，NaY/perlite 样品片状颗粒在 5～10 μm，膨胀珍珠岩自身几十微米的大片状颗粒在晶化后期已经观察不到。珍珠岩自身的 BET 比表面积仅为 10 m^2/g，晶化一定时间后，比表面积可达 626 m^2/g，比表面积的增加主要来自 Y 沸石的贡献。孔体积尤其是微孔体积的增加，更是直接证明了 Y 沸石的生成。样品的总表面积与微孔比面积之差接近 90 m^2/g，外表面积较高，也说明样品中 Y 沸石的晶粒较小。更有意思的是，普通 NaY 的振实密度为 0.400 g/cm^3，显著高于 NaY/perlite 复合材料的振实密度（0.293 g/cm^3），这是由珍珠岩的特性带来的特点。

---

* 通信作者：申宝剑，E-mail：baojian@cup. edu. cn；电话：13601399095。国家自然科学基金（U1462202）。

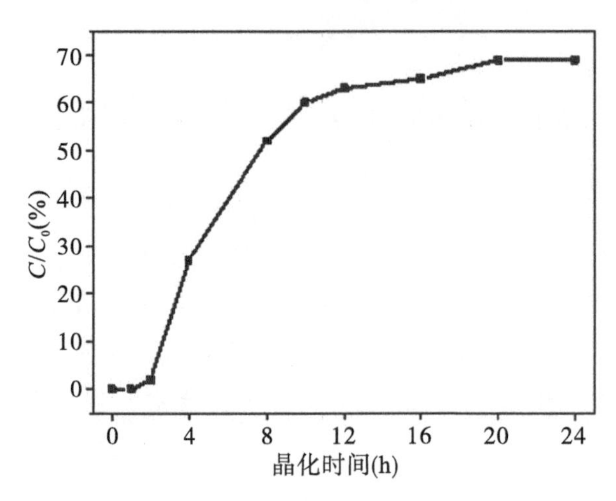

图 1　NaY/珍珠岩复合材料的晶化曲线

图 2　NaY/perlite 复合材料的 SEM 电镜图片

改性后得到的改性复合材料 USY/perlite 样品的形貌几乎没有变化。Y 沸石仍旧以准纳米大小晶粒生长在膨胀珍珠岩表面，而且 Y 沸石晶粒能够稳定分布，没有发生团聚。而且，DSC 表征证明由于这种附晶生长模式，导致 Y 沸石的耐热和水热稳定性都明显增加。

以 USY/perlite 复合材料和 USY 为主剂制备的催化剂催化裂化评价结果表明，以重油（VGO）为原料时，与催化剂 Cat（USY）相比，催化剂 Cat（USY/perlite）具有更低的干气、焦炭产率，油浆产率虽然提高了 1.6wt.%，但是柴油产率相当，汽油产率提高了 4.6wt.%。NH_3-TPD 表征结果证明 USY/perlite 不论是弱酸还是强酸酸量都低于 USY。说明其优异的催化活性和选择性源于 USY/perlite 更为发达的介孔结构。该复合材料良好的扩散性能弥补了其酸性的不足，提高了 VGO 的转化率，降低了汽油的二次裂化和焦炭的生成。这些数据表明，该类材料在催化领域中有广阔的应用前景。

石蜡/膨胀石墨复合相变储热材料的制备和性能

王　洋[1,2]　田云峰[1,2]　曾　萍[1,2]　姜凌艺[1,2]　李　珍[1,2*]
（1. 中国地质大学（武汉）材料与化学学院，武汉 430074；
2. 中国地质大学（武汉）纳米矿物材料及应用
教育部工程研究中心，武汉 430074）

随着能源危机的出现，能源的有效利用越来越受到人们的关注。热能储存技术是一种行之有效的提高能源利用率、保护环境的重要技术，使用相变材料储热可节能和提高效率[1,2]。在各种相变储热介质中，石蜡具有相变潜热大、固—液相变过程容积变化小、热稳定性好、无过冷现象和价格低廉等优点，成为中、低温储热技术中应用广泛的相变储热材料。但是单纯的石蜡相变材料导热性能差、储热慢，在相变温度上为液体从而容易渗漏[3]。本研究采用熔融共混法制备石蜡/不同粒径膨胀石墨复合相变储热材料（图 1），对样品进行 XRD、FT－IR、SEM、DSC 和 LFA 表征分析，研究了不同粒径膨胀石墨的质量比例对复合相变储热材料性能的影响。结果表明：随着小粒径膨胀石墨含量的增加，复合相变储热材料的热扩散系数先增大后减小（图 2 和图 3）。在大小粒径膨胀石墨质量比例为 9：1 时，石蜡充分利用了大小粒径膨胀石墨的镶嵌式空间结构，复合相变材料的热扩散系数为 1.964×10^{-6} m^2/s，比纯石蜡提高了 22 倍，相变潜热为 144.2 J/g。

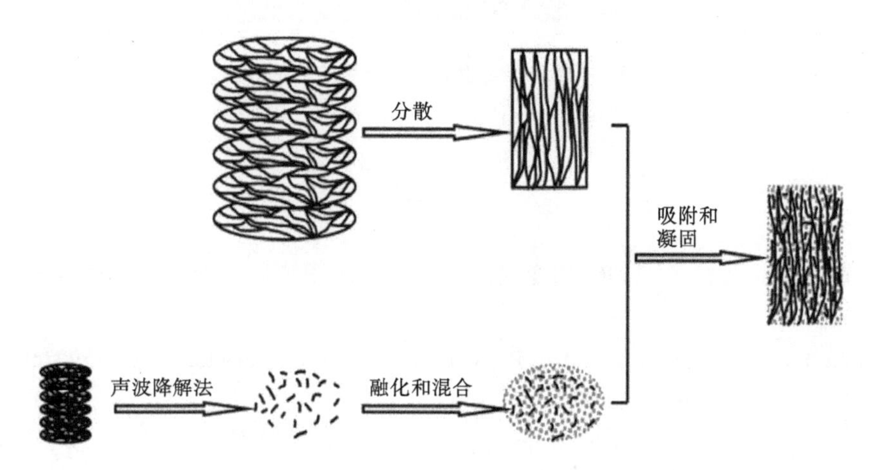

图 1　制备 EG/石蜡复合相变储热材料流程示意

* 通信作者：李珍；E-mail：zhenli@cug. edu. cn；手机号：13588066098。

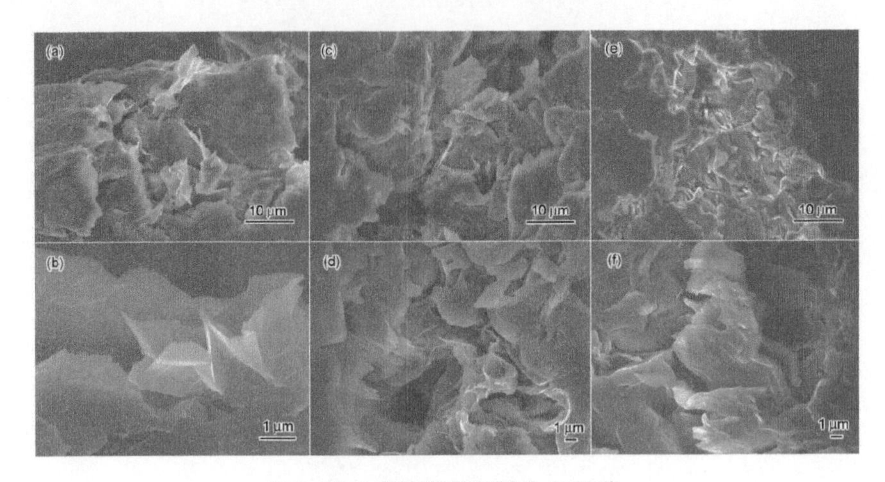

图 2　复合相变储热材料的 SEM 像

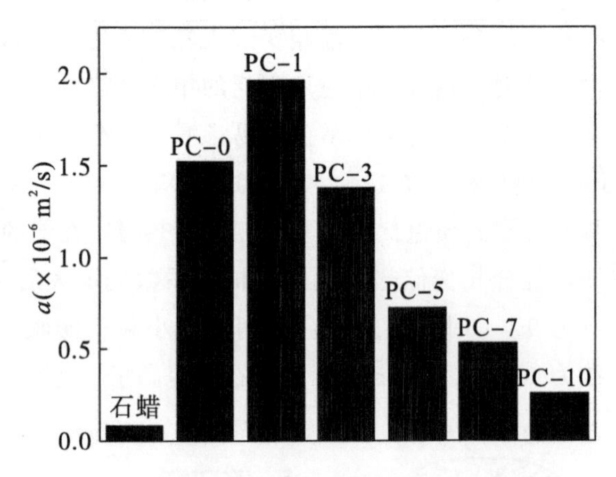

图 3　石蜡和复合相变储热材料的热扩散率

参考文献

[1] 张寅平，胡汉平，孔祥冬. 1996. 相变储热—理论与应用［M］. 合肥：中国科技大学出版社，339.

[2] Pielichowska K，Pielichowski K. 2014. Phase change materials for thermal energy storage［J］. Progress in Materials Science，65，67.

[3] 戴琴，周莉，朱月，等. 2014. 改善石蜡相变材料导热性能的研究进展［J］. 当代化工，43（7）：1257.

$CaCl_2 \cdot 6H_2O$/硅藻土复合相变储能材料制备与性能

邓　勇　杨紫娟　李金洪*

（非金属矿物与固废资源材料化利用北京市重点实验室，

中国地质大学（北京）材料科学与工程学院，北京 100083）

无机水合盐属于中低温相变储能材料，具有合适相变温度、储能密度大和潜热值高等优点，但也存在相变泄露、过冷和导热率低等问题。本研究以 $CaCl_2 \cdot 6H_2O$（CCH）为相变材料，提出利用硅藻土的多孔结构吸附定形制备复合相变储能材料（ss－PCM），添加成核剂 $SrCl_2 \cdot 6H_2O$（SCH）有效降低冷度，并添加石墨强化传热。研究结果表明，硅藻土对 CCH 封装容量为 65wt%，1wt%的 SCH 使 ss－PCM 的过冷度几乎消除（0.3 ℃），10wt%的石墨使 ss－PCM 导热率由 0.56 W/（m·K）提高到 0.95 W/（m·K），并阐明了过冷抑制机理和强化传热机制。添加成核剂和石墨后，ss－PCM 依然保持了较高储热能力：熔化温度为 29.1 ℃，熔化焓为 110.1 J/g，凝固温度为 28.8 ℃，凝固焓为 108.7 J/g。经过 200 次熔化-凝固相变循环后，ss－PCM 化学性质不发生变化，熔化和凝固焓分别降低了 0.5%和 0.2%。

相变储能材料是一种能够通过周围环境温度调节自身相变吸收环境中热量或将自身储存热量释放出来的新型功能材料，具有储能密度大、温度恒定和过程易控制等优点，可解决能量在时间和空间上不匹配的矛盾，在建筑节能、太阳能光热转换和航空航天等领域有广泛应用。水合盐相变材料在建筑节能实际应用中存在以下问题[1]：（1）相变过程形态不稳定。造成建筑基体材料的腐蚀，导致力学性能变差，增加安全隐患。（2）过冷度大。较大过冷度会导致水合盐相变材料不能在应用温度范围内结晶放热。（3）导热率低。导热率直接影响相变材料与外界环境进行热交换的能力，只有较高的导热率才能实现储能调温控制环境温度波动的效果。本研究针对以上问题，制备了过冷抑制和强化传热的 ss－PCM。

实验采用的原料主要有饱和 $CaCl_2$溶液、SCH 和石墨等。复合相变材料制备方法为熔融浸渍法。主要仪器有 SEM（HITACHI S－4800）、DSC（214 Polyma，NETZSCH）、导热系数仪（XIATECH TC 3000E）和温度数据记录仪（TOPRIE TP720）等。

添加成核剂 SCH 后，ss－PCM 的过冷大幅减小，在 SCH 添加量为 1wt%时，ss－PCM 的过冷度降低至 0.3 ℃，几乎消除（图 1）。成核剂降低了以界面能为主要障碍的成核位垒，降低了成核所需能量，为相变材料结晶提供了优先成核位置，因而有效降低了过冷度。

随着石墨添加量（x）的增加，ss－PCM 导热率（y）呈线性增加，线性拟合方程为：$y = 3.8822x + 0.5601$（$R^2 = 0.9997$），当添加 10wt%石墨后，ss－PCM 导热率提高了 70%（图 2）。

* 通信作者：李金洪；E-mail：jinhong@cugb. edu. cn；国家自然科学基金（U1607113）、青海省科技计划项目（2017-HZ-805）资助项目。

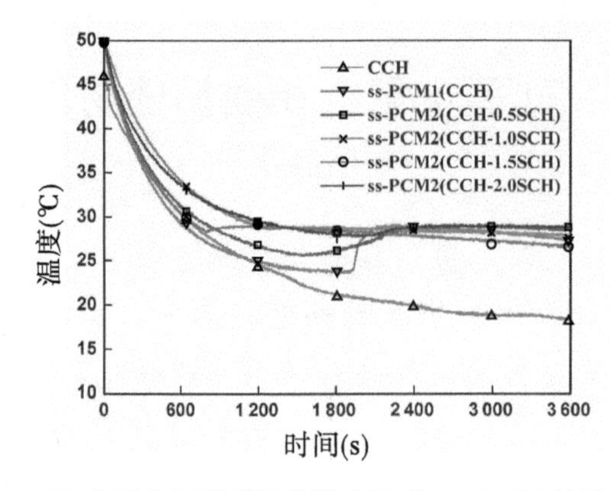

图 1　CCH 和添加不同质量分数 SCH 的 ss－PCM 的过冷曲线

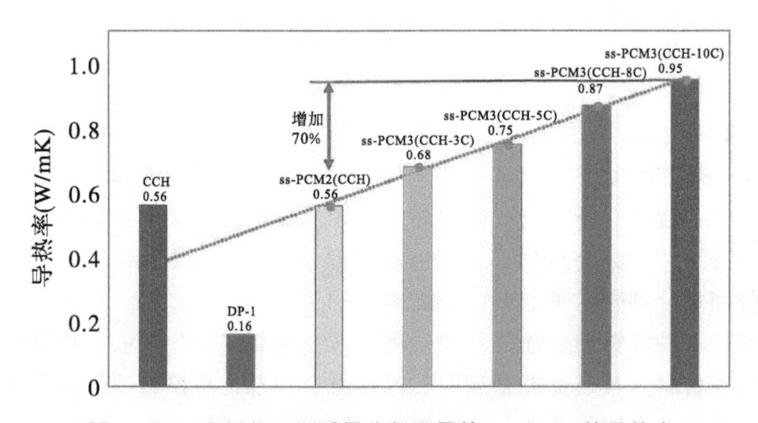

图 2　CCH 和添加不同质量分数石墨的 ss－PCM 的导热率

由于合适的相变温度、较低的过冷及改善的传热性能，制备的 ss－PCM 在建筑节能领域有较大应用前景。

参考文献

［1］Deng Y，Li J H，Deng Y X，et al. 2018. Supercooling suppression and thermal conductivity enhancement of $Na_2HPO_4 \cdot 12H_2O$/expanded vermiculite form-stable composite phase change materials with alumina for heat storage［J］. ACS Sustainable Chemistry & Engineering，6：6792-6801.

g－C_3N_4在聚光条件下的光催化还原CO_2制CH_4行为

周贤机 高志宏 卢晗锋 张泽凯*

（浙江工业大学大学化学工程学院，杭州 310032）

氮化碳（CNs）是一种较为廉价的非金属材料，拥有许多特殊的物化性质，如独特的半导电性、碱性、高硬度、高化学稳定性和高机械稳定性等。氮化碳是氮掺入碳阵列中形成，这不仅提供了一个带隙不超过 2.7 eV 的电子结构，也有助于增强其场发射、光催化和碱催化、碳捕获和能量储存等性能。这为其在许多领域的应用，如催化、储氢、二氧化碳捕集和储存、水处理、太阳能转化以及特殊污染物检测等打下了基础。如文献报道，CNs 材料的性质主要是由其结构、组成和结晶度决定，典型的氮化碳材料可能具有 5 种不同的结构，包括 1 个二维石墨碳氮化物C_3N_4（g－C_3N_4）和 4 个三维 CNs 结构，即 a－C_3N_4、b－C_3N_4、cubic－C_3N_4和 pseudocubic－C_3N_4等。在这些结构中，g－C_3N_4由于合成条件温和、结构稳定、物化性质高度可调、应用范围广泛而受到许多研究人员的关注[1,2]。

近年来，g－C_3N_4一个代表性的应用是作为光催化还原CO_2制CH_4的催化剂。作为一个一石多鸟的反应，光催化还原CO_2制CH_4可以同时达到减少大气中CO_2排放，实现碳循环和太阳能转化与储存等目标，是能源和环境领域的重要课题之一。常用的光催化还原CO_2催化剂主要为半导体材料如TiO_2等，其缺点在于带隙较宽，只能利用短波区的太阳光线；而 g－C_3N_4带隙相对较窄，可以吸收利用可见光，从而提高太阳光的吸收效率，有望成为一种良好的光催化还原CO_2催化剂。

在光催化反应过程中，光还原CO_2的反应速率与光强直接相关，有一种观点认为：

$$-r_{CO_2}=kI^a\frac{P_{H_2O}P_{CO_2}}{(1+K_{H_2O}P_{H_2O}+K_{CO_2}P_{CO_2})^2}$$

这就意味着，提高光强可以提高反应速率。为此，本文采取聚光方式提高入射光强，并讨论其对 g－C_3N_4催化还原CO_2行为的影响。

g－C_3N_4以三聚氰胺作为前驱体制备，称取所需质量三聚氰胺放入坩埚中，在马弗炉中进行热聚合。样品在N_2气氛中以 5 ℃/min 的速率从室温加热至 550 ℃并保持 4 h 得浅黄色样品即为 g－C_3N_4。使用前将 g－C_3N_4放入模具并压成不同厚度和尺寸的薄片。

反应在装有高透射率石英玻璃的釜式反应器中进行。光源为 300 W 高压 Xe 灯（CEL－HXF300，AuLight）。光强约为 177 mW/cm^2。用菲涅耳透镜（Mylens Co.，Shenzhen）作为聚光镜聚光。光线照射穿透玻璃和菲涅尔透镜并垂直照射到催化剂表面。通过改变透镜和催化剂之间的距离可以调节聚光比。CO_2和水通过反应器上的开口注入，产物采用气相色谱分析。

* 通信作者：张泽凯；E-mail：zzk@zjut. edu. cn；手机号：13968043221。

图 1 显示了 g - C_3N_4催化还原 CO_2的 CH_4产率。为了确认 CH_4实际上来源于 CO_2，首先进行了在反应器中无 CO_2并且仅含有纯 N_2和水的空白实验，以确保没有由 g - C_3N_4的杂质产生 CH_4。由图 1 可得，自然入射光下未经预处理的 g - C_3N_4的活性较差，CH_4产率仅为 0.21 μmol·g^{-1}·h^{-1}。通过聚光（聚光比＝10）可将 CH_4产量提高至约 2.49 μmol·g^{-1}·h^{-1}。产物曲线在 6 h 内近似于直线上升，表明催化剂在此反应条件下性能稳定。此外，在空气氛围中通过聚光 1 h 进行预处理，g - C_3N_4的 CH_4产率增加至约 3.39 μmol·g^{-1}·h^{-1}。采取聚光对催化剂预处理有利于 g -C_3N_4的 CO_2光还原性能的进一步提高。进一步对不同聚光比下的反应速率进行分析，可以得到光强的反应级数。

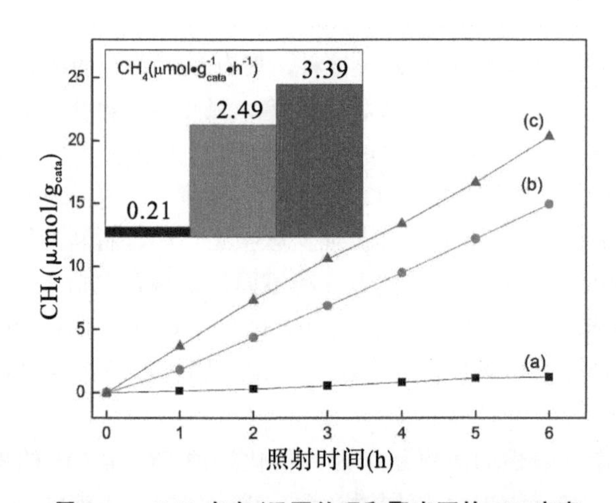

图 1　g - C_3N_4在有/无预处理和聚光下的 CH_4产率

总之，g - C_3N_4在光催化还原 CO_2领域具有较好的应用前景，而聚光技术可望促进其在该领域的进一步发展。

参考文献

［1］Lakhi K S，Park D H，Al-Bahily K，et al. 2017. Mesoporous carbon nitrides：synthesis，functionalization，and applications［J］. Chem. Soc. Rev.，46：72 - 101.

［2］Li D，Abanades S，Chen Y F，et al. 2018. Enhanced activity of TiO_2 by concentrating light for photoreduction of CO_2 with H_2O to CH_4［J］. Catalysis Communications，156：648 - 659.

Pd/膨润土催化剂的制备及液相甲醇选择氧化性能

季生福* 覃荣现　刘建芳　穆金城
（北京化工大学化工资源有效利用国家重点实验室，北京 100029）

甲醇是一种基本化工产品，是重要的有机化工原料。近几年来，随着煤化工的发展，国内甲醇行业也呈现着快速发展的局势，国内甲醇市场呈现供大于求的局面，因此促进甲醇下游产品的开发具有重要的意义。甲醇可以转化为众多的具有更高附加值的化学品，如甲醇经过一步反应可合成甲醛[1]、甲酸[2]、二甲醚[3]等，同时，甲醇又可与一步反应产物甲醛和甲酸在酸催化作用下经缩合或酯化反应进一步得到甲缩醛或甲酸甲酯。其中，甲酸甲酯被誉为万能的化学中间体，是当前 C1 化学发展的热点之一，作为 C1 化学的中间纽带越来越受到重视。传统工业上合成甲酸甲酯的方法主要有甲醇酯化法、甲醇羰基法、甲醇脱氢法、合成气直接合成等两步法，两步法的工艺路线长、成本高，因此甲醇在双功能催化剂上一步合成甲酸甲酯引起了人们的高度重视。甲醇一步氧化法制甲酸甲酯有气相氧化和液相氧化两种，气相法存在着原料甲醇浓度低的缺点，且需要用到和循环大量的惰性气体，温度也不易控制。液相甲醇氧化法具有更大的潜在研究价值，而液相甲醇选择氧化制备甲酸甲酯的关键是制备出同时具有氧化性和酸性的双功能催化剂。

Pd 催化剂对很多催化反应具有很好的催化效果，将贵金属 Pd 负载到具有酸性位点的膨润土载体上，Pd 为甲醇氧化反应提供氧化性位点，膨润土载体能为反应提供路易斯酸性位点。膨润土（Bentonite）的主要成分是蒙脱石（Montmorillonite），其化学式为 Al_2O_3·［$4Si_2O_2$］·nH_2O，是一类天然硅铝酸盐黏土矿物质，具有吸附、离子交换等多种特性，其优异的特性为其在催化领域的应用提供了可能性，同时膨润土在国内储量大、价格低，非常适合作为催化剂做载体使用，有很好的利用前景和经济效益。

本文以膨润土为载体，采用浸渍法制备了不同 Pd 负载量的负载型催化剂，对制备的催化剂进行了 XRD 等表征分析，评价了不同 Pd 负载量下的 Pd/Bentonite 催化剂对甲醇液相氧化一步合成甲酸甲酯的反应性能。催化剂在原料甲醇的质量浓度为 8 g/L、反应温度为 150 ℃、O_2压力为 2 MPa、反应时间为 5 h 的条件下，以膨润土为载体的不同 Pd 负载量的催化剂对甲醇选择氧化制甲酸甲酯的反应性能如表 1 和图 1 所示。由图 1 可以看出，随着 Pd 负载量的增加，甲醇的转化率随之升高，甲酸甲酯的选择性反而随之降低。这是因为随着 Pd 负载量的增大，催化剂拥有更多的氧化性位点，有利于甲醇的氧化，因此甲醇转化率提高。随着 Pd 负载量的增加，载体的酸性位点部分被覆盖，导致酯化步骤被削弱，使中间产物不能及时转化成甲酸甲酯而生成了其他副产物，因此甲酸甲酯的选择性下降。由表 1 可见，在相同条件反应条件下，2wt％Pd/Bentonite 的反应效果最好，甲醇转化率达到 60.25％，甲酸甲酯收率为 24.89％。

---

*通信作者：季生福；E-mail：jisf@mail. buct. edu. cn；手机号：18046545619。国家自然科学基金（21573015）资助项目。

表 1　不同 Pd 负载量的 Pd/膨润土催化剂催化甲醇选择氧化制甲酸甲酯反应的性能

实验	催化剂	甲醇转化率（%）	甲酸甲酯选择性（%）	甲酸甲酯收率（%）
1	膨润土	32.65	17.15	5.59
2	1%Pd/膨润土	42.91	54.15	23.23
3	2%Pd/膨润土	60.25	41.32	24.89
4	3%Pd/膨润土	65.23	34.01	22.18

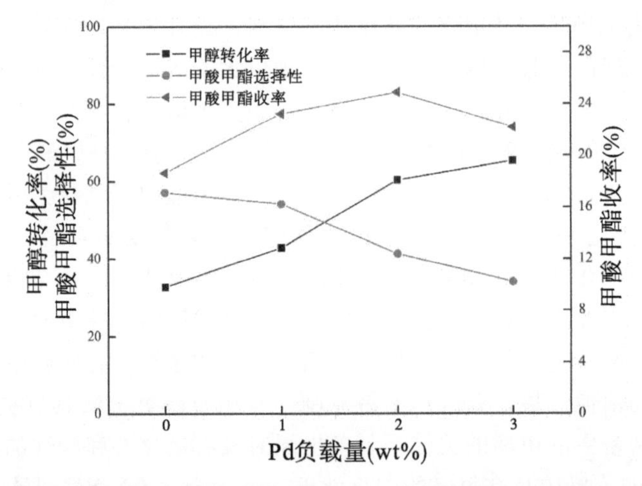

图 1　不同 Pd 负载量的 Pd/膨润土催化剂催化甲醇选择氧化制甲酸甲酯反应的性能

参考文献

[1] Koivikko N，Laitinen T，Ojala S. 2011. Formaldehyde production from methanol and methyl mercaptan over titania and vanadia based catalysts [J]. Applied Catalysis B Environmental，103（1）：72－78.

[2] Hansen J A，Ehara M，Piecuch P. 2013. Aerobic oxidation of methanol to formic acid on Au8－：benchmark analysis based on completely renormalized coupled-cluster and density functional theory calculations [J]. Journal of Physical Chemistry A，117（40）：10416－10427.

[3] Solyman S M，Betiha M A. 2014. The performance of chemically and physically modified local kaolinite in methanol dehydration to dimethyl ether [J]. Egyptian Journal of Petroleum，23（3）：247－254.

高岭石纳米管的热稳定性

孙世平[1]　刘静豪[1,2]　许红亮[1*]　范二闯[1]　邵　刚[1]　王海龙[1]
（1. 郑州大学材料科学与工程学院，郑州 450001；
2. 宇通客车股份有限公司工艺部，郑州 450061）

我国拥有丰富的高岭土资源，但目前多用于陶瓷、造纸、涂料等行业，产品附加值较低，因此，开发具有功能特性、高附加值的高岭土产品并研究其物理性能，对于拓展高岭土的应用领域具有十分重要的意义。

本研究以天然高岭石为主要原料，首先，通过化学插层、超声等协同剥片技术制备片状高岭石，进而制得片状高岭石/DMSO插层复合体，然后将其与甲醇进行溶剂热反应制得高岭石/甲醇插层复合体，再以高岭石/甲醇插层复合体为前驱体、与CTAC的甲醇溶液于100 ℃进行溶剂热反应24 h制备出高岭石纳米管（KNTs）[1]。将KNTs分别在300～800 ℃煅烧2 h，得到煅烧KNTs样品。综合采用X射线衍射仪、扫描电子显微镜、红外光谱仪、综合热分析系统、比表面积及孔径分析仪等仪器、设备，研究了高岭石原料、KNTs及其煅烧产物的物相、显微结构、比表面积等随煅烧温度变化的规律，从而评价KNTs的热稳定性。

图1为高岭石原料、KNTs及其不同温度煅烧后的SEM图。原料高岭石为结晶良好的晶体，粒径为3 μm以下，呈现出典型的层状结构。所制KNTs为管状结构，形貌和结构比较完整，管径为10～30 nm，长度为500～1 000 nm。400 ℃煅烧后，KNTs的衍射峰仍然存在，但500 ℃煅烧后则基本消失；KNTs的DSC－TG分析显示，其在460 ℃有1个吸收峰并伴有较大的失重。因此，KNTs的结构羟基于460 ℃脱除并转变为偏高岭石。煅烧温度对KNTs的形貌和管状结构影响不大，800 ℃煅烧后仍保持着较完整的管状结构。

图1　高岭石原料（a）、高岭石纳米管（b）和高岭石纳米管在800 °C煅烧后（c）的SEM图

*通信作者：许红亮；E-mail：xhlxhl@zzu. edu. cn；手机号：13939022916。河南省高等学校重点科研项目计划（15A430010）、河南省科技攻关计划（182102210005）资助项目。

KNTs 的比表面积为 57.41 m^2/g，孔容 0.263 cm^3/g，平均孔径为 18.34 nm。随着煅烧温度的升高，KNTs 的比表面积和孔容而逐步增大，500 ℃煅烧时达到最大，此后随煅烧温度的升高而逐步减小；KNTs 平均孔径的变化规律则与此相反。KNTs 在 500 ℃煅烧后，其比表面积、孔容、平均孔径分别达到了 91.74 m^2/g、0.312 cm^3/g、13.95 nm。

以天然高岭石为原料，采用溶剂热法可制得高岭石纳米管。高岭石纳米管在 300～800 ℃煅烧时，晶体结构于 460 ℃受到破坏，形貌和管状结构受煅烧温度的影响不大，但高岭石纳米管的比表面积、孔容、平均孔径则随煅烧温度的升高而改变。这一研究结果为调控高岭石纳米管的比表面积、孔容和孔径等提供了思路，也为其在高温环境下应用于材料、环境治理及化工等领域提供了理论依据。

参考文献

［1］Xu H L，Jin X Z，Chen P，et al. 2015. Preparation of kaolinite nanotubes by a solvothermal method［J］. Ceramics International，41：6463－6469.

磷矿渣粉石灰石粉复合掺合料对骨料碱活性的抑制效果

王　珩[1,2,3]　刘伟宝[1,2,3]　陆采荣[1,2,3]　梅国兴[1,2,3]　戈雪良[1,2,3]　杨　虎[1,2,3]
（1. 南京水利科学研究院，南京 210024；
2. 水利部水工新材料工程技术研究中心，南京 210024；
3. 水文水资源与水利工程科学国家重点实验室，南京 210029）

在优质粉煤灰日益不足的大环境下，磷渣粉石灰石粉复合掺合料（简称 PL）作为替代粉煤灰的一种新型混凝土掺合料，其主要物理力学性能与Ⅱ级粉煤灰接近，但其对于骨料的碱活性抑制效果尚需研究。碱–骨料反应的破坏性不仅要从反应本身（骨料活性、碱含量）来预测，还需要考虑骨料粒径分布、湿度、水泥品种和数量、水胶比、掺合料[1]。掺合料不但可以与碱结合，减少其对骨料的破坏，还可以改善孔结构，降低碱离子迁移的能力，而迁移能力又与孔结构有一定关系[2]。

试验选用石英玻璃砂作为活性骨料，在保证有效碱含量相同的前提下，分析不同水胶比试件的碱骨料反应膨胀率及浆体孔溶液中离子浓度。采用中热硅酸盐水泥、水泥加粉煤灰、水泥加 PL 掺合料作为胶凝材料制作砂浆试件，水胶比分别为 0.35、0.45、0.55，掺合料用量分别为 25％、45％、65％，并使用 NaOH 补碱至胶凝材料的 1％。试件拆模后放置 38 ℃养护箱中隔水养护，24 h 后测基长，然后测量基长之后的 7、14、28、56 d 的膨胀率。

不同水胶比试件 28 d 龄期膨胀降低率见图 1。在水胶比为 0.45 时，与纯水泥基准组相比，膨胀率的降低率最小，其原因是该水胶比下的孔结构的致密程度既不足以阻挡碱离子的迁移，又不能对膨胀应力进行很好的释放。0.45 水胶比时膨胀率试验结果见图 2，可以发现，纯水泥（基 2）的膨胀率最大，掺用粉煤灰的试件膨胀率最小，而掺用 PL 掺合料的试件，膨胀率居中，且随着掺量增加而降低。各试件膨胀率随着龄期增加而增加，28 d 后趋于稳定。由于石英玻璃的高活性，在 56 d 后，试件表面出现龟裂纹破坏，故停止测量。

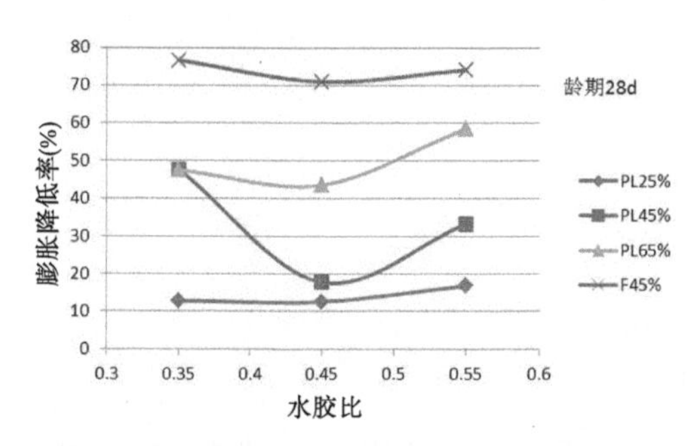

图 1　不同掺量下膨胀降低率与水胶比的关系

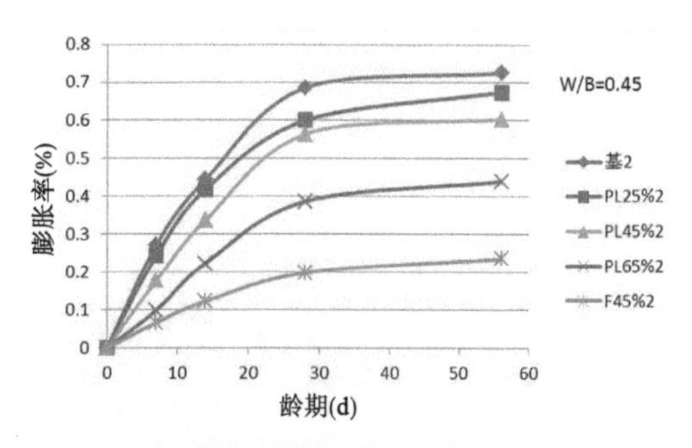

图 2　石英玻璃砂浆试件膨胀率（水胶比 0.45）

采用0.3～0.6 mm石英玻璃砂和无活性河砂制作砂浆试件，其水胶比为0.55，PL掺合料掺量为45%，粉煤灰掺量为30%，补碱至胶材1%。试件拆模后，浸泡于38 ℃自来水中养护至需要的龄期，随后进行孔溶液榨取，采用ICP测试各试件R^+（Na^+、K^+）浓度，结果见图3。

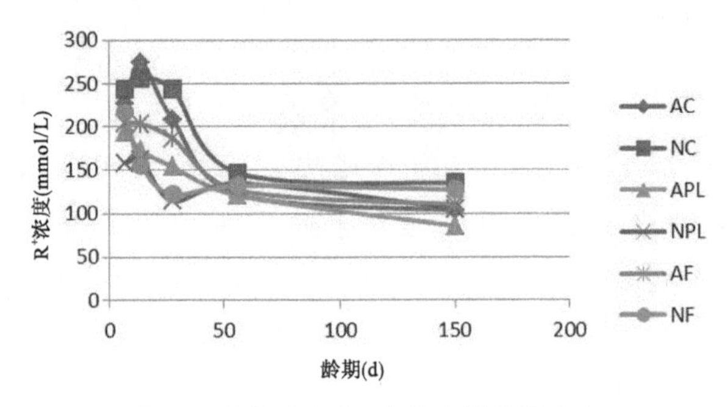

图3　砂浆孔溶液R^+浓度随龄期变化

注：A、N分别表示活性、非活性骨料；C、PL、F分别表示纯水泥、PL掺合料、粉煤灰

试件膨胀率与孔溶液中R^+相关性并不强，14 d时，纯水泥试件R^+浓度有上升段，但掺合料试件没有，与文献[3]一致。随着龄期延长，R^+（Na^+、K^+）的浓度呈降低状态，且在56 d后趋于稳定，掺有PL掺合料组在后期仍持续下降，表明掺合料在持续吸收R^+。

从膨胀率试验和孔溶液分析来看，由于粉煤灰属于酸性掺合料，在相同掺量下PL掺合料没有粉煤灰对膨胀率的抑制效果好，但PL掺合料在掺量达到45%以上时，对碱活性也有较好的抑制效果。

参考文献

［1］Smolizyk H G. 1975. Investigation on the fiffusion of Na^+ ion in concrete［A］. Proceedings of symposium on alkali-aggregate reaction preventive measures［C］. Reykjavik，Iceland.

［2］Hornain H，Thuret B，Guedon-Dubied S，et al. 1996. Influence of aggregates and mineral additives on the composition of the pore solution［A］. Proceedings of the 10th international conference on alkali-aggregate in concrete［C］. Melbourne，Australia.

［3］Duchesne J，Bérubé M A. 1994. The effectiveness of supplementary cementing materials in suppressing expansion due to ASR：Another look at the reaction mechanisms，Part 2：Pore solution chemistry［J］. Cement and Concrete Research，24（2）：221－230.

石墨粒度对锂离子电池电化学性能的影响

王巧平　白云山*　马红竹*

（陕西师范大学化学化工学院，西安 710119）

锂离子电池与传统电池相比，嵌锂电位低而且平坦，大部分嵌锂容量分布在0～0.20 V，可以保证锂离子电池高且平稳的工作电压[1-3]。嵌入型电极理论分析结果表明，电极活性材料的颗粒尺寸、比表面积、电极厚度和孔隙度对电极的电化学性能均有明显的影响[4,5]。文献报道，天然石墨的粒度与分布对锂离子电池的初始充放容量有较大的影响，而对其首次效率影响相对较小[6]。新疆某白云石矿中含有大量的鳞片石墨，矿产资源丰富，为了进行有效利用，通过一定的方法提取精制，进一步探究该石墨的粒度对锂离子电池电化学性能的影响。

采用不同的溶剂，探究研磨方式对石墨粉的形貌、粒度及厚度的影响，进一步探究其对锂离子电池电化学性能的影响。实验采用的鳞片石墨（BG）来自于新疆某白云石矿。经实验探究，采用球磨与胶体磨（JM－F65）相结合，乙醇作溶剂，以蒸馏水为沉降剂，用简单沉降法对研磨后的石墨粉进行粒度分级。将样品、乙炔黑与聚偏氟乙烯（PVDF）按照8∶1∶1的质量比例涂膜，干燥，以锂片（天津中能锂业）为对电极，组装成电池，采用高精度电池测试系统（VOO486，深圳）、电化学工作站（CHI660，上海）测其电化学性能，利用X射线衍射仪与环境扫描电子显微镜表征结构与形貌。

与进口石墨对比，实验室精制石墨表面平整，具有良好的层状结构，有利于锂离子的脱嵌，从而减少了自由电子流动的阻碍因素。采用简单的沉降法对研磨后的石墨进行分级，利用Nano Measurer软件计算平均粒径（表1）。通过首次充放电比容量、不可逆容量和循环性能等研究，发现BG3～BG4（石墨粒度15～25 μm）的电化学性能更好，比容量达到365 mA·h·g^{-1}；不可逆容量小，循环稳定性好，具有相对来说比较优异的性能（图1）。虽然最近几年报道了很多有高比容量的负极材料，但在商业化应用中石墨类碳材料仍占据着巨大的市场，对其改性研究还是很有必要的，但实验的不足之处是机械研磨的效率不高，所以今后仍需继续改进，从而实现大规模的生产应用。

表1　BG1～BG6的平均粒径统计

名称	BG1	BG2	BG3	BG4	BG5	BG6
平均粒径（μm）	7.66	11.86	15.46	20.34	25.05	33.51

* 通信作者：白云山，E-mail：baiys@snnu.edu.cn；马红竹，E-mail：hzmachem@snnu.edu.cn。手机号：13636819611。钙镁分离及石墨提取中试技术（1204070143）。

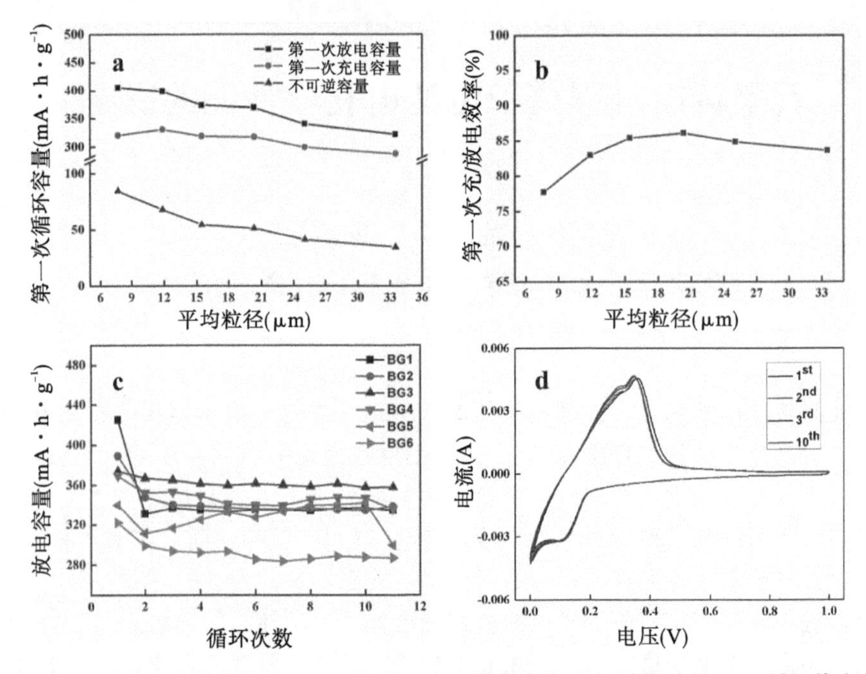

图 1　石墨粒度对首次充放电性能（a)、效率（b)、循环性能（c）的影响及 BG4 循环伏安曲线（d)

参考文献

［1］Chen J T，Zhou H H，Chang W B，et al. 2003. Effect of particle size on the properties of graphite anode materials with lithium imbedded［J]. Journal of physical chemistry，19（3）：278－282.

［2］Li W，Yang Z，Jiang Y，et al. 2014. Crystalline red phosphorus incorporated with porous carbon nanofibers as flexible electrode for high performance lithium-ion batteries［J]. Carbon，78：455－462.

［3］Wang L，He X，Li J，et al. 2012. Nano-structured phosphorus composite as high-capacity anode materials for lithium batteries［J]. Angewandte Chemie International Edition，124（36）：9168－9171.

［4］Shao-Yang H E，Zeng J B，Jiang F M. 2015. Numerical reconstruction and characterization analysis of microstructure of lithium-ion battery graphite anode［J]. Journal of Inorganic Materials，30（9）：906－912.

［5］Roy P，Srivastava S. 2015. Nanostructured anode materials for lithium ion batteries［J]. Journal of Materials Chemistry A，3（6）：2454－2484.

［6］Li Z，Huang J，Liaw B Y，et al. 2014. A review of lithium deposition in lithium-ion and lithium metal secondary batteries［J]. Journal of Power Sources，254（254）：168－182.

磁性蒙脱石纳米粒子固定纤维素酶

吴琳梅　曾庆虎　朱登勇　汪　涵　夏觅真*

（安徽医科大学生命科学学院，合肥 230000）

纤维素酶催化降解纤维素，是获得燃料乙醇和化学品的重要途径。采用固定化技术对纤维素酶进行固定，可提高其稳定性，使其能够重复使用[1]，因此纤维素酶的固定化是国内外研究的热点，其中载体被认为是决定固定化酶效果的重要因素[2]。在众多载体材料中，蒙脱石呈层状结构，具有较大的比表面积、较强离子交换能力和吸附能力，具有非均匀性的电荷分布、较好的生物兼容性和机械强度，是固定化酶的良好载体[3]。近年来，将蒙脱石制成磁性蒙脱石纳米粒子，利用外加磁场，可将其从反应体系中分离，极大拓展了蒙脱石材料在固定化酶上的应用[4]。但目前磁性蒙脱石固定纤维素酶的研究鲜有报道，固定化条件及机理仍不清楚。因此，本文采用共沉淀法制备磁性蒙脱石纳米粒子，并以其为载体，探究吸附法固定纤维素酶。

磁性蒙脱石纳米粒子的制备[5]：$FeCl_2 \cdot 6H_2O$ 与 $FeCl_2 \cdot 4H_2O$ 按照摩尔比 $Fe^{2+}:Fe^{3+}=1:1.8$ 的比例加入 100 mL 蒸馏水中，于 80 ℃下用 1 mol/L NaOH 滴加至 pH＝11，加入一定量的蒙脱石反应 1.5 h，50％乙醇洗至中性，干燥备用；纤维素酶的固定：20 mL 5 mg/mL 纤维素酶溶液中加入 0.05 g 载体，40 ℃下吸附 2 h，磁性分离，280 nm 下测定吸光度；酶活力测定：固定化酶和游离酶分别配制成 1 mg/mL 的溶液，加入 3 倍体积的 0.5％ CMC－Na 溶液，50 ℃下摇床水浴振荡30 min，固液分离，DNS 显色，530 nm 下测定吸光度。酶活回收率＝（固定化酶活力/加入游离酶总活力）×100％。

表 1　Fe_3O_4与蒙脱石的比例对纤维素酶吸附和活性的影响

Fe_3O_4：Mt（质量比，g/g）	1：1	1：2	1：3	纯 Mt
纤维素酶吸附量（mg/g）	860.8	867.4	635.0	704.6
酶活回收率（％）	51.88	49.29	45.66	44.95

磁性 Fe_3O_4能够促进蒙脱石对纤维素酶的吸附，增加固定化酶的活力，且当磁性 Fe_3O_4与蒙脱石的比例为 1：1 时，纤维素酶的吸附量和活性最大；但随着蒙脱石含量的进一步增加，纤维素酶吸附量和活性急剧减小（表 1）。当吸附达到饱和时，载体的载酶量和固定化酶活力最大（图 1a）；纤维素酶的分子运动会受到温度的影响，温度越高分子运动速度越快，导致吸附量上升，但过高的温度使酶变性，进而失活（图 1b）；随着吸附时间的增加，载体上酶的吸附量增

* 通信作者：夏觅真；E-mail：xiamizhenna@163. com；电话：0551-65160393。安徽医科大学博士科研启动基金（0807016101）资助项目。

加，然而随着吸附时间的延长，大量纤维素酶聚集，导致酶活性中心相互靠近相互干扰，无法发挥催化活性（图 1c）；纤维素酶的等电点 pI 在 5.5 左右，当 pH<pI 时，酶带正电，可通过静电吸引固载到蒙脱石负电荷层板上，增加吸附量，但随着 pH 值逐渐靠近 pI，酶电荷逐渐减弱至中性，与蒙脱石的静电作用减弱，吸附量降低（图 1d）。因此，磁性蒙脱石纳米粒子固定纤维素酶的最佳条件：纤维素酶的浓度为 5 mg/mL，吸附温度为 40 ℃，反应时间为 1 h，溶液 pH 值为 4.8。

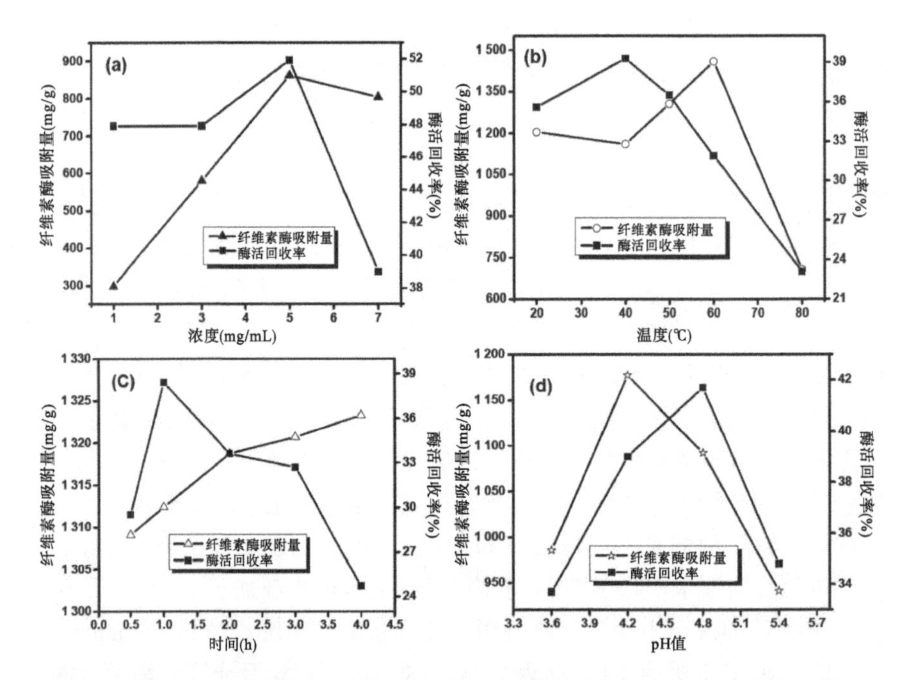

图 1　磁性 Fe_3O_4 与蒙脱石比例为 1∶1 时纤维素酶浓度（a）、吸附温度（b）、吸附时间（c）以及溶液 pH 值（d）对纤维素酶的吸附和固定化酶活性的影响

参考文献

［1］Mateo C，Palomo J M，Fernandez-Lorente G，et al. 2007. Improvement of enzyme activity，stability and selectivity via immobilization techniques［J］. Enzyme & Microbial Technology，40（6）：1451－1463.

［2］Gokhale A A，Lee I. 2012. Cellulase immobilized nanostructured supports for efficient saccharification of cellulosic substrates［J］. Topics in Catalysis，55（16－18）：1231－1246.

［3］An N，Zhou C H，Zhuang X Y，et al. 2015. Immobilization of enzymes on clay minerals for biocatalysts and biosensors［J］. Applied Clay Science，114：283－296.

［4］Chen L，Zhou C H，Fiore S，et al. 2016. Functional magnetic nanoparticle/clay mineral nanocomposites：preparation，magnetism and versatile applications［J］. Applied Clay Science，127－128：143－163.

［5］Chang J L，Ma J C，Ma Q L，et al. 2016. Adsorption of methylene blue onto Fe_3O_4/activated montmorillonite nanocomposite［J］. Applied Clay Science，132－140.

分散剂对水煤浆流变性能的影响及作用机理

葛　新[1]　赵　亮[2]　陈新志[2]　钱　超[2*]
（1. 江南大学化学与材料工程学院，无锡 214122；
2. 浙江大学化学工程与生物工程学院浙江省化工
高效制造技术重点实验室，杭州 310027）

将等离子裂解煤的固体产物制成水煤浆，对不同条件下所表现出的流变学性能进行了系统研究，主要考察分散剂对于固体产物制备水煤浆的流变性能影响。分别对比了 3 种离子型分散剂（萘磺酸钠甲醛缩聚物、木质素磺酸钠、马来酸钠—丙烯酸钠共聚物）和 1 种非离子型分散剂（磺酸盐月桂醇聚氧乙烯醚）作为分散剂时固体产物制备水煤浆流变性能，发现无分散剂加入时，该体系是典型的“剪切变稀型”流体。当固含量为 50%时，分散剂的加入使得体系的流变性质变为拥有峰值黏度的特殊状态，而当固含量上升至 55%时，该现象消失。以萘磺酸钠甲醛缩聚物为分散剂时，效果最优，并且随着固含量的提升，相同剪切速率下黏度也逐渐增大，并且当固含量超过 55%后，黏度上升速率非常快。不同分散剂体系的水煤浆黏—温关系大体相似，均为随温度升高，体系黏度下降，但下降幅度与速率不同。使用分散剂后，体系的稳定性改善，经过静置沉降实验 12 d 后产生清液明显减少至 0.5%，前后表观黏度值的变化范围也得到了降低。

氢等离子体裂解煤制取乙炔过程得到的固体产物约占原料煤粉质量的 60%以上，并具有含碳量高、粒径小和含硫低等特点[1,2]。因此，将其制成水煤浆[3]具有很可观的经济与环境效益。但目前其分散剂匹配性研究较少，因此研究分散剂对固体产物制备的水煤浆流变性能的影响及作用机理具有重要的意义。

本文将氢等离子体裂解煤制取乙炔工艺所得到的固体产物作为研究对象，将其制备成为水煤浆并进行了系统的流变学分析。主要考察了 4 种不同类型分散剂对于固体产物制备的水煤浆流变性能的影响及作用机理。

（1）对于固含量为 50%的固体产物水煤浆，引入不同分散剂会造成体系成为拥有峰值黏度的特殊流体，同时分散剂降低体系黏度的效果在固含量上升至 55%时就消失了，甚至出现了反效果（图 1）。

（2）在萘磺酸钠甲醛缩聚物为分散剂的作用下，固体产物制备的水煤浆的表观黏度随着固含量的上升而逐渐增大，当固含量超过 55%时，黏度上升速率非常快（图 1）。

（3）不同种类分散剂对于固体产物水煤浆的黏—温关系的影响基本相同，均为随温度升高，体系黏度下降，但下降幅度与速率不同（图 2）。

---

*通信作者：钱超；E-mail：qianchao @ zju. edu. cn；手机号：13606602125。国家重点研发计划项目（2016YFB0301800）、国家自然科学基金（21606104，21606104）和浙江省化工高效制造技术重点实验室开放课题（No. ACEMT-17-03）资助项目。

(4) 分散剂改善了固体产物制备的水煤浆稳定性，经过静置沉降实验 12 d 后产生清液明显减少至 0.5%，前后表观黏度值的变化也得到了降低。

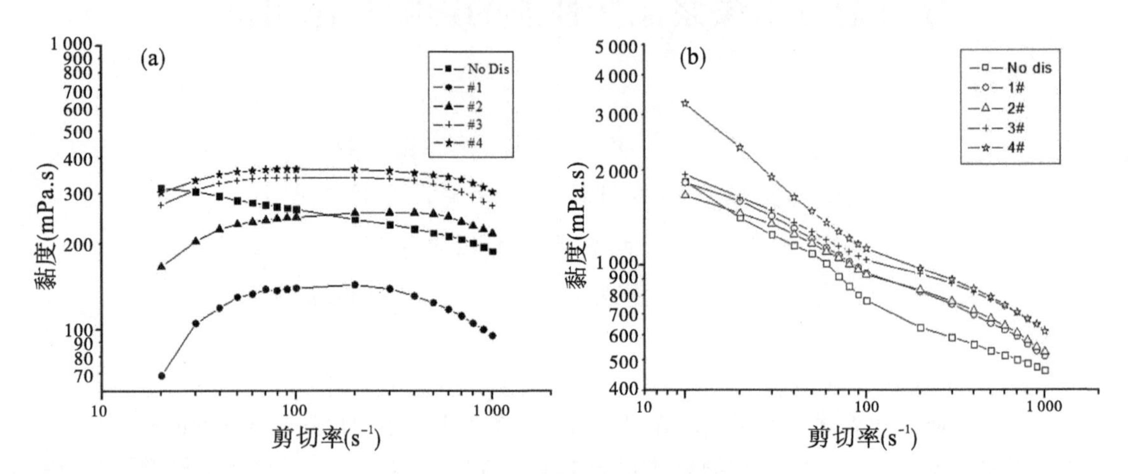

图 1 分散剂对体系表观黏度的影响

注：a. 固含量 50%；b. 固含量 55%

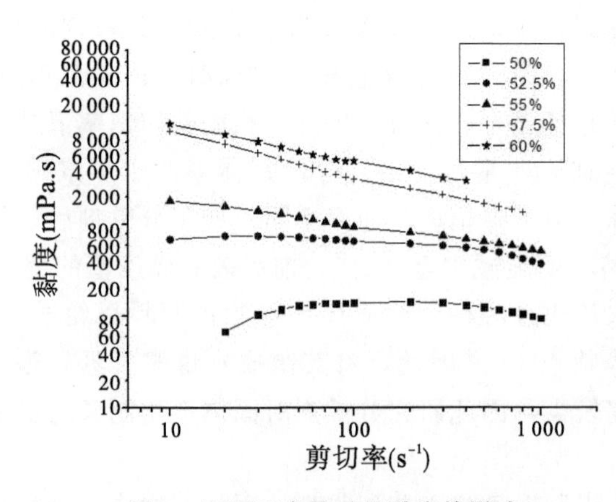

图 2 不同固含量对于黏度的影响

参考文献

[1] Zhang X, Qian C, Chen X, et al. 2013. Cleaning the residue of plasma pyrolysis of coal through alkali-acid treatment with recycle of alkali [J]. Energy Sources Part a-Recovery Utilization and Environmental Effects, 35 (20): 1939 - 1945.

[2] Chen X, Ge X, Zhang X, et al. 2015. Preparation of activated carbon from the residue of plasma pyrolysis of coal by steam activation [J]. Energy Sources Part a-Recovery Utilization and Environmental Effects, 37 (4): 440 - 446.

[3] Chen X, Zhao L, Zhang, X, et al. 2012. An investigation on characteristics of coal-water slurry prepared from the solid residue of plasma pyrolysis of coal [J]. Energy Conversion and Management, 62: 70 - 75.

Mo－V/黏土催化剂催化甘油脱水氧化过程

姜雪超[1]　吴书涛[1]　童东绅[1]　俞卫华[1]　周春晖[1,2*]
（1. 浙江工业大学化学工程学院，杭州 310014；
2. 青阳非金属矿研究院，青阳 242800）

由于环境、生态问题以及化石燃料资源的减少，导致对绿色和清洁能源的需求增加，推动了以生物基化学品为原料的新一代化学工艺的发展[1]。过去十多年来使用生物基原料生产精细化学品受到科学和工业界越来越多的关注。甘油是一种生物基平台化学品；以甘油三脂的形式存在动植物体中。工业上，以甘油三脂为原料的皂化制皂、水解制脂肪酸和酯交换生产生物柴油等过程均会产生甘油副产物[2]。此外，甘油还可以通过微生物发酵淀粉、酶催化淀粉转化、氢解山梨醇和催化分解木质纤维素等途径获得[3]。近年来，生物柴油（biodiesel）每年全球需求量不断增长，在 2023 年其产量预计将会达 4. 0×10^{10} L。工业上，生物柴油主要通过酯交换法获取，生产过程中每生产 1 吨生物柴油伴生约 0. 1 吨甘油，目前成为甘油主要来源之一[4]。

甘油，化学名为丙三醇（1,2,3－propanetriol，$C_3H_5(OH)_3$），为最简单三元醇，由碳链上带有羟基的 3 个碳链组成。常温下为无色无味黏稠且易吸湿性液体[5]。在 20 ℃下，甘油物理化学性质：密度 1.26 g/cm，黏度 1.5 Pa·s，熔点 18.2 ℃，沸点 290 ℃，闪点 160 ℃，热能 4.32 kcal/g，表面张力 64 mN/m，温度系数－0.059 8 mN/（m），为强极性物质，易溶于水和醇，微溶于常见有机溶剂，不溶于烃。甘油可用于化妆品、食品及动物饲料等，这些行业的需求量难以解决大宗工业过程副产甘油的有效利用。诸多研究人员正积极探索和开发甘油转化高附加值化学品新工艺，如催化甘油脱水制丙烯醛、蒸汽重整制氢气氧化生成二羟基丙酮、醚化合成缩聚甘油及与羧酸酯化脱水生成甘油脂等，但这些反应尚处于实验室探索或工业研究阶段。相对而言，近十年来甘油脱水氧化连串反应生产丙烯酸成为研究人员的新研究热点。丙烯酸作为重要化工原料，用于制备涂料、塑料及橡胶等产品，丙烯酸需求量在 2025 年预计高达 9.0×10^9 kg。目前，工业中丙烯酸主要由 C3 烷烯烃在有氧环境中经两步氧化制取。如果能成功开发甘油脱水氧化连串反应制丙烯酸工艺，将会显著降低 C3 烷烃依赖，同时降低能耗和减少 CO_x 等温室气体排放。此外，还能有效解决过剩甘油引起柴油工业发展受限难题。开发高效的、抗失活的催化剂是实现甘油脱水氧化连串反应制丙烯酸工业化过程的核心问题[5]。

本实验将钼和钒的前驱体（钼酸铵和偏钒酸铵）按照不同比例溶解在去离子水中，通过浸渍法制备 Mo－V/蒙脱石催化剂。在连续流动固定床多相催化微型反应器中进行催化剂（0.5 g，20～40 目）脱水氧化反应性能评价。

酸改性的蒙脱石和酸改性的蒙脱石负载后 Mo－V 氧化物的样品的 XRD 图中仍呈现蒙脱石的

* 通信作者：周春晖；E-mail：chc. zhou@aliyun. com；clay@zjut. edu. cn；手机号：13588066098。国家自然科学基金（21373185，41672033）资助项目。

特征衍射峰（图 1）；随着焙烧温度的提高，蒙脱石层状结构的 001 晶面衍射特征峰以及 020 晶面衍射特征峰依然存在，然而 001 晶面衍射峰在负载 Mo－V 后向着低角度偏移，可能的原因是负载后的催化剂的层间距变大。此外，经过焙烧的蒙脱石在 26.60°处的 SiO_2 的特征衍射峰强度增加，反映出蒙脱石结构的无序性增加，相对使得晶体 SiO_2 晶面衍射增强。

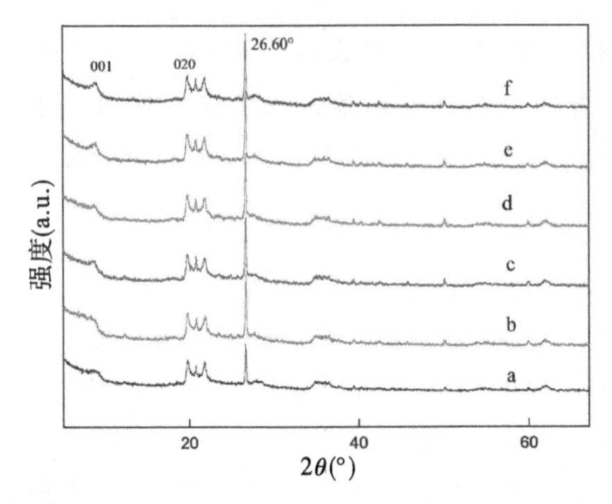

图 1　Mo－V/蒙脱石在不同焙烧温度下的 XRD 衍射图

注：a. H－MT－450；b. Mo－V/H－MT－300；c. Mo－V/H－MT－350；
d. Mo－V/H－MT－400；e. Mo－V/H－MT－450；f. Mo－V/H－MT－500（℃）

参考文献

[1] Zhou C H，Beltramini J N，Lin C X，et al. 2011. Selective oxidation of biorenewable glycerol with molecular oxygen over Cu-containing layered double hydroxide-based catalysts [J]. Catalysis Science & Technology，1：111－122.

[2] Kong P S，Aroua M K，Wan M A W D. 2016. Conversion of crude and pure glycerol into derivatives：A feasibility evaluation [J]. Renewable & Sustainable Energy Reviews，63：533－555.

[3] Zhou C H，Zhao H，Tong D S，et al. 2013. Recent advances in catalytic conversion of glycerol [J]. Catalysis Reviews，55：369－453.

[4] Zhou C H，Beltramini J N，Fan Y X，et al. 2008. Chemoselective catalytic conversion of glycerol as a biorenewable source to valuable commodity chemicals [J]. Chemical Society Reviews，37：527－549.

[5] Jiang X C，Zhou C H，Tesserd R，et al. 2018. Coking of catalysts in catalytic glycerol dehydration to acrolein. Industrial & Engineering Chemistry Research. DOI：10. 1021/acs. iecr. 8b01776.

碱激发在煅烧煤矸石粉体材料活性评价中的应用

刘　朋　王爱国* 孙道胜　管艳梅　李　燕　胡普华
（安徽建筑大学安徽省先进建筑材料重点实验室，合肥 230022）

煤矸石是采煤和选煤过程中产生的固体废弃物，约占全国固体废弃物总量的 20%。历年来我国煤矸石堆积量已超过 50 亿吨，而年利用率仅为 60%左右[1]，如何合理利用煤矸石并提高其附加值亟待解决。煤矸石中含有的黏土质矿物在高温下煅烧后会发生相变产生活性，使得煅烧煤矸石可作为活性混合材添加到水泥中[2]。而煤矸石活性的高低是影响其在水泥工业中高效应用的关键。

目前对于活化煤矸石活性评价常用的方法有强度评价法[3]、火山灰性试验法[4]、维卡法[5]、现代分析测试方法[6]等，但是这些方法或呈现结果慢，或操作复杂，或难以定量活化煤矸石的活性。而碱激发地质聚合物具有“敏感性高”“强度来源单一”“强度发展迅速”等特点，能够很好地避免上述问题。本文在碱激发胶凝材料的研究基础上，使用改性水玻璃作为激发剂激发煅烧煤矸石的活性，探究了碱激发煅烧煤矸石地质聚合物的强度与其活性间的关系，并与其他活性评价法进行了对比，确定了煅烧煤矸石粉体材料活性的碱激发快速评价法。

煤矸石取自淮北某煤矿，水泥为江南小野田生产的 P·Ⅱ 52.5 型水泥。激发剂采用模数 1.2、固含量 40%的自配水玻璃。取粒径小于 9.5 mm 原状煤矸石，分别在 550、650、750、850、950 ℃下煅烧，然后通过振动磨粉磨并过 200 目方孔筛。碱激发实验液固比为 0.5，采用 20 mm×20 mm×80 mm 的模具，试样入模振捣后连同模具密封并在 40 ℃下养护至指定龄期，测试其强度。参照国家标准分别进行强度比、火山灰性、化学结合水含量测试。分别对样品做 XRD、IR、SEM 测试。

XRD、IR 结果表明（图 1），原状煤矸石经煅烧后其所含的高岭石矿物脱羟基转变为偏高岭石，使得其活性大大增强。SEM 显示（图 1），原状煤矸石表面结构较为致密，呈现出“片层状”结构；经煅烧后，煤矸石表面变得“疏松多孔”，“片层状”结构几乎完全被破坏，这是高岭石矿物原有的晶体结构被破坏的表现。

火山灰性试验结果表明（图 2），经 650 ℃及 750 ℃煅烧后的煤矸石样品的总碱度及 CaO 含量最低，表明该温度煅烧下煤矸石火山灰活性较强。碱激发地质聚合物结果表明（表 1），随着煅烧温度的升高，地质聚合物强度呈现出先升高后降低的趋势。经 750 ℃煅烧后的煤矸石制备的地质聚合物强度最高，1、3 d 强度分别可达 74、98.4 MPa，与上述火山灰性试验结果相符。

碱激发快速评价法很好地避免了其他方法的缺陷，其操作简单，价格低廉，能迅速得到试验结果，并定量材料活性大小，可作为一种快速评价煅烧煤矸石活性的方法。经 750 ℃煅烧的煤矸石具有较强的火山灰活性，将其制备成活性混合材用于水泥基材料中具有良好的经济效益和环境效益。

* 通信作者：王爱国；E-mail：wag3134@126. com；手机号：13955146169。国家自然科学基金（51778003）、高校优秀中青年骨干人才国内外访学研修项目（gxfxZD2016134）、安徽省科技攻关计划项目（1301042127）、安徽省高等教育人才项目（皖教高［2014］11 号文）。

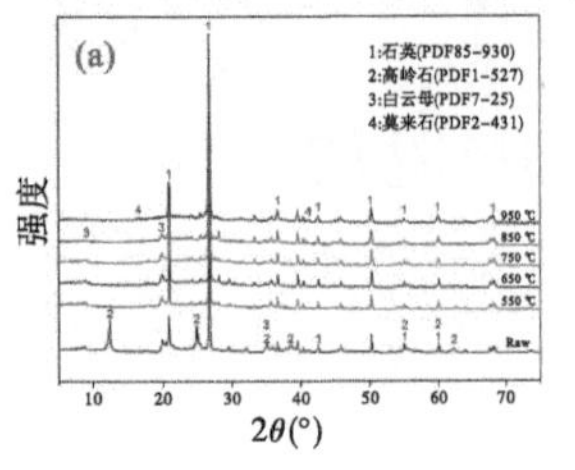

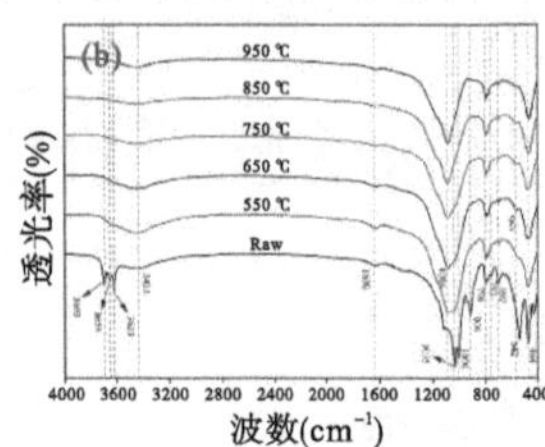

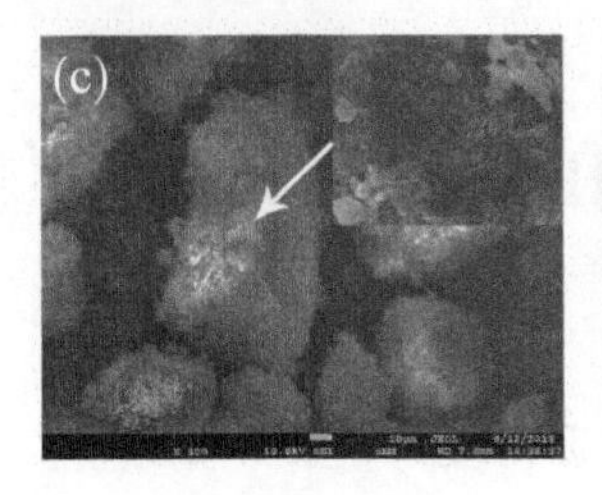

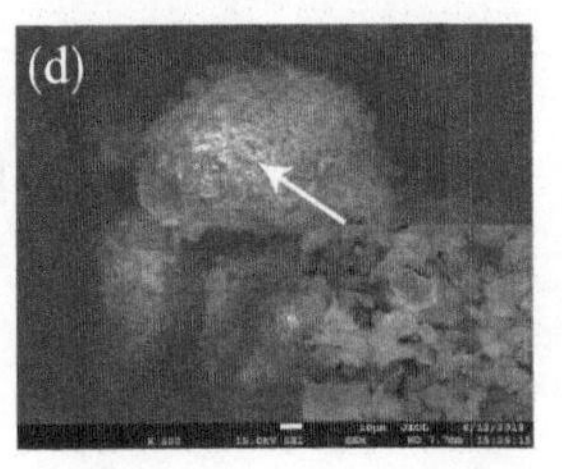

图 1　煅烧温度对煤矸石结构的影响

注：a. XRD；b. IR；c. SEM－Raw；d. SEM－750 ℃

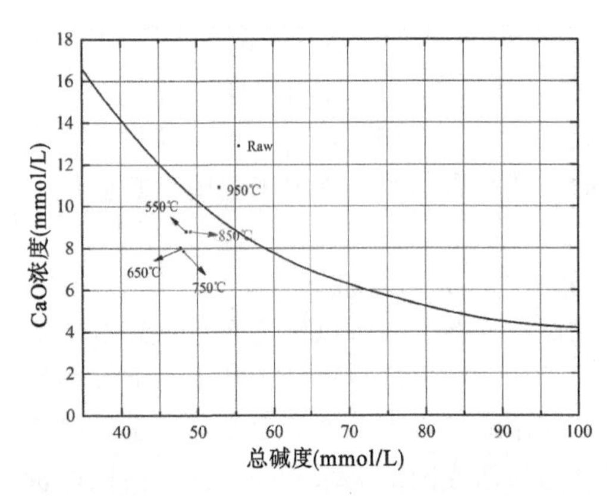

图 2　煅烧温度对煅烧煤矸石火山灰活性的影响

表 1　煅烧温度对地质聚合物力学性能的影响

编号	煅烧温度（℃）	弯曲强度（MPa）		抗压强度（MPa）	
		1 d	3 d	1 d	3 d
G0	Raw	/	/	/	/
G1	550	6.0	8.0	60.9	82.3
G2	650	7.1	8.0	69.3	91.0
G3	750	7.6	8.4	74.0	98.4
G4	850	0.5	1.7	7.2	31.3
G5	950	/	/	/	/

参考文献

[1] GuoY，Zhao Q，Yan K，et al. 2014. Novel process for alumina extraction via the coupling treatment of coal gangue and bauxite red mud [J]. Industrial & Engineering Chemistry Research，53 (11)：4518－4521.

[2] Guo W，Zhu J P，Li D X，et al. 2010. Early hydration of composite cement with thermal activated coal gangue [J]. Journal of Wuhan University of Technology (Materials Science Edition)，25 (1)：162－166.

[3] Zhang C S. 2006. Pozzolanic activity of burned coal gangue and its effects on structure of cement mortar [J]. Journal of Wuhan University of Technology (Materials Science Edition)，21 (4)：150－153.

[4] Li D X，Song X Y，Gong C C，et al. 2006. Research on cementitious behavior and mechanism of pozzolanic cement with coal gangue [J]. Cement & Concrete Research，36 (9)：1752－1759.

[5] Cao Z，Cao Y D，Dong H J，et al. 2016. Effect of calcination condition on the microstructure and pozzolanic activity of calcined coal gangue [J]. International Journal of Mineral Processing，146：23－28.

[6] Li C，Wan J H，Sun H H，et al. 2010. Investigation on the activation of coal gangue by a new compound method [J]. Journal of Hazardous Materials，179 (1)：515－520.

矿物加工外循环撞击流反应器的强化微观混合性能

佘启明[1]　周春晖[2*]

（1. 黄山学院化学化工学院，黄山 245041；

2. 浙江工业大学化学工程学院，杭州 310032）

撞击流（简记 IS）的概念最初由前苏联 Elperin[1] 提出，早期主要应用于气固相体系，是一类特殊的可以显著强化相间传递的流动结构。近年来，人们开始关注撞击流促进微观混合的特性及其应用潜力，并努力将这一特性应用于矿物加工分离工艺，如液相、液固相反应和反应—沉淀或结晶等过程[2,3]。撞击流反应器正是利用撞击流促进微观混合的特性而设计，其内部特殊的流动状况使撞击流反应器的传热、传质效率得到了明显强化，在矿物加工工程中的矿物分离方面具有良好的应用前景。

本研究提出一种外循环撞击流反应器，采用循环回路外置的方式避免回流强化混合，对于悬浮颗粒较大的矿物加工、矿物浮选、矿物结晶沉淀等需要强化微观混合的场合可显著提高微观混合效果，提高矿物加工效率。

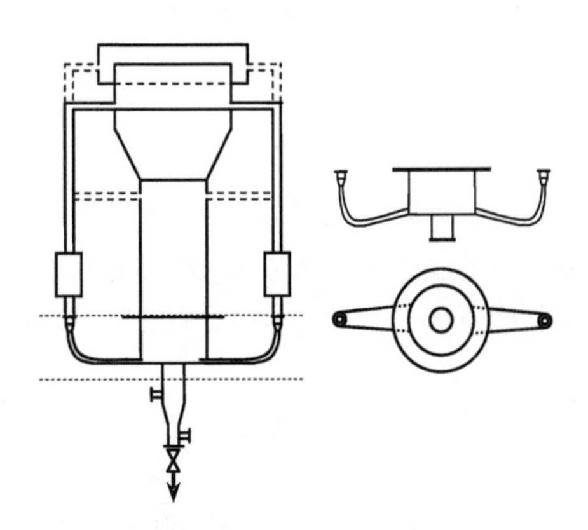

图 1　外循环撞击流反应器撞击段结构

如图 1 所示，撞击段结构为 304 不锈钢材质，左右 2 个外循环喷口相对布置并各与水平面呈一定角度上扬，与外部循环管路间使用卡箍联接，方便拆卸和更换新型撞击结构，外循环回路可根据反应器上部不同接口进行拆卸更换，从低到高共 3 种循环方式；中部直筒型部分采用螺栓联接，保证装置整体结构稳定性，适用于矿物浮选、强化吸附等矿物加工场合。

本研究所用主要实验装置为自行设计的外循环撞击流反应器（ECISR）微观混合实验装置

* 通信作者：周春晖；E-mail：qyimcn@126. com；手机号：13588066098。安徽省教育厅自然科学研究项目（KJHS2016B09）资助。

（图 2），撞击段结构如图 1 所示，目前对于反应器内的微观混合研究主要采用化学方法，应用平行竞争反应体系考察微观混合对化学反应的影响。本研究采用 Yu 和 Bourne 等[4]所提出的酸碱中和与氯乙酸乙酯水解平行竞争反应体系来研究 ECISR 内的微观混合状况，在注满反应器的清水部分由电磁流量计 5_3、循环泵 6_3以一定的流量经淘析腿循环；大部分由循环位置 B 出反应器，经电磁流量计 5_1（5_2）、循环泵 6_1（6_2），由撞击段的喷口高速流入反应器，形成撞击并循环。NaOH 溶液经蠕动泵 2 由进料管 3 注入反应器 4 的不同进料位置。进料完毕后，从取样口 7 处取样进行色谱分析。实验结束，打开出口阀门 8 排放。

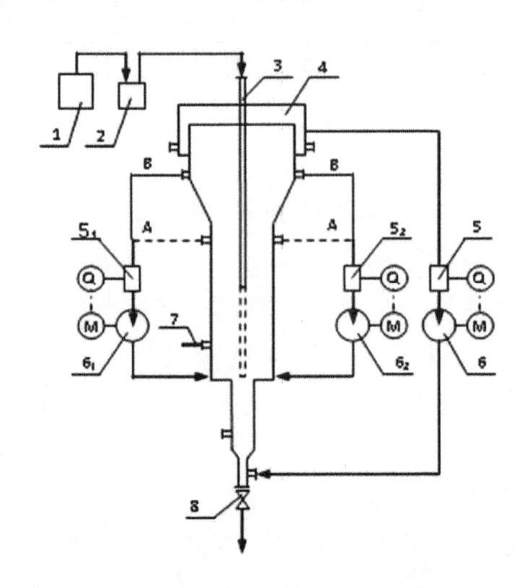

图 2　外循环撞击流反应器（ECISR）微观混合实验装置

注：1——NaOH 溶液；2——蠕动泵；3——进料管；4——撞击流反应器；5——电磁流量计；6——循环泵；7——取样口；8——出口阀门；A、B——循环位置；M——电动机执行机构；Q——电磁流量计

通过气相色谱分析对表征微观混合性能的产物分布的离集指数 X_Q进行了系统的研究，分别考察了 ECISR 内进料时间、进料位置、循环流量对产物分布的影响，实验结果表明：撞击区上部（400 mm）X_Q最低，当进料时间超过 2 400 秒时 X_Q值已不再随进料时间延长而降低；X_Q在外循环流量达 2 000～5 000 L/h 时急剧下降，维持在 0.12～0.19，微观混合效果良好。

参考文献

[1] Tamir A. 1996. 撞击流反应器——原理和应用［M］. 伍沅，译. 北京：化学工业出版社.

[2] 薛松，吴明，徐志高，等. 2012. 撞击流-活性炭吸附法制备氧化钇超细粉体［J］. 稀有金属，36（3）：440－445.

[3] 胡立舜，王兴军，于广锁，等. 2008. 撞击流反应器用于甲醇合成反应［J］. 化工学报，59（5）：1136－1142.

[4] Yu S Y. 1993. Micromixing and parallel reactions［J］. Swiss（PhD thesis），Swiss Federal Institute of Technology Zurich.

沸石—炭复相材料吸附氨氮的性能

吴小贤[1,2*] 李春生[1,2] 陈玲霞[1,2] 潘方珍[1,2]
（1. 浙江省地质矿产研究所，杭州 310007；
2. 国土资源部黏土矿物重点实验室，杭州 310007）

许多非金属矿物材料因其独特的结构而具有良好的吸附和离子交换性能，且储量大，价格低，对环境无污染，是一类环境友好、很有发展前景的优质廉价吸附剂，被广泛应用于污水治理等行业[1-3]。目前非金属矿物材料在天然河道、湖泊等水处理过程应用时主要直接以粉体或高温烧结成陶粒形式使用。当水体中投入大量粉体吸附材料时，这些吸附材料难以回收，最终沉积到水底，造成了该水域淤泥增加和二次脱附污染。而高温烧结成陶粒存在着能耗高，同时由于高温烧结作用破坏或部分破坏了非金属矿物本身结构特点而降低了吸附性能。因此，制备低能耗绿色可回收利用的非金属矿物水处理材料，对非金属矿产资源高效利用及扩大其在环保领域应用具有重大意义和市场前景。

本研究以浙江省缙云县斜发沸石粉为主要原料（粒度小于 200 目），添加木屑、淀粉等有机物为碳化结合剂，通过造粒、非氧化性低温热处理制备得到了粒度为 0.5～1 cm 的沸石—炭复相材料水处理剂颗粒，研究了不同碳化结合剂含量（5%～20%）、不同热处理温度（400～700 ℃）等对不同氨氮浓度污水的吸附性能影响。氨氮吸附实验在 250 mL 锥形瓶中进行，准确称取不同条件制备的沸石—炭复相材料水处理剂颗粒 2 g 放入锥形瓶内，并量取 200 mL 配制好浓度的氨氮溶液加入锥形瓶，开始氨氮吸附实验。根据设定吸附时间按时选取上清液进行氨氮浓度测试，计算氨氮吸附量。

采用德国布鲁克 D8 ADVANCE X 射线衍射仪对沸石粉及制备得到的沸石—炭复合水处理剂颗粒物相组成进行分析；采用 WEW - 50B 微机液压万能试样机对沸石—炭复相材料水处理剂颗粒强度进行测试；采用德国卡尔・蔡司的∑IGMA 型场发射扫描电镜对原料进行显微形貌观察；采用环境标准《水质氨氮的测定纳氏试剂分光光度法》（HJ 535—2009）对吸附试验过程中污水中氨氮含量进行检测。

研究发现，随着热处理温度升高，沸石—炭复相水处理剂颗粒强度逐渐增加，400 ℃热处理后颗粒强度为 2.8 MPa，700 ℃热处理后颗粒强度为 5.3 MPa。随着碳化结合剂从 5%增加到 20%，样品强度从 2.8 MPa 下降到 1.0 MPa。不同温度热处理后得到的沸石—炭复相材料 XRD 分析得到主要矿物相跟沸石原矿基本一致，为斜发沸石和丝光沸石，同时伴生矿物有少量石英。对沸石—炭复相材料水处理剂颗粒断口形貌观察，发现颗粒内部为疏松多孔结构，断口中有大量微米级炭结构与沸石粉体颗粒表面相互连接（图 1）。沸石—炭复合水处理剂颗粒强度获得机理主要是颗

*通信作者：吴小贤；E-mail：xiaoxian8411@163. com；手机号：18268835253。浙江省公益技术应用研究（2017C33006）资助项目。

粒中亚微米及微米尺度的片状、纤维/管状、蜂窝状碳结构通过粘接、包裹和限位作用将沸石粉体颗粒固定住，形成颗粒强度。

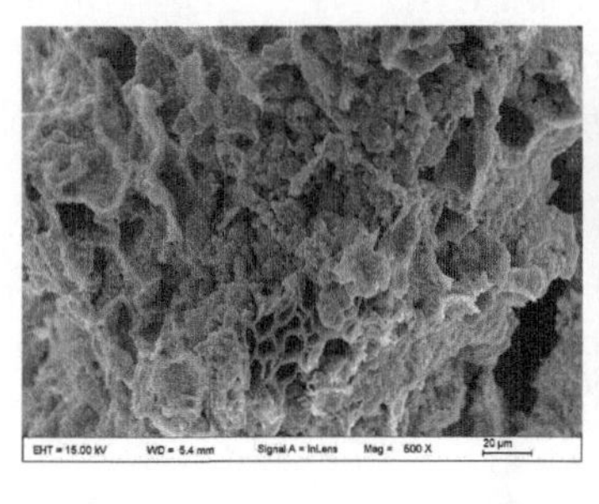

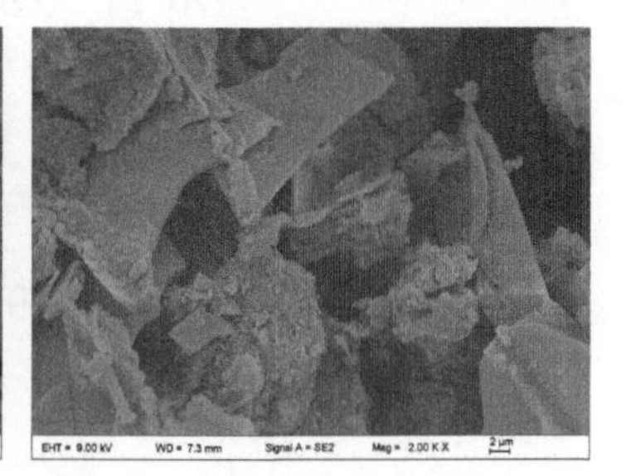

图 1　沸石—炭复合水处理剂颗粒中碳形貌结构

通过氨氮吸附实验可以表明，本研究制备的沸石—炭复相材料水处理剂颗粒具有较好的吸附氨氮性能。实验得出：5%碳化结合剂添加量样品有最优氨氮吸附性能；400 ℃热处理样品吸附速率最优，起始浓度为 15 mg/L 氨氮溶液被 400 ℃热处理制备样品（5%碳化结合剂）吸附 24 h 后氨氮溶液浓度降为 4.69 mg/L，氨氮吸附量为 1.03 mg/g；吸附 100 h 后氨氮溶液浓度降为 1.35 mg/L，氨氮吸附量为 1.36 mg/g，氨氮去除率为 90.97%。

氨氮吸附实验发现，相同质量条件下沸石—炭复相材料水处理剂颗粒在吸附速率上弱于沸石粉体，但在够吸附时间下对氨氮的吸附量同沸石粉体吸附量接近。本研究创新性地采用炭结合方式通过在非氧化性气氛下低温热处理将非金属矿物粉体造粒成在水中应用时具有一定强度的可回收颗粒复相材料，一方面利用了炭的结合强度来实现造粒，一方面生成的活性炭本身有很高的孔隙率及比表面积，不会降低非金属矿物粉体颗粒的吸附性能，甚至能增强其吸附净水性能。

参考文献

[1] 黄晓鸣，潘敏，陈天虎，等. 2016. 天然斜发沸石吸附去除水中氨氮机理研究［J］. 矿物学报，36（3）：371 - 376

[2] Liu H B，Peng S C，Shu L，et al. 2013. Effect of Fe_3O_4 addition on removal of ammonium by zeolite NaA［J］. Journal of Colloid and Interface Science，390：204 - 210.

[3] Lin L，Lei Z F，Wang L，et al. 2013. Adsorption mechanisms of high-levels of ammonium onto natural and NaCl-modified zeolites［J］. Separation and Purification Technology，103：15 - 20.

MgAlLa 水滑石基复合氧化物的制备及光催化性能

徐敏虹* 潘国祥　王永亚　郭玉华　伍　涛
（湖州师范学院材料工程系，湖州 313000）

近年来，国内外文献报到了水滑石及水滑石基复合氧化物在有机污染物降解中的应用[1]。研究发现，经过高温焙烧而得到的复合氧化物比水滑石具有更好的光催化活性[2]。另一方面，因为稀土元素电子结构的特殊性，常被用作各类材料的重要掺杂元素。将稀土元素掺杂进水滑石中制得水滑石前驱体，最后再经高温焙烧有望形成高催化活性的复合氧化物。因此，用稀土元素掺杂水滑石形成复合氧化物光催化剂具有重要的研究意义和实用价值。

本文通过共沉淀法制得一系列不同摩尔比的 Mg/Al/La 的水滑石前驱体（MgAlLa - LDHs），经 600 ℃焙烧得到 Mg - Al - La 复合氧化物（MgAlLa - LDOs）。通过 XRD、SEM、FT - IR 和 UV - Vis 等技术对样品的晶体结构、表面形貌和光催化活性进行表征。并在可见光照射下，通过测定亚甲基蓝（MB）溶液的光催化降解率来评价 MgAlLa - LDOs 的光催化活性。

采用共沉淀法制备以 NO_3^- 为层间阴离子的 MgAlLa - LDHs。并置于 600 ℃的马弗炉煅烧5 h，得到 MgAlLa - LDOs。Mg^{2+}、Al^{3+}、La^{3+} 摩尔比为 3∶0.5∶0.5，具体表示为 $Mg_3Al_{0.5}La_{0.5}$ - LDOs。

以 MB 溶液模拟污染废水，称取约 0.05 g $Mg_3Al_{0.5}La_{0.5}$ - LDOs 置于烧杯中，再加入 25 mL 5 mg/L MB 溶液，超声分散 2 min。以 150 W 金属钠光灯灯模拟可见光照射，每隔 10 min 取样，离心，分离沉淀。得到的上层清液用 UV - 2600 紫外分光光度计（日本岛津）测试其吸光度。

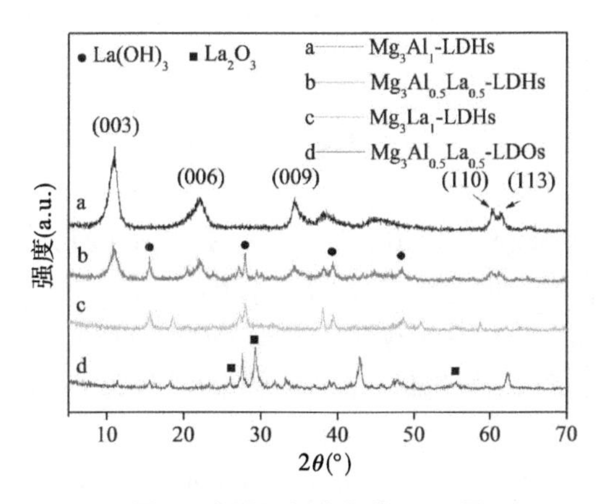

图 1　水滑石衍生物的 XRD 图

* 通信作者：徐敏虹；E-mail：xumh123@163. com；手机号：13665758929。浙江省自然科学基金（12345678）资助项目。

图 1 为水滑石衍生物的 XRD 谱图。对比 Mg_3Al_1-LDHs 和 $Mg_3Al_{0.5}La_{0.5}$-LDHs 的 XRD 谱图可见，在 $2\theta=11.7°$、22.5°、34.5°、60.5°和 62.4°附近都出现了衍射峰，表明成功地合成了水滑石样品[3]。而 c 谱线则没有表现出水滑石的特征衍射峰，表明当 La^{3+} 离子全部取代 Al^{3+} 离子时，合成的物质不是水滑石。在 $Mg_3Al_{0.5}La_{0.5}$-LDOs 的 XRD 谱图中可见 600 ℃焙烧下，$Mg_3Al_{0.5}La_{0.5}$-LDHs 的特征峰消失，说明合成的水滑石前驱体的层状结构崩塌，形成了 Mg-Al-La 复合氧化物，并且相对应的特征峰强而峰型尖锐，则证明结晶度良好。

$Mg_3Al_{0.5}La_{0.5}$-LDOs/H_2O_2对 MB 的光催化如图 2（a）所示，在 664 nm 处有一强吸收峰，且随着光催化反应时间的延长，MB 溶液的最大吸收峰逐渐减弱。当光催化到 1 h 后，吸收峰不明显，这归因于 MB 光催化降解过程中其发色基团的不饱和共轭键被打断[4]。图 2（b）为在 $Mg_3Al_{0.5}La_{0.5}$-LDOs/H_2O_2反应体系中加入不同捕获剂对 MB 光催化降解的影响。由图 2（b）可知，不加捕获剂光催化反应 1 h 后 MB 剩余率为 0.11%。在体系中加入对苯醌（BQN）、EDTA 时，MB 的降解效果明显下降，而加入异丙醇（IPA）捕获剂时，MB 的降解速度明显快于未加反应体系。这证明了在 $Mg_3Al_{0.5}La_{0.5}$-LDOs/H_2O_2反应体系中，$\cdot O_2^-$ 和 h^+ 是光催化的主要活性种，而 · OH 的作用较小。

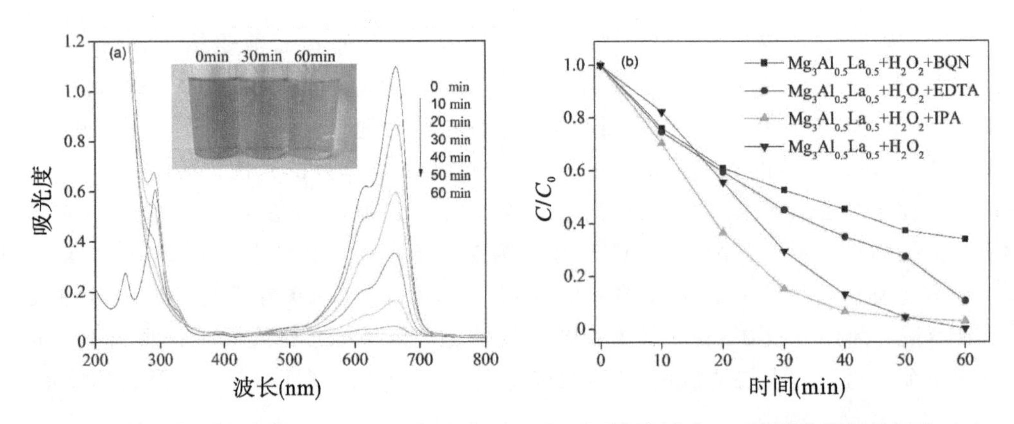

图 2 MB 降解过程中的 UV-Vis 吸收光谱（a）和不同捕获剂对 MB 光催化降解的影响（b）

参考文献

[1] Mora M，López M I，Jiménez-Sanchidrián C，et al. 2010. Ca/Al mixed oxides as catalysts for the meerwein-ponndorf-verley reaction [J]. Catalysis Letters，136（3-4）：192-198.

[2] Xu X，Xie L，Li Z，et al. 2013. Ternary MgO/ZnO/In_2O_3，heterostructured photocatalysts derived from a layered precursor and visible-light-induced photocatalytic activity [J]. Chemical Engineering Journal，221（4）：222-229.

[3] Pang X，Ni Z M，CAO F，et al. 2012. Hydrogen production from aqueous-phase reforming of ethylene glycol over Ni/Sn/Al hydrotalcite derived catalysts [J]. Applied Clay Science，58（4）：108-113.

[4] Zhang T，Oyama T，Aoshima A，et al. 2001. Photooxidative n-demethylation of methylene blue in aqueous tio2 dispersions under UV irradiation [J]. Journal of Photochemistry & Photobiology A Chemistry，140（2）：163-172.

月桂酸/多孔碳化木复合相变材料的制备及性能

杨志伟　邓　勇　李金洪*

（非金属矿物与固废资源材料化利用北京市重点实验室，
中国地质大学（北京）材料科学与工程学院，北京 100083）

月桂酸是一种常见的有机相变材料，具有优异的化学稳定性、热稳定性和高潜热（175～220 J/g），经常作为复合相变材料的封装材料，据报道月桂酸与高岭石、海泡石、膨胀蛭石等多孔材料制备的复合相变材料虽有许多优异的性能，但同时具有诸多缺点，例如基体对月桂酸的相变行为的限制作用及不令人满意的封装量（30wt.％～70wt.％）等。本研究制备了一种新的封装基体——多孔碳化木，用以封装月桂酸制备月桂酸/多孔碳化木复合相变材料。研究表明：月桂酸的封装率高达 81.1wt.％，复合相变材料的相变潜热约为 178.2 J/g。非等温结晶结果表明，月桂酸的活化能（－333.4 kJ/mol）高于复合相变材料的活化能（－170.47 kJ/mol），而半结晶时间低于复合相变材料，表明多孔碳化木基体显著提高了月硅酸的成核速率，但略微限制了其晶体生长。热循环测试、FTIR 和 TGA 结果表明月桂酸/多孔碳化木复合相变材料显示出优异的热可靠性、化学相容性和热稳定性。

由于化石燃料能源短缺和能源利用率低，能源危机已引起全球关注。热能储存（TES）成为合理有效利用现有能源的有效方法之一。对于现有的储热方法，潜热储存因其高储热能力和优异的相变行为，已在各个领域得到广泛应用和研究[1,2]。相变材料（PCM）是实现潜热储存的最有效介质，已应用于太阳能储存、绝缘服装、余热回收、建筑节能等。目前相变材料应用于建筑节能方面较为广泛，多为矿物基复合相变材料，本文力求寻找一种新的基体材料，使其具有复合相变材料的优异性能，同时克服普遍存在的基体对相变材料相变行为的限制作用的问题，进一步增大相变材料的储热能力，进而应用于更广泛的领域。本研究针对以上问题，制备了一种新型的复合相变材料—月桂酸/多孔碳化木复合相变储能材料。

实验采用的原料主要有氢氧化钠溶液、亚硫酸钠溶液和桐木等，复合相变材料制备方法为真空熔融浸渍法，主要仪器有 SEM（HITACHI S－4800）、DSC（214 Polyma，NETZSCH）和导热系数仪（XIATECH TC 3000E）等。

DSC 结果表明：月桂酸/多孔碳化木复合相变材料的潜热高达 178.2 J/g，且多孔碳化木的碳化程度对基体的封装性能影响不大，计算的潜热略高于实验潜热（图 1），表明多孔碳化木对月桂酸的相变行为几乎没有限制作用。

* 通信作者：李金洪；E-mail：jinhong@cugb. edu. cn；国家自然科学基金（U1607113）、青海省科技计划项目（2017-HZ-805）资助项目。

从图 2（a）可以看出，多孔碳化木具有规则的孔道结构、大量的孔体积，具有作为封装基体的特性；图 2（b）中孔道几乎全部封装满月桂酸，没有泄露，表明多孔碳化木具有优异封装性能。

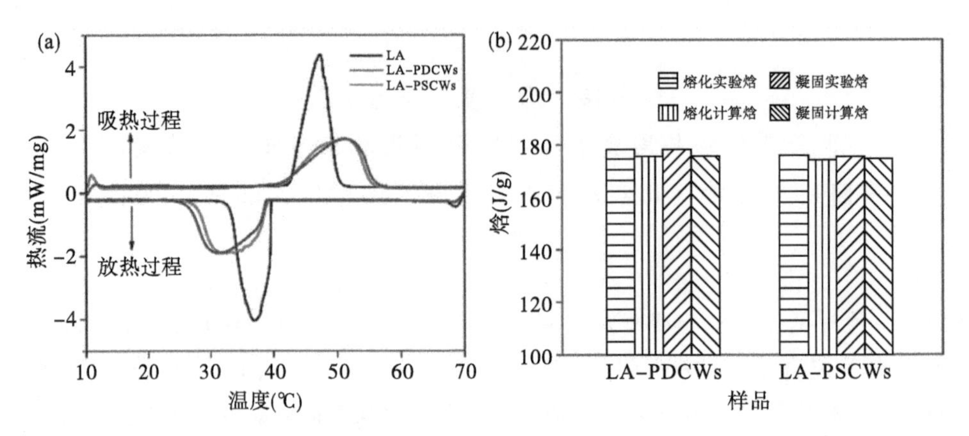

图 1　LA 和 LA/PCWs ss－CPCM DSC 曲线（a）及实验和计算的焓的比较（b）

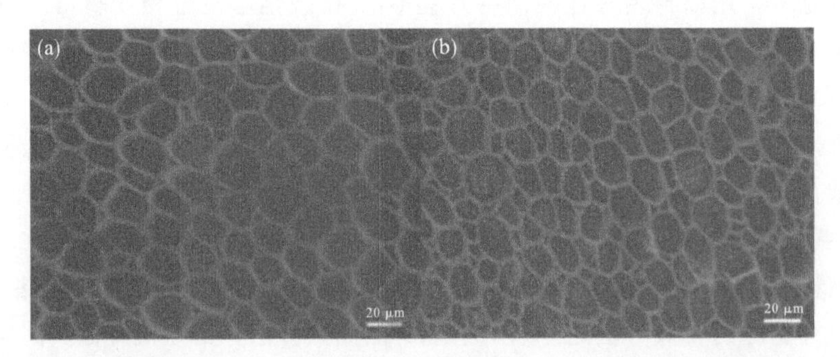

图 2　处理后的多孔碳化木基体（a）和封装后的月桂酸/多孔碳化木复合相变材料（b）电镜图

本研究制备的复合相变材料由于其低密度、优异的热可靠性、热稳定性和储热能力，可广泛应用于家具、建筑等相变储能领域。

参考文献

［1］Al-Maghalseh M，Mahkamov K. 2018. Methods of heat transfer intensification in PCM thermal storage systems：Review paper［J］. Renewable & Sustainable Energy Reviews，92：62－94.

［2］Khan Z，Khan Z，Ghafoor A. 2016. A review of performance enhancement of PCM based latent heat storage system within the context of materials，thermal stability and compatibility［J］. Energy Conversion & Management，115（5）：132－158.

梯度化富含介孔结构沸石分子筛的设计与表征

王承栋　李金洪*

（非金属矿物与固废资源材料化利用北京市重点实验室，

中国地质大学（北京）材料科学与工程学院，北京 100083）

梯度化介孔沸石结合了高酸度和热稳定性，与传统沸石类似，具有极高的传质性能。模板化，脱铝和脱硅是常见的用于形成附加介孔的策略[1]。其中碳纳米粒子和中尺度阳离子聚合物一般用模板法合成。脱铝和脱硅是基于优先从沸石骨架中提取铝或硅的方法，在沸石合成后通过酸或碱进行二次处理。通过改变碱或酸的浓度、处理的持续时间和温度，可以形成不同沸石孔道的类型。实验成功合成了梯度化富含介孔结构沸石，其介孔直径分布在 3～5 nm。

实验选用的是 NaY 型沸石，与 X 型的沸石相比拥有更大的硅铝比。现代工业对沸石分子筛的酸性和水热稳定性有了更高的要求，因此实验利用酸碱调节沸石的酸性且对其进行硅烷化来提高它的表面稳定性，同时通过二次造孔法实现对沸石孔道结构的调控，希望能在确保沸石的结晶性不被明显破坏的情况下在沸石晶体内部引入介孔结构，特别是具有三维连通性质的介孔或者是大孔结构，提高扩散效率，进而有效地提高沸石的催化活性。

本实验选取的材料有 NaY 型分子筛、NaOH、$(NH_4)_3PO_4 \cdot 3H_2O$、$C_6H_5CH_3$、HCl、三甲基氯硅烷。通过粉末 X 射线衍射对沸石进行物相分析，利用扫描电子显微镜对沸石微观形貌进行观察，通过核磁共振和红外光谱对沸石内部 Al 原子的所处位置结构变化以及表面沸石所接的键的类型进行观测，最后使用氮气脱附来验证孔的大小和形貌。

从图 1a 可以看出，经过酸洗处理后，四组图像中都出现了回滞环，说明沸石内部已经出现介孔结构。二次造孔法通过脱硅作用将骨架硅溶解。骨架 Al 被认为在溶解机理中具有“孔隙导向”作用，因为负电荷的 AlO_4^- 单元可以防止从框架中提取相邻的硅原子，所以硅铝比对梯度化孔道的形成也有极大影响。实验通过对沸石进行碱处理，在脱硅的同时调控硅铝比，达到一个良好的梯度化孔道结构。图 2 为 HY 型沸石经过酸洗碱洗后 N_2 吸附脱附曲线，说明在二次造孔法制得的沸石中出现介孔，通过观察孔径分布图可以发现碱洗后孔道孔径主要还是集中在 3～5 nm，不同的热处理温度和热处理时间对沸石孔道的形成有显著影响。

实验通过离子交换及高温热处理等方式来对 NaY 型沸石的酸度进行改性，未进行改性的沸石直接进行酸浸，其结晶度大幅降至 32.04%，而改性沸石在酸浸后结晶度反而有 5%～8% 等的提升，说明离子交换、高温热处理等一系列改性措施对酸浸沸石起到保护作用。高温脱铝法破坏了沸石中的铝氧八面体，使沸石中产生游离态的铝氧五面体，创建了非均匀的孔道结构，为之后的酸洗提供了缺陷条件。随着热处理时间的增加，沸石内部 Al 的量出现变化，在随后的酸洗碱

* 通信作者：李金洪；E-mail：jinhong@cugb. edu. cn；中央高校基本科研业务费优秀教师基金（53200859163）资助项目。

洗中，不同处理时间导致其比表面积的变化程度有所差异，时间越长，其比表面积变化越大。在温度为 873 K 条件下，当热处理时间为 3 h 时，碱洗前后表面积从 317.62 m^2/g 减小为 303.02 m^2/g，处理时间升高到 6 h 时，碱洗前后表面积从 449.17 m^2/g 减小为 342.10 m^2/g。

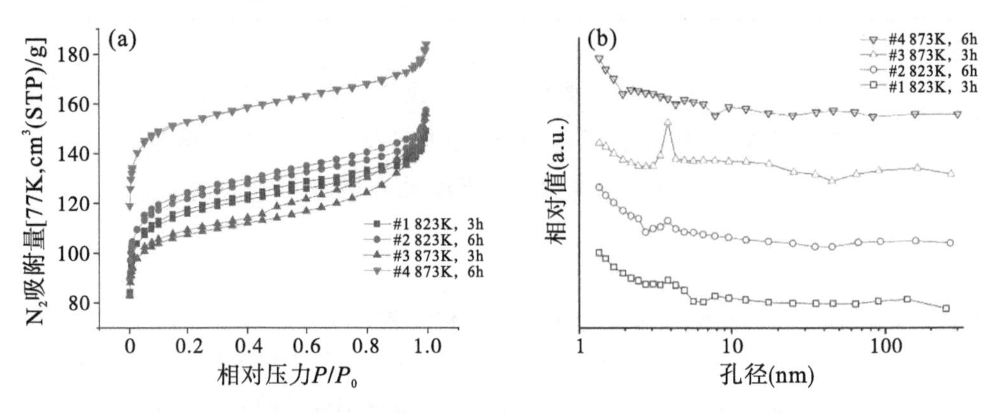

图 1　HY 型沸石经过酸洗后 N_2 吸附脱附曲线（a）及孔径分布（b）

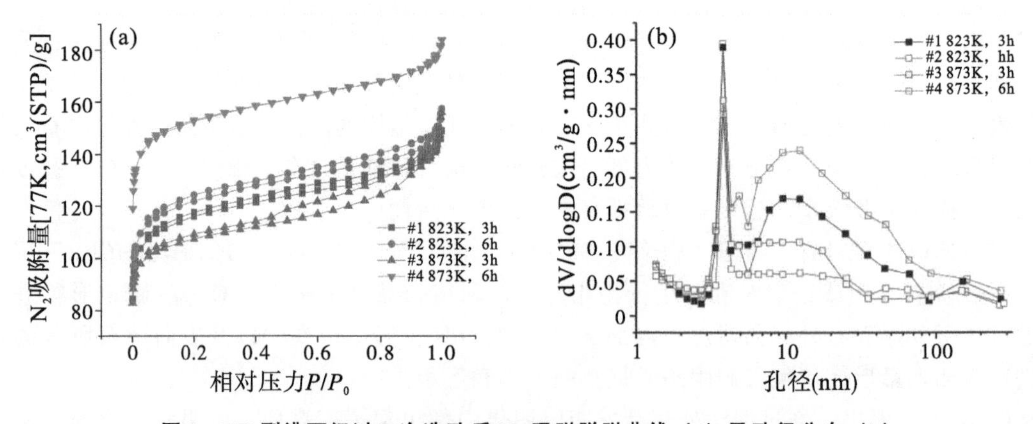

图 2　HY 型沸石经过二次造孔后 N_2 吸附脱附曲线（a）及孔径分布（b）

参考文献

［1］Feliczak-Guzik A. 2017. Hierarchical zeolites：Synthesis and catalytic properties［J］. Microporous & Mesoporous Materials，259：33 - 45.

硅酸钙合成方法和新应用

夏淑婷[1]　周春晖[1,2*]

（1. 浙江工业大学化学工程学院，杭州 310014；

2. 青阳非金属矿研究院，青阳 242800）

硅酸钙材料具有多种结构形式，主要包括 C－S－H（水化硅酸钙）、活性硅酸钙、硬硅钙石、多孔硅酸钙等。C－S－H 是一种无定形的物质，在常温条件下呈凝胶态，比表面积大、孔径均匀、无毒环保，是硅酸盐水泥成分中主要水化产物；也可以作为污水重金属吸附材料；活性硅酸钙晶体呈针状结构，表面经过改性处理以后，常用于造纸、橡胶等领域；硬硅钙石晶体结构为纤维状或针状，耐热温度高，常用作保温和阻燃材料。多孔硅酸钙材料具有密度小、强度高、耐高温、耐腐蚀以及具有较高的表面积等优点，常用作光催化剂和药物的载体，细轻质硅酸钙也可以作为造纸填料[1-6]。硅酸钙板由白水泥、胶水、玻璃纤维复合而成的多元材料，因为防火、防潮、隔音、隔热等性能，常用作建筑材料（图 1）。

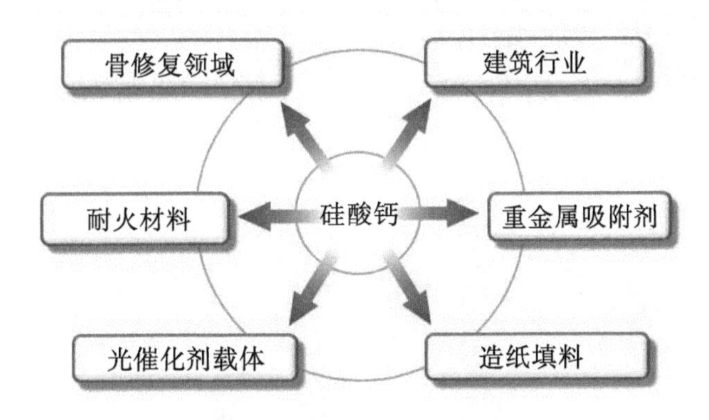

图 1　硅酸钙的主要应用领域

在人工合成中，通常用溶液反应法或静态/动态水热合成水化硅酸钙、沉淀法合成多孔硅酸钙、微乳液法或化学沉淀法制备纳米硅酸钙、模板法合成介孔硅酸钙。

不同钙硅比的硅酸钙矿物因为晶体结构中键长分布和配位多面体链接方式的不同，导致硅酸盐矿物晶体结构的化学性质及水化活性方面有较大差异。例如，在光电或半导体材料中，需要没有水化活性，提高器件的使用寿命；在建筑材料中，希望加速硅酸盐矿物的水化速率，从而达到节能减排的要求。

因为硅酸钙优良的生物相容性，且能与多种化学物质形成复合材料，应用范围随之变广。研究表明硅酸钙粉体或陶瓷具有很好的生物活性和诱导沉积类骨羟基磷灰石层的能力，且将硅酸钙喷涂至骨修复钛合金表面形成涂层，该涂层仍具有良好的生物活性。同时水化硅酸钙在环保行业应用也很广泛，不仅可以吸附污水中的氨氮，改性后还能吸附重金属离子。但由于C－S－H是一种超细粉末，在酸性环境会发生溶解，因而在处理酸性重金属废水时容易损耗，需要通过改性增强其耐酸性能，硅酸钙—壳聚糖聚合物能更有效地吸附污水中的重金属离子。

硅酸钙也可以利用工业废料制得，将粉煤灰中非晶态 SiO_2 化学提取转化成超细轻质多孔的硅酸钙填料时，用于纸张填料。硅酸钙存在难分散、易絮聚等缺点，在制作保温材料时往往要加入分散剂、胶黏剂以达到使用要求，而植物纤维则可有效改进以上问题。在针叶木浆纤维的细胞腔壁以及分丝帚化形成的网络空间内原位合成硅酸钙，有望应用于抄造文化用纸的工业生产上。

参考文献

[1] Hao F，Qin L，Liu J. et al. 2018. Assessment of calcium sulfate hemihydrate-tricalcium silicate composite for bone healing in a rabbit femoral condyle model [J]. Materials Science & Engineering C.

[2] Santos R L，Horta R B，Pereira J，et al. 2018. Alkali activation of a novel calcium-silicate hydraulic binder with $CaO/SiO_2=1.1$ [J]. Journal of the American Ceramic Society.

[3] 徐鹏，刘忠，王成海，等. 2017. 纤维原位合成硅酸钙及其造纸性能 [J]. 天津科技大学学报，32（3）：45－49.

[4] 王倩倩，李晓冬，沈晓冬. 2017. 硅酸钙矿物的晶体结构 [J]. 南京工业大学学报（自然科学版），39（1）：39－45.

[5] 刘立华，杨刚刚，王易峰，等. 2016. 模板法合成介孔硅酸钙及其对重金属离子的吸附性能 [J]. 环境化学，35（9）：1943－1951.

[6] 谭蔚，张志豪，刘丽艳，等. 2016. 半晶态硅酸钙材料结晶状态对过滤分离的影响 [J]. 天津大学学报：自然科学与工程技术版，49（3）：248－252.

改性黑滑石填充氯丁橡胶及补强机理

许子帅[1]　汤庆国[1,2*]　王　菲[1,2]　梁金生[1,2]
（1. 河北工业大学能源与环保材料研究所，天津 300130；
2. 河北工业大学材料科学与工程学院，天津 300130）

在橡胶制品的生产中，为降低成本，减少环境污染和对化石能源的依赖，人们正在寻求炭黑填料的替代品。其中，具有天然纳米结构和纳米粒径的黏土矿物成为首选，如杜玉龙等[1]用协同改性凹凸棒土填充三元乙丙橡胶，当填料添加量为 100 分时，其拉伸强度达到 18.57 MPa。肖诚斌等[2]用氯丁橡胶溶液插层蒙脱石，当反应时间为 72 h，温度为 65 ℃，蒙脱石填量为 4%时，制备出剥离型氯丁橡胶/蒙脱土复合材料的拉伸强度 22.0 MPa。可见，表面有机化改性的黏土矿物对各种橡胶具有优异补强性能。黑滑石是滑石形成过程中，因碳或有机碳，以及铁、锰等致色成分浸染或混入而形成的黑色或灰黑色的滑石矿物（$3MgO \cdot 4SiO_2 \cdot H_2O$），具有滑石矿物的惰性、绝缘性、热稳定性、流变性等，但颜色缺陷限制了其在造纸、涂料等领域中的应用[3-5]。本研究以改性黑滑石为补强填料，探讨改性黑滑石对氯丁橡胶的补强机理及作为密封材料的适应性。

黑滑石产自江西省广丰县，其中滑石占 90%以上，石英约 5%，及有机碳，化学成分：SiO_2 为61.04%、CaO 为 2.14%、MgO 为 27.24%、C 为 2.0%。原矿经烘干、粉碎、改性备用；氯丁橡胶 DCR－36（日本电气公司）；改性剂为钛酸酯偶联剂 NDZ－201；其他助剂分别为 2,2,4－三甲基－1,2－二氢化喹啉聚合体（RD）和 4,4′－二异辛基二苯胺（4010NA）、石蜡、N－环己基－2－苯骈噻唑次磺酰胺（CZ）、氧化镁、氧化锌、硫磺和无水乙醇。

力学性能按国标 GB/T 528—2009、GB/T 529—2008 测试，结合傅里叶红外光谱、X 射线衍射分析和扫描电镜分析、动态机械性能分析等研究晶体结构的变化及其补强机理。

（1）比较炭黑 N330（C70）、超细黑滑石（H70）和改性黑滑石（GH70）为填料，添加量均为 70 分时，复合氯丁橡胶性能的变化见表 1。由表 1 可知，改性黑滑石复合氯丁橡胶的拉伸强度和撕裂强度最大。H70 和 GH70 的扯断伸长率均超过 1400%，成为高弹性复合橡胶。

表 1　复合橡胶的力学性能

样品	100%（MPa）	300%（MPa）	拉伸强度（MPa）	扯断伸长率（MPa）	撕裂强度（$N \cdot mm^{-1}$）
C70	1.54	3.723	13.71	653.5	41.09
H70	2.49	3.97	14.62	1495	49.13
GH70	2.25	3.66	16.56	1487	54.91

* 通信作者：汤庆国；E-mail：qingguo_tang@163. com；手机号：13132097129。

(2) 改性黑滑石及补强氯丁橡胶的傅里叶红外光谱分析见图 1；加工工艺对复合硫化胶中黑滑石晶体结构的影响见图 2。由图 1 可知，改性黑滑石及补强氯丁橡胶表面官能团的变化。由图 2 可知，橡胶的混炼、硫化及冷冻加工均能够使黑滑石的晶体结构发生变化，使黑滑石（001）晶面衍射峰变宽，强度降低，意味着橡胶的黏合力对层状黑滑石形成一定的剥层，完整黑滑石内核总量减少；在 $2\theta=4.3°$处出现新的衍射峰，意味着插层作用使滑石的晶面间距由 0.94 nm 增加到 1.97 nm。说明改性黑滑石与橡胶基体产生了很好的界面亲和性。

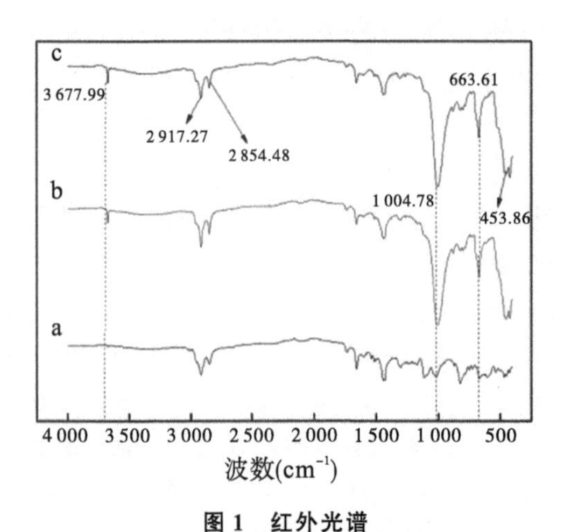

图 1　红外光谱

注：a. 氯丁橡胶；b. 黑滑石复合氯丁橡胶；c. 改性黑滑石复合氯丁橡胶

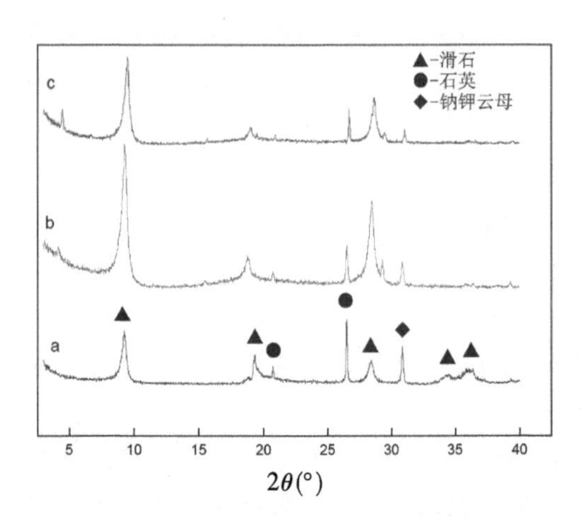

图 2　改性黑滑石与硫化胶、冰冻硫化胶的 XRD

注：a. 改性黑滑石；b. 即时硫化胶；c. 冰冻硫化胶

以改性黑滑石为填料补强氯丁橡胶，探讨改性剂用量和工艺条件对复合橡胶性能的影响。结果表明，当改性黑滑石添加量为 70 分时，制备出扯断伸长率达 1487％的高弹性复合氯丁橡胶，其拉伸强度、撕裂强度、扯断伸长率分别比炭黑填充氯丁橡胶提高 20.8％、33.6％和 127.5％，可以满足密封材料对环境适应性的特殊要求。

参考文献

[1] 杜玉龙，汤庆国，王菲，等. 2013. 协同改性凹凸棒石对补强三元乙丙橡胶性能的影响［J]. 硅酸盐学报，41（1）：71－75.

[2] 肖诚斌，张泽朋，刘建辉，等. 2006. 氯丁橡胶/蒙脱石纳米复合材料的研制［J]. 材料导报，（S1）：208－210.

[3] Palaniandy J S. 2011. Comparative study of water-based and acid-based sonications on structural changes of talc［J]. Applied Clay Science，51（4）：399－406.

[4] Dellisanti F，Valdrè G，Mondonico M. 2009. Changes of the main physical and technological properties of talc due to mechanical strain［J]. Applied Clay Science，42（3－4）：398－404.

[5] 任叶叶，张俭，严俊，等. 2015. 应用 X 射线衍射—红外光谱等技术研究滑石在机械力研磨中的形貌和晶体结构变化及影响机制［J]. 岩矿测试，34（2）：181－186.

解淀粉芽孢杆菌对 Re（Ⅶ）和 Se（Ⅳ）在膨润土中的扩散影响

王永亚[1]　赵帅维[2]　潘国祥[1]　徐敏红[1]　伍　涛[1*]
（1. 湖州师范学院工学院，湖州 313000；
2. 中国辐射防护研究院，合肥 213456）

高庙子膨润土具有渗透率低、阳离子交换容量高和储量大等优点，是我国高放废物地质处置库的候选回填材料。它的表面带负电荷，由于阴离子排斥效应，很难阻滞长寿命放射性裂片元素 $^{79}SeO_3^{2-}$ 和 $^{99}TcO_4^-$ 的扩散。因此，在高放废物地质处置安全评价研究中，这些放射性阴离子的扩散行为是研究的重点内容[1,2]。处置库中含有大量的芽孢杆菌，可以通过风化、分解和转化改变膨润土的结构和性能，从而改变回填材料对核素的阻滞能力[3]。也可以通过螯合作用和还原作用，改变核素的扩散行为[4]。

本文采用贯穿扩散法研究解淀粉芽孢杆菌对 Re（Ⅶ）［模拟 ^{99}Tc（Ⅶ）］和 Se（Ⅳ）［模拟 ^{79}Se（Ⅳ）］在高庙子膨润土中的扩散行为。采用扩散参数拟合程序（FDP）对数据进行处理，得到了 Re（Ⅶ）和 Se（Ⅳ）的有效扩散系数和有效孔隙率。最后采用阴离子排斥模型，研究压实膨润土的空隙分布与压实密度的关系，进而了解解淀粉芽孢杆菌如何影响 Re（Ⅶ）和 Se（Ⅳ）的扩散行为。

采用贯穿扩散法（Through-diffusion method）[2]，所用的实验装置由 1 个蠕动泵、1 个扩散池（内置压实膨润土，密度为 1 600 kg/m^3）、1 个高浓度储液瓶和 1 个低浓度储液瓶组成，分别与扩散池的两侧相连。

在高浓度储液瓶和低浓度储液瓶中装入除了 Re（Ⅶ）和 Se（Ⅳ）之外的溶液，平衡 5 个星期。接着将解淀粉芽孢杆菌、Re（Ⅶ）或 Se（Ⅳ）加入高浓度储液瓶中，在蠕动泵的带动下，Re（Ⅶ）或 Se（Ⅳ）在扩散池的一侧形成高浓度的液膜，将会很快穿透膨润土块，扩散到另一侧的低浓度储液瓶中，每隔一段时间更换低浓度储液瓶，采用电感耦合等离子体光谱仪测量 Re（Ⅶ）或 Se（Ⅳ）的浓度，得到 Re（Ⅶ）或 Se（Ⅳ）的通量和累积扩散量随时间变化的实验数据。

采用贯穿扩散法获得 Re（Ⅶ）和 Se（Ⅳ）的累计扩散总量和扩散通量与扩散时间的实验数据，结合 FDP 软件对实验数据进行处理，获得有效扩散系数和有效孔隙率，结果见表 1。由于阴离子排斥效应，Re（Ⅶ）不能进入膨润土的 TOT 层间空隙，它不吸附在膨润土的表面。而 Se（Ⅳ）在本实验条件下为 $HSeO_3^-$，能够与膨润土表面形成外层络合物。解淀粉芽孢杆菌存在下，提高了 $HSeO_3^-$ 的分配系数[5]，降低了有效扩散系数和表观扩散系数。

目前，常用阴离子排斥模型（AMP）描述放射性阴离子的扩散机理[6]，阴离子主要通过膨润

* 通信作者：伍涛；E-mail：twu@zjhu. edu. cn；手机号：13655729651。浙江省自然科学基金（LY18B070006）资助项目。

土的颗粒间空隙向外扩散。该模型是在膨润土的孔隙模型基础上建立的，所用参数见表 2。图 1 为不同压实密度条件下，Re（Ⅶ）在膨润土中的空隙分布。可以看出，解淀粉芽孢杆菌不影响膨润土的扩散双电层孔隙率，但是增加了膨润土的 TOT 层间孔隙率，使得 Re（Ⅶ）的有效孔隙率降低，由于膨润土的层间距变化不明显，对 Re（Ⅶ）的有效和表观扩散系数影响不大。

表 1　芽孢杆菌存在下 Re（Ⅶ）和 Se（Ⅳ）在压实膨润土中的扩散参数

元素	条件	C_0 (mg/L)	D_e （× $10^{-11}m^2/s$）	D_a （× $10^{-11}m^2/s$）	α （—）	K_d （× $10^{-4}m^3/kg$）
	1 000	180 ±10	6.13 ±0.35	18.7 ±0.4	0.33 ±0.02	/
	1 400	180 ±10	3.26 ±0.18	18.3 ±0.2	0.19 ±0.01	/
	1 600	180 ±10	2.09 ±0.12	13.3 ±0.1	0.16 ±0.01	/
Re（Ⅶ）	1 800	180 ±10	0.29 ±0.02	2.9 ±0.1	0.10 ±0.01	/
	1 300*	1 900 ±150	5.3 ±0.4	19.0 ±2.0	0.28 ±0.02	/
	1 600*	1 900 ±150	1.80 ±0.1	10.0 ±2.0	0.18 ±0.01	/
	1 800*	600 ±50	0.30 ±0.02	2.1 ±0.2	0.14 ±0.01	/
	1 000	630 ±20	3.82 ±0.13	3.78 ±0.23	1.01 ±0.02	3.82 ±0.53
	1 400	630 ±20	2.56 ±0.09	2.17 ±0.13	1.18 ±0.06	4.89 ±0.64
	1 600	630 ±20	1.03 ±0.04	0.93 ±0.05	1.21 ±0.06	5.07 ±0.67
Se(Ⅳ)	1 800	630 ±20	0.46 ±0.02	0.34 ±0.07	1.35 ±0.18	5.95 ±0.57
	1 300*	10 000 ±200	5.4 ±0.3	7.8 ±0.5	0.69 ±0.03	1.4
	1 600*	8 000 ±200	2.3 ±0.2	3.4 ±0.4	0.68 ±0.05	1.8
	1 800*	16 000 ±200	0.45 ±0.02	0.7 ±0.05	0.62 ±0.03	1.7

注：* 表示没有解淀粉芽孢杆菌

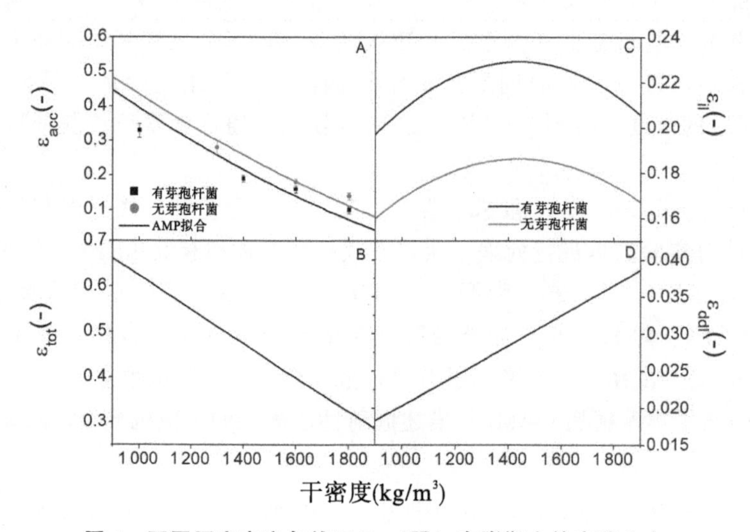

图 1　不同压实密度条件下 Re（Ⅶ）在膨润土的空隙分布

表 2　APM 拟合参数

条件	A_{ext} (m^2/g)	A_{int} (m^2/g)	n_c (—)	m	A (nm)	B (nm)	f_{ddl} (—)	f_{dens} (—)
有芽孢杆菌	25.8	74.9	30	0.754	0.523	0.905	1	0.8
无芽孢杆菌	25.8	74.9	30	0.754	0.532	0.905	1	0.65

参考文献

[1] Tsai T L，Tsai S C，Shih Y H，et al. 2017. Diffusion characteristics of HTO and TcO_4^- -^{99}Tc in compacted Gaomiaozi (GMZ) bentonite [J]. Nuclear Science and Techniques，28.

[2] Wu T，Wang Z，Wang H，et al. 2017. Salt effects on Re (Ⅶ) and Se (Ⅳ) diffusion in bentonite [J]. Applied Clay Science，141：104－110.

[3] Zhu Y，Li Y，Lu A，et al. 2011. Study of the interaction between bentonite and a strain of Bacillus mucilaginosus [J]. Clays and Clay Minerals，59：538－545.

[4] Mishra R R，Prajapati S，Das J，et al. 2011. Reduction of selenite to red elemental selenium by moderately halotolerant Bacillus megaterium strains isolated from Bhitarkanika mangrove soil and characterization of reduced product [J]. Chemosphere，84：1231－1237.

[5] Wu T，Wang H，Zheng Q，et al. 2014. Diffusion behavior of Se (Ⅳ) and Re (Ⅶ) in GMZ bentonite [J]. Applied Clay Science，101：136－140.

[6] Idiart A，Pękala M. Models for Diffusion in Compacted Bentonite. Swedish Nuclear Fuel and Waste Management Company：SKB TR－16－06，2016；Vol. SKB TR－16－06，p SKB TR－16－06.

黏土矿物对西图则氏假单胞菌降解原油的影响

李　磊[1]　万云洋[1*]　罗　娜[1]　何欣月[1]　刘　源[1]　穆红梅[1]　张　越[2]

(1. 中国石油大学（北京），油气资源与勘探国家重点实验室，
油气污染防治北京市重点实验室，北京 102249；
2. 中海石油气电集团有限责任公司，北京 100028)

假单胞菌属由于能够产生鼠李糖脂生物表面活性剂而在微生物提高石油采收率（MEOR）[1]领域得到应用。其降解原油过程和代谢产物已经得到广泛研究，但是由于忽视了储层中黏土矿物的影响，真实储层原油成分的变化与室内降解实验存在差异。本文综合使用原油族组分分析和气相色谱—质谱分析（GC－MS）探讨西图则氏假单胞菌（*Pseudomonas stutzeri*）在有无黏土(47.4%和0%)条件下原油降解产物的组分变化。实验结果表明：黏土组的饱和烃浓度高于无黏土组(74.47%>61.65%)，而其余组分（芳烃、非烃和沥青质）低于无黏土组；GC－MS数据支持该结果，具体表现在黏土组的63种（63/64）饱和烃都高于无黏土组（图1a）；黏土对西图则氏假单胞菌降解芳烃的影响较为复杂：抑制萘、一甲基萘、菲、一甲基菲、二甲基菲、三甲基菲、一乙基菲、芴、甲基芴、䓛和芘的降解，促进二甲基萘、三甲基萘、四甲基萘、甲基二苯并呋喃、一甲基联苯、二甲基联苯、三甲基联苯和三芳甾烷的降解（图1b）。黏土抑制西图则氏假单胞菌降解饱和烃，这可能是吸附作用导致的；而在芳环存在时，黏土对不同芳烃展现出不同的作用，特征离子可证明黏土对同一系列的芳烃影响规律一致，这可能是微生物对特定化学键的作用[2]，也可能是催化作用所致。

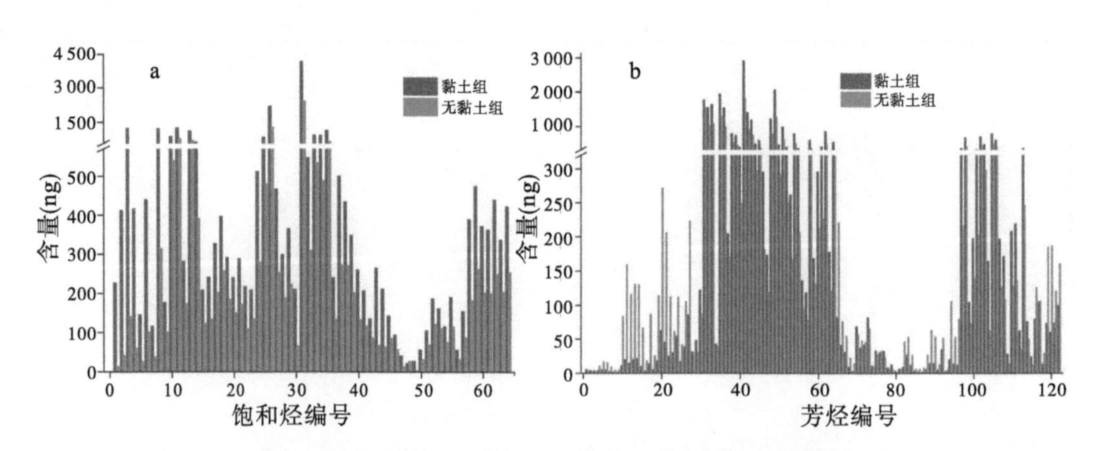

图1　降解后黏土组和无黏土组的饱和烃和芳烃含量对比

* 通信作者：万云洋；E-mail：wanyunyang@cup.edu.cn；电话：010-89732230。国家科技重大专项(2016ZX05050011)、国家自然科学基金（41373086）、科技新星与领军人才培养（Z161100004916033）资助项目。

图 1a 中：饱和烃编号 1. C15 -二环倍半萜烷；2. C15 -二环倍半萜烷；3. 8β（H）-补身烷；4. C15 -二环倍半萜烷；5. C16 -二环倍半萜烷；6. C15 -二环倍半萜烷；7. C16 -二环倍半萜烷；8. 8β（H）-升补身烷；9. 13β（H），14α（H）- C19 三环萜烷；10. 13β（H），14α（H）- C20 三环萜烷；11. 13β（H），14α（H）- C21 三环萜烷；12. 13β（H），14α（H）- C22 三环萜烷；13. 13β（H），14α（H）- C23 三环萜烷；14. 13β（H），14α（H）- C24 三环萜烷；15. 13β（H），14α（H）- C25 三环萜烷（R）；16. 13β（H），14α（H）- C25 三环萜烷（S）；17. 13β（H），14α（H）- C26 三环萜烷（R）；18. C24 -四环萜烷；19. 13β（H），14α（H）- C26 -三环萜烷（S）；20. 13β（H），14α（H）- C28 三环萜烷（R）；21. 13β（H），14α（H）- C28 三环萜烷（S）；22. 13β（H），14α（H）- C29 三环萜烷（R）；23. 13β（H），14α（H）- C29 三环萜烷（S）；24. 18α（H）- 22，29，30 三降藿烷（Ts）；25. 17α（H）- 22，29，30 三降藿烷（Tm）；26. 17α，21β（H）- 30 降藿烷（C29H）；27. 18α，21β（H）- 30 降新藿烷（C29Ts）；28. C30 重排藿烷；29. 17β（H），21α（H）- 30 降莫烷；30. 18α（H）-奥利烷；31. 17α（H），21β（H）-藿烷（C30H）；32. 17β（H），21α（H）-莫烷（C30M）；33. 17α（H），21β（H）- 30 升藿烷（22S）；34. 17α（H），21β（H）- 30 升藿烷（22R）；35. 伽马蜡烷；36. 17β（H），21α（H）- 30 升莫烷（22S＋22R）；37. 17α（H），21β（H）- 30，31 二升藿烷（22S）；38. 17α（H），21β（H）- 30，31 二升藿烷（22R）；39. 17α（H），21β（H）- 30，31，32 三升藿烷（22S）；40. 17α（H），21β（H）- 30，31，32 三升藿烷（22R）；41. 17α（H），21β（H）- 30，31，32，33 四升藿烷（22S）；42. 17α（H），21β（H）- 30，31，32，33 四升藿烷（22R）；43. 17α（H），21β（H）- 30，31，32，33，34 五升藿烷（22S）；44. 17α（H），21β（H）- 30，31，32，33，34. 五升藿烷（22R）；45. C21 - 5α（H）-孕甾烷；46. C22 - 5α（H）-升孕甾烷；47. 13β（H），17α（H）-重排胆甾烷（20S）；48. 13β（H），17α（H）-重排胆甾烷（20R）；49. 13α（H），17β（H）-重排胆甾烷（20S）；50. 13α（H），17β（H）-重排胆甾烷（20R）；51. 5α（H），14α（H），17α（H）-胆甾烷（20S）；52. 5α（H），14β（H），17β（H）-胆甾烷（20R）；53. 5α（H），14β（H），17β（H）-胆甾烷（20S）；54. 5α（H），14α（H），17α（H）-胆甾烷（20R）；55. 24 -乙基- 13β（H），17α（H）-重排胆甾烷（20R）；56. 24 -乙基- 13α（H），17β（H）-重排胆甾烷（20S）；57. 24 -甲基，5α（H），14α（H），17α（H）-胆甾烷（20S）；58. 24 -甲基，5α（H），14β（H），17β（H）-胆甾烷（20R）；59. 24 -甲基，5α（H），14β（H），17β（H）-胆甾烷（20S）；60. 24 -甲基，5α（H），14α（H），17α（H）-胆甾烷（20R）；61. 24 -乙基，5α（H），14α（H），17α（H）-胆甾烷（20S）；62. 24 -乙基，5α（H），14β（H），17β（H）-胆甾烷（20R）；63. 24 -乙基，5α（H），14β（H），17β（H）-胆甾烷（20S）；64. 24 -乙基，5α（H），14α（H），17α（H）-胆甾烷（20R）

图 1b 中：1. 萘；2. 2 -甲基萘；3. 1 -甲基萘；4. 2，6 -＋2，7 -二甲基萘；5. 1，3 -＋1，7 -二甲基萘；6. 1，6 -二甲基萘；7. 1，4 -＋2，3 -二甲基萘；8. 1，5 -二甲基萘；9. 1，2 -二甲基萘；10. 1，3，7 -三甲基萘；11. 1，3，6 -三甲基萘；12. 1，4，6 -＋1，3，5 -三甲基萘；13. 2，3，6 -三甲基萘；14. 1，2，7 -＋1，6，7 -三甲基萘；15. 1，2，6 -三甲基萘；16. 1，2，4 -三甲基萘；17. 1，2，5 -三甲基萘；18. 1，4，5 -三甲基萘；19. 1，3，5，7 -四甲基萘；20. 1，3，6，7 -四甲基萘；21. 1，4，6，7 -＋1，2，4，6 -＋1，2，4，7 -四甲基萘；22. 1，2，5，7 -＋1，3，6，8 -四甲基萘；23. 2，3，6，7 -四甲基萘；24. 1，2，6，7 -四甲基萘；25. 1，2，3，7 -四甲基萘；26. 1，2，3，6 -四甲基萘；27. 1，2，5，6 -＋1，2，3，5 -四甲基萘；28. 1，2，3，5，7 -五甲基萘；29. 1，2，3，6，7 -五甲基萘；30. 1，2，3，5，6 -五甲基萘；31. 菲；32. 3 -甲基菲；33. 2 -甲基菲；34. 甲基菲；35. 9 -甲基菲；36. 1 -甲基菲；37. 3 -乙基菲；38. 2 -乙基菲＋9 -乙基菲＋ 3，6 -二甲基菲；39. 1 -乙基菲；40. 2，6 -＋2，7 -＋3，5 -二甲基菲；41. 2，10 - ＋ 1，3 - ＋ 3，10 - ＋ 3，9 -二甲基菲；42. 1，6 -＋ 2，9 - ＋ 2，5 -二甲基菲；43. 1，7 -二甲基菲；44. 2，3 -二甲基菲；45. 4，9 - ＋ 4，10 - ＋ 1，9 -二甲基菲；46. 1，8 -二甲基菲；47. 1，2 -二甲基菲；48. 1，3，6 -＋1，3，10 -＋2，6，10 -三甲基菲；49. 1，3，7 -＋2，6，9 -＋2，7，9 -三甲基菲；50. 1，3，9 -＋2，3，6 -三甲基菲；51. 1，6，9 -＋1，7，9 -＋2，3，7 -三甲基菲；52. 1，3，8 -三甲基菲；53. 2，3，10 -三甲基菲；54. 三甲基菲；55. 1，6，7 -三甲基菲；56. 1，2，6 -三甲基菲；57. 1，2，7 -＋1，2，9 -三甲基菲；58. 1，2，8 -三甲基菲；59. 芴；60. 3 -甲基芴；61. 2 -甲基芴；62. 1 -甲基芴；63. 4 -甲基芴；64. C2 -芴；65. 二苯并噻吩；66. 4 -甲基二苯并噻吩；67. 2 -＋3 -甲基二苯并噻吩；68. 1 -甲基二苯并噻吩；69. 4 -乙基二苯并噻吩；70. 4，6 -二甲基二苯并噻吩；71. 2，4 -二甲基二苯并噻吩；72. 2，6 -二甲基二苯并噻吩；73. 3，6 -二甲基二苯并噻吩；74. 2，8 -二甲基二苯并噻吩；75. 2，7 -＋3，7 -二甲基二苯并噻吩；76. 1，4 -＋1，6 -＋1，8 -二甲基二苯并噻吩；77. 1，3 -＋3，4 二甲基二苯并噻吩；78. 1，7 -二甲基二苯并噻吩；79. 2，3＋1，9 -二甲基二苯并噻吩；80. 1，2 -二甲基二苯并噻吩；81. 二苯并呋喃；82. 4 -甲基二苯并呋喃；83. 2 -＋3 -甲基二苯并呋喃；84. 1 -甲基二苯并呋喃；85. 3 -甲基联苯；86. 4

-甲基联苯；87. 3-乙基联苯；88. 3,5-二甲基联苯；89. 3,3'-二甲基联苯；90. 3,4'-二甲基联苯；91. 4,4'-二甲基联苯；92. 3,4-二甲基联苯；93. 3,5,3'-三甲基联苯；94. 3,5,4'-三甲基联苯；95. 3,4,3'-三甲基联苯；96. 3,4,4'-三甲基联苯；97. 萤蒽；98. 芘；99. 苯并［a］芴；100. 苯并［b］芴；101. 2-甲基芘；102. 4-甲基芘；103. 1-甲基芘；104. 苯并［a］蒽；105. 屈；106. 3-甲基屈；107. 2-甲基屈；108. 4-甲基屈；109. 6-甲基屈；110. 1-甲基屈；111. 苯并［b］荧蒽；112. 苯并［k］荧蒽；113. 苯并［e］芘；114. 苯并［a］芘；115. 苝；116. C20 三芳甾烷；117. C21 三芳甾烷；118. C26，三芳甾烷（20S）；119. C26，＋C27，三芳甾烷（20R＋20S）；120. C28，三芳甾烷（20S）；121. C27，三芳甾烷（20R）；122. C28，三芳甾烷（20R）

参考文献

［1］李磊，穆红梅，万云洋，等. 2016. 新疆彩南油田彩 9 区块原位微生物分析及提采潜力预测［J]. 中国石油第六届化学驱提高采收率年会论文集，北京：石油工业出版社，277－285.

［2］Du W D，Wan Y Y，Zhong N N，et al. 2011. Status quo of soil petroleum contamination and evolution of bioremediation［J]. Petroleum Science，8（4）：502－514.

Continuous Synthesis of Nanominerals in Supercritical Water

Aymonier Cyril
（CNRS，Univ. Bordeaux，ICMCB，UPR 9048，F－33600 Pessac，France）

Supercritical fluids-based technologies are developed for more than 40 years. Years after years the sub-and supercritical fluids route finds new applications in the field of materials by design from organics to inorganics through carbon-based materials[1]. This continuous method is fast (few tens of seconds), sustainable and scalable and gives access to high quality nanostructured materials with unique physico-chemical properties; this means which can not be obtained with other synthetic methods[2]. More recently, supercritical continuous fluids technology, especially using water as solvent, has been investigated as an alternative and scalable approach for clay mineral processing[3-6].

This presentation proposes to introduce first the specific properties of supercritical fluids, with a focus on the ones of water, followed by the description of the principle of the processing of materials in supercritical fluids with the associated technologies[7,8]. The need to open the black box of the near-and supercritical fluids processes brought us to develop numerous tools for in situ characterizations to have a better insight into hydrodynamics, thermodynamics, chemistry and nucleation & growth[9]. This will be also described knowing that this is a key issue to better understand but also to better control.

We have proposed the first proof of the synthesis in few tens of seconds of clay minerals, namely talc, in a continuous process[3]. This innovative route offers the possibility to obtain a range of nanominerals differing in their degree (s) of crystallinity and size just by adjusting synthesis time and/or temperature. Beyond the control of these characteristics, this synthetic talc exhibits unique properties as its hydrophilicity. This hydrophilic character of the synthetic talc has conducted to the formulation of the first fluid talc filler. More recently, we went one-step forward with the demonstration of the possibility to prepare highly crystalline geominerals in just a few seconds again but under thermodynamically metastable conditions. This will be illustrated with the synthesis of the torbermorite mineral which is not abundant in nature but very interesting in the construction industry[4,5]. As a result of the supercritical continuous hydrothermal synthesis, highly crystalline fibrillar tobermorite can be obtained in just a few seconds under thermodynamically metastable conditions at 400 ℃ and 23.5 MPa.

In addition to chemistry, it is also possible to play with the process to develop multifunctional materials. Instant one-pot synthesis of functional LDH has been developed using a subcritical continuous hydrothermal multiple process for the preparation of i) pristine LDH with different compositions by varying the cation (Mg, Ni, or Zn) and the anion (CO_3, NO_3) and ii) functional LDH by varying the

E-mail：cyril. aymonier@icmcb. cnrs. fr.

functionalization agent, for example large/long organic molecules (hybrid LDH), enzymes (bio-hybrid LDH) or still inorganic metal /oxide nanocrystals (inorganic-LDH)[6].

The benefits of the sub-and supercritical continuous hydrothermal route include not only better performances for advanced applications but also environmental issues associated with the synthesis process. We will also stress challenges towards the transfer at the industrial scale of this technology in a near future.

References

[1] Padmajan S, Poulin P, Aymonier C. 2016. Prospects of supercritical fluids in realizing graphene based functional materials [J]. Advanced Materials, 28: 2663 - 2691.

[2] Roig Y, Marre S, Cardinal T, et al. 2011. Synthesis of exciton luminescent ZnO nanocrystals using continuous supercritical microfluidics [J]. Angewandte Chemie International Edition, 50: 12071 - 12074.

[3] Dumas A, Claverie M, Slostowski C, et al. 2016. Fast geomimicking using chemistry in supercritical water [J]. Angewandte Chemie International Edition, 55 (34): 9795 - 10149.

[4] Diez-Garcia M, Gaitero J J, Dolado J S, et al. 2017. Ultra-fast tobermorite supercritical hydrothermal synthesis under thermodynamically metastable conditions [J]. Angewandte Chemie International Edition, 56: 1 - 6.

[5] Diez-Garcia M, Gaitero J J, Santos I, et al. 2018. Supercritical hydrothermal flow synthesis of xonotlite nanofibers [J]. J. Flow Chem., DOI: 10. 1007/s41981 - 018 - 0012 - 7.

[6] Pascu O, Marre S, Cacciuttolo B, et al. 2017. Instant one pot preparation of functional layered double hydroxides (LDH) via a continuous hydrothermal approach, manuscript in preparation [J]. ChemNanoMat, 3 (9): 614 - 619.

[7] Aymonier C, Loppinet-Serani A, Reveron H, et al. 2006. Review of supercritical fluids in inorganic materials science [J]. Journal of Supercritical Fluids, 38: 242 - 251.

[8] Aymonier C, Philippot G, Erriguible A, et al. 2018. Playing with solvents in supercritical conditions and the associated technologies for advanced materials by design [J]. J. Supercrit. Fluids, 134: 184 - 196.

[9] Philippot G, Bojesen E, Elissalde C, et al. 2016. Insights into $BaTi_{1-y}Zr_yO_3$ ($0\leqslant y\leqslant 1$) synthesis under supercritical fluid conditions [J]. Chemistry of Materials, 28 (10): 3391 - 3400.

Clay Minerals as Industrial Adsorbents: A Review

Riccardo Tesser* Vincenzo Russo Rosa Turco Rosa Vitiello
Martino Di Serio
(University of Naples "Federico Ⅱ", Naples, 90126, Italy)

Adsorption is one among the best processes for water treatments because of its significant advantages if compared with other more energy-consuming separation techniques. In the assessment of suitable adsorption solid materials, clays minerals have recently assumed a key role because they are materials abundantly present, cheap and easy to be adapted to many adsorption applications. In the literature, they have been successfully used as adsorbents for heavy metals removal from water both with recovery or for environmental purposes[1]. Clays minerals, in their natural or chemically modified form, have been also extensively used in waste water plant treatments for removing organic molecules, pesticide residues, heavy metals or dyes. This wide variety of possible application fields has motivated the search of high efficiency and low-cost adsorbents as a potential alternative to activated carbon. Some characteristics of clay minerals such as high specific surface area, ion exchange properties, absence of toxicity, layered structure and surface chemistry, have promoted many studies based on these materials as adsorption media.

The use of clay minerals as adsorbents for removing organic pollutants or dyes is of particular interest[2,3], introduced in water streams from different industries like pharmaceutical, paper and pulp, tannery, bleaching and textile. A huge number of dyes, in fact, are currently used in many manufacturing processes and, according to some estimations[2], a significant amount of these substances (an average of 5%—20%) enters directly in the environment through wastes.

Others interesting and promising fields that could be considered, in which clay minerals can been employed in adsorption applications, are removal of fluorine from water, heavy metals from soil, antibiotics residue from water or soil.

In any adsorption application, it is of fundamental importance the theoretical modeling approach, both for what concern the batch of the continuous operation. The correct design of industrial adsorption equipment involves a detailed knowledge on the following topics: (i) adsorption mechanism, (ii) adsorption equilibrium and (iii) adsorption kinetics. A considerable number of models have been proposed and reviewed in the literature for the description of the mentioned aspects[4] and some of them have been proved very efficient in the description of experimental data. Such as example, in Fig. 1, the agree-

* Corresponding author: Riccardo Tesser; E-mail: riccardo. tesser@unina. it; Mobile Phone No.: +39-3295613464.

ment obtained is reported for the adsorption kinetics of methylene blue onto silica with different amount of adsorption material.

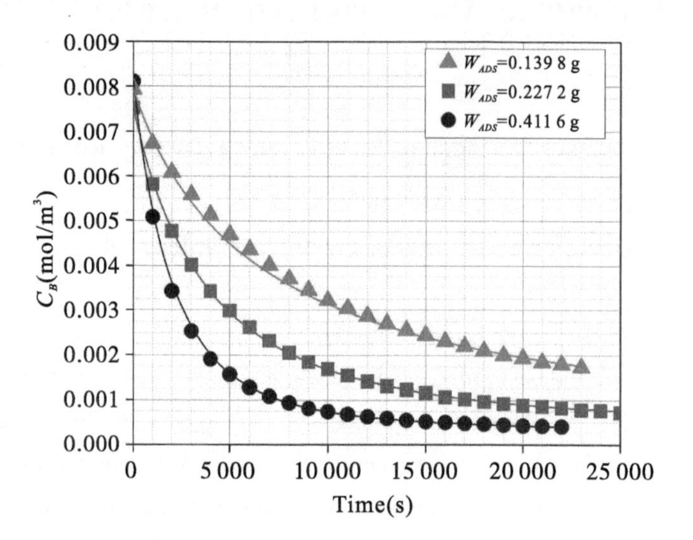

Fig. 1 Methylene blue adsorption kinetics[4]

In the present paper clay minerals, in natural or in modified form, will be reviewed in the applications as cheap, efficient and promising sorbent material. The review is devoted to both clays applications in adsorption operations and to the related data modelling for batch and continuous devices.

References

[1] Burakov A E, Galunin E V, Burakova I V, et al. 2018. Adsorption of heavy metals on conventional and nanostructured materialsfor wastewater treatment purposes: A review [J]. Ecotoxicology and Environmental Safety, 148: 702 - 712.

[2] Kausar A, Iqbal M, Javed A, et al. 2018. Dyes adsorption using clay and modified clay: A review [J]. Journal of Molecular Liquids, 256: 395 - 407.

[3] Adeyemo A A, Adeoye I O, Bello O S. 2017. Adsorption of dyes using different types of clay: a review [J]. Appl Water Sci, 7 (2): 543 - 568.

[4] Russo V, Trifuoggi M, Di Serio M, et al. 2017. Fluid-solid adsorption in batch and continuous processing: A review and insights into modeling [J]. Chemical Engineering & Technology, 40 (5): 799 - 820.

Hydration Mechanisms of Geosynthetic Clay Liners

Asli Acikel[1]　Will Gates[2*]　Abdelmalek Bouazza[1]
(1. Monash University, Melbourne, 3800, Australia;
2. Deakin University, Burwood, 3125, Australia)

Geosynthetic clay liners (GCLs) are important components in most modern engineered barriers for containing wastes as well as associated liquids and gases generated during the service life of the barrier. To be effective as barriers to liquids and gases the bentonite component within geosynthetic clay liners (GCLs) must be hydrated to >80% gravimetric water content (GWC)[1]. Recent experiments on hydration of GCLs, and the bentonite extracted from the same GCL[2], indicate that the smectite component of the bentonite controls the vapor-phase hydration of GCLs below ≈30% GWC. However, when deployed in the field at their as-manufactured moisture content of ≈ 10% GWC, subsoil mineralogy strongly influences the final hydration state reached, which is often <30% GWC[3] and well below the hydration state required for effective barrier properties. The composite structure of GCLs, in combination with the properties of the bentonite used (granulated vs powdered), has recently been shown[4] to strongly influence GCL hydration above its water entry value (WEV) from a given subgrade. While the mineralogy and particle size of the subgrade soil largely controls its resulting AEV[4] during drying, the micro-pore structure of the bentonite controls the WEV of the wetting GCL[2,4,6,7]. This paper presents recent experimental results that better GCL hydration from subgrade soils, taking into account the mineralogical, microstructural and geotechnical properties of the bentonite and the subgrade soil. Redistribution of water between the GCL and subgrade soils depends on their characteristic water retention curves and the initial suction values of both media (Fig. 1). In Fig. 1, Zone 2 represents conditions in which only the subgrade soil can maintain capillary connections, but in Zone 3 neither the GCL nor the subgrade can maintain capillary connections. In the absence of capillary connections between GCL and the subgrade, a GCL hydrating from the subgrade will remain wholly within Zones 2 and 3[4]. A GCL will take up water above its WEV (and therefore enter Zone 1) only where good capillary connectivity occurs between the GCL and the subgrade, and where the subgrade is able to release water by capillary action (i. e., the subgrade initially resides in Zone 1). To obtain optimal wetting of a GCL from subgrade soils, it is necessary to design subgrades to maximize the difference between the WEV of the GCL and the initial suction value of the subgrade. For example if using subgrades composed of fine sands, the subgrade should be easier to pre-hydrate to a suction value lower than the suction value of the target GWC of the GCL.

* Will P. Gates; E-mail: will. gates@deakin. edu. au; +61 3 9246 8373.

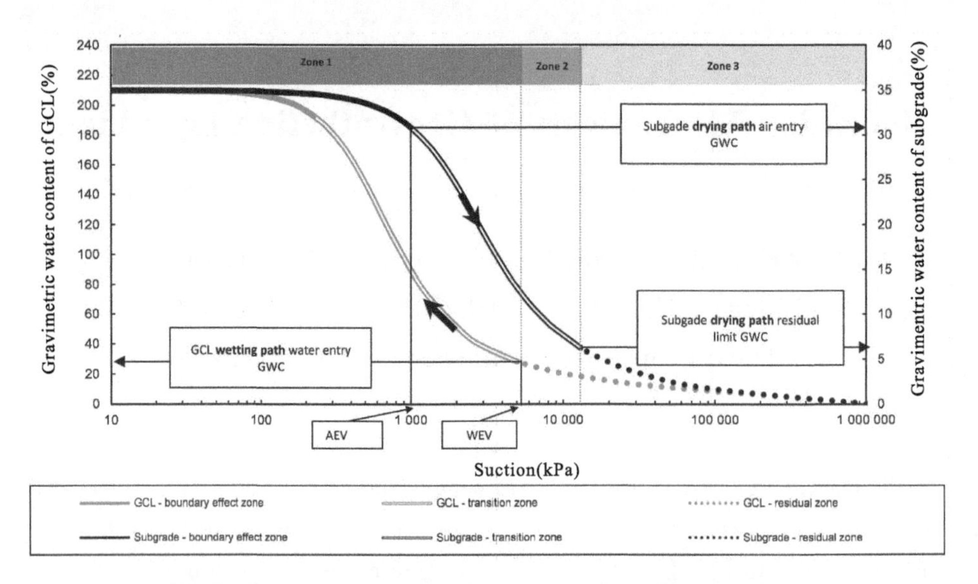

Fig. 1 Redistribution of water depends on the relative differences of water entry value of the wetting GCL and the air entry value of the drying subgrade (after [5])

References

[1] Rouf Md, Bouazza A, Singh R M, et al. 2016. Gas flow unified system for sequential meaasurment of gas diffusion and gas permeability of partially hydrated geosynthetic clay liners [J]. Canadian Geotechnical Journal, 53 (6): 1000 - 1012.

[2] Carnero-Guzman G G, Gates W P, Bouazza A, et al. 2018. Using neutron spectroscopy to measure soil water retention at high suction values [J]. Canadian Geotechnical Journal, In Press (July 2018).

[3] Bouzza A, Ali M A, Gates W P, et al. 2017. New insight on geosynthetic clay liner hydration: The key role of subsoil mineralogy [J]. Geosynthetics International, 24: 139 - 150.

[4] Acikel A S, Gates W P, Singh R M, et al. 2018. Insufficient initial hydration of GCLs from subgrades: Factors and causes [J]. Geotextiles and Geomembranes, In Press (July, 2018).

[5] Acikel A S, Gates W P, Singh R M, et al. 2018. Time dependent unsaturated behaviour of geosynthetic clay liners [J]. Canadian Geotechnical Journal, In Press, June 2018.

[6] Gates W P, Aldridge L P, Carnero-Guzman G G, et al. 2017. Water desorption and absorption isotherms of sodium montmorillonite: A QENS study [J]. Applied Clay Science, 146: 97 - 104.

[7] Gates W P, Dumedah G, Bouazza A. 2018. Micro X-ray visualization of the interaction of geosynthetic clay liner components after partial hydration [J]. Geotextiles and Geomembranes, 46: 739 - 747.

Advance in Layered Double Hydorxide for Sustainable Environment Protection

Guang Ren Qian*　Jia Zhang　Ji Zhi Zhou　Xiu Xiu Ruan　Yun Feng Xu
Dan Chen　Jian Yong Liu
(Shanghai University, Shanghai 200444, China)

The layered double hydroxides (LDHs), known as hydrotalcite-like materials or as anionic clays, are a large group of natural and synthetic materials readily produced when suitable mixtures of metal salts are exposed to base. They consist of layers, containing the hydroxides of two different kinds of metal cations and possessing an overall positive charge, which is neutralized by the incorporation of exchangeable anions. Accordingly, the layered double hydroxide was developed in environmental application for heavy metals and anions removal in wastewater.

Recently, our group reported heavy metals such as Zn, Cu, Cr, and Ni removal via LDH formation in the wastewaters[1-3] (Fig. 1). The efficient removal of heavy metal led to the purification of the wastewater and the promising hydroxide catalyst with heavy metals for organic compound degradation. Such strategy was applied in most wastewaters for heavy metal recycle, which provided the novel approach to the treatment of electroplating wastewater (Fig. 2).

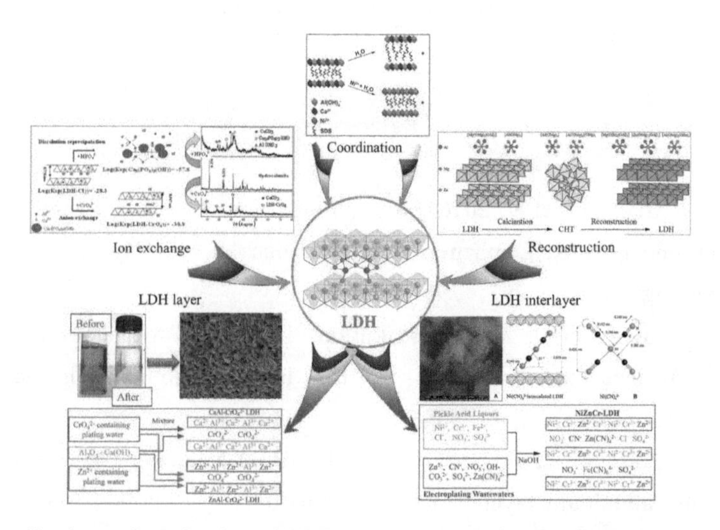

Fig. 1　The application and mechanism of LDH in the removal of heavy metals and organic pollutants

* Corresponding author: Guang Ren Qian; E-mail: grqian@shu. edu. cn; Mobile Phone No.: ++86-21-66137746.
The National Nature Science Foundation of China: No. 21477071, No. 91543123 et al.

Moreover, the anions exchanging of LDH was used in the remediation of soil with several contaminant anions, such as chromate, selenate, arsenate, phosphate. A series of CaAl, CaFe, MgFe - LDH was developed, which showed the efficient purification of anions in soil.

For the advance in application of LDH, we loaded an exfoliated NiFe - LDH nanosheet on the graphene for electrochemical water splitting[4]. Our work provides the insight into LDH in the environmental application.

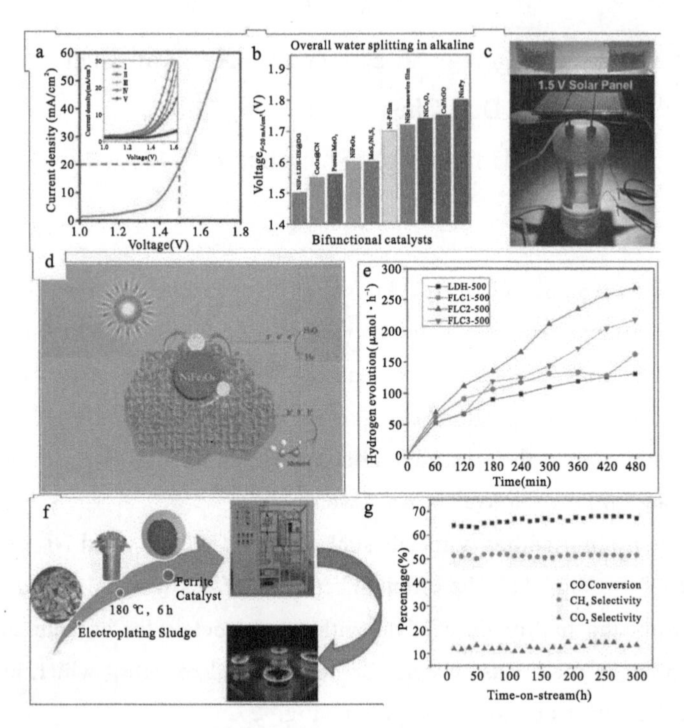

Fig. 2 The application and mechanism of LDH in OER, HER and Fischer-tropsch synthesis

References

[1] Ruan X, Chen Y, Chen H, et al. 2016. Sorption behavior of methyl orange from aqueous solution on organic matter and reduced graphene oxides modified Ni - Cr layered double hydroxides [J]. Chemical Engineering Journal, 297: 295 - 303.

[2] Zhou J, Su Y, Zhang J, et al. 2015. Distribution of OH bond to metal-oxide in Mg3 - xCaxFe - layered double hydroxide (x = 0 - 1. 5): Its role in adsorption of selenate and chromate [J]. Chemical Engineering Journal, 262: 383 - 389.

[3] Chen H, Qian G, Ruan X, et al. 2016. Removal process of nickel (Ⅱ) by using dodecyl sulfate intercalated calcium aluminum layered double hydroxide [J]. Applied Clay Science, 132: 419 - 424.

[4] Jia Y, Zhang L, Gao G, et al. 2017. A heterostructure coupling of exfoliated Ni - Fe hydroxide nanosheet and defective graphene as a bifunctional electrocatalyst for overall water splitting [J]. Advanced Materials, 29 (17): 1700017.

Enhancing Oxidative Capability of Ferrate (Ⅵ) for Oxidative Destruction of Phenol in Water through Intercalation of Ferrate (Ⅵ) into Layered Double Hydroxide

Ji Zhi Zhou　Jian Zhong Wu　Xin Huang　Ming Qi Zhang
Wei Kang Shu　Jia Zhang　Guang Ren Qian*
(Shanghai University, Shanghai 200444, China)

Chemical oxidative processes represent a major step in water treatment for elimination of various pollutants. Among different chemical oxidants, ferrate (FeO_4^{2-}) has attracted special attention due to a high redox potential and little formation of byproducts. Ferrate (Ⅵ) has been extensively studied for chemical oxidation of a broad range of organic contaminants in water[1]. Despite, the self-decay of Fe (Ⅵ) to Fe oxides led to the over-consuming of ferrate[2]. To solve this issue, ferrate intercalated Ca/Al-layered double hydroxide (Ferrate - LDH) was developed for the high efficiency of organic contaminate degradation.

Ferrate - LDH materials were synthesized via coprecipitation method (Fig. 1). For example, $CaCl_2$ and $AlCl_3 \cdot 6H_2O$ were dissolved in deionized water to form Solution A. NaOH dissolved in deionized water, and then the solution was added potassium ferrate to prepare Solution B. Solution A was quickly poured into Solution B to form a suspension. The suspension was stirred at 10 ℃ for 4 hours under a N_2 - bubbledatmosphere. After mixing, the solid products were centrifuged, rinsed, freeze-dried. The dried samples were grinded and stored.

Ferrate - LDH was synthesized and characterized. Ferrate was stably present in the LDH interlayers, in agreement with results of the Density Functional Theory calculation. The oxidative capability of Ferrate - LDH was examined in terms of the mineralization of phenol in water. The Ferrate - LDH could achieve up to 86.8% utilization efficiency during oxidative destruction of phenol in water (pH=6.5, TOC=38.3 mg/L), advantageous over direct ferrate addition that only achieved 12.6% utilization efficiency (Fig. 2). A slower evolution of dioxygen (a final product of Ferrate (Ⅵ) self-decay) was observed in the Ferrate-LDH water system, suggesting that the LDH structure inhibited Ferrate (Ⅵ) self-decay. Characterization of the LDH products before and after oxidation of phenol revealed that ferric (hydr) oxides capable of surface catalyzing Ferrate (Ⅵ) self-decay were formed on the LDH surface, not inside the LDH interlayers, suggesting that ferric (Ⅲ) was repelled from the LDH interlayers. Iso-

* Corresponding author: Guang Ren Qian; E-mail: grqian@shu.edu.cn; Mobile Phone No.: ++86-21-66137746.
The National Science Foundation Project of China (No. 51678351, No. 21707087).

lation of Fe（Ⅵ）present in the LDH structure from these active iron products may be responsible for the inhibited Ferrate（Ⅵ）self-decay when Ferrate－LDH was dosed to water. This study demonstrates that the intercalation of ferrate in LDH represents a promising approach to more efficiently and economically utilizing Ferrate（Ⅵ）for the elimination of water pollutants.

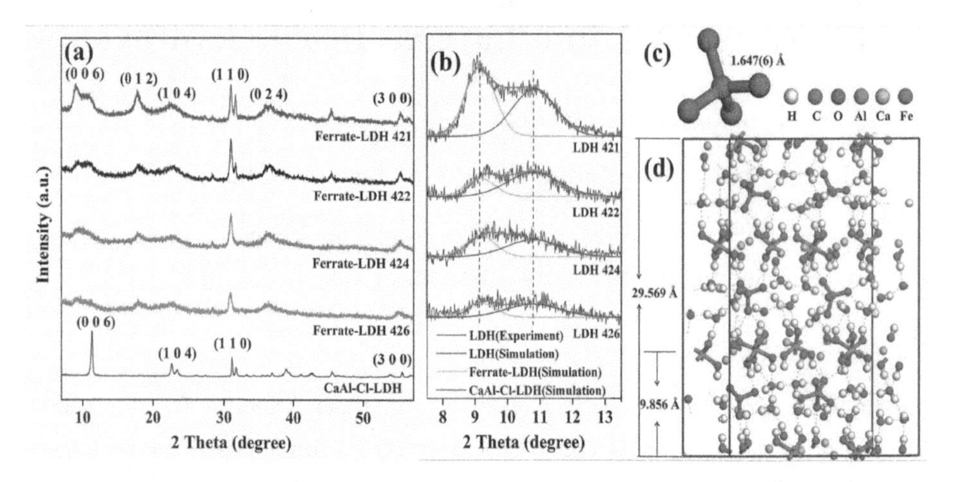

Fig. 1　XRD patterns of Ferrate-LDHs and geometric structure of CaAl－Ferrate－LDH

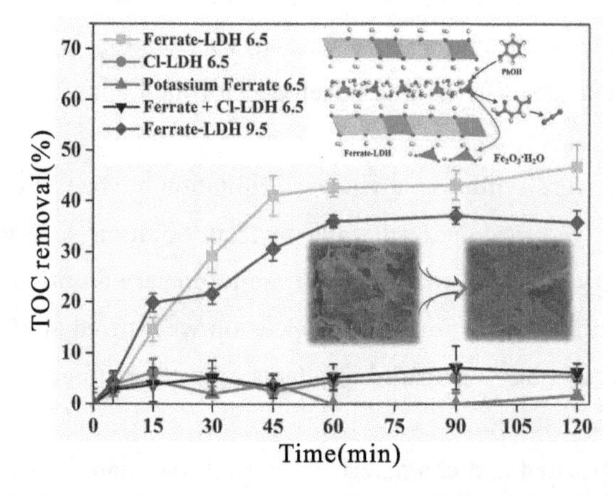

Fig. 2　Removals of phenol TOC after addition ferrate－LDH，Cl－LDH，potassium ferrate and potassium ferrate ＋ Cl－LDH; SEM patterns of ferrate－LDH before and after removal on phenol

References

［1］Lee Y，Kissner R，Von G U. 2014. Reaction of ferrate（Ⅵ）with ABTS and self-decay of ferrate（Ⅵ）：kinetics and mechanisms［J］. Environmental Science & Technology，48（9）：5154－5162.

［2］Jiang Y，Goodwill J E，Tobiason J E，et al. 2015. Effect of different solutes，natural organic matter，and particulate Fe（Ⅲ）on ferrate（Ⅵ）decomposition in aqueous solutions［J］. Environmental Science & Technology，49（5）：2841－2848.

Morphological Characteristics of Indoor Dust in Building Material Markets

Ling Li Zhou*
(Institute of Atmospheric and Environmental Sciences, Goethe-University, Frankfurt am Main, 60438, Germany)

Because indoor dust originates from different sources, its composition varies greatly, and dust particle sizes could range from nanometers to millimeters in the same location. Increasing studies have shown that indoor dust is a composite reservoir of pollutants from various sources and a significant source of exposure to these chemicals to humans[1,2]. However, to date, few studies focus on morphological characteristics of indoor dust and the association with chemical abundance in different dust size fraction.

In this study, indoor dust samples were collected from different building material markets in the Rhine/Main area (Germany) by use of vacuum cleaner. These dust samples were separated into three sub-fractions: the coarse particle size fraction (F1, $<150-200$ μm), the medium size fraction (F2, $<63-50$ μm) and the finest size fraction (F3, <63 μm). These fractions were investigated by optical microscopic under transmitted light and transmitted polarizing light (Leica DMRX POL, Wetzlar, Germany) to identify the most abundant particles and solvent extracts of the dust size fractions were analyzed with gas chromatography-mass spectrometry (Thermo Scientific, Dreieich, Germany) to quantify eleven organophosphate flame retardants (OPFRs).

As shown in the Fig. 1, the dust size fraction F1 (Fig. 1a and 1b) and F2 (Fig. 1c and 1d) of sample B4 from a building material market contained many polyurethane foam fragments, fiber-like materials, plant debris, organic particles with irregular shape (ash), but few mineral particles. In contrast, F3 (Fig. 1e and Fig. 1f) is more likely composed of mineral material particles showing a similar geometrical shape with clay and soil, but less of fiber-like materials and polyurethane foam fragments.

The concentrations of OPFRs in indoor dust from building material markets show that $\sum$OPFRs in F1 (153 μg/g) and F2 (196 μg/g) were higher than those observed F3 (88 μg/g). The distribution of the $\sum$OPFRs in different dust size fractions is not agreement with generally increasing concentrations of pollutants together with decreasing particle sizes because of the increasing surface areas of particles. This could be due to the fact that OPFR-treated insulation materials and polyurethane foams easily tear with the formation of larger fragments, and OPFRs remained bound to particles of the original polymer ma-

* Corresponding author: Ling Li Zhou; E-mail: zhou@iau. uni-frankfurt. de; Mobile Phone No.: +49-1744161143.

trix in indoor dust from building material markets. This result characterized the materials in different particle size fractions and provides insights into the origins of OPFRs in indoor dust.

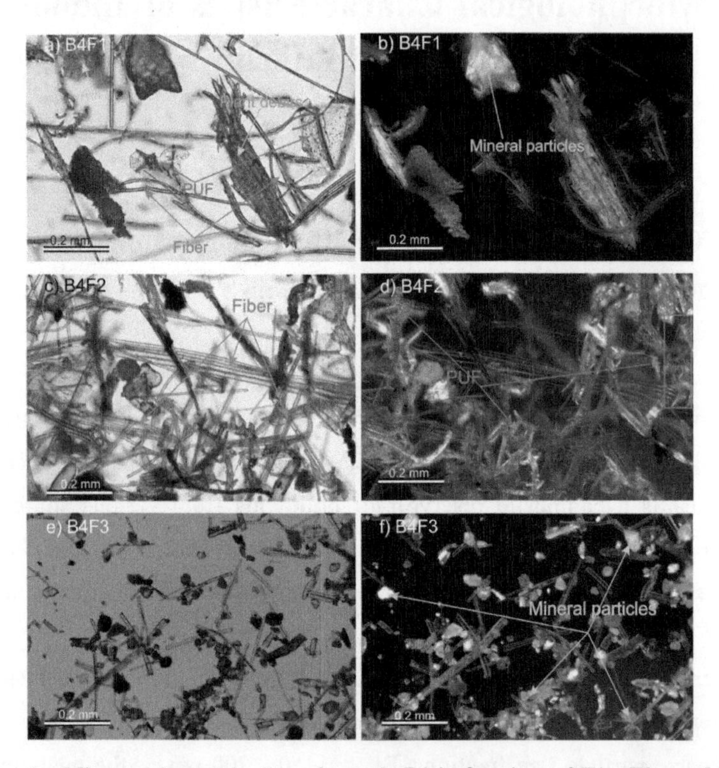

Fig. 1 Microscopic examination of sample B4 in fractions of F1, F2, and F3 under transmitted light (a, c, e) and transmitted polarizing light (b, d, f)

References

[1] van der Veen I, de Boer J. 2012. Phosphorus flame retardants: properties, production, environmental occurrence, toxicity and analysis [J]. Chemosphere, 88: 1119 - 1153.

[2] Zhou L, Hiltscher M, Püttmann W. 2017. Occurrence and human exposure assessment of organophosphate flame retardants in indoor dust from various microenvironments of the Rhine/Main region, Germany [J]. Indoor Air, 27: 1113 - 1127.

Photocatalytic Reduction of NO_x over $La_{1-x}Pr_xCoO_3$/ Attapulgite Nanocomposites

Ke Nian Wei[1,2]　Shi Xiang Zuo[1]　Xia Zhang Li[1]　Chao Yao[1*]

(1. School of Petrochemical Engineering, Changzhou University, Changzhou 213164, China;

2. Key Laboratory for Soft Chemistry and Functional Materials, Nanjing University of Science and Technology, Ministry of Education, Nanjing 210094, China)

A series of $La_{1-x}Pr_xCoO_3$/attapulgite ($La_{1-x}Pr_xCoO_3$/ATP) nanocomposites prepared by a facile sol-gel method were applied to photo-SCR at low temperature[1,2] (Fig. 1). XRD, SEM, TEM, NH_3-TPD, H_2-TPR, XPS, and DFT were used to characterize the structures, morphologies and photocatalytic activities of samples (Fig. 2). The photo-SCR of $La_{1-x}Pr_xCoO_3$/ATP nanocomposites was effectively evaluated using NO_x. The results of photo-SCR indicated that when the Pr doped was 0.5, the NO_x conversion and N_2-selectivity were the best in the reaction temperature of 150—250 ℃ compared with other sample due to the $PrCoO_3$ phase co-precipitates and forms a coherent heterojunction of $PrCoO_3/La_{0.5}Pr_{0.5}CoO_3$ on ATP. The $La_{0.5}Pr_{0.5}CoO_3$/ATP catalyst demonstrates the best performance reaching as high as 92% of NO_x conversion rate in the low temperature.

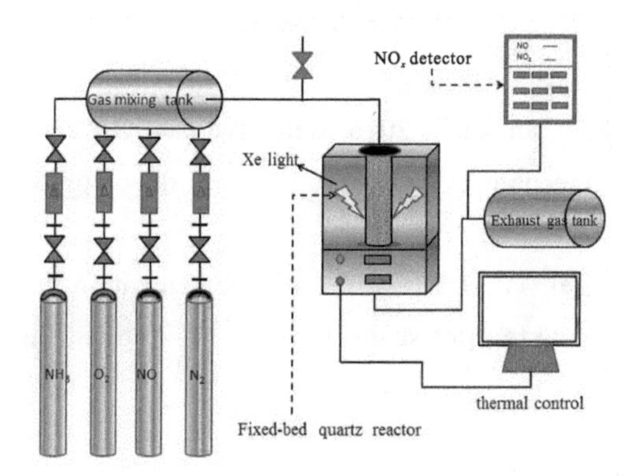

Fig. 1　The schematic diagram of photo - SCR of NO_x

---

* Corresponding author: Chao Yao; E-mail: yaochao420@163.com; Mobile Phone No.: ++86-13815053502. Grants or funding sources: National Science Foundation of China (51674043, 51702026).

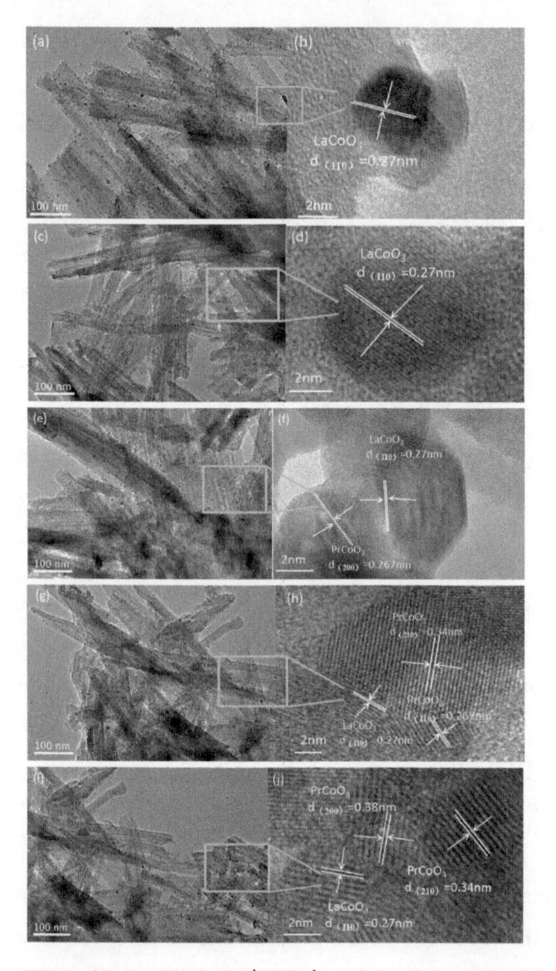

Fig. 2　TEM of $La_{1-x}Pr_xCoO_3$/ATP（x=0.1，0.3，0.5，0.7，0.9）

References

［1］Yamamoto A，Teramura K，Tanaka T. 2016. Selective catalytic reduction of NO by NH_3 over photocatalysts（Photo-SCR）：mechanistic investigations and developments ［J］. The Chemical Record，16：2268－2277.

［2］Yu J C，Nguyen V H，Lasek J，et al. 2016. NO_x abatement from stationary emission sources by photo-assisted SCR：Lab-scale to pilot-scale studies ［J］. Applied Catalysis A：General，523：294－303.

Star-shaped Polylactide on Promoting the Exfoliation of Clay with Improved Fire Retardancy and Mechanical Properties

Xin Wen* Xue Cheng Chen Mijowska Ewa

(Nanomaterials Physicochemistry Department, West Pomeranian University of Technology Szczecin, 70 - 311, Szczecin, Poland)

Polymer/clay nanocomposites have attracted a lot of attention in recent years due to their potential applications and fundamental academic interest[1]. It has been proved that the formation of exfoliated clay instead of intercalated clay is desirable to significantly improve the properties of the resulting nanocomposites. Therefore, extensive efforts have been made to obtain exfoliated clay for fabricating high-performance polymer nanocomposites. In our previous work[2], we found that star-shaped polylactide (SPLA) could promote the exfoliation of clay via annealing treatment. Herein we add 15wt% and 30wt% of SPLA (three-arm star-shaped PLA, Mn=124 500; PDI=1.09) into PLA (4032D from Natureworks) /clay (MMT 1.34TCN from Nanocor) mixtures, respectively. The dispersion of clay and the related performances of PLA/clay nanocomposites were investigated.

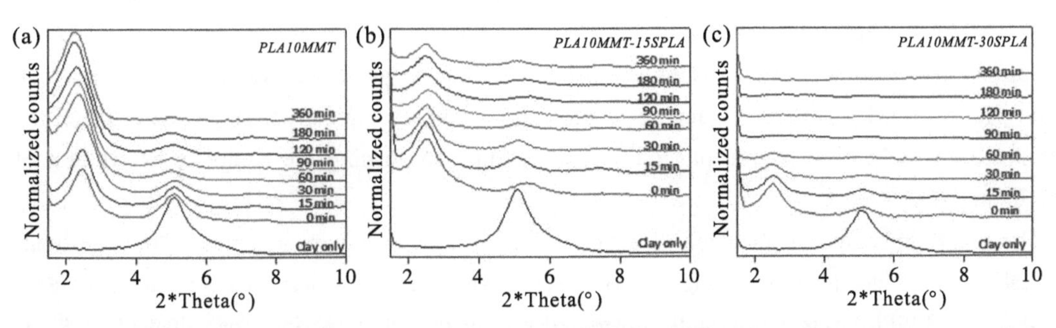

Fig. 1 XRD for linear polylactide/clay nanocomposites (PLA10MMT) (a), PLA10MMTwith 15wt% star-shaped polylactide (PLA10MMT - 15SPLA) (b) and PLA10MMTwith 30wt% star-shaped polylactide (PLA10MMT - 30SPLA) (c) annealed for various periods of time at 140 ℃

The PLA-clay samples were annealed at 140 ℃ with different time and analyzed by XRD. Compared to the original clay, the position of the (001) basal plane diffraction peak in PLA10MMT shifted to a lower angle (Fig. 1a), implying the formation of intercalated structure. For PLA10MMT-15SPLA, the (001) peak shifted to a lower angle, and its intensity became much weaker, meaning that partly exfoli-

* Corresponding Author, E-mail: Xin. Wen@zut. edu. pl.

ated structure of clay was achieved (Fig. 1b). As shown in Fig. 1c, very interestingly, the (001) peak in PLA10MMT-30SPLA gradually decreased with increasing annealing time, and after continuous annealing for 90 min, it completely disappeared, which indicates the formation of a completely exfoliated state of the clay.

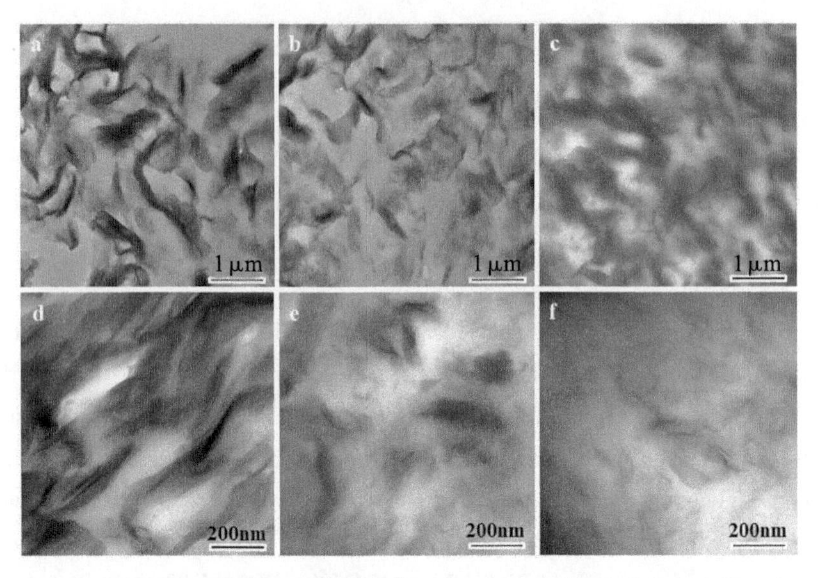

Fig. 2 TEM images of PLA10MMT (a, d), PLA10MMT - 15SPLA (b, e) and PLA10MMT - 30SPLA (c, f) after annealing at 140 ℃ for 360 min

The dispersed state ofclay after annealing treatment for 360 minutes was further examined by TEM. In PLA10MMT system, numerous intercalated parallel platelets were detected (Fig. 2a and 2d). In contrast, the PLA10MMT - 15SPLA exhibited smaller-sized aggregates, and some separated platelet were clearly visible (Fig. 2b and 2e). Differently, the large tactoids completely disappeared in PLA10MMT - 30SPLA, but widely uncorrelated platelets were present, confirming the complete exfoliation of clay (Fig. 2c and 2f).

The fire retardancy and mechanical properties of PLA/clay nanocomposites were investigated by cone calorimeter and tensile testing, respectively. The peak of heat release rate (PHRR) for PLA10MMT - 30SPLA was significantly decreased due to its better barrier effect than other PLA samples. Moreover, PLA10MMT - 30SPLA displayed higher tensile strength and elongation at break, indicating an improvement on mechanical performances.

References

[1] Kiliaris P, Papaspyrides C D. 2010. Polymer/layered silicate (clay) nanocomposites: An overview of flame retardancy [J]. Progress in Polymer Science, 35: 902 - 958.

[2] Yao K, Wen X, Tan H, et al. 2013. Insight on the striking influence of the chain architecture on promoting the exfoliation of clay in a polylactide matrix during the annealing process [J]. Soft Matter, 9: 10891 - 10898.

Highly Efficient and Low-cost Solar Steam Generation via Bilayered Attapulgite

Juan Jia　Cheng Jun Wang　Jian Li Zhang
Yue Yue Yang　Wei Dong Liang*
(Lanzhou University of Technology, Lanzhou 730050, China)

Solar steam generation has been attracting wide attention for boosting the evolution of solar-energy-harvesting technology[1]. In the past few years, a variety of materials have been used to enhance solar absorption effectively[2,3]. However, there are still the challenges that remain in connection to the material costs, complicated fabrications[4]. Here, we used the three processes to prepare the bilayered attapulgite. Firstly, the crosslinking agents (PVA and PVP) were added into the poly (acrylamide) /attapulgite composites[5], and then freeze-dried by refrigerated (−18 ℃) for 10 h to get the attapulgite gel (AG), finally, the AG were treated by a simple flame in order to prepare the bilayered attapulgite gel (BEA).

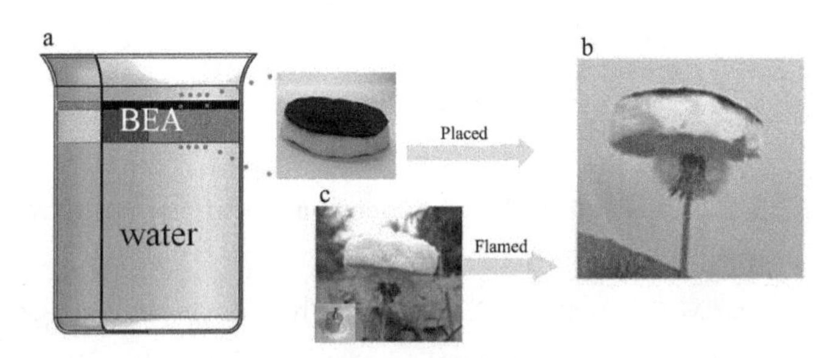

Fig. 1　The schematic and photograph of the BEA

Fig. 1a is a schematic of the bilayered attapulgite-based steam generation, the BEA is consisted of the carbonization layer and the thermal isolation layer, the carbonization layer (F-AG) can enable efficient solar absorption and the AG layer can provide a sufficient water supply and prevent heat transfer in exactly the same time. Fig. 1b shows the BEA with a weight of 100 g stand on the top of a dandelion, and the AG stand on the top of a dandelion is demonstrated in Fig. 1c. We can know that the BEA has the excellent mechanical stability, thermostability, light absorption, pore structures from Fig. 2. We measured the evaporation rates of water under a constant solar illumination. The corresponding energy

* Corresponding author: Wei Dong Liang; E-mail: davidlucas@163.com; Mobile Phone No.: ++86-13619 365531.
The authors are grateful to NSFC (Grant No. 51462021).

efficiency（η）for solar steam generation of the BEA can be analyzed using the following equation[2]：$\eta = mh_{LV}/q_i C_{opt}$. The energy efficiencies of the BEA can achieve up to 85% at the 1sun. The bilayered attapulgite（BEA）not only has pore structures，but also has low density and thermal conductivity（0.07 $W \cdot m^{-1} \cdot K^{-1}$），so the BEA can self-float on the surface of water and transport water. As a result，the BEA has achieved 1.2 $kg \cdot m^{-2} \cdot h^{-1}$ under 1 sun illumination and enable 85% solar-to-vapor efficiency. The new design reveals that the natural inorganic minerals can be fabricated as solar steam generation more than provides inspiration for the high-performance solar steam conversion devices.

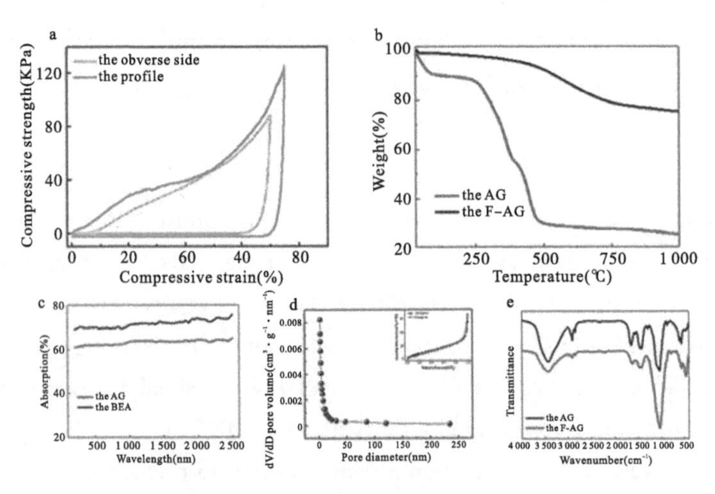

Fig. 2　Properties characterization

References

[1] Li X Q，Lin R X，Ni G，et al. 2018. Three-dimensional artificial transpiration for efficient solar waste-water treatment [J]. National Science Review，5：70－77.

[2] Yin Z，Wang H M，Jian M Q，et al. 2017. Extremely black vertically aligned carbon nanotube arrays for solar steam generation [J]. ACS Applied Materials Interfaces，9：28596－28603.

[3] Jiang Q，Tian L，Liu K K，et al. 2016. Bilayered biofoam for highly efficient solar steam generation [J]. Advanced Materials，28：9400－9407.

[4] Xue G B，Liu K，Chen Q，et al. 2017. Robust and low-cost flame-treated wood for high-performance solar steam generation [J]. ACS Applied Materials Interfaces，9：15052－15057.

[5] Li A，Wang A，Chen J. 2010. Studies on poly（acrylic acid）/attapulgite superabsorbent composites. Ⅱ. Swelling behaviors of superabsorbent composites in saline solutions and hydrophilic solvent-water mixtures [J]. Journal of Applied Polymer Science，94：1869－1876.

第六章

非金属矿与养殖、农业和土壤

黏土矿物促进生物质碳化的作用研究

李贵黎[1,2]　夏淑婷[1,2]　童东绅[1]　周春晖[1,2*]
（1. 浙江工业大学化学工程学院，杭州 310014；
2. 青阳非金属矿研究院，青阳 242800）

黏土矿物是含水硅酸盐化合物，具有由硅氧四面体和铝氧八面体组成的层状结构。它们广泛存在于各类地质体中，也是土壤的主要组成部分，影响着土壤的结构和性能。黏土矿物具有比表面积大、空隙多以及极性强等特征。特殊的晶体结构赋予了黏土矿物较强的吸附性、脱水、膨胀、收缩和离子交换性能等。生物碳生物质（如竹子、松木、稻草和玉米芯等）或生物质衍生的有机化合物（如纤维素、淀粉和蔗糖等）在 300～700 ℃下于缺氧或绝氧环境中经高温热裂解后生成的一种多孔固体碳质材料。将纤维素材料热分解成具有高能量密度的水合物是另一种选择。作为一类功能材料，生物碳在农业和环境应用中显示出潜在或已发现的实际用途，生物碳可作为土壤改良剂、肥料缓释载体及二氧化碳封存剂等[1]。向土壤中添加生物碳以修复受污染的土壤[2]，增强土壤水的保留并减少营养损失。木质纤维素生物质是非常丰富的可再生资源之一，显示出作为化石燃料补充剂的可持续燃料的巨大潜力。黏土矿物和木质纤维素之间的相互作用研究有助于了解土壤中生物量的演变和自然界中化石燃料的地质形成[3]。许多研究已经对纤维素的液化、热解和气化进行了尝试以产生燃料或中间化学品。生物质碳化是一个复杂的化学化工过程，受温度、时间、压力、催化剂等的影响[4]。将生物质转化为生物炭可能是长期碳封存的有效方式。催化剂的加入降低了碳化反应所需的活化能。黏土矿物可通过美拉德反应、多酚理论、选择性保存和吸附保护等机制，经配体交换、疏水相互作用、阳离子桥作为催化剂和吸附剂来固定有机物质[5]。黏土矿物也可作为催化剂通过黏土表面上的 Lewis 和 Bronsted 酸位点，将木质纤维素催化裂解成生物碳[6]。但是关于生物碳的类型、黏土矿物的转化、生物质碳化的机理尚不明确。本实验利用黏土矿物蒙脱石、伊利石与生物质纤维素、木质素为原料，使用水热碳化法在不同的温度、时间条件下进行了黏土矿物存在下的生物质碳化实验；通过傅立叶变换红外光谱、粉末 X 射线衍射、热重分析等对实验样品进行了表征。

黏土矿物蒙脱石与生物质碳化反应生成生物碳（图 1）；黏土矿物伊利石与生物质碳化反应生成生物碳（图 2）在一定的温度和热解时间条件下，黏土矿物的加入对生物质碳化生物碳的类型有影响；在达到一定的温度和压力下，黏土矿物蒙脱石和伊利石能够通过伊蒙混合层发生转化；在温度为 100～150 ℃范围时纤维素、木质素为不完全碳化；随着温度的升高，液相产物中还原性糖、糠醛、有机酸等的生成量逐渐增加；当温度升高到 175 ℃及以上时纤维素、木质素完全碳化生成生物碳，同时还原性糖、糠醛、有机酸的生成量逐渐减少。

* 通信作者：周春晖（1970-）；E-mail：chc. zhou @ aliyun. com；电话：13588066098。国家自然科学基金（41672033）。

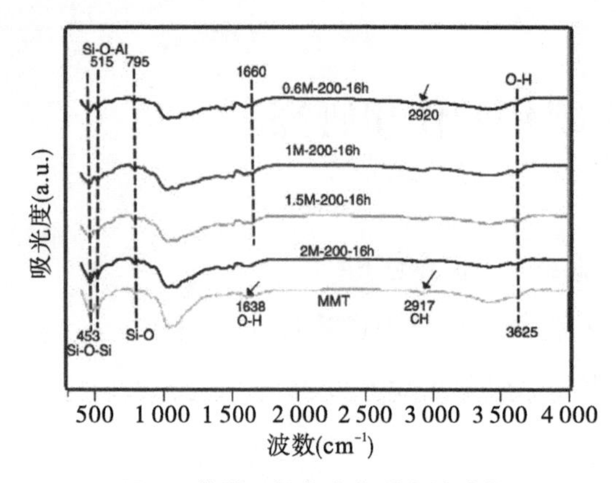

图 1 蒙脱石催化生物质水热碳化反应固体产物的 FT-IR 图

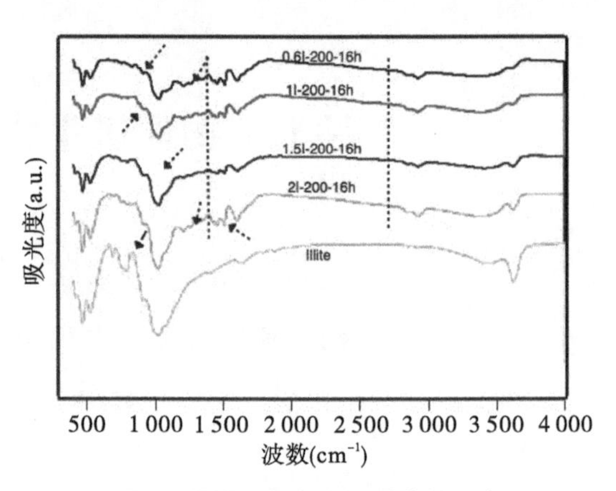

图 2 伊利石催化生物质碳化反应固体产物的 FT-IR 图

参考文献

[1] Zhou C H, Xia X, Lin C X, et al. 2011. Catalytic conversion of lignocellulosic biomass to fine chemicals and fuels [J]. Chemical Society Reviews, 40: 5588-5617.

[2] Wu L M, Zhou C H, Keeling J, et al. 2012. Towards an understanding of the role of clay minerals in crude oil formation, migration and accumulation [J]. Earth-Science Reviews, 115 (4): 373-386.

[3] Wu L M, Zhou C H, Tong D S, et al. 2014. Novel hydrothermal carbonization of cellulose catalyzed by montmorillonite to produce kerogen-like hydrochar [J]. Cellulose, 21 (4): 2845-2857.

[4] Wu L M, Tong D S, Li C S, et al. 2016. Insight into formation of montmorillonite-hydrochar nanocomposite under hydrothermal conditions [J]. Applied Clay Science, 119: 116-125.

[5] Ahmad M, Rajapaksha A U, Lim J E, et al. 2014. Biochar as a sorbent for contaminant management in soil and water: a review [J]. Chemosphere, 99: 19-33.

[6] Chen L, Chen X L, Zhou C H, et al. 2017. Environmental-friendly montmorillonite-biochar composites: facile production and tunable adsorption-release of ammonium and phosphate [J]. Journal of Cleaner Production, 156: 648-659.

超微粉碎凹凸棒石对肉鸡生长和养分利用的影响

杜明芳　张瑞强　王　坤　何青芬　温　超　周岩民*

（南京农业大学动物科技学院，南京 210095）

凹凸棒石具有多种生物学功能，应用于饲料中可改善动物生产性能、提高饲料养分利用率、吸附有毒有害物质。随着凹凸棒石颗粒粒径的减小，其比表面积增大，吸附性能提高[1]。目前，动物生产中所用凹凸棒石以常规粉碎产品（200 目）为主，且肉鸡饲料以 1%添加量为宜[2]。而有关日粮中添加超微粉碎凹凸棒石对肉鸡的影响尚未见报道。为此，本试验通过在日粮中添加超微粉碎凹凸棒石，研究其对肉鸡生长性能和养分表观利用率的影响，为凹凸棒石在饲料中的合理应用提供参考依据。

本试验采用超微粉碎分级机对常规粉碎凹凸棒石进行粉碎，收集粒度为 1 000 目的超微粉碎凹凸棒石。选取 1 日龄爱拔益加肉鸡 192 只，随机分成 3 组，每组 8 个重复，每个重复 8 只，分别饲喂基础日粮和在基础日粮中添加 1%常规粉碎凹凸棒石、1%超微粉碎凹凸棒石的试验日粮。于试验 42 d 对肉鸡进行空腹称重，并统计肉鸡试验期间的耗料量，计算肉鸡试验期 1～42 d 的平均日增重（ADG）、平均日采食量（ADFI）和料重比（F∶G）。于试验 38～40 d 收集肉鸡新鲜粪便样品，测定饲料及粪样中的有机物、粗蛋白、粗脂肪和酸不溶灰分的含量，计算各种养分的表观利用率。

由表 1 可见，与对照组相比，常规粉碎凹凸棒石对肉鸡生长性能无显著影响（$P>0.05$），而超微粉碎凹凸棒石显著降低了肉鸡试验期的 F∶G（$P<0.05$）。黏土类矿物所含的矿物元素在消化液作用下释放，被机体吸收利用并参与重要激素、酶的合成，从而促进机体生长[3]；同时，凹凸棒石可有效吸附霉菌毒素、有毒重金属等，降低有毒有害物质的毒性作用，并可增加肠道食糜的黏性以及延长食糜在消化道的滞留时间，从而改善动物对饲料的利用，提高生产性能[4]。凹凸棒石超微粉碎后比表面积增大，吸附性和黏性增强，促进了矿物元素的释放、有毒有害物质的吸附以及机体对饲料的利用。因此，超微粉碎凹凸棒石改善肉鸡生长性能的效果优于常规粉碎凹凸棒石。

由表 2 可见，与对照组相比，日粮添加常规粉碎凹凸棒石对肉鸡养分表观利用率虽有所提高，但差异不显著（$P>0.05$），而超微粉碎凹凸棒石显著提高了肉鸡有机物和粗脂肪表观利用率（$P<0.05$）；且与常规粉碎凹凸棒石组相比，超微粉碎凹凸棒石组有机物表观利用率亦显著提高（$P<0.05$）。凹凸棒石具有较强的吸附性和黏滞性，饲料中添加凹凸棒石可增加食糜黏性，延长食糜在肠道中的消化时间，有利于营养物质的消化吸收。随着颗粒粒度的减小，凹凸棒石的吸附性和黏性增强，更有利于养分利用。本试验中，超微粉碎凹凸棒石可显著提高肉鸡养分表观利用

* 通信作者：周岩民；E-mail：zhouym6308@163. com；手机号：13805162705。凹凸棒石黏土功能性饲料产品研发基金（201501）资助项目。

率，而常规粉碎凹凸棒石则无此效果，表明超微粉碎凹凸棒石提高肉鸡对饲料养分利用的作用更强。

本研究结果表明，日粮添加1%超微粉碎凹凸棒石可降低肉鸡F∶G，提高肉鸡对有机物和粗脂肪的表观利用率，改善肉鸡生长性能和养分利用率；超微粉碎凹凸棒石在肉鸡饲料中的应用效果优于常规粉碎凹凸棒石。

表1　超微粉碎凹凸棒石对肉鸡生长性能的影响

项目[1]	对照组	常规粉碎组	超微粉碎组	标准误	P值
ADG（g/d）	58.56	59.49	60.99	0.67	0.384
ADFI（g/d）	104.76	105.35	104.67	1.10	0.965
F/G（g∶g）	1.79^{a}	1.77^{ab}	1.72^{b}	0.01	0.032

注：同行数据肩标字母不同表示差异显著（P＜0.05）；[1] ADG表示平均日增重，ADFI表示平均日采食量，F∶G表示料重比

表2　超微粉碎凹凸棒石对肉鸡养分表观利用率的影响（%）

项目	对照组	常规粉碎组	超微粉碎组	标准误	P值
有机物	60.12^{b}	62.35^{b}	72.51^{a}	1.50	＜0.001
粗脂肪	50.56^{b}	55.30^{ab}	68.73^{a}	2.78	0.016
粗蛋白	44.95	43.22	54.00	2.55	0.149

注：同行数据肩标字母不同表示差异显著（P＜0.05）

参考文献

[1] Berhane T M，Levy J，Krekeler M P S，et al. 2016. Adsorption of bisphenol A and ciprofloxacin by palygorskite-montmorillonite：Effect of granule size，solution chemistry and temperature [J]. Applied Clay Science，132-133：518－527.

[2] Chen Y P，Cheng Y F，Li X H，et al. 2016. Dietary palygorskite supplementation improves immunity，oxidative status，intestinal integrity，and barrier function of broilers at early age [J]. Animal Feed Science and Technology，219：200－209.

[3] Slamova R，Trckova M，Vondruskova H，et al. 2015. Clay minerals in animal nutrition [J]. Applied Clay Science，51：395－398.

[4] Safaeikatouli M，Boldaji F，Dastar B，et al. 2012. The effect of dietary silicate minerals supplementation on apparent ileal digestibility of energy and protein in broiler chickens [J]. International Journal of Agriculture and Biology，14：299－302.

沸石对肉鸡生长和肌肉抗氧化能力的影响

曲恒漫　陈跃平　程业飞　李　俊　赵宇瑞　周岩民*
（南京农业大学动物科技学院，南京 210095）

抗生素应用于饲料中可促进动物生长、改善动物健康，但抗生素的使用易导致耐药细菌产生等公共健康隐患。减少或禁止使用抗生素已成为必然趋势[1]，肌肉组织氧化损伤能够导致组织中脂质过氧化并产生大量自由基，影响肌肉品质[2]，因此，迫切需要开发更加安全、高效的改善动物健康、促生长并提高动物产品质量的产品。沸石具有独特的吸附、离子交换、催化及分子筛等特性[3]，应用于饲料中可吸附病原菌和霉菌毒素等有害物质，改善动物肠道健康，提高动物生产性能[3]，增强动物抗氧化和抗病能力[2]，具有一定的替代抗生素潜力。为此，本文通过沸石对肉鸡生长性能和肌肉抗氧化能力的影响，研究其作为抗生素替代品的可行性，以便为沸石在肉鸡饲料中的合理应用提供理论参考依据。

试验选取 144 只体重相近的 1 日龄爱拔益加（AA）肉鸡随机分为 3 组，每组 6 重复，每重复 8 只鸡，对照组饲喂基础日粮，试验组为抗生素组和沸石组，分别饲喂添加 50 mg/kg 金霉素和 10 g/kg沸石的试验日粮。试验期为 42 d。分别于 21、42 d 对饲料及肉鸡进行称重，计算肉鸡生长性能，并于 42 d 进行采样，取左侧胸肌肌肉进行肌肉抗氧化能力检测。

由表 1 可知，与对照组相比，添加抗生素增加了肉鸡 1～42 d 平均日增重（$P<0.05$）；添加沸石增加了 1～21 d 的平均日增重（$P<0.05$）和平均日采食量（$P<0.05$）；抗生素和沸石组 1～21 d 料重比均显著降低（$P<0.05$）。沸石表现出的促生长效果可能是其具有较强的吸附能力，可吸附肠道有害物质，改善了肠道微生态环境，延长饲料在肠道中停留的时间，进而提高了营养物质的利用。因此，沸石可在一定程度上替代抗生素促进动物生长。

由表 2 可知，与对照组相比，添加抗生素提高了胸肌超氧化物歧化酶活性（$P<0.05$）；添加沸石提高了胸肌谷胱甘肽过氧化物酶活性（$P<0.05$），降低了丙二醛含量（$P<0.05$），且显著低于抗生素组。日粮添加沸石或抗生素均提高了肉鸡胸肌抗氧化能力，可能是由于抗生素可抑制有害菌增殖，而沸石可以吸附有害物质，从而减少毒素在肌肉中的沉积，由此可减少毒素对肌肉造成氧化损伤。

综上所述，沸石替代饲料中的抗生素，能够改善肉鸡的生长性能，提高肌肉抗氧化能力，在一定程度上具有替代抗生素的潜力。

* 通信作者：周岩民；E-mail：zhouym@njau. edu. cn；手机号：13805162705。

表 1 沸石对肉鸡生长性能的影响

时间（d）	项目	对照组	抗生素组	沸石组	标准误	P 值
	平均日增重（g/d）	26.44^{b}	28.27^{ab}	29.96^{a}	0.529	0.017
1～21	平均日采食量（g/d）	40.56^{b}	42.11^{ab}	44.60^{a}	0.629	0.024
	料重比（g/g）	1.56^{a}	1.49^{b}	1.51^{b}	0.014	0.017
	平均日增重（g/d）	72.57	73.36	69.48	1.199	0.432
22～42	平均日采食量（g/d）	151.18	153.66	141.99	2.225	0.093
	料重比（g/g）	2.12	2.07	2.01	0.025	0.214
	平均日增重（g/d）	49.59^{b}	52.70^{a}	49.40^{b}	0.477	0.001
1～42	平均日采食量（g/d）	95.15	95.58	90.94	0.920	0.079
	料重比（g/g）	1.91	1.83	1.84	0.016	0.120

注：同行数据肩标不同字母表示差异显著（$P<0.05$）

表 2 沸石对肉鸡胸肌抗氧化能力的影响

项目	对照组	抗生素组	沸石组	标准误	P 值
SOD（U/mg protein）	47.50^{b}	55.44^{a}	53.46^{ab}	1.347	0.028
MDA（nmol/mg protein）	0.84^{a}	0.75^{a}	0.54^{b}	0.043	0.011
GSH（mg/g protein）	2.73	3.06	3.13	0.122	0.402
GSH－Px（U/mg protein）	1.79^{b}	2.22^{ab}	2.81^{a}	0.167	0.033

注：同行数据肩标不同字母表示差异显著（$P<0.05$）；SOD 表示超氧化物歧化酶，MDA 表示丙二醛，GSH 表示谷胱甘肽，GSH－Px 表示谷胱甘肽过氧化物酶

参考文献

[1] Wegener H C. 2006. Antibiotics in animal feed and their role in resistance development [J]. Current Opinion in Microbiology，6：439-445.

[2] Brigita H，Mislav Đ，Mirela P，et al. 2017. Antioxidative status and meat sensory quality of broiler chicken fed with XTRACT© and zeolite dietary supplementation [J]. Pakistan Journal of Agricultural Sciences，54：897－902.

[3] Papaioannou D，Katsoulos P D，Panousis N，et al. 2005. The role of natural and synthetic zeolites as feed additives on the prevention and/or the treatment of certain farm animal diseases：A review [J]. Microporous and Mesoporous Materials，84：161－170.

固相载锌凹凸棒石对河蟹生长和肠道菌群的影响

张瑞强[1]　姜　滢[2]　温　超[1]　陈跃平[1]　刘文斌[1]　周岩民[1*]
（1. 南京农业大学动物科技学院，南京 210095；
2. 江苏金康达集团，盱眙 211700）

锌（zinc，Zn）是动物生长必需的微量元素之一，参与动物机体的生长发育、生殖、免疫等生理过程[1]。凹凸棒石（palygorskite，Pal）是一种具有链层状纤维晶体结构的含水富镁铝硅酸盐黏土矿物，具有较大的比表面积、优良的离子交换和吸附性能[2]。固相载锌凹凸棒石（Zn－Pal）是利用固相离子熔融法在一定条件下将 Zn 负载于 Pal 而制成的一种含金属离子无机抗菌剂[3]。

研究表明，Zn－Pal 能够作为一种新型锌源添加剂应用于动物饲料中，改善动物生长和肠道微生物区系[3,4]。但对水产动物的影响鲜见报道。因此，本试验旨在研究 Zn－Pal 对中华绒螯蟹生长和肠道菌群的影响，以便为 Zn－Pal 在水产养殖中的应用提供理论参考。

试验用凹凸棒石由江苏神力特生物科技股份有限公司提供，载锌凹凸棒石参考 Yan 等[3]的方法由实验室制得，Zn 含量为 28.19 mg/g。中华绒螯蟹由江苏金康达集团提供。将 432 只健康、规格一致的中华绒螯蟹随机分为 4 组，每组 6 个重复，每个重复 18 只，分别饲养在 24 个循环桶中，对照组饲喂无外源 Zn 添加剂的基础日粮，试验组分别饲喂基础日粮添加 0.5、1 和 2 g/kg 的 Zn－Pal 的试验日粮，试验为期 8 周。

由表 1 可知，Zn－Pal 具有提高中华绒螯蟹末均重、增重率和特定生长率的趋势（$0.1>P>0.05$）。与对照组相比，Zn－Pal 试验组饵料系数数值上分别降低了 8.62%、8.62%、16.62%。这可能是由于 Zn 作为多种机体酶的辅助因子，能够改善机体蛋白质、碳水化合物和能量代谢，进而提高中华绒螯蟹生长发育[1]。

由表 2 可知，Zn－Pal 能够降低中华绒螯蟹肠道食糜大肠杆菌数量（Linear，$P=0.034$）。与对照组相比，日粮中添加 1 g/kg Zn－Pal 降低了中华绒螯蟹肠道食糜大肠杆菌数量（$P<0.05$）。Zn 对细菌和真菌具有抗菌作用，其主要机制包括：（1）锌离子通过库伦作用破坏细菌细胞膜结构和功能，从而使细菌死亡；（2）过量锌离子进入微生物内部，能够催化微生物体内自由基生成，造成微生物氧化应激而死亡[3]。此外，Zn 离子在动物肠道中可缓慢释放，表现出持久的抗菌效果，改善肠道菌群组成。

由上述结果可知，Zn－Pal 作为一种新型无机抗菌剂应用于动物饲料中，可在一定程度上提高中华绒螯蟹的生长性能，改善肠道菌群结构。

* 通信作者：周岩民；E-mail：zhouym6308@163. com；手机号：13805162705。

表 1 固相载锌凹凸棒石对中华绒螯蟹生长性能的影响

组别[1]	对照组	固相载锌凹凸棒石（g/kg）			SEM	P 值	
		0.5	1	2		Linear	Quadratic
初均重（g）	16.05	16.18	16.06	15.88	0.11	0.534	0.493
末均重（g）	36.57	37.67	44.75	42.16	1.58	0.096	0.552
增重率（%）	127.84	133.48	178.39	164.57	9.51	0.070	0.597
特定生长率（%day^{-1}）	1.43	1.50	1.80	1.72	0.07	0.055	0.542
饵料系数（g/g）	3.25	2.97	2.97	2.71	0.18	0.339	0.979

注：[1]增重率＝（末均重－初均重）/初均重* 100%；特定生长率＝［ln（末均重）－ln（初均重）］/天数* 100%；饵料系数＝总耗料量/（末总重－初总重）

表 2 固相载锌凹凸棒石对中华绒螯蟹肠道食糜菌群影响（log CFU/g）

组别	对照组	固相载锌凹凸棒石（g/kg）			SEM	P 值	
		0.5	1	2		Linear	Quadratic
大肠杆菌	5.22^{a}	4.83^{ab}	4.38^{b}	4.57^{ab}	0.13	0.034	0.210
气水单胞菌	5.83	5.40	5.01	5.35	0.14	0.153	0.172
乳酸菌	3.99	4.14	3.37	3.79	0.13	0.222	0.579

注：同行肩标无相同字母表示差异显著（$P<0.05$）

参考文献

［1］Halver J E，Hardy R W. 2002. Fish nutrition［M］. New York：（Third edition）Academic press.

［2］Galan E. 1996. Properties and applications ofpalygorskite-sepiolite clays［J］. Clay Minerals，31：443－453.

［3］颜瑞. 2016. 固相载锌凹凸棒石黏土对肉鸡锌生物利用率及免疫调节机制的研究［D］. 南京：南京农业大学.

［4］Yang W L，Chen Y P，Cheng Y F，et al. 2016. An evaluation of zinc bearing palygorskite inclusion on the growth performance，mineral content，meat quality，and antioxidant status of broilers［J］. Poultry science，95：878－885.

基于尖晶石结构特征的锌/铬污染土壤结构化固定机制

吴 非[1,2,3] 刘承帅[1,2*] 吕亚辉[2] 廖长忠[2] 高 庭[1,2,3] 马胜寿[2] 李芳柏[2]
(1. 中国科学院地球化学研究所环境地球化学国家重点实验室，贵阳 550081；
2. 广东省生态环境技术研究所广东省农业环境综合治理重点实验室，广州 510650；
3. 中国科学院大学，北京 100049)

城市中工厂搬迁遗留场地的重金属污染问题正日益成为制约棕地可持续开发利用的主要因素[1]，这种情况在中国尤其突出。因为快速的城市化对土地资源提出了多重要求，同时公众对居住环境越来越关注[2]。处理重金属污染的有效途径之一是通过热处理将重金属转化为稳定的尖晶石结构[3]。土壤污染一般是多种重金属复合污染[4]。与单一金属污染相比，土壤中的多种重金属具有交互作用并表现出综合的生态效应，使得土壤修复难度增加[5]。而以前的研究大多关注单一重金属污染土壤的修复。因此，该研究的目的是研究多种重金属复合污染土壤的高温处理方法，并通过浸出实验探讨烧结体的长期稳定性。

固化实验中选用 ZnO 和 Cr_2O_3 模拟锌铬污染土壤，Zn 和 Cr 的摩尔比为 1∶2。将 ZnO 和 Cr_2O_3 混合物在球磨机中充分混合；将混合后的样品在压片机中以 320 MPa 的压力压制成圆柱状体；将圆柱体置于马弗炉中在目标温度（700～1 300 ℃）下煅烧 3 h。自然冷却至室温后，将烧结体破碎研磨成粉末状。采用 XRD 表征烧结体物相，并进行毒性浸出实验。烧结体中重金属浸出实验采用美国环保局 Method 1311 Toxicity Characteristic Leaching Procedure（TCLP）标准流程。以 pH＝2.9 的乙酸缓冲液为浸出液，反应总时间为 0.75～21 d。

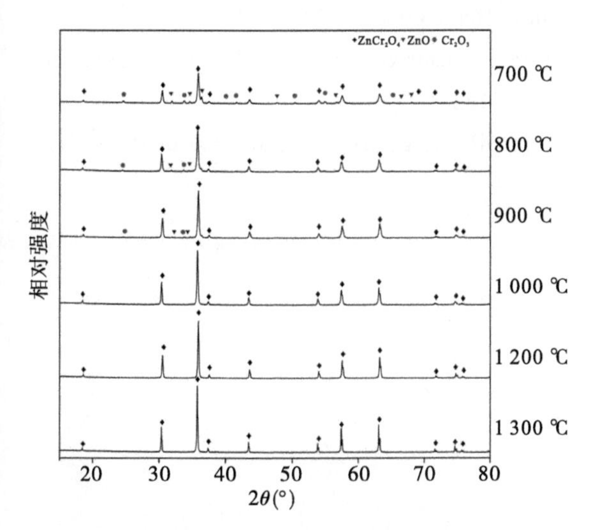

图 1 ZnO 和 Cr_2O_3 混合高温煅烧后烧结体的 XRD 图

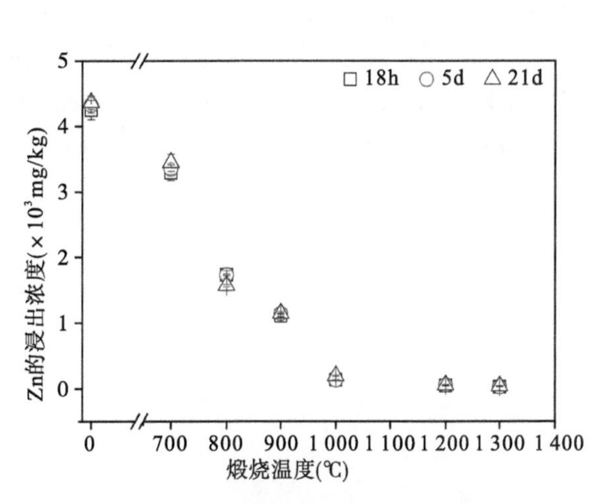

图 2 样品煅烧前后锌的浸出浓度

XRD 结果显示（图 1），锌和铬同时固定到 $ZnCr_2O_4$尖晶石结构中，作为尖晶石结构分子组分，且 $ZnCr_2O_4$是烧结体中的主要物相。随着温度的升高，$ZnCr_2O_4$的特征峰强度逐渐增加。当煅烧温度超过 1 000 ℃时，ZnO 和 Cr_2O_3的特征峰完全消失。因此从该研究结果可以得出，高温处理能有效固定金属氧化物，烧结后的重金属以结构态形式固定于尖晶石结构中。

高温处理前后，样品中锌的浸出浓度如图 2 所示。煅烧前，锌的浓度达到 4 000 mg/kg。随着煅烧温度的升高，锌的浸出浓度从 3 500 mg/kg 降低到 50 mg/kg，比烧结前降低了 80%以上。有报道称，在酸性环境中的质子消耗过程中主要是质子—金属交换[6]，所以在没有烧结的样品和低温（700～900 ℃）烧结体中 ZnO 与 H^+ 反应产生了大量 Zn^{2+} 。

尖晶石是一独特的矿物结构，其结构通式为 XY_2O_4。尖晶石中的分子键结合紧密，键能高，在酸性环境中具有很强的稳定性。因此，将 ZnO 和 Cr_2O_3充分混合压实后，在一定温度下煅烧，烧结体中形成了 $ZnCr_2O_4$尖晶石结构。浸出实验结果显示，金属的浸出浓度大大降低，实现了对污染土壤重金属的固定化，并排除了二次污染的可能性。由于煅烧所得的烧结体中重金属的浸出风险极低，并可在酸性环境下长期稳定存在。因此，获得的烧结体可作为再生材料用于建筑材料、市政工程等，是一种有效的固废资源化利用处置技术。

参考文献

［1］Ren W，Xue B，Geng Y，et al. 2014. Inventorying heavy metal pollution in redeveloped brownfield and its policy contribution：Case study from Tiexi District，Shenyang，China ［J］. Land Use Policy，38：138 - 146.

［2］Sun L，Geng Y，Sarkis J，et al. 2013. Measurement of polycyclic aromatic hydrocarbons（PAHs）in a Chinese brownfield redevelopment site：The case of Shenyang ［J］. Ecological Engineering，53：115 - 119.

［3］Tang J，Su M，Zhang H，et al. 2018. Assessment of copper and zinc recovery from MSWI fly ash in Guangzhou based on a hydrometallurgical process ［J］. Waste Management.

［4］Fonseca B，Figueiredo H，Rodrigues J，et al. 2011. Mobility of Cr，Pb，Cd，Cu and Zn in a loamy sand soil：A comparative study ［J］. Geoderma，164（3 - 4）：232 - 237.

［5］Lombi E，Zhao F J，Dunham S J，et al. 2001. Phytoremediation of heavy metal-contaminated soils ［J］. Journal of Environmental Quality，30（6）：1919 - 1926.

［6］Snellings R. 2015. Surface chemistry of calcium aluminosilicate glasses ［J］. Journal of the American Ceramic Society，98（1）：303 - 314.

磁性 Fe_3O_4 纳米环对牛血清蛋白的吸附性能

刘小楠[1,2]　张友魁[3]　李金山[2*]

（1. 西南科技大学环境友好能源材料国家重点实验室，绵阳 621010；

2. 中国工程物理研究院化工材料研究所，绵阳 621000；

3. 中国科学技术大学国家同步辐射实验室，合肥 230029）

磁性纳米粒子可以使蛋白质的分离更加高效和迅速，随着对蛋白质研究的深入，磁性纳米粒子分离蛋白质的研究变得日益活跃，且研究多使用超顺磁纳米颗粒。然而涡旋磁性纳米环应用于蛋白质分离与检测的研究鲜有报道[1,2]。现有的研究发现，MNRs 在用于蛋白质的检测时由于其在临床的强磁场环境中稳定性较差，使得其在蛋白质的检测过程中灵敏度很低，应用性能有待改进。血浆中含有的可溶性蛋白质主要是血清蛋白，在血清蛋白中牛血清蛋白与人的血清蛋白具有很大的相似性，且具有很好的水溶解性、稳定性、低成本、易得到而备受人们关注，因而选取牛血清蛋白进行分离实验。

配制 $FeCl_3 \cdot 6H_2O$、NaH_2PO_4 和 Na_2SO_4 三种混合溶液，然后将混合溶液转移到反应釜中，并保持在 220 ℃条件下反应 48 h。将所得粉体转移到管式气氛炉进行热处理，在氢氩混合气氛下，360 ℃保温 5 h 得到黑色粉末样品备用。将牛血清蛋白标准溶液、去离子水和考马斯亮蓝溶液加入小称量瓶后，轻轻摇匀，静置 5 min 后进行紫外—可见光分光光度测量，并绘制标准曲线。该实验采用磷酸盐缓冲溶液进行不同 pH 值、不同浓度 BSA 溶液的配制。

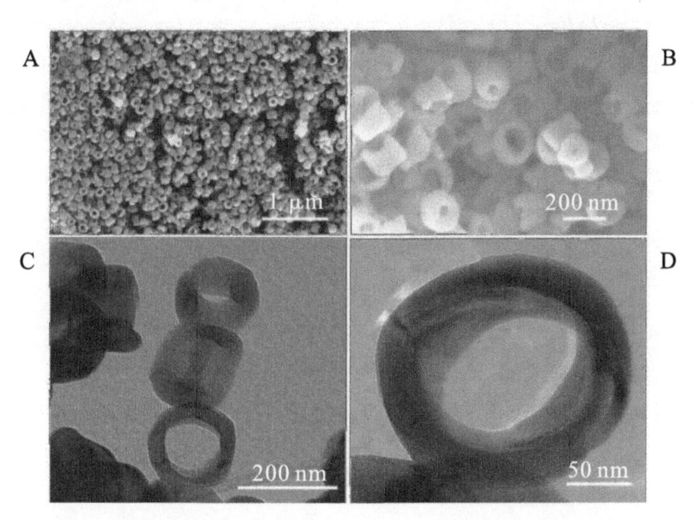

图 1　纳米环状氧化铁 SEM（A，B）和 TEM（C，D）照片

* 通信作者：李金山；E-mail：liuxiaonan@swust. edu. cn；手机号：13547116168。中国博士后科学基金面上项目、四川省博士后科研项目特别资助、四川省科学技术厅面上项目、四川省教育厅重点项目（17ZA0401）资助项目。

利用磁性 Fe_3O_4纳米环对 BSA 进行吸附性能研究。实验证明，在 pH=5、BSA 溶液初始浓度为 2.6 mg·mL^{-1}、吸附时间为 3 h 的条件下，磁性 Fe_3O_4纳米环对 BSA 的吸附达到平衡（图 1、2）。通过实验数据和公式，计算得出磁性 Fe_3O_4纳米环对 BSA 的最大吸附量为 325.2 mg·g^{-1}，高于据报道所知的磁性 Fe_3O_4纳米颗粒对 BSA 的吸附量 250 mg·g^{-1}。磁性 Fe_3O_4纳米环吸附能力的增强存在原因可能是：一方面，磁性 Fe_3O_4纳米环具有较高的比表面积；另一方面，磁性 Fe_3O_4纳米环特殊的环形腔体结构可使部分蛋白质分子进入环腔内，更容易吸附固定 BSA；再者，受到其特有的涡旋磁畴特性影响，从而有效增强了吸附效果。

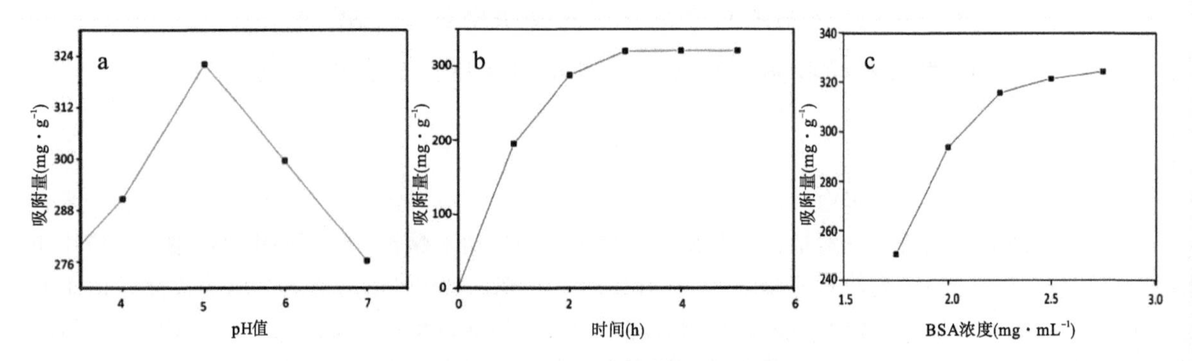

图 2　磁性 Fe_3O_4 纳米环对 BSA 吸附性能分析

注：a. pH 值对 BSA 吸附量的影响；b. 吸附时间对 BSA 吸附量的影响；c. 不同浓度 BSA 对吸附量的影响

参考文献

［1］Liu X N，Zhu F H，Wang W，et al. 2016. Synthesis of single-crystalline iron oxide magnetic nanorings as electrochemical biosensor for dopamine detection［J］. International Journal of Electrochemical Science，11：9696－9703.

［2］Liu X N，He G，Li J S，et al. 2017. The synthesis and mechanism of（001）orientated hematite nanorings：A combined theoretical and experimental investigation［J］. NANO，12（9）：72－82.

川西平原还田秸秆 DOM 对矿物细颗粒吸附 SMX 的影响

曾　丹[1,2]　王　彬[1,2*]　黎　明[1,2]　朱静平[1]　谌　书[1,2]　白英臣[3]

（1. 西南科技大学环境与资源学院，绵阳 621010；

2. 西南科技大学，固体废物处理与资源化教育部重点实验室，绵阳 621010；

3. 中国环境科学研究院，环境基准与风险评估国家重点实验室，北京 100012）

溶解性有机质（dissolved organic matter，DOM）是一类混合物，通常指环境中可以被水或稀盐溶液提取，对污染物环境地球化学行为具有显著影响。研究发现，无机矿物对离子型 OCs（如抗生素）的影响与有机质相当，甚至更高[1]。

本文以矿物细颗粒吸附 SMX 为基础，为探明川西平原还田秸秆 DOM 对矿物细颗粒吸附磺胺甲恶唑（sulfamethoxazole，SMX）的影响机理，研究了矿物细颗粒吸附 SMX 的动力学过程以及 DOM 对此过程的影响，比较分析了矿物颗粒吸附前后的傅里叶变换红外光谱（fourier transform infrared spectroscopy，FI-IR）特征。

图 1 为矿物颗粒吸附 SMX 的动力学拟合过程。准一级动力学模型对 3 种矿物细颗粒的拟合存在较大偏差，而拟二级动力学模型的 r_{adj}^2 在 0.969～0.986，双室一级动力学模型的 r_{adj}^2 在 0.994～0.998。可见，双室一级动力学更加适用于描述该吸附动力学过程。

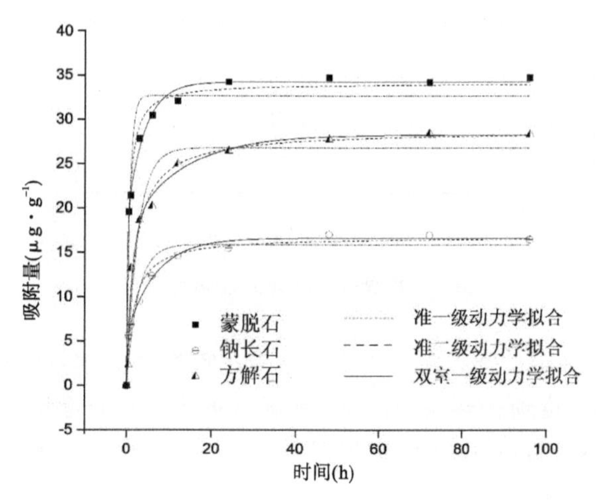

图 1　矿物颗粒吸附 SMX 的动力学过程

* 通信作者：王彬；E-mail：greenworldwb@swust.edu.cn；手机号：18121898065。国家自然科学基金项目（41403081），四川省应用基础研究计划项目（2015JY0168），西南科技大学高端引进人才项目（13zx7126），绵阳市科技计划项目（15S-02-1），西南科技大学教学改革与研究项目（14xn0019）。

图 2 反映了秸秆 DOM 对矿物颗粒吸附 SMX 的影响。投加 DOM 后，矿物颗粒对 SMX 的吸附量明显增加；其中，蒙脱石和钠长石对 SMX 的吸附量增长较显著而方解石较弱，其增量分别为 28.94、28.34、2.40 $\mu g \cdot g^{-1}$，这与 3 种矿物颗粒的化学组分以及颗粒的表面活性有关。

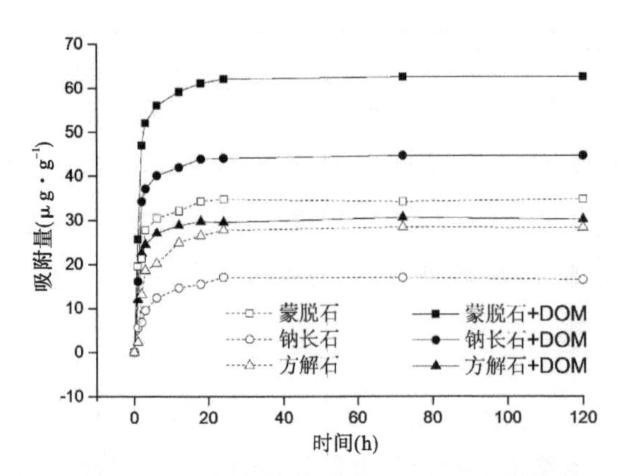

图 2　秸秆 DOM 对矿物颗粒吸附 SMX 的影响

图 3 呈现了蒙脱石的红外光谱学特征。吸附后的蒙脱石较吸附前在波数 3 700、1 600、1 000 cm^{-1}附近的尖锐吸收峰和在波数 3 600～3 000 cm^{-1}的宽吸收带明显减弱，且与 DOM 作用下波谱基本一致，DOM 作用下吸附量的提升与蒙脱石溶出 Al^{3+} 与 DOM 结合使得吸附位点增加有关。

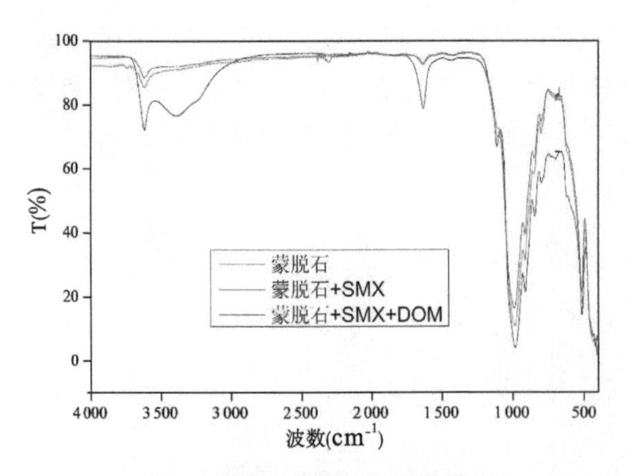

图 3　蒙脱石的红外光谱学特征

该研究表明，SMX 在 3 种矿物表面的吸附动力学过程符合双室一级动力学模型，DOM 作用下矿物颗粒的吸附量不同程度地得到提升，而 3 种矿物的模型拟合度均有所降低。红外光谱分析得出，DOM 在蒙脱石、钠长石对 SMX 的吸附行为过程中有影响，而方解石几乎无影响。

参考文献

[1] Schwarzenbach R P, Westall J. 1981. Transport of nonpolar organic compounds from surface water to groundwater. Laboratory sorption studies [J]. Environmental Science and Technology, 15 (11): 1360 - 1367.

方解石细颗粒与金黄色葡萄球菌的近尺寸作用

董发勤[1,2*]　周世平[1,2]　周　青[1]　李　帅[1]　代群威[1,2]　边　亮[1,2]　邓建军[3]
（1. 西南科技大学环境与资源学院，绵阳 621010；
2. 固体废物处理与资源化教育部重点实验室，绵阳 621010；
3. 四川绵阳四〇四医院，绵阳 621000）

近年来，雾霾污染问题频发，且有愈演愈烈的趋势。因此，针对大气可吸入颗粒污染物的研究成为热点。已有研究发现，大气颗粒物主要由矿物细颗粒组成，同时伴随有大量的细菌体共同悬浮于空气当中[1,2]。可以推测，通过研究分析大气中矿物细颗粒与细菌体各自特性及二者共存状态、相互作用机制，可以为雾霾污染防治提供重要的理论依据。本文在研究矿物细颗粒与细菌体相互作用的基础上[3]，通过实验和模拟计算进一步探讨了两者之间在分子水平的相互作用机理。

实验矿物细颗粒选用最常见的方解石颗粒，购自上海一基实业有限公司，为优级纯，经湿法研磨提纯，D_{50}为 3.973 μm。实验用细菌体选择金黄色葡萄球菌（SA），取自绵阳市四〇四医院。矿物颗粒与细菌体形貌变化与存在状态通过 SEM 观察。采用选择性强的荧光分光光度法检测体系及胞内活性氧自由基（ROS）的变化。采用分子力学构建方解石颗粒与 SA 的团簇模型。采用蒙特卡洛计算方解石细颗粒与 SA 相互作用过程中的吸附能变化。

实验结果表明，SA 液体培养体系中添加方解石颗粒可以加速葡萄糖（GLU）的消耗，同时方解石颗粒溶蚀现象显著，且与 SA 细胞明显黏附，菌体形态出现破裂现象。方解石颗粒与 SA 作用24 h 后，均会导致菌胞内活性氧自由基升高，且呈现浓度—时间剂量效应（图 1）。由于活性氧自由基升高，超氧化物歧化酶开始清除自由基，其活性均有所下降。由于矿物细颗粒的加入导致细菌通透性发生变化，乳酸脱氢酶释放，其活性有所上升（图 2）。

蒙特卡洛的计算结果表明，SA 生物分子在方解石表面完成了稳定的吸附，吸附能为－2 736.17～－1 818.71 kcal·mol^{-1}。通过分子动力学弛豫和优化，方解石与生物分子作用距离处在2.4～2.8Å，且方解石的钙离子和碳酸根离子均参与作用。具体表现为：Ca 离子主要作用于氨基酸的羧基上羰基氧原子，磷脂的磷酸根基团（磷氧双键）、磷酸甘油酯的酯键（O 原子），磷脂中脂肪酸甘油酯的酯键（O 原子）等。碳酸根离子主要作用于氨基酸的氨基上的氢、羧基上的氢、亚甲基上的氢，以及磷脂上磷酸根基团羟基上的氢等。

---

* 通信作者：董发勤；E-mail：fqdong@swust. edu. cn；联系电话：0816-6089013。国家自然科学基金（41602033、41130746、41572025、41472046），四川省应用基础研究项目重点项目（2018JY0426）。

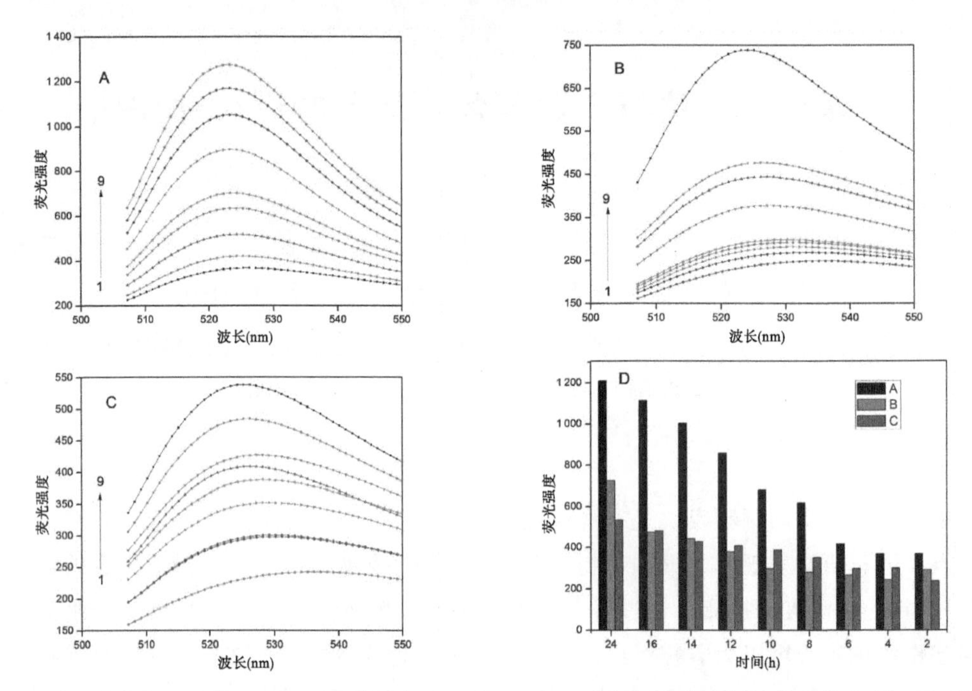

图 1　方解石培养基不同时间（A）、纯培养基不同时间（B）、方解石与金黄色葡萄球菌作用不同时间（C）的滤液荧光光谱以及不同时间各个体系 ROS 荧光强度（D）

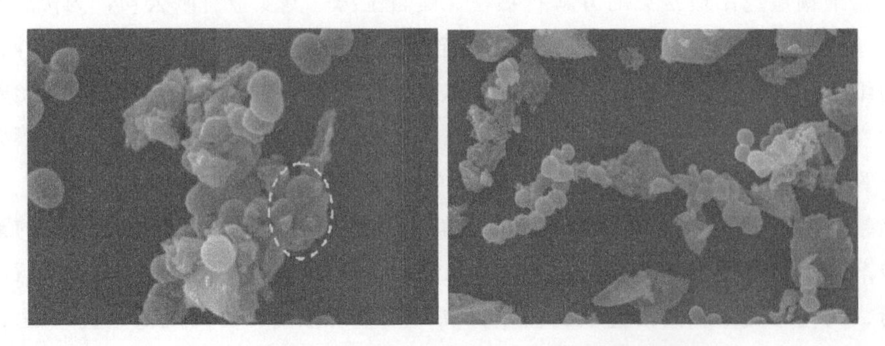

图 2　方解石与金黄色葡萄球菌作用后的 SEM 结果

参考文献

[1] Rattigan O V，Felton H D，Bae M S. 2010. Multi-year hourly PM2. 5 carbon measurements in New York：Diurnal，day of week and seasonal patterns [J]. Atmospheric Environment，21 (2)：76 - 82.

[2] Yang M J，Rodger P M，Harding J H. 2009. Molecular dynamics simulations of peptides on calcite surface [J]. Mol Simulat，35 (7)：547 - 53.

[3] Zhao Y L，Dong F Q，Dai Q W，et al. 2017. Variation of preserving organic matter bound in interlayer of montmorillonite induced by microbial metabolic process [J]. Environmental Science and Pollution Research，1 - 8.

生物炭—铁锰尖晶石复合材料对锑镉污染土壤的钝化效果

汪玉瑛[1,2]　计海洋[3]　吕豪豪[1,2]　刘玉学[1,2]　何莉莉[1,2]　杨生茂[1,2*]

（1. 浙江省农业科学院环境资源与土壤肥料研究所，杭州 310021；

2. 浙江省生物炭工程技术研究中心，杭州 310021；

3. 浙江师范大学化学与生命科学学院，金华 321004）

锑（Sb）作为不可再生资源在世界上占有重要地位。过量的锑进入地表，不仅造成地表土壤和水体的重金属污染，还能干扰降低人体内酶的活性，是一种具有潜在毒性和致癌性的元素。镉（Cd）是国际癌症研究机构和世界卫生组织确认的一种致癌物质，会导致人体骨损伤、痛痛病等使骨骼变形或骨折，最终导致死亡，更为严重的是，锑常与镉（Cd）、砷（As）等重金属共存，进一步加剧了生态与健康风险。因此，我国锑、镉复合污染土壤的修复与治理亟待研究解决。

生物炭作为一类新型环境功能材料在温室气体减排、农业土壤改良、农作物增产提质等方面具有巨大的应用潜力，成为近年来环境修复的研究热点；但由于生物炭表面主要是带负电荷的官能团，对阴离子的吸附效果较差。因此，需要通过改性手段活化生物炭表面性质。我们前期以茶叶枝条为原料，通过表面负载的方法制备生物炭—铁锰尖晶石纳米复合材料，并探究其对废水中Sb（Ⅲ）和Cd（Ⅱ）的吸附性能影响及作用机理[1]。结果表明：生物炭—铁锰尖晶石纳米复合材料对锑、镉的最大吸附量均显著高于未改性生物炭，且锑、镉的吸附存在着协同作用。然而，迄今为止，铁锰基生物炭复合材料对锑镉复合污染土壤的修复报道甚少，铁锰基生物炭复合材料对锑镉在土壤中的赋存形态、迁移转化规律及降低锑镉生物有效性机制还缺乏相关的研究。

本研究以茶叶枝条为原料，通过表面负载的方法制备生物炭—铁锰尖晶石纳米复合材料，探明不同用量的生物炭及生物炭—铁锰尖晶石复合材料对土壤理化性质、土壤锑镉污染的钝化效果及黑麦草中锑镉含量的影响。

供试生物炭为茶叶枝条生物炭及铁锰尖晶石—生物炭复合材料。供试作物为黑麦草。将10目筛的风干供试土壤装盆，每盆1 000 g，然后将生物炭及铁锰尖晶石—生物炭复合材料按质量百分比0.5%、1%和2%的添加量加入盆中（S2～S7），充分混匀，种植黑麦草。同时，以未添加生物炭的污染土壤（S1）及原土样（S0）作为对照。试验设个8处理，3次重复。试验在保持田间持水量60%情况下进行，60 d后取样分析。

实验结果表明，施加生物炭及铁锰尖晶石改性生物炭对土壤理化性质有显著的影响。与不施生物炭的对照相比，施用生物炭可显著提高污染土壤pH值、有效磷、速效钾、全氮和有机质含量，且随添加量增加而幅度增大。添加生物炭及改性生物炭后，黑麦草的株高和干物质量均提高，而且改性生物炭较未改性生物炭提高幅度增大（表1）。同时，改性生物炭有效降低了污染

*通信作者：杨生茂；E-mail：yangshengmao@263. net；手机号：15067126622。浙江省重点研发计划项目（2015C03020）、浙江省自然科学基金项目（LQ17D020002）资助项目。

土壤及植物中的有效态锑镉含量，而未改性生物炭降低对镉的钝化效果好于锑（图 1）。另外，改性生物炭处理的黑麦草中重金属含量明显降低。可见，改性生物炭显著降低了土壤重金属锑镉的生物有效性。因此，生物炭—铁锰尖晶石纳米复合材料是一种潜在的高效土壤钝化剂。

表 1　收获的黑麦草株高和干物质量

处理	株高（cm）	干物质量（$g \cdot pot^{-1}$）
S0	41.00±1.00 a	2.04±0.24 a
S1	16.67±1.15 f	0.39±0.14 f
S2	21.00±1.00 e	1.02±0.14 e
S3	30.33±1.53 d	1.51±0.12 cd
S4	31.00±1.00 d	1.56±0.09 bcd
S5	33.67±1.53 c	1.44±0.09 d
S6	35.67±1.53 bc	1.74±0.11 bc
S7	36.33±1.15 b	1.78±0.10 b

注：不同小写字母表示处理间差异达 0.05 显著水平

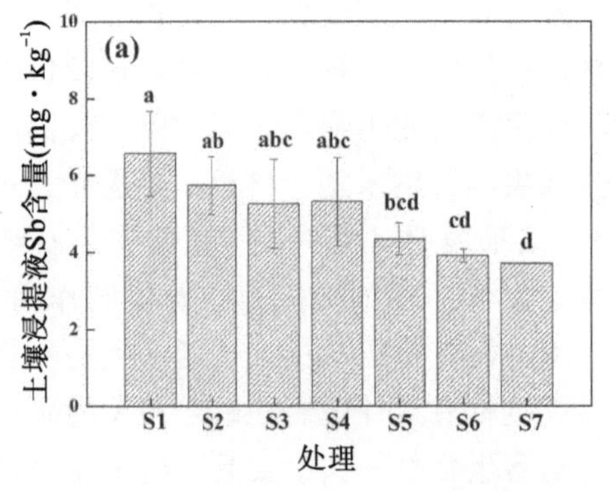

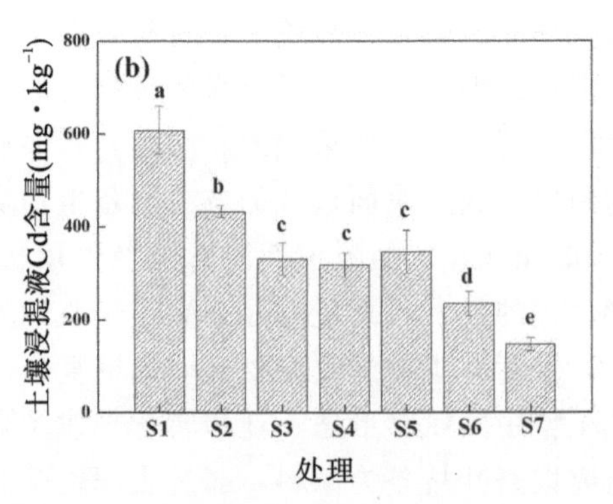

图 1　土壤浸提液中锑（a）和镉（b）含量

注：不同小写字母表示处理间差异达 0.05 显著水平

参考文献

[1] Wang Y Y，Ji H Y，Lu H H，et al. 2018. Simultaneous removal of Sb（Ⅲ）and Cd（Ⅱ）in water by adsorption onto a $MnFe_2O_4$- biochar nanocomposite [J]. RSC Advances，8：3264 - 3273.

竹炭降低小麦吸收积累土壤镉的作用

倪　幸[1]　黄旗颖[1,2]　叶正钱[1*]
（1. 浙江农林大学环境与资源学院，杭州 310032；
2. 浙江省金华市土肥站，金华 321000）

小麦是易积累重金属镉（Cd）的主要粮食作物，在 Cd 污染农田种植会引起小麦籽粒 Cd 污染，危害人体健康。施用生物炭可有效降低农田土壤有效态 Cd 含量，抑制植物对 Cd 的吸收，从而保障农产品安全，因而在土壤修复和消减重金属食品污染风险方面潜力巨大[1,2]。本文以竹炭为生物炭材料，通过盆栽试验探讨 Cd 胁迫条件下竹炭对小麦吸收积累土壤 Cd 的影响。

材料与方法：采集 0～20 cm 耕层的 Cd 污染土壤，土壤 pH＝6.0，有机质 46.4 g/kg，Cd 3.3 mg/kg。经添加外源 Cd（按 2 mg/kg 添加），老化 1 个月后获得 Cd 胁迫土壤。供试生物质炭为市售竹炭粉，pH＝9.41，未检测出 Cd。按质量 0%、0.1%、1%、5%的比例将竹炭与 Cd 胁迫土壤混匀，播种小麦，在小麦成熟期取植物和土壤样品进行分析测定。

结果与讨论：低用量竹炭（0.1%、1%）对土壤 pH 值、有机质及有效态 Cd 含量的影响小，但是在高用量（5%）时，土壤 pH 值、有机质含量都显著提升，相反，土壤有效态 Cd 水平显著下降（表 1），竹炭可能是通过升高土壤 pH 值，增加土壤有机质含量，改变土壤重金属的形态来实现的[3]。

表 1　不同竹炭用量对土壤性质的影响

竹炭用量（%）	土壤 pH 值	土壤有机质（g/kg）	土壤有效镉（mg/kg）
0	6.00 b	42.71 c	3.15 a
0.1	5.94 b	44.23 c	3.09 a
1	6.11 b	52.57 b	3.11 a
5	6.43 a	91.49 a	2.90 b

注：同列数据后小写字母不同表示不同处理间差异显著（$P<0.05$）

小麦植物生长和产量受竹炭用量的影响较小，总体表现为低用量（0.1%、1%）促进、高用量（5%）抑制小麦的生长，但处理间无显著差异。受竹炭对土壤性质及对小麦生长的影响，随着竹炭施用量的增加，小麦植株各器官 Cd 含量呈先增加后减少的趋势，0.1%竹炭施用量的 Cd 含量最高，而 5%竹炭施用量的 Cd 含量最低，显著低于其他各处理（表 2）。综上，低量竹炭的

* 通信作者：叶正钱；E-mail：yezhq@zafu. edu. cn；手机号：18989866187。浙江省科技厅（2018C03028）资助项目。

施用促进了小麦植株对 Cd 的吸收，而高量竹炭的施用可有效降低小麦对 Cd 的吸收，该研究结果与其他生物炭类似研究一致[4,5]。

表 2　不同竹炭用量对小麦植株各部位器官 Cd 含量的影响（mg/kg）

竹炭用量（%）	根	茎鞘	剑叶	其余叶	麦壳	籽粒
0	4.58 b	1.81 b	1.54 ab	2.76 a	0.51 ab	0.60 a
0.1	5.27 a	2.37 a	2.65 a	3.09 a	0.77 a	0.68 a
1	4.52 b	1.78 b	1.54 ab	2.81 a	0.52 ab	0.58 a
5	3.01 c	1.39 b	0.93 b	1.99 b	0.32 b	0.39 b

注：同列数据后小写字母不同表示不同处理间差异显著（$P<0.05$）

结论：竹炭的施用提高了土壤 pH 值和有机质含量，降低了土壤有效态 Cd 含量，但竹炭对土壤 Cd 有效性的作用与施用量密切相关，只有在高量（5%）施用时才显著降低土壤 Cd 的有效性，减少 Cd 在小麦籽粒中的积累。

参考文献

[1] Yang X，Liu J，Mcgrouther K，et al. 2016. Effect of biochar on the extractability of heavy metals (Cd，Cu，Pb，and Zn) and enzyme activity in soil [J]. Environmental Science & Pollution Research，23：974－984.

[2] Park J H，Choppala G K，Bolan N S，et al. 2011. Biochar reduces the bioavailability and phytotoxicity of heavy metals [J]. Plant & Soil，348：439－451.

[3] Beesley L，Moreno-Jiménez E，Gomez-Eyles J L，et al. 2011. A review of biochars' potential role in the remediation，revegetation and restoration of contaminated soils [J]. Environmental Pollution，159：3269－3282.

[4] 张晗芝，黄云，刘钢，等. 2010. 生物炭对玉米苗期生长、养分吸收及土壤化学性状的影响 [J]. 生态环境学报，19：2713－2717.

[5] 周建斌，邓丛静，陈金林，等. 2008. 棉秆炭对镉污染土壤的修复效果 [J]. 生态环境，17：1857－1860.

碳纳米管/Fe_3O_4复合材料构建及微波降解抗生素

田林涛 刘世媛 吕国诚* 廖立兵*
（中国地质大学（北京），材料科学与工程学院，北京 100083）

抗生素是可以杀死或抑制微生物（细菌、真菌、病毒、古细菌、原生动物和微藻）生长的抗菌剂类型。随着人类和兽医中抗生素的使用越来越多，其对土壤和水生态系统的污染问题不容忽视。其中，抗生素水污染是非常严重的环境问题，目前大多数处理方法，如化学氧化、生物降解、膜生物反应器、光催化降解等难以实现抗生素完全降解并且会产生有害的副产物。与其他方法相比，微波废水处理快速、高效以及环保，微波诱导降解污染物的方法可以实现更高效更彻底的抗生素去除，且没有明显的副作用。微波吸收材料在微波诱导降解污染物中起着重要作用。因此，开发高效催化材料具有潜在的应用价值[1-4]。

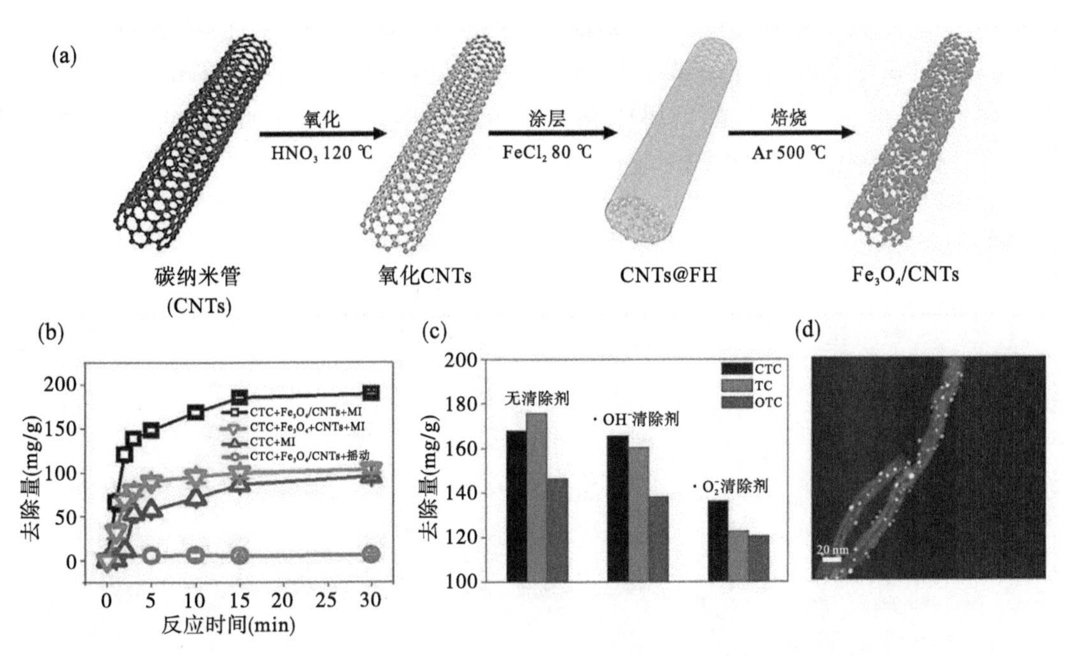

图 1 Fe_3O_4/CNTs 的合成过程示意图（a）、Fe_3O_4/CNTs 为催化剂的微波诱导降解金霉素的降解效果（b）、不同捕获剂对降解效果的影响（c）及 Fe_3O_4/CNTs 的 Mapping 图像（d）

*通信作者：吕国诚；E-mail：guochenglv@cugb. edu. cn；廖立兵；E-mail：lbliao@cugb. edu. cn。国家重点科技攻关计划（2017YFB0310704）、国家自然科学基金青年基金（51604248）资助项目。

在该研究中我们设计了一种新型的微波诱导催化剂——Fe_3O_4/CNTs，通过简单的制备方法控制 Fe_3O_4纳米颗粒（<10 nm）在碳纳米管上均匀生长。该复合材料具有多孔结构和高比表面积，可促进与污水之间的相互作用。Fe_3O_4和 CNTs 两种材料的介电损耗和磁损耗的结合以及良好的阻抗匹配使该复合材料具有较强的微波吸收能力。在微波作用下，该复合材料表面可以产生很多“热点”，这些材料表面的高温热点促进了吸附在材料表面的抗生素的快速氧化降解。结果显示（图 1），Fe_3O_4/CNTs 在微波作用下对金霉素、四环素和土霉素均具有较好的降解效果，其中 CTC 的去除量达到 185 mg/g，其降解率（k＝0.037 6）远高于先前报道。添加自由基捕获剂的实验结果表明，微波作用下微波诱导催化剂在水溶液中产生的羟基自由基（·OH）与超氧自由基（$\cdot O_2^-$）也有助于抗生素的去除。同时通过质谱和离子色谱分析证实，金霉素的最终降解产物是无害的小分子 CO_2和 NO_3^-，没有其他副产物。该研究结果表明，Fe_3O_4/CNTs 复合材料是一种性能优异的微波诱导氧化催化剂，与其他有机物降解方法相比，以 Fe_3O_4/CNTs 为催化剂的微波诱导氧化降解可以更高效、彻底地去除水中的抗生素。

参考文献

［1］Baquero F，Martínez J L，Cantón R. 2018. Antibiotics and antibiotic resistance in water environments［J］. Current Opinion in Biotechnology，19：260－265.

［2］Jian X，Wu B，Wei Y F，et al. 2016. Facile synthesis of Fe_3O_4/GCs composites and their enhanced microwave absorption properties［J］. Applied Materials & Interfaces，8：6101－6109.

［3］Zhang Z H，Xu Y，Ma X P，et al. 2012. Microwave degradation of methyl orange dye in aqueous solution in the presence of nano－TiO_2－supported activated carbon（supported－TiO_2/AC/MW）［J］. Journal of Hazardous Materials，209－210：271－277.

［4］Lai T，Lee C C，Huang G L，et al. 2008. Microwave-enhanced catalytic degradation of 4－chlorophenol over nickel oxides［J］. Applied Catalysis B－Environmental，78：151－157.

第七章

非金属矿纳米技术与生物医药健康

过氧化氢刺激响应埃洛石纳米管基复合材料合成与应用

张海磊* 温 昕 武永刚 巴信武
（河北大学化学与环境科学学院，保定 071002）

埃洛石纳米管是一种天然纳米管状材料，具有较大的长径比和空腔体积、出色的生物相容性以及低廉的售价，在催化、生物医药等领域具有较大的应用前景[1,2]。过氧化氢是一种重要的活性氧，其在生命活动和环境保护方面起到重要作用。在生命活动中，过氧化氢扮演着免疫标志的角色。因此，实现对过氧化氢快速、灵敏的检测以及构建过氧化氢响应型释药体系，对工业生产、医疗保健和人体健康均具有十分重要的意义。

我们通过芳基硼酸对埃洛石纳米管进行表面改性，制备了“埃洛石纳米管—芘”复合物和“埃洛石纳米管—荧光素”复合物；利用芳基硼酸酯中的B—C键可在过氧化氢作用下发生断键的机制，将两种改性产物开发为可对过氧化氢定量检测的荧光探针，其中“埃洛石纳米管—芘”复合物在过氧化氢作用下发生荧光衰减，即“turn off”效应；“埃洛石纳米管—荧光素”复合物在过氧化氢作用下发生荧光增强，即“turn on”效应。上述两种复合物用于检测过氧化氢具有良好的灵敏性和专属性，其中“埃洛石纳米管—荧光素”复合物可利用细胞内源性过氧化氢，实现长效荧光染色的效果[3,4]。

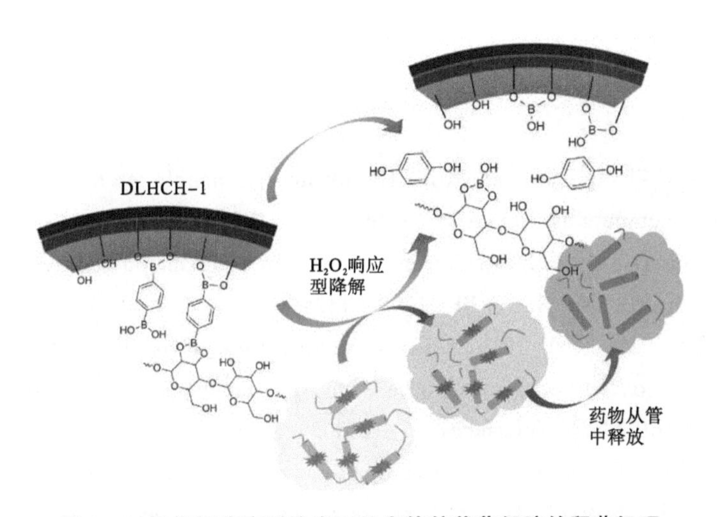

图1 过氧化氢响应型埃洛石纳米管基载药凝胶的释药机理

我们进而利用已建立的芳基硼酸对埃洛石纳米管改性的方法，通过1,4-苯二硼酸对埃洛石

*通信作者：张海磊；E-mail：zhanghailei@hbu. edu. cn；手机号：15930920929。国家自然科学基金（Nos. 21274037，21474026）、河北省教育厅基金（QN2018052）资助项目。

纳米管进行处理，制得含有活性芳基硼酸基团的苯二硼酸改性埃洛石纳米管。进而以苯二硼酸改性的埃洛石纳米管为交联剂，以可压性淀粉为基材，合成了埃洛石纳米管基天然多糖水凝胶。该凝胶在具有过氧化氢响应性释药的同时，可良好地抑制药物的“突释”效应（图 1）[5]。

我们进一步利用芳基硼酸对埃洛石纳米管的改性方法及原子转移自由基聚合法，制备了聚离子液体接枝改性埃洛石纳米管（图 2），有望拓展其在电池材料领域的应用，以期利用其长径比较高的优势制备具有各向异性的导体材料。

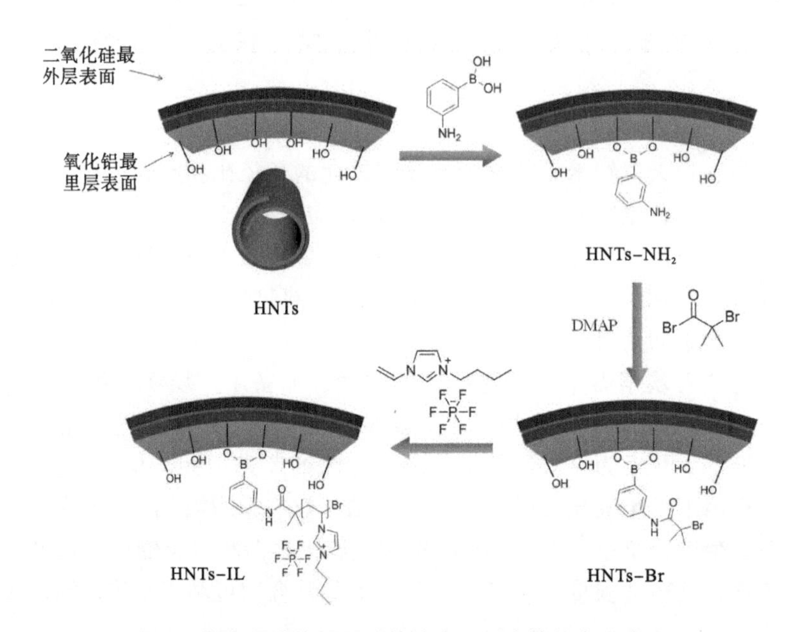

图 2　聚离子液体接枝改性埃洛石纳米管的合成路线

参考文献

[1] Lvov Y，Wang W，Zhang L，et al. 2016. Halloysite clay nanotubes for loading and sustained release of functional compounds [J]. Adv. Mater.，28：1227 - 1250.

[2] Liu M，Jia Z，Jia D，et al. 2014. Recent advance in research on halloysite nanotubes-polymer nanocomposite. Prog. Polym. Sci.，39：1498 - 1525.

[3] Zhang H，Ren T，Ji Y，et al. 2015. Selective modification of halloysite nanotubes with 1 - pyrenylboronic acid：A novel fluorescence probe with highly selective and sensitive response to hyperoxide [J]. ACS Appl. Mater. Interfaces，7：23805 - 23811.

[4] Dong J，Zhao Z，Liu R，et al. 2017. Investigation of a halloysite-based fluorescence probe with a highly selective and sensitive “Turn-on” response upon hydrogen peroxide [J]. RSC Adv.，7：55067 - 55073.

[5] Liu F，Bai L，Zhang H，et al. 2017. Smart H_2O_2 - responsive drug delivery system made by halloysite nanotubes and carbohydrate polymers [J]. ACS Appl. Mater. Interfaces，9：31626.

还原性含铁黏土抗菌机理

夏庆银[1]　王　曦[1]　董海良[1,2*]　曾　强[1]
（1. 中国地质大学（北京）生物地质与环境地质国家重点实验室，北京 100083；
2. 迈阿密大学（美国），牛津市，俄亥俄州，美国）

前人研究表明，自然界中大量存在的黏土混合体系对某些致病菌具有抑制作用[1,2]。还原性含铁黏土（RIC），尤其是伊利石—蒙脱石矿物，是天然抗菌黏土的重要组成部分[3,4]。Williams等[5]研究指出，混合黏土矿物在酸性条件下溶解、释放出溶解态 Fe^{2+}，进入细胞内催化自由基反应，是其杀菌的基本原理。然而，纯的含铁黏土矿物的杀菌作用尚未被证实，且黏土矿物中普遍存在结构 Fe（Ⅱ）［矿物晶格中的 Fe（Ⅱ）］，在杀菌过程中的作用还有待研究。

本实验体系中，将晶格中结构铁被还原的黏土矿物绿脱石（rNAu－2）与大肠杆菌（*E. coli*）混合并暴露在空气中。结果表明，rNAu－2 的杀菌作用与体系 pH 值和结构 Fe（Ⅱ）的浓度有关。pH 值为 6 时，2 g/L［5.6 mM 结构 Fe（Ⅱ）］的 rNAu－2 在 24 h 内可以完全杀死 *E. coli*（～10^8 cells/mL），而 pH 值为 7 和 8 时 rNAu－2 几乎没有杀菌作用。加入·OH 捕获剂或在无氧条件下 rNAu－2 的杀菌效果显著下降，这表明 rNAu－2 的杀菌作用与其结构 Fe（Ⅱ）在空气中氧化产生·OH 相关。原位的成像结果表明，*E. coli* 菌体两端的心磷脂（Cardiolipin，CL）被氧化破坏，同时活性氧（Reactive oxygen species，ROS）和溶解态的 Fe^{2+} 在菌体两端聚集区和被氧化破坏的心磷脂区域高度重叠。

我们的研究表明黏土矿物不仅仅可以通过酸解释放溶解态 Fe^{2+} 进行杀菌，其晶格中结构态 Fe（Ⅱ）在杀菌过程中还发挥着极其重要的作用。黏土矿物结构 Fe（Ⅱ）是菌体外产生·OH 的主要来源，这些·OH 氧化心磷脂改变了菌体的通透性，促进溶解态 Fe^{2+} 进入细菌体内从而杀死细菌。

由于菌体和黏土表面的负电性，二者间的库伦排斥可能影响黏土杀菌的效果[6]。为加强黏土杀菌效果，利用壳聚糖（Chitosan）插层改性的表面带正电的 rNAu－2（rC－NAu－2）分别对革兰氏阴［大肠杆菌（*E. coli*）］阳性菌［金黄色葡萄球菌（S. aureus）］进行实验。结果表明，含 3 mM 结构 Fe（Ⅱ）的 rC－NAu－2 在 pH 值为 6 的条件下就可以在 24 h 内完全杀死 *E. coli*（～10^8 cells/mL）。即使在 pH 值为 7 和 8 的条件下，含 3 mM 结构 Fe（Ⅱ）的 rC－NAu－2 也能够在 24 h 内杀死 99％的 *E. coli*（2 个数量级）。含 4.5 mM 结构 Fe（Ⅱ）的 rC－NAu－2 在 pH 值为 6 的条件下可以在 24 h 内杀死 90％以上的 S. aureus，而同样浓度的 rNAu－2 对 S. aureus 而言，则检测不到明显杀菌效果。

我们的研究结果证实前人提出的抗菌模型的合理性，证实了 ROS 在攻击细胞膜上心磷脂的

*通信作者：董海良；E-mail：dongh@cugb. edu. cn；手机号：15910914282。国家自然科学基金（41572328）资助项目。

重要性，伴随着 ROS 的攻击，可溶性 Fe^{2+} 进入胞内，随后引发循环芬顿（Fenton）作用，从而导致胞内的蛋白质受损。本研究首次证明结构 Fe（Ⅱ），尤其是在近中性 pH 值条件下，是・OH 的主要来源，并证实心磷脂在黏土抗菌过程中的重要作用，从而改善之前的模型。基于这些新发现，我们将黏土矿物杀菌的有效 pH 值范围扩大至对人类皮肤温和的中性条件，对其进一步产业化发展具有推动作用。目前的实验结果指明我们下一步的研究工作应该将重心放在整个生理生化过程，尤其是菌株在 ROS 等氧化压力下的基因及代谢表达变化，同时关注胞内 ROS 累计至细菌死亡过程中细胞具体的理化参数变化。

参考文献

[1] Williams L B，Haydel S E. 2010. Evaluation of the medicinal use of clay minerals as antibacterial agents [J]. International Geology Review，52：745 - 770.

[2] Williams L B. 2017. Geomimicry：harnessing the antibacterial action of clays [J]. Clay Minerals，52：1 - 24.

[3] Williams L B，Haydel S E，Giese R F，et al. 2008. Chemical and mineralogical characteristics of French green clays used for healing [J]. Clays and Clay Minerals，56：437 - 452.

[4] Morrison K D，Underwood J C，Metge D W，et al. 2014. Mineralogical variables that control the antibacterial effectiveness of a natural clay deposit [J]. Environmental Geochemistry and Health，36：613 - 631.

[5] Williams L B，Metge D W，Eberl D D，et al. 2011. What makes a natural clay antibacterial? [J]. Environmental Science & Technology，45：3768 - 3773.

[6] Singh R，Dong H，Zeng Q，et al. 2017. Hexavalent chromium removal by chitosan modifiedbioreduced nontronite [J]. Geochimica et Cosmochimica Acta，210：25 - 41.

埃洛石纳米管在生物医学领域中的应用

刘明贤* 郑静琪 吴 帆 赵秀娟 张 军 周长忍
（暨南大学化学与材料学院，广州 510632）

埃洛石纳米管（Halloysite nanotubes，HNTs）是自然界中形成的黏土纳米材料，具有独特的中空管状结构，化学式为 $Al_2Si_2O_5$（OH）$_4$ · H_2O，是 1∶1 型硅酸盐矿物。HNTs 管内径尺寸是 10～30 nm，外径尺寸为 40～70 nm，而长度为 0.2～1 μm，在世界各地都有沉积（图 1）。HNTs 具有高比表面积、强吸附能力、良好的水分散性、优异的生物相容性、活泼的表面基团的优势。HNTs 纳米粒子展现出作为药物载体、创面修复、组织工程支架和生物检测等领域的应用潜力[1]。

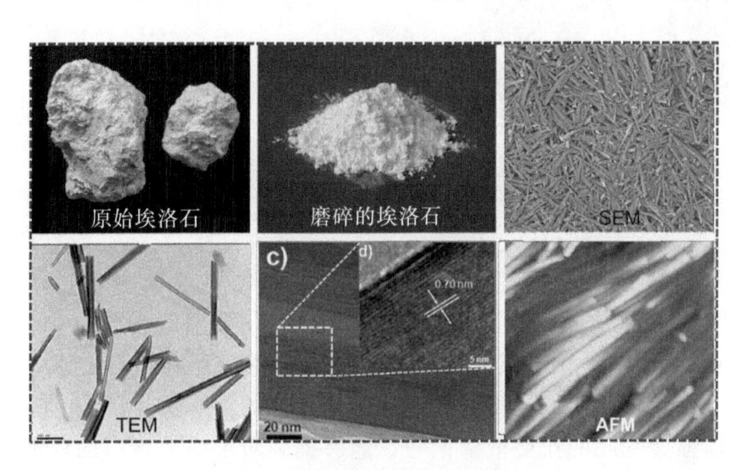

图 1　埃洛石的外观和微观形态

系统研究了大分子修饰后的 HNTs 作为姜黄素、阿霉素（DOX）等药物载体的可行性，探究了其体外和体内治疗乳腺癌的效果[2,3]。修饰后的 HNTs 具备带正电荷的表面和良好的血液相容性，可以负载药物促进 MCF－7 细胞的凋亡。由于 HNTs 类似于碳纳米管的独特针状结构，其可以直接穿透细胞膜或通过内吞进入细胞，通过线粒体和细胞核双重损伤协同机制来增强药物的抗癌效率。通过原位注射纳米药物到 4T1 肿瘤小鼠的肿瘤部位进行抑瘤实验。发现其可以显著地抑制肿瘤的生长，小鼠全部存活至 60 d 以上，并且小鼠的主要脏器并无损伤。功能化 HNTs 在临床抗癌中提供了新可能。

利用聚苯乙烯磺酸钠改性埃洛石，有效地改善了其在水中分散稳定性。之后巧妙利用毛细管、夹片受限空间、球板受限空间等方式将埃洛石进行干燥过程取向排列，利用重力、摩擦力和

* 通信作者：刘明贤；E-mail：liumx@jnu. edu. cn；手机号：13560412830。国家自然科学基金（No. 51502113）资助项目。

表面张力的力学平衡原理，进行多次的粘附—脱粘过程，最终获得了在多种基底上形成的纳米阵列条纹，条纹中的纳米管也呈现一定程度的取向排列，而且在偏光下显示出规则的马耳他十字现象，非常类似聚合物球晶和液晶的光学折射现象。通过改变分散液的浓度、干燥温度、剪切力条件等，可调控形成的图案结构。制备的埃洛石规则条带结构表面可以捕获循环肿瘤细胞，也能作为引导细胞取向的基底[4]（图 2）。

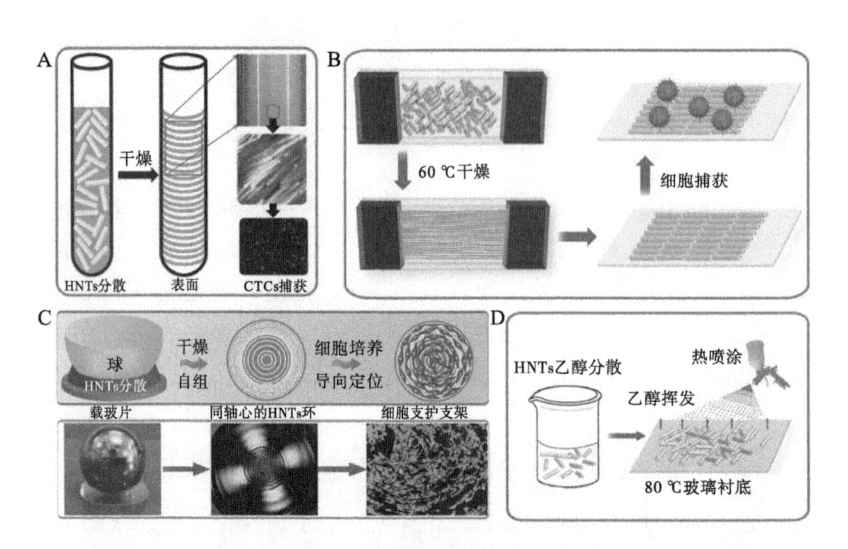

图 2　在不同的受限空间中埃洛石的组装及有序结构形成和应用

注：A. 玻璃管；B. 狭缝；C. 球—板；D. 热喷涂

由于纳米埃洛石和癌细胞之间存在强相互作用，该团队研究发现，HepG－2、MCF－7、Neuro－2a、A549 等癌细胞在埃洛石粗糙表面上的捕获率均高于 80%。与 EpCAM 抗体偶联后可将癌细胞的捕获率在 3 h 内提高到 92%。表面可从加入肿瘤细胞的血液中以及临床癌症病人血液高效率捕获并杀死癌细胞，从而成为早期诊断和病情监控的纳米诊疗平台。进一步研究发现，埃洛石能够数量级地增强金基底的拉曼散射信号，能够提高对牛血清蛋白的检测能力。这表明埃洛石在多种生物检测领域的应用潜力，从而可能实现重大疾病的生物信号体外检测和靶向治疗。

参考文献

[1] Liu M，Jia Z，Jia D，et al. 2014. Recent advance in research on halloysite nanotubes-polymer nanocomposite. Prog. Polym. Sci.，39：1498－1525.

[2] Liu M，Chang Y，Yang J，et al. 2016. Functionalized halloysite nanotube by chitosan grafting for drug delivery of curcumin to achieve enhanced anticancer efficacy [J]. Journal of Materials Chemistry B，4（13）：2253－2263.

[3] Yang J，Wu Y，Shen Y，et al. 2016. Enhanced therapeutic efficacy of doxorubicin for breast cancer using chitosan oligosaccharide-modified halloysite nanotubes [J]. ACS applied materials & interfaces，8（40）：26578－26590.

[4] Liu M，He R，Yang J，et al. 2016. Stripe-like clay nanotubes patterns in glass capillary tubes for capture of tumor cells [J]. ACS applied materials & interfaces，8（12）：7709－7719.

钙磷序贯释药纳米递送系统治疗肝肿瘤多药耐药

王　琪[1,2]　董　阳[1]　朱为宏[2]　段友容[1*]

（1. 上海交通大学医学院附属仁济医院上海市肿瘤研究所，上海 200032；

2. 华东理工大学化学与分子工程学院，先进功能材料重点实验室，上海 200032）

肿瘤细胞的多药耐药性（MDR）产生已经成为肝癌化疗的主要障碍。三磷酸腺苷（ATP）依赖型外排泵能够将进入肿瘤细胞内的药物大量排出细胞，从而大大降低肿瘤内有效药物浓度，极大降低药物作用效果，从而引起肿瘤耐药。因此，如何有效抑制外排泵对药物的外排作用，是治疗肝肿瘤细胞耐药的关键因素。

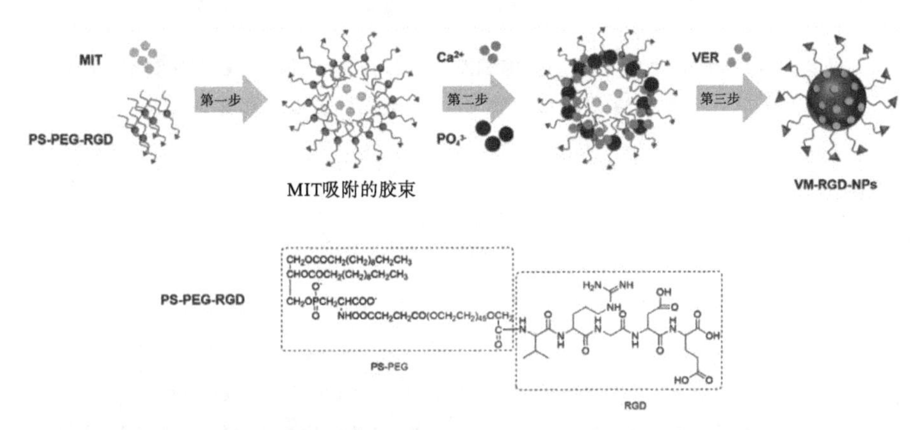

图 1　治疗多药耐药肝肿瘤的壳核结构纳米粒 VM－RGD－NPs 的制备路线[1]

注：第一步为 RGD－PS－PEG 自组装；第二步为钙离子和磷酸根离子吸附于胶束表面；第三步为逆转耐药试剂 VER 吸附于磷酸钙表面的孔隙中

针对上述思路，本研究设计了一种多功能的、可生物降解的、肿瘤靶向的、核壳结构纳米载体（RGD－NPs）用于耐药型肝癌的治疗（图 1）。磷酸钙壳（可诱导 pH 值触发式的溶酶体逃逸），磷脂酰丝氨酸—聚乙二醇核（PS－PEG，可实现纳米粒的体内长循环）和主动靶向配体的 RGD 肽（可引导纳米粒主动靶向至肿瘤组织）。耐药抑制剂（维拉帕米 VER）和化疗药物（米托蒽醌，MIT）分别封装在纳米粒的外壳层和内核层中制备得到载药纳米粒 VM－RGD－NPs。纳米粒的壳核结构使其包载的壳层和核层药物逐级释放，从而实现了两种药物的协同作用，极大减弱了多药耐药肿瘤细胞对抗肿瘤药物 MIT 的外排效应。此外，在溶酶体的低 pH 值条件下，磷酸钙解离，触发了纳米粒从溶酶体中逃逸。同时，优化 MDR 肿瘤细胞对纳米粒的摄取能力和摄取途

* 通信作者：段友容，手机号：13818706255。

径，极大地提高了细胞内有效化疗药物浓度，提高了疗效。在该药物递送系统中，PEG 延长了体内循环时间，RGD 肽明显提高了纳米粒对多药耐药肿瘤的亲和力[2]。

VM－RGD－NPs 将传统的同时施与耐药抑制剂核化疗药物的 MDR 治疗策略，通过将两者分别包载进纳米粒的壳层和核层的方法改进成为序贯给药的治疗方式。并且，VM－RGD－NPs 对耐药肝癌细胞的治疗作用也得到了有力的证明，RGD－NPs 对 MDR 肝癌有显著的协同治疗作用，为 MDR 肿瘤的治疗提供了一种新的治疗途径。

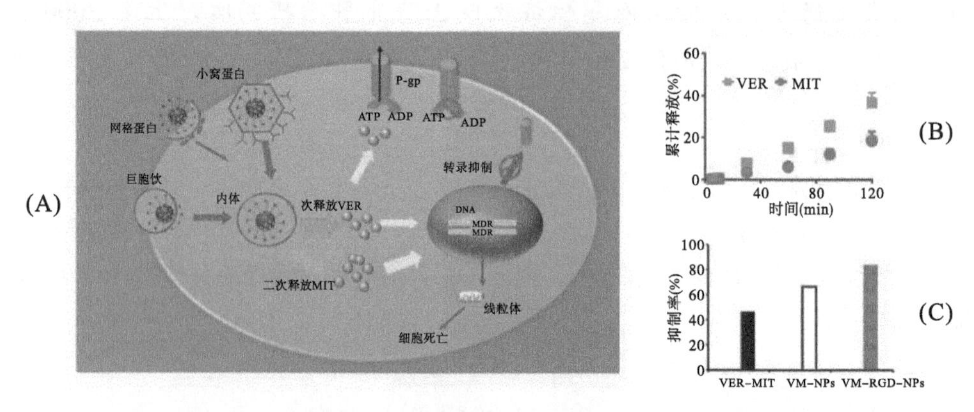

图 2　VM－RGD－NPs 在耐药肿瘤细胞内的作用路径（A)、耐药抑制剂 VER 和化疗药物 MIT 的释放曲线（B）及耐药肝肿瘤的抑制率（C)

参考文献

[1] Wang Q，Zhang Z R，Duan Y R. 2015. Targeted polymeric therapeutic nanoparticles：Design and interactions with hepatocellular carcinoma [J]. Biomaterials，56：229－240.

[2] Wang Q，Zhu W H，Zhang Z R，et al. 2018. Multifunctional shell-core nanoparticles for treatment of multidrug resistance hepatocellular carcinoma [J]. Adv. Funct. Mater.，1706124.

负载磷酸钙脂质体的水凝胶构建及骨修复

程若昱 崔文国*

（上海市中西医结合防治骨与关节病损重点实验室，上海市伤骨科研究所，上海交通大学医学院附属瑞金医院，上海 200025）

骨缺损是指骨的结构完整性被破坏。创伤、感染、肿瘤以及各种先天性疾病等是导致骨缺损的主要原因。目前，治疗骨缺损的方法主要有自体骨移植、异体骨移植、组织工程技术等[1]。而在组织工程技术之中，水凝胶作为支架中重要的一类而被广泛应用于各项研究[2]。水凝胶是一种高分子网络体系，性质柔软，在溶液体系中可保持一定的形状并且可以吸收大量的液体，被广泛应用于工业、农业、医疗等领域[3]。但是，单纯的水凝胶材料在医疗领域中只能起到支架的作用，却难以实现加快恢复骨缺损部位诱导和再生。纳米磷酸钙在自然界骨组织的形成过程中起到了至关重要的作用。虽然骨的组成类型大不相同，但是其组成成分中的无机成分大多为纳米磷酸钙[4]。纳米磷酸钙可赋予骨以良好的机械性能以及生物活性。此外，在生物体内纳米磷酸钙可在调控下进行定向自组装从而形成特殊的矿化结构[5]。CaP（calcium phosphate）纳米粒作为纳米磷酸钙中的一员，因其良好的生物相容性、骨诱导性以及药物传递作用而被广泛应用于药物治疗[6]。但是 CaP 纳米颗粒因其粒径仅为纳米级别，难以在局部提供一定的力学支撑作用从而无法作为支架使用。因此，本研究旨在构建一种将药物与无机粒子相结合的多功能骨诱导水凝胶，不仅可以在局部起到支架的作用，还可在局部持续释放药物以及无机纳米颗粒，从而更好地促进缺损部位的修复。

制备同时负载有机药物维生素 D_3（VD_3）以及无机磷酸钙纳米粒的脂质体（CaP－Lip），并通过动态光散射粒度分析仪、透射电子显微镜、扫描电子显微镜、高效液相色谱仪对其粒径、电位、形态学以及体外释放特性进行考察；进一步将上述脂质体与甲基丙烯酸酐修饰的明胶（GelMA）水凝胶相结合制备复合水凝胶（VD_3&CaP－Lip@Gel），并通过扫描电子显微镜、高效液相色谱仪、CCK－8 试剂盒、细胞活死染色、碱性磷酸酶（ALP）活力测定、ALP 染色、茜素红 S 染色、茜素红定量分析，对该复合水凝胶的形态学、体外释放特性、体外矿化能力、细胞相容性、细胞黏附性以及体外促成骨能力进行考察。

本研究所构建的 VD_3&CaP－Lip@Gel 体系，实现了不同类型药物于同一载体中的负载，使得促成骨药物 VD_3 可实现在水凝胶体系中的均匀分散以及长达 17 d 的体外持续释放；并且 CaP 的引入赋予了复合水凝胶体系以生物矿化的能力，在 SBF 环境中该复合体系可促进生物矿化过程，在水凝胶孔洞中可观察到明显的结晶形成，如图 1 所示。细胞相容性实验结果显示，该复合材料有着良好的细胞相容性以及细胞黏附性；由于 CaP 以及 VD_3 均有着较为明显的促成骨分化作用，因此，VD_3&CaP－Lip@Gel 相较于阴性对照组、阳性对照组以及 GelMA 组，有着明显的体

* 通信作者：崔文国（教授/博导），研究方向：骨组织再生材料。E-mail：wgcui80@hotmail. com。

外促成骨分化能力。

综上所述，良好的生物相容性以及体外促成骨诱导能力、简单的制备方法、制备材料来源广泛这些优点都使得该双效骨诱导性有机-无机复合水凝胶为拓宽水凝胶在骨修复以及组织工程中的应用提供一种有前景的策略。

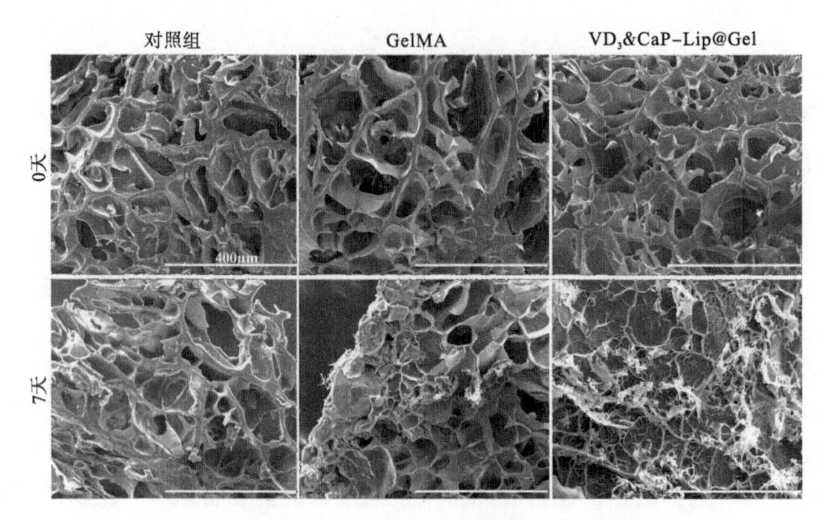

图 1 VD_3 &CaP - Lip@Gel 水凝胶体外矿化结果

参考文献

[1] Drosse I, Volkmer E, Capanna R, et al. 2008. Tissue engineering for bone defect healing: an update on a multi-component approach [J]. Injury, 39 (2): 9 - 20.

[2] Luo Y, Shoichet M S. 2004. A photolabile hydrogel for guided three-dimensional cell growth and migration [J]. Nat. Mater., 3 (4): 249 - 253.

[3] Billiet T, Vandenhaute M, Schelfhout J, et al. 2012. A review of trends and limitations in hydrogel-rapid prototyping for tissue engineering [J]. Biomaterials, 33 (26): 6020 - 6041.

[4] Schmitz J P, Hollinger J O, Milam S B. 1999. Reconstruction of bone using calcium phosphate bone cements: A critical review [J]. J Oral Maxillofac Surg., 57 (9): 1122 - 1126.

[5] Tadic D, Epple M. 2004. A thorough physicochemical characterisation of 14 calcium phosphate-based bone substitution materials in comparison to natural bone [J]. Biomaterials, 25 (6): 987 - 994.

[6] Olton D, Li J, Wilson M E, et al. 2007. Nanostructured calcium phosphates (NanoCaPs) for non-viral gene delivery: Influence of the synthesis parameters on transfection efficiency [J]. Biomaterials, 28 (6): 1267 - 1279.

二氧化硅包裹磁性—荧光多功能纳米材料用于循环肿瘤细胞检测

陈景瑶[1]　黄　鑫[1]　乐文俊[1,2]　陈炳地[1*]
（1. 同济大学医学院，上海 200092；
2. 东方医院再生医学研究所，上海 200120）

近 30 年来临床开始转向在循环血中寻找癌症标记物。如针对前列腺癌的 PSA，针对卵巢癌的 CA－125，和一般性标记物 AFP、CEA 等，但这些血化验的准确性、可靠性、特异性和敏感度都很有限，假阳性及假阴性的发生率很高。且这些指标缺少针对多种癌症的广谱性。

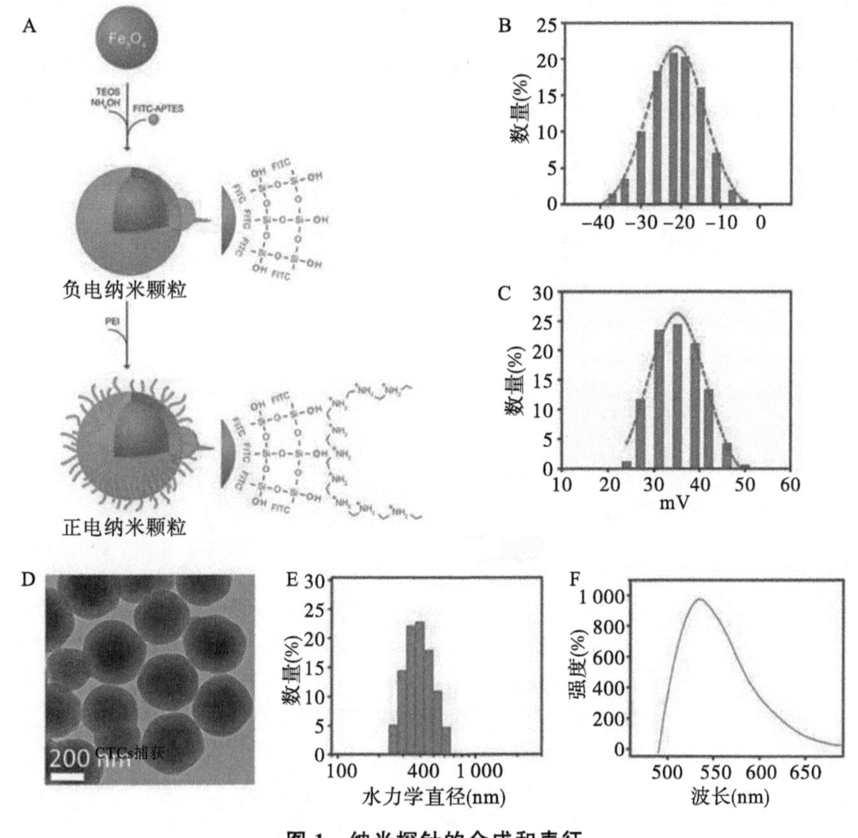

图 1　纳米探针的合成和表征

* 通信作者：陈炳地；E-mail：inanochen@tongji. edu. cn；手机号：18721796963。国家自然科学基金（81772285）资助项目。

目前，最新也最受临床医生关注的癌症早期检测技术是能够直接寻找循环血液中癌细胞的技术（CTC 技术）。这些癌细胞是从固定癌病灶上脱落下来的，然后被体液流动带入循环血，这同时也是癌症转移的重要机制。因此，近 10 年如何在循环血液中寻找捕捉癌细胞在抗癌研究领域里就变成越来越火的科研话题，对癌症的诊断、治疗及预后监测均有着重要的指导意义和实用价值，然而，目前采用普通免疫方法检测因为普遍缺少优异的癌症标记物而困难重重。

本项目技术发明人之一的崔征教授通过十几年来对抗癌新疗法的研究发现几乎所有的有代谢活性癌细胞都具有如超高度使用葡萄糖通过无氧酵解产能的特点，由此产生癌细胞表面生理特性的特殊变化，从而导致癌细胞与正常细胞的显著差异。基于该癌细胞表面生理特性变化的发现，崔教授独创性提出广谱性 CTC 检测技术全新理念；根据本团队多年开发成熟的合成多功能纳米材料的技术平台设计出全新的多功能纳米颗粒。经过大量系统性细胞学实验验证，全新纳米颗粒对癌细胞有超高效的特异性的吸附力，同时又有极强的富集作用、细胞荧光成像及量化能力（图 1）。

该技术有望实现以下检测目标：在 2 h 以内，在 2 mL 循环血中检测富集到循环肿瘤细胞，且适用于大多数种类的癌细胞的快速、准确检测。该技术成本低、高灵敏、快速简便且具有普适性。经过广泛调研，该技术特别适合用于癌症的大范围筛查、早期检测以及癌症病人治疗方案指导，预后监测，复发可能性评价。

本课题组开发了一种带静电、荧光和超顺磁性纳米探针用于靶向癌细胞而不使用任何分子生物标记物。所有 22 个随机选择的不同器官的癌细胞系均可与带正电的纳米探针特异性结合（图 2）。

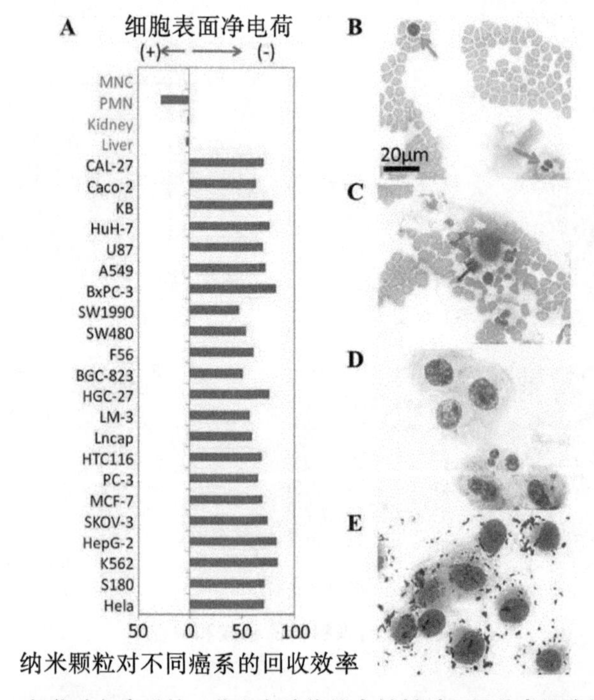

图 2 二氧化硅包裹磁性—荧光多功能纳米材料对不同种类细胞的捕获

血红素—氧化还原石墨烯/硫堇修饰金纳米粒子双重信号放大CaMV35S核酸传感器

严伍文　谢晶琦　操小栋　叶永康*
（合肥工业大学食品科学与工程学院，合肥 230009）

当前世界上多数国家管理部门认为必须对转基因食品进行具体评估，这不论对转基因产品进行标识管理，转基因食品的安全性评价，对转基因与非转基因原料的分别输送，或是对转基因原料和食品的检测都是必不可少的。由此可见，寻找快速有效、准确可靠检测食品中转基因成分的分析方法具有十分重要的意义[1]。通常，聚合酶链式反应（PCR）方法由于其高灵敏度和高稳定性而成为转基因生物检测的首选方法[2]，但其存在如操作复杂、要求严苛、运行成本和时间成本高等一些不利因素。生物传感器方法因其操作简便、价廉而日益受到关注[3]。石墨烯（GO）纳米材料由于其固有的性质，比如提供高的比表面积装载生物分子的平台，以及促进的高导电性生物分子和表面之间的电子转移，在过去几年中被广泛应用于生物传感器的构建。

本文构建了一种基于血红素—还原氧化石墨烯（hemin-rGO）和金纳米粒子（AuNPs）修饰的玻碳电极（AuNPs/hemin-rGO GCE）的新型电化学核酸生物传感器用于检测转基因花椰菜花叶病毒35S（CaMV35S）启动子的目标片段（tDNA）。

实验部分：花椰菜花叶病毒（CaMV）35S启动子作为DNA靶序列。所有的合成寡核苷酸序列均购自上海生工，包括捕获序列（pDNA，S1）、互补序列（cDNA，S2）、单碱基错配序列（Mis-1）、两碱基错配序列（Mis-2）和三碱基错配序列（Mis-3）。制备hemin-rGO后将其修饰在预处理过的玻碳电极上（GCE），再电沉积预先制备AuNPs，再通过Au-S修饰S1单链，通过杂交反应与不同浓度的S2杂交后，最后与硫堇（Thi）修饰AuNPs温浴15 min，在最优化条件下测定DPV信号，并与各错配序列相对照。

图1左显示典型的具有轻微皱纹的片状形状，表明rGO可以提供大的比表面积。图1中所示，hemin-rGO纳米片与rGO相比没有显著差异，表明石墨烯片在水溶液中分散良好。从图1右中可以看到大量明显的颗粒状纳米结构，表明AuNPs已成功制备。

为了实现传感器的良好电化学性能，优化了磷酸盐缓冲液pH值、硫堇浓度、AuNPs反应时间、杂交温育时间等影响因素。制作的DNA传感器是否能够检测出真实样品的敏感性和特异性，用阳性真实样品和阴性真实样品作为检测目标，得到的实验结果见图2。

这种三明治型电化学核酸生物传感器，当tDNA浓度的对数和示差脉冲极谱法（DPV）电流值在$1\times10^{-16}\sim1\times10^{-10}$ M范围内呈线性关系，最低检出限是9.45×10^{-17} M，实现对超低含量目标DNA的检测。该传感器具有良好的选择性、稳定性和重复性，并可用于实际PCR样品的检测。

* 通信作者：叶永康；E-mail：yongkang. ye@hfut. edu. cn；手机号：18919663650。国家自然科学基金（31772099）、安徽省自然科学基金（1508085MC47）资助项目。

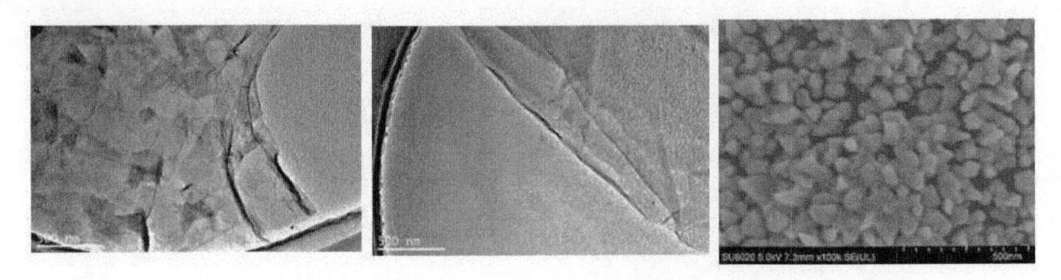

图 1　rGO（左）、hemin - rGO（中）和 AuNPs（右）的 TEM 图像

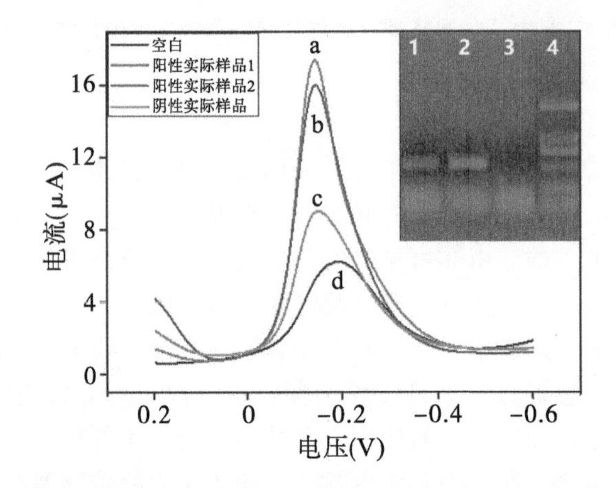

图 2　阳性实际样品（a，b）、阴性实际样品（c）和空白（d）的 PCR 产物响应的 DPV 峰值电流

注：插图为 PCR 扩增产物的凝胶电泳图。泳道 1、2 为阳性实际样品，泳道 3 为阴性实际样品，泳道 4 为 DL2000 DNA 标记从上到下依次是 2 000、1 000、750、500、250 和 100 bp

参考文献

[1] 张文珠，史剑浩，肖铭. 2015. 转基因产品分析方法的研究进展 [J]. 分析测试技术与仪器，21（4）：191 - 198.

[2] Sun Y，Nguyen N T，Kwok Y C. 2008. High-throughput polymerase chain reaction in parallel circular loops using magnetic actuation [J]. Analytical Chemistry，80（15）：6127 - 6130.

[3] Huang D，Liu H，Zhang B，et al. 2009. Highly sensitive electrochemical detection of sequence-specific DNA of 35S promoter of cauliflower mosaic virus gene using CdSe quantum dots and gold nanoparticles [J]. Microchimica Acta.，165（1 - 2）：243 - 248.

生物绿色仿生合成的 CuS@BSA 纳米颗粒在光热/MRI 诊疗一体化中的应用

褚中运　王智明　贾能勤*
（上海师范大学化学系，上海 200234）

肿瘤的诊断与治疗一直是广大科学研究人员广泛关注的问题，将磁共振成像（MRI）与光热疗法（PTT）相结合，可使癌症治疗诊断更加高效和准确，从而实现多功能化的一步多法。

在本课题中，基于生物矿化的理论基础，以牛血清白蛋白 BSA 为生物模板[1,2]，通过简单的一锅法在绿色环保、经济安全、环境友好的实验条件下合成了平均粒径约为 16.5 nm 的 CuS@BSA 纳米粒子（图 1）[1]。合成的 CuS 颗粒表面包覆有丰富的 BSA，在后期的研究中可以利用 BSA 表面含有的丰富官能团（如氨基、羧基、羟基等）实现多样化的修饰[2]。通过稳定性测试发现，CuS@BSA 可以在 pH＝4.0～7.0 的溶液环境下稳定存在，在不同的溶液体系，如 PBS、血清、细胞培养基中同样可以稳定存在，这为后续的生物应用提供了良好的稳定性条件。进一步的研究发现 CuS@BSA 具有极强的近红外吸收特性和磁共振成像能力，因此，它可以用作癌症光热治疗和 MRI 诊断的多功能化纳米载体。

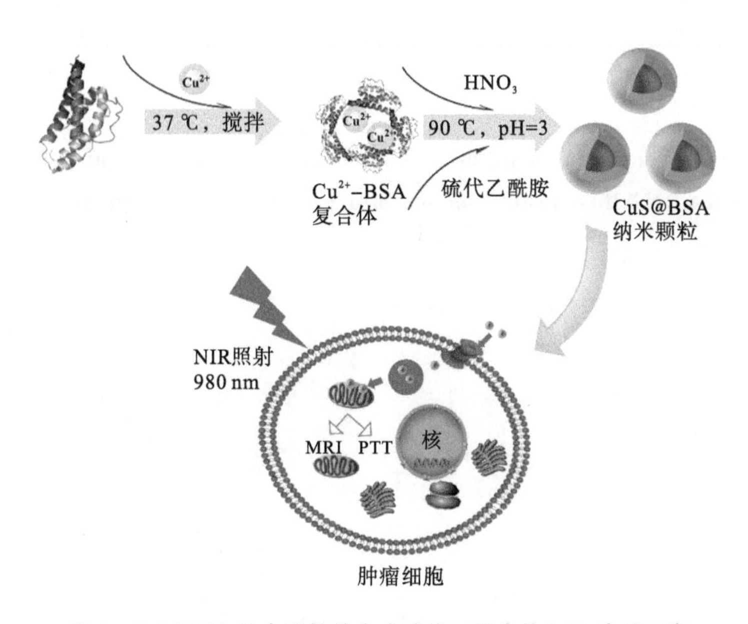

图 1　CuS@BSA 纳米颗粒的合成路线以及光热/MR 应用示意

* 通信作者：贾能勤；E-mail：nqjia@shnu. edu. cn；手机号：13818195398。

通过进一步的体外毒性研究表明，CuS@BSA 纳米粒子具有非常低的生物毒性。基于其良好的 980 nm 近红外吸收能力（图 2），通过一定功率的光热孵育发现 CuS@BSA 可以作为光热材料达到非常明显的肿瘤细胞杀伤效果。此外，与临床上广泛使用的 MRI 试剂—马根维显（Magnevist，$r1=3.13\ mM^{-1}\cdot s^{-1}$）相比，CuS@BSA 还表现出一定的弛豫值（$r1=0.26\ mM^{-1}\cdot s^{-1}$），这为 CuS 纳米粒子在分子影像方面的应用提供了新的思路。

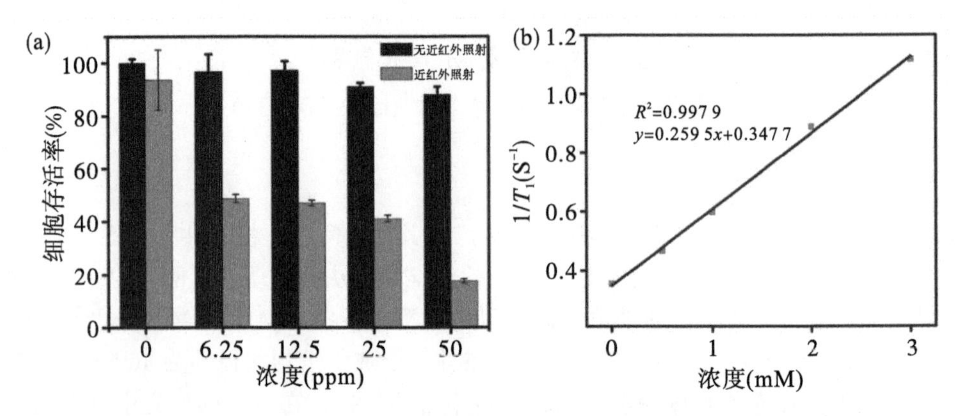

图 2　(a) A549 细胞与 CuS@BSA 共同孵育 24 h 后，在 980 nm 近红外照射下的细胞杀伤效果（$1.57\ w\cdot cm^{-2}$）；(b) CuS@BSA 的 T1 弛豫效率

参考文献

[1] Chu Z Y，Wang Z M，Chen L N，et al. 2018. Combining magnetic resonance imaging with photothermal therapy of CuS@BSA nanoparticles for cancer theranostics [J]. ACS Appl. Nano Mater.，1：2332 - 2340.

[2] Wang Z M，Wu H，Shi H Y，et al. 2016. A novel multifunctional biomimetic Au@BSA nanocarrier as a potential siRNA theranostic nanoplatform [J]. J. Mater. Chem. B，4：2519 - 2526.

体内微环境诱导的多孔硅荧光行为用于肿瘤成像

沃芳洁　金　尧　崔瑶轩　李乐昕　邬建敏*

（浙江大学化学系分析化学研究所，杭州 310058）

手术切除是癌症很有效的临床治疗手段之一。然而在临床操作中，肿瘤和邻近组织之间的边界不清，对于肿瘤切缘的确定造成了障碍[1]。为解决这一问题，近红外（NIR）光手术引导系统在近几年起着重要作用并引起了人们的广泛关注[2]。随着纳米技术的发展，纳米材料由于其原有的特性，在生物成像中显示出巨大的潜力。与其他材料相比，荧光多孔硅（LuPSi）的一个显著优点是其荧光行为与表面化学之间的高度相关性，容易受到外部环境氧化水平和酸碱度的影响。碱性基团可加快将 Si—H 键氧化成 Si—OH 键的化学过程，从而加速 LuPSi 的荧光激活和衰减过程[3]。前期研究显示，肿瘤组织内的 pH 值低于癌旁组织中的 pH 值[4]，因此将 LuPSi 微粒注射入肿瘤组织及癌旁组织中，其在癌旁组织中将经历一个比肿瘤组织中显著加快的荧光动力学过程。癌旁组织中的 LuPSi 微粒将首先被激活，使肿瘤区域呈现一个负成像状态，随着时间的推移，肿瘤组织中的 LuPSi 也接着被激活，而癌旁组织中的 LuPSi 则开始衰减并最终淬灭，导致荧光成像由负成像转变为正成像状态（图 1）。通过这种方法，我们可以区分肿瘤和邻近组织之间的边界。

LuPSi 使用电化学刻蚀法制备，在氢氟酸和乙醇的混合电解液中对硼掺杂的单晶硅片进行［100］晶面上的阳极氧化，然后将多孔片层从硅基底上剥离，并用超声波破碎。将得到的多孔硅微粒浸泡于 PBS 中进行荧光激活。

将 LuPSi 微粒分散于 pH 值为 6.5 或 7.5 的 PBS 中，37 ℃下孵育。在紫外光激发下，利用光谱仪检测不同时间点的荧光光谱。为了增加 LuPSi 的荧光稳定性，使用 3-（三甲基甲硅基）甲基丙烯酸丙酯（结构式如图 2a）对其进行表面修饰。

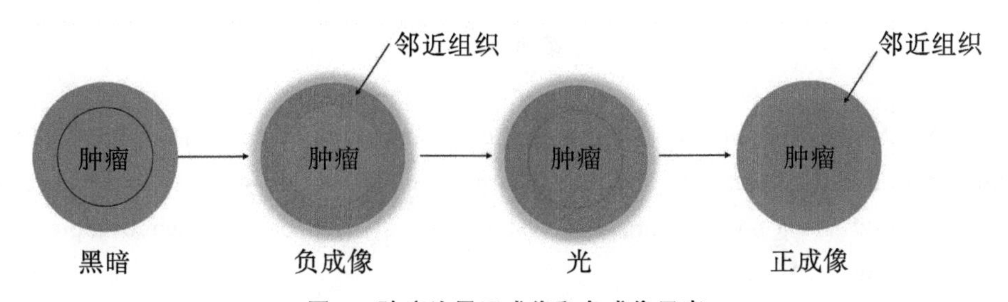

图 1　肿瘤边界正成像和负成像示意

图 2b 显示，LuPSi 在不同 pH 值的缓冲液中表现出不同的发光动力学。未经修饰的 LuPSi 在

* 通信作者：邬建敏；E-mail：wjm-st1@zju. edu. cn；手机号：13958135820。中国博士后科学基金（507300-X91805）资助项目。

pH 值为 6.5 和 7.5 的 PBS 中荧光衰减时间分别为 5 h 和 3 h，而经 3－（三甲基甲硅基）甲基丙烯酸丙酯修饰后的 LuPSi 的荧光强度增加，且衰减时间则延长到 14 h 和 8 h。因此，表面修饰 3－（三甲基甲硅基)甲基丙烯酸丙酯可改善 LuPSi 的荧光性能，增加其在不同 pH 值中的衰减时间差，使其可在长达 8 h 的时间内保持较高的相对荧光强度比（图 2c），该结果有利于提高体内应用中的肿瘤区域与癌旁组织的信噪比。该课题利用 LuPSi 材料本身的荧光特性，将其应用于肿瘤成像中，是一次对 LuPSi 生物医学应用领域的拓展，也为肿瘤边缘的术中判定提供一项新的解决策略。

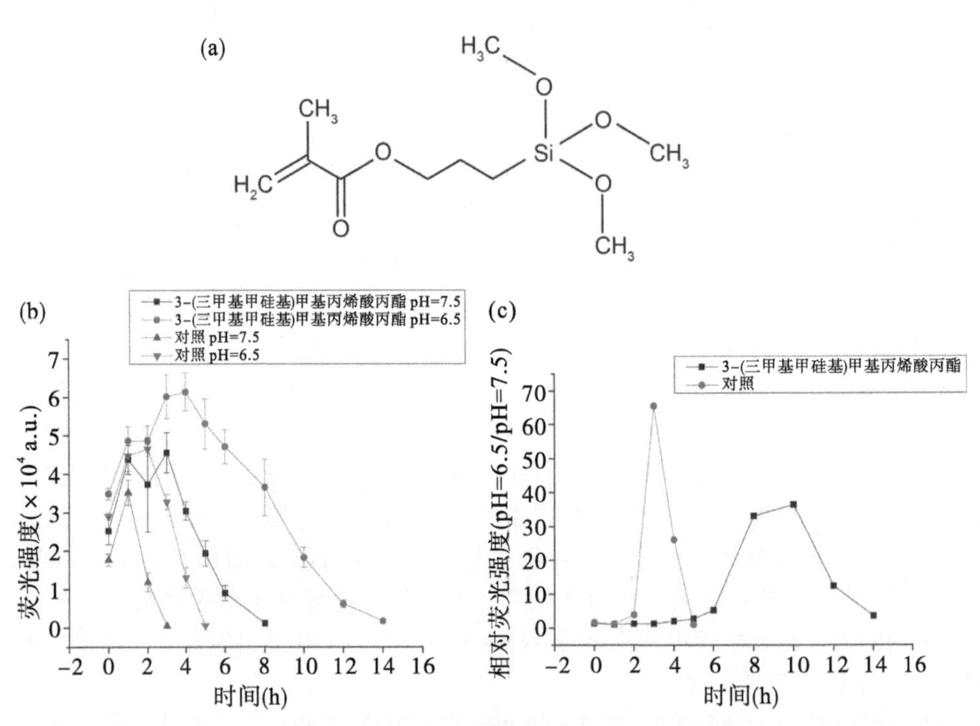

图 2　LuPSi 的修饰及其荧光行为

注：a. 表面修饰的化合物结构式；b. LuPSi 在 PBS 中的荧光变化；c. LuPSi 在不同 pH 值条件下的相对荧光强度比

参考文献

［1］Hiroshima Y，Maawy A，Metildi C A，et al. 2014. Successful fluorescence-guided surgery on human colon cancer patient-derived orthotopic xenograft mouse models using a fluorophore-conjugated anti-CEA antibody and a portable imaging system［J］. Journal of Laparoendoscopic & Advanced Surgical Technique，24：241－247.

［2］Vahrmeijer A L，Hutteman M，van der Vorst J R，et al. 2013. Image-guided cancer surgery using near-infrared fluorescence［J］. Nature Reviews Clinical Oncology，10：507－518.

［3］Chen X S，Wo F J，Jin Y，et al. 2017. Drug-porous silicon dual luminescent system for monitoring and inhibition of wound infection［J］. ACS Nano，11：7938－7949.

［4］Xu L，Fidler I J. 2000. Acidic pH-induced elevation in interleukin 8 expression by human ovarian carcinoma cells［J］. Cancer Research，60：4610－4616.

非金属矿生物医药材料的开发和应用

鲍康德[1,2]　周春晖[3,4]　姜程曦[1*]

（1. 温州医科大学药学院，温州 325035；

2. 池州市九华山黄精研究所，池州 247100；

3. 浙江工业大学化学工程学院，杭州 310014；

4. 青阳非金属矿研究院，青阳 242804）

全世界非金属矿物多达 3 500 余种，主要应用于化工、机械、能源、汽车、轻工、食品加工、冶金、建材等产业上。由于分析测试技术、设备等的限制，对非金属矿物的组成及含量、微量元素的种类和赋存状态、有关矿物晶体结构及晶体化学特点以及水溶速度、浸出及吸附等物化性能认知不足，使得其在传统医药行业的应用主要集中在直接入药（如蒙脱石等）和药物载体等辅料（如高岭土等）等方面[1]。随着科技进步，近年来呈现出由传统中药为主兼向西药发展、以载体等辅料为主兼向主料发展、以外用为主兼向内用发展的趋势。非金属矿药物以其作用缓和持久、疗效稳定、无副作用等优点，在国内外新型高效医药产品领域日益凸显其重要地位[1,2]。

药用非金属矿物是生药中不可再生的重要资源，包括天然矿物（多数可供药用，如朱砂）、矿物的加工品（如芒硝）和动物及其骨骼的化石（如龙骨）[3]。矿物药的种类虽少，但在临床上应用颇广，它们代表了现代药用非金属矿物开发利用的先进水平，如以蒙脱石为主要成分，治疗胃炎、肠道炎和急慢性腹泻等的特效药。随着经济、技术的飞速发展，人们日益关注非金属矿物的药用研究和开发利用。非金属矿物的研究应用已经从外科临床进展到内科临床，以及人体保健等多个领域[2,3]。

与化工学上按矿物原料的化学成分和结构（分为 11 类），或者按材料的功能进行分类（分为 9 类）不同[4]，医药学上通常以矿物中对药效起重要作用的阳离子为依据，这对矿物药进行分类和应用研究有诸多便利[2]（详见表 1）。

和其他药物活性成分一样，非金属矿物材料必须同时具备“安全性、有效性、可控性”才可能成为药品。良好的临床疗效和投资回报是非金属矿物新药研发立项的重要依据。影响项目进程与投资收益的国家宏观政策分析是开发非金属矿物药首要的评估要素，必须对影响目标治疗领域、目标品种市场的医药产业和临床用药管理政策因素等进行全面正确解读。对某一适应症/功能主治的医学现状分析；在研品种遴选与未来竞争对手等研发进展分析，并对关键技术要有前瞻性；当前竞争态势与未来市场策略的市场格局分析，以及对市场潜力预测与投资风险评估的综合分析是不可或缺的要素。

非金属矿物在生物医药领域的开发新空间值得关注，主要有处方药，如含特定非金属微量元

* 通信作者：姜程曦；E-mail：jiangchengxi@126. com；手机号：18969715696。国家中医药管理局中药标准化建设项目（ZYBZH - Y - SC - 40）。

素的抗肿瘤、抗病毒、抗高血压等药物；非处方药物，如以蒙脱石为主要成分、添加了少量助剂等可治疗胃炎、肠道炎和急慢性腹泻；保健品，以非金属矿物为原料的保健用品较普遍，如美容品和化妆品等；现代农业：选用于改良土壤，抗菌生根、保湿抗旱等[2]。

表 1 按药效活性阳离子分类的非金属矿物

矿物类别	常用非金属矿物
钾化合物类	硝石（KNO_3）
钠化合物类	芒硝（$Na_2SO_4 \cdot 10H_2O$）、玄明粉（Na_2SO_4）、硼砂（$Na_2[B_4O_5(OH)_4] \cdot 8H_2O$）、大青盐（NaCl）等
钙化合物类	石膏（$CaSO_4 \cdot 2H_2O$）、寒水石（$CaCO_3$）、龙骨［$CaCO_3$、$Ca_3(PO_4)_2$等］、紫石英（CaF_2）等
镁化合物类	滑石［$Mg_3(Si_4O_{10})(OH)_2$］等
铝化合物类	白矾［$KAl(SO_4)_2 \cdot 12H_2O$］、赤石脂［$Al_4(Si_4O_{10})(OH)_8 \cdot 4H_2O$］等
锌化合物类	炉甘石（$ZnCO_3$）等
铁化合物类	赭石（Fe_2O_3）、磁石（Fe_3O_4）、自然铜（FeS_2）、皂矾（绿矾）（$FeSO_4 \cdot 7H_2O$）等
铜化合物类	胆矾（$CuSO_4 \cdot 5H_2O$）、铜绿等
铅化合物类	铅丹（Pb_3O_4）、密陀僧（PbO）等
砷化合物类	雄黄（As_2S_2）、雌黄（As_2S_3）、信石（As_2O_3）等
硅化合物类	白石英、玛瑙、浮石（SiO_2）、青礞石等
铵化合物类	白硇砂（NH_4Cl）等
其他类	硫黄（S）、琥珀等

现代科学技术的进步和分析检测设备的发明推动着药理学研究的不断深入。药用非金属矿物以矿物离子（胶体）具有吸附病菌、病毒的功能，并藉此调节肌体组织细胞膜电荷平衡，调节新陈代谢；微量元素对酶的激活作用等药理药效活性显示出全新用途。利用某些非金属矿物本身所具有的特殊结构及其物化性能，如矿物颗粒的表面带电性、吸附性、膨胀性、胶结性和润滑性等，可吸附各种病菌，对其药理学研究已达到分子水平。随着选矿提纯和纳米加工等新技术的应用，作为绿色医药或化妆品的原辅料，非金属矿物将随着对其本质认识的深入在人类医疗保健事业中发挥更大作用。

参考文献

[1] 李萍. 2005. 生药学［M］. 北京：中国医药科技出版社.
[2] 谭琦，冯安生，刘新海，等. 2015. 中国非金属矿产资源领域发展现状与趋势［J］. 矿产保护与利用，4：52－56.
[3] 国家药典委员会. 2015. 中国药典（2015 年版）［M］. 北京：中国医药科技出版社.
[4] 周春晖，俞卫华，童东绅，等. 2011. 非金属矿材料及产品标准汇编［M］. 北京：中国质检出版社（中国标准出版社）.

石墨烯量子点在杀菌中的应用研究

吴　颖　章泽飞　郭昱良　桂　馨*
（同济大学生命科学与技术学院，上海 200092）

由于石墨烯和脂质分子之间强烈的分散相互作用，纳米片层石墨烯可以渗透到细胞膜中，并且提取出大量的磷脂，从而抑制细胞和细菌的生长[1]。然而，大尺寸的石墨烯及其衍生物的抗菌能力有限，所以需要更多的手段来提高抗菌的活性。其中零维石墨烯纳米片，即石墨烯量子点（graphene quantum dots，GQDs），作为石墨烯家族的重要一员，目前在生物医学领域中的应用侧重于药物输送、肿瘤治疗和生物检测，在抗菌方面的研究并不多见。

GQDs 可以通过切割大尺寸的石墨烯得到，所以 GQDs 的粒径更小，从而获得了比石墨烯更大的比表面积，相同质量下可以装载更多的药物；而且不同于最常见的石墨烯和氧化石墨烯，GQDs 不仅能够高效地将光能转换为热能，同时还能产生活性氧，过量的活性氧会破坏细胞膜、蛋白质以及 DNA 从而杀伤细胞。可见，GQDs 无需装载光敏剂，即可达到光热和光动力的协同杀伤效果，而这两种效应的协同可以更大程度地杀伤细胞和细菌。

基于此，本课题利用 655 - nm 激光诱导 GQDs 产生大量的热量和活性氧，达到协同杀伤大肠杆菌的目的（图 1）。

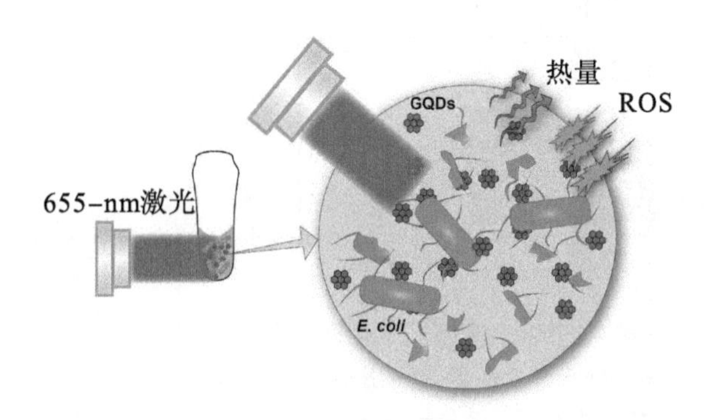

图 1　实验示意

以石墨粉为原料合成了 GQDs。结果显示，制备的 GQDs 在溶液中的分布均匀，粒径大小主要集中在 8～14 nm。

在 655 - nm 激光诱导下，该 GQDs 迅速将光能转换为热能，同时产生活性氧。在被激光照射 5 min 后，0.6 mg/mL GQDs 水溶液平均升温幅度超过 25 ℃，相对活性氧水平从最初的 2.6±0.9 上升到 18.5±6.2，表明 GQDs 是高效的光热和光动力材料。

* 通信作者：桂馨，女，工程师，E-mail：guixin@tongji. edu. cn。

在证实了单独的 1 mg/mL GQDs 处理不影响细菌活性之后，考察了 655 - nm 激光照射下，GQDs 对细菌的杀伤情况。研究发现，当 GQDs 浓度为 0.1 mg/mL 时，激光照射 20 min 后，可以观察到大肠杆菌生长被抑制的趋势；随着浓度增大到 0.2 mg/mL 时，发现大肠杆菌的生长受到了显著的抑制；而浓度为 0.5 mg/mL 时，平板里的大肠杆菌菌落数量仅为 2 位数；更加突出的是，0.6 mg/mL GQDs 在激光照射 20 min 后可以完全杀灭大肠杆菌，相比之下，单独的激光照射对大肠杆菌的生长无明显影响（图 2a 和 2b）。证明一定浓度的 GQDs 辅以激光照射是一种有潜力的杀菌方法。

进一步考察了 GQDs 杀灭细菌的有效时间。发现 0.6 mg/mL GQDs 在低功率密度的 655 - nm 激光照射 15 min 的情况下，就可以将大肠杆菌完全杀灭（图 2c 和 2d），证明在激光照射下的 GQDs 可以短时、高效、可控地杀灭细菌。

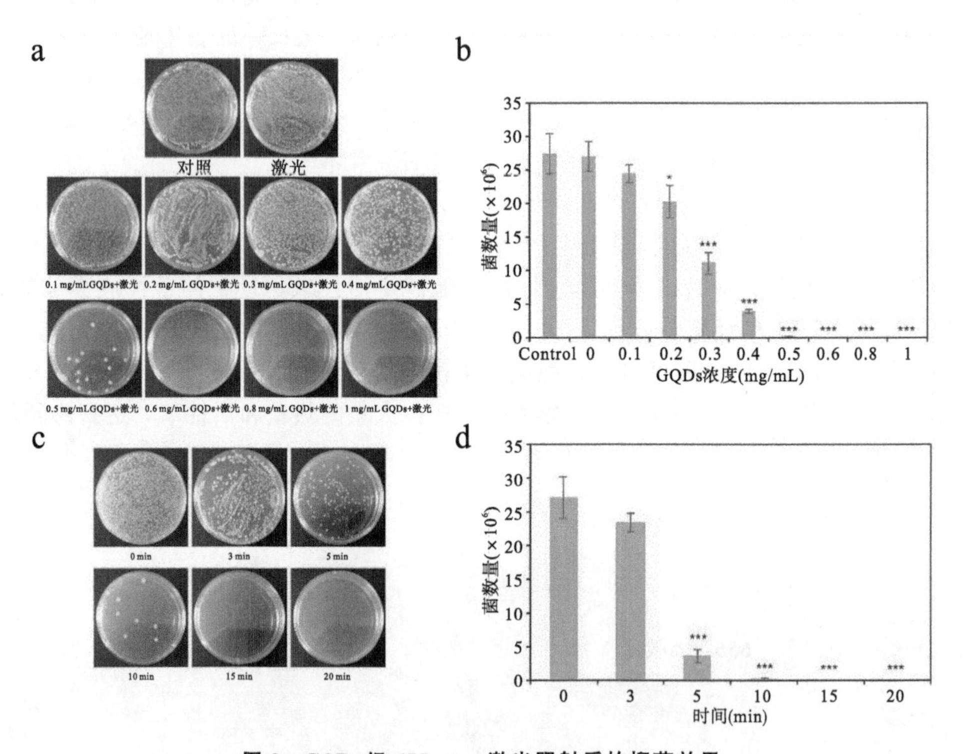

图 2　GQDs 经 655 - nm 激光照射后的抑菌效果

注：a. 不同浓度的 GQDs 经 20 min 655 - nm 激光照射后的抑菌效果（定性分析）；b. 不同浓度的 GQDs 经 20 min 655 - nm 激光照射后的抑菌效果（定量分析）；c. 0. 6 mg/mL 的 GQDs 经不同时间 655 - nm 激光照射后的抑菌效果（定性分析）；d. 0.6 mg/mL GQDs 经不同时间 655 - nm 激光照射后的抑菌效果（定量分析）

本研究利用 655 - nm 激光诱导 GQDs 产生的光热效应和光动力效应，用于杀菌的研究，对于推动 GQDs 在生物医学领域中的应用以及扩展杀菌的新方法均具有重要意义。

参考文献

[1] Tu Y S，Lv M，Xiu P，et al. 2013. Destructive extraction of phospholipids from *Escherichia coli* membranes by graphene nanosheets [J]. Nature Nanotech.，8：594 - 601.

聚谷氨酸薄膜—羧基化多壁碳纳米管修饰玻碳电极测定有机磷及其含酶果蔬清洗盐活性研究

夏子豪 操小栋 叶永康*

（合肥工业大学食品科学与工程学院，合肥 230009）

我国是一个农业大国，农业不仅是中国最基础的产业，同时也保障着我国人民的生活与工作。但是由于大量不规范的农药使用，我国的很多农产品都受到了严重的农药污染。目前，解决农药的环境残留问题，采用生物酶降解法是一条较为合适、有效的途径。近年来，市面上出现了很多生物酶类果蔬清洗剂，如比亚酶果蔬清洗液、该娅生物酶清洗液等，它们不同于传统的表面活性剂内清洗剂，而是依靠酶的降解作用去除果蔬表面残留的农药，去除效果较传统的表面活性剂内清洗剂有很大的提高[1]。食盐也具有杀菌消毒的作用，同时能帮助去除果蔬表面的农药残留，将农药降解酶与表面活性剂清洗液结合起来，开发一种复合型农药清洗液，将会大大降低农产品中农药残留的危害。

本文以甲基对硫磷（MP）为实验对象，利用聚谷氨酸薄膜与羧基化多壁碳纳米管共同修饰的玻碳电极，采用线性伏安法在最优条件下对 MP 进行了快速灵敏的检测，通过测定有机磷降解酶降解前后的底液中 MP 的量，为探究有机磷降解酶的最适活性条件提供了有效手段。

实验部分：用 0.05 μm 氧化铝粉末多玻碳电极进行打磨、抛光至镜面。然后将抛光后的玻碳电极置于硝酸水溶液（$V_{硝酸}$ ∶ $V_{水}$ = 1∶1）、丙酮以及超纯水中分别超声 3 min。并用氮气吹干，置于干燥封闭的环境中，室温下存放。用 pH=7 的 0.1 mol/L PBS 来配制 0.01 mol/L 谷氨酸溶液。将处理后的玻碳电极浸入 0.01 mol/L 谷氨酸溶液中，在 0～2.0 V 电位范围内循环扫描 10 圈（扫描速度 0.1 V/s）[2]。取 5 μL 均匀分散的羧基化多壁碳纳米管壳聚糖溶液，滴加在晾干后的聚谷氨酸薄膜修饰的电极上，置于密闭干燥的环境中晾干，待用。

图 1 为电化学传感器对 MP 富集作用示意图。由于聚合谷氨酸薄膜修饰的电极对 MP 的富集作用，在线性伏安扫描（LSV）曲线上出现 1 个还原峰（图 2），当在谷氨酸薄膜上滴加多壁碳纳米管修饰后，对 MP 的富集作用进一步提升，LSV 曲线上的 MP 还原峰的峰电流更大，所以采用聚谷氨酸薄膜—多壁碳纳米管共同修饰的电极能够很好地富集溶液中的 MP，并放大电化学信号。

为了实现传感器的良好电化学性能，优化了多壁碳纳米管修饰量与电解质溶液的 pH 值。在最佳测定条件下，检测的线性范围为 0.05～1.0 μg/mL（R^2 = 0.989），检测限为 5 ng/mL（S/N=3）。方法简单方便，灵敏度良好。

* 通信作者：叶永康；E-mail：yongkang. ye @ hfut. edu. cn；手机号：18919663650。企业委托项目（W2017JSKF0147）资助项目。

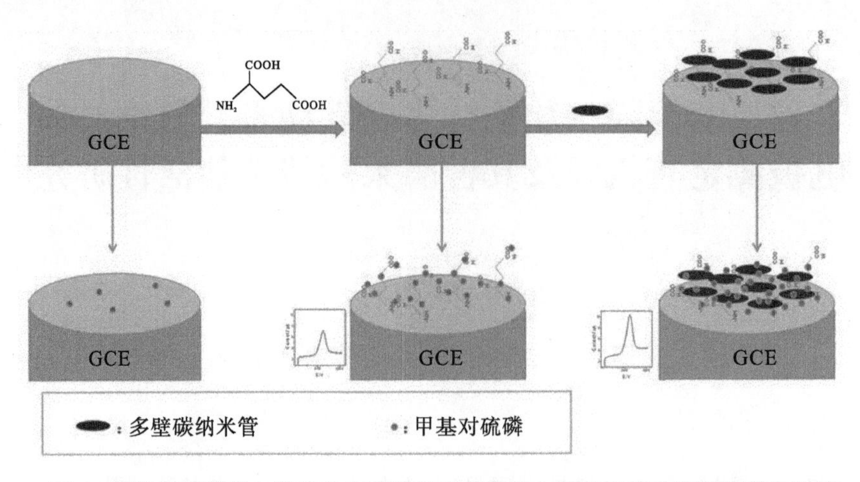

图 1　聚谷氨酸薄膜—羧基化多壁碳纳米管修饰电极对甲基对硫磷的富集作用

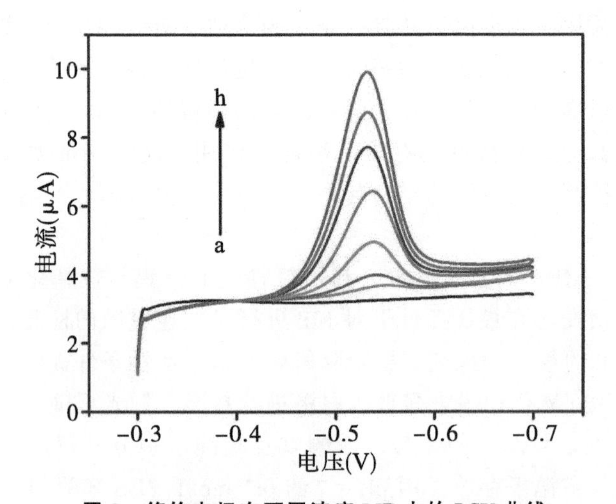

图 2　修饰电极在不同浓度 MP 中的 LSV 曲线

注：MP 浓度为 0、0.05、0.1、0.2、0.4、0.6、0.8、1.0 μg/mL。

扫描条件：支持电解质为 pH＝7 的 0.1 mol/L 的 PBS，扫速为 0.1 V/s

实验表明，有机磷降解酶在 pH＝8、温度为 30 ℃、最适的 NaCl 浓度为 0.234 g/L 的条件下，加入优化后表面活性剂，具体成分配方为酶液 50％、NaCl 2.34％、环糊精 0.146％、蔗糖脂肪酸酯（SE）0.146％、水 47.368％。合适的降解时间为 20～25 min。

参考文献

［1］刘建利. 2010. 有机磷农药降解酶的研究进展［J］. 广东农业科学，37（1）：60－64.

［2］王春燕，由天艳，田坚. 2011. 聚谷氨酸修饰电极同时检测对苯二酚和邻苯二酚［J］. 分析化学，39（4）：528－533.

氧化还原酶调控功能凝胶的制备及医学应用

王启刚

（同济大学化学科学与工程学院，上海 200092）

我们开发系列氧化还原酶实现温和自由基聚合用于高分子凝胶的制备[1]。常见的氧化还原酶包括辣根过氧化物酶、葡萄糖氧化酶、胺氧化酶、氨基酸氧化酶等，它们通过氧化拔氢、还原脱羟基和氧化脱羧等不同单电子转移途径形成引发自由基。酶促自由基聚合成胶相对于常用的热、射线及超声等自由基引发成胶手段具有具有聚合条件温和、成胶时间可控、生物相容性等优点。因此，酶催化聚合及氧化交联可用于高分子复合凝胶的温和制备和打印成型[2]（图 1）。在此基础上，我们进一步开发了含酶的宏观凝胶及纳米凝胶，实现 ROS 响应性生物酶诱导显像和治疗研究[3]。

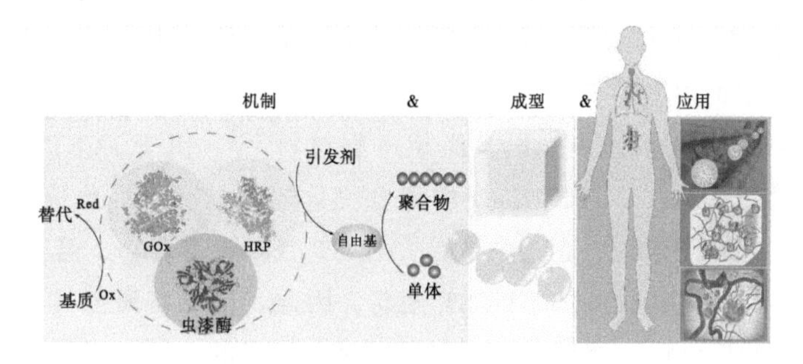

图 1　氧化还原酶调控凝胶的制备及应用示意

首先我们制备多种/串联生物酶共担载的纳米凝胶和可注射凝胶体系，在体内循环中各生物组分均不会产生生物毒性作用，当到达肿瘤等病变组织部位高水平的 ROS 响应性发生酶催化反应。肿瘤原位的酶诱导产生的具有杀伤作用的单线态氧组分会诱导肿瘤细胞产生更多的双氧水和超氧自由基，在功能酶作用下发生类氧化爆发，最终实现安全高效的病变微环境响应性治疗。

参考文献

[1] Wang X，Chen S S，Wu D B，et al. 2018. Oxidoreductase initiated radical polymerizations to design hydrogel and micro/nanogel：Mechanism，molding and applications [J]. Advanced Materials，30：1705668.

[2] Wei Q C，Xu M C，Liao C A，et al. 2016. Printable hybrid hydrogel by dual enzymatic polymerization with superactivity [J]. Chemical Science，7：2748－2752.

[3] Wang X，Niu D C，Li P，et al. 2015. Dual-enzyme-loaded multifunctional hybrid nanogel system for pathological responsive ultrasound imaging and T－2－weighted magnetic resonance imaging [J]. ACS Nano，9：5646－5656.

具有 LCST－UCST 温敏性的磁性纳米聚集体

王春尧　迟　海　袁　华　袁伟忠
（同济大学材料科学与工程学院，上海 201804）

磁性纳米粒子具有独特的物理与化学性能因而在催化、磁成像、药物控释等领域具有重要研究价值。纳米四氧化三铁制备方法简单、磁性能突出，因而更加受到关注。但四氧化三铁纳米粒子易于团聚、腐蚀与氧化。在四氧化三铁纳米粒子表面进行修饰聚合物或表面活性剂有利于磁性纳米粒子的分散稳定。但这些方法仍不能阻止磁性纳米粒子的腐蚀氧化。因此，需要采用更加有效的方法稳定与保护四氧化三铁纳米粒子。生物医用的磁性杂化纳米粒子应包括化学稳定、生物相容、强磁性和低磁滞行为。一般而言，二氧化硅包覆的磁性四氧化三铁 $Fe_3O_4@SiO_2$ 能够符合这些要求[1]。二氧化硅层不仅可以保护磁核免于氧化，而且能够提供活性位点进行聚合物功能化。具有 LCST 和 UCST 的温敏性聚合物能够对温度作出响应，因而非常有望用于生物医用材料，包括药物控释、智能生物活性表面以及组织工程材料[2,3]。由于 LCST 与 UCST 具有相反的温敏行为，因而能够实现 LCST－UCST 转变的聚合物就非常具有研究价值[4-6]。聚甲基丙烯酸-N,N-二甲氨基乙酯（PDMAEMA）是一种典型的 LCST 型温敏聚合物，但通过与 1,3-丙磺酸内酯反应，得到的季铵化的 PDMAPS 具有 UCST 性能。因此，制备与研究具有 LCST－UCST 转变的以 $Fe_3O_4@SiO_2$ 为核的聚合物—无机杂化材料是非常有意义的。

实验部分：$FeCl_2 \cdot H_2O$，Aldrich，$FeCl_6 \cdot H_2O$，油酸，Triton X－100，hexyl alcohol，TEOS，2-溴异丁酰溴（BiBB），PMDETA，1,3-丙磺酸内酯直接使用。DMAEMA 单体通过碱性氧化铝柱除去阻聚剂。溴化亚铜（CuBr）分别用乙酸与乙醇处理。测试仪器包括全反射傅里叶红外分光光度计、紫外—可见分光光度计、透射电子显微镜、动态光散射、磁振仪等。纳米四氧化三铁采用共沉淀法制备；而最终的 $Fe_3O_4@SiO_2$－g－PDMAEMA 以 $Fe_3O_4@SiO_2$－Br 为引发剂，通过 ATRP 制备。

图 1 表明，$Fe_3O_4@SiO_2$－g－PDMAEMA 呈现出了典型的 LCST 性能，其相转变温度约为 40 ℃，当 PDMAEMA 转变为 PDMAPS 后，$Fe_3O_4@SiO_2$－g－PDMAPS 则呈现出 UCST 行为，其相转变温度约为 30 ℃。DLS 结果符合上述结论。图 2 为纳米粒子、纳米聚集体的 TEM 照片以及自组装和温敏性示意图。尺寸较为均一的磁性四氧化三铁被包覆在二氧化硅中。表面接枝的聚合物呈现出了温度响应性。

结论：$Fe_3O_4@SiO_2$－g－PDMAEMA 杂化磁性材料通过 $Fe_3O_4@SiO_2$－Br 引发 DMAEMA 的 ATRP 反应得到。通过季铵化反应，$Fe_3O_4@SiO_2$－g－PDMAPS 很容易制备得到。其自组装体以 $Fe_3O_4@SiO_2$ 为核，PDMAEMA/PDMAPS 为壳。LCST－UCST 得以实现。在生物医药、智能纳米材料领域具有潜在的应用。

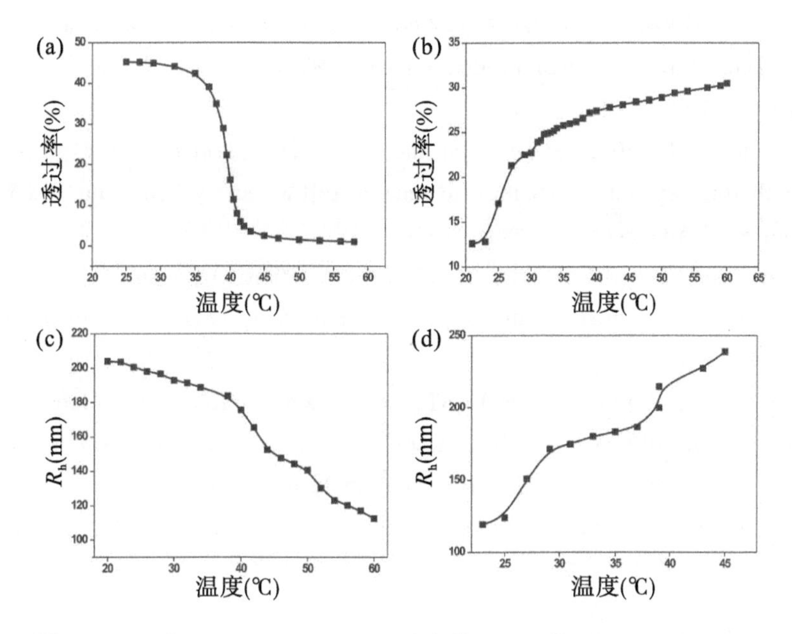

图 1　Fe_3O_4@SiO_2-g-PDMAEMA（a）和 Fe_3O_4@SiO_2-g-PDMAPS（b）的透过率曲线以及 Fe_3O_4@SiO_2-g-PDMAEMA（c）和 Fe_3O_4@SiO_2-g-PDMAPS（d）基于温度的流体力学半径（R_h）

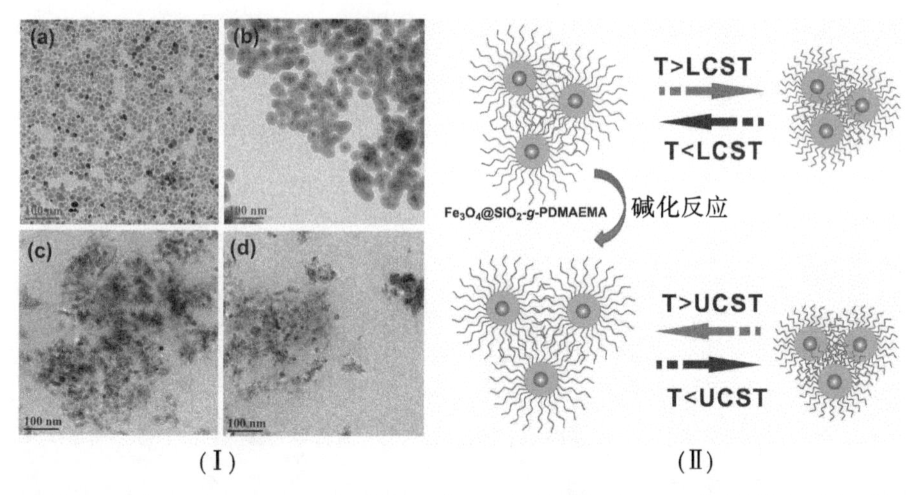

图 2　（Ⅰ）Fe_3O_4（a）、Fe_3O_4@SiO_2（b）、Fe_3O_4@SiO_2-g-PDMAEMA（c）在 25 ℃时的透射电子显微镜（TEM）照片；（Ⅱ）Fe_3O_4@SiO_2-g-PDMAEMA 和 Fe_3O_4@SiO_2-g-PDMAPS 杂化聚集体 LCST-UCST 的温度响应性示意图

参考文献

[1] Deng Y H，Qi D W，Deng C H，et al. 2008. Superparamagnetic high-magnetization microspheres with an Fe_3O_4@SiO_2 core and perpendicularly aligned mesoporous SiO_2 shell for removal of microcystins [J]. Journal of the American Chemical Society，130：28-29.

[2] Wu G，Chen S C，Zhan Q，et al. 2011. Well-defined amphiphilic biodegradable comb-like graft copolymers：Their unique architecture-determined LCST and UCST Thermoresponsivity [J]. Macromolecules，44：999 - 1008.

[3] Yuan W Z，Chen X N. 2016. Star-shaped and star-block polymers with a porphyrin core：from LCST-UCST thermoresponsive transition to tunable self-assembly behaviour and fluorescence performance [J]. RSC Advances，6：6802 - 6810.

[4] Qiu X P，Korchagina E V，Rolland J，et al. 2014. Synthesis of a poly（N-isopropylacrylamide）charm bracelet decorated with a photomobile α - cyclodextrin charm [J]. Polymer Chemistry，5：3656 - 3665.

[5] Hou L，Wu P Y. 2015. Comparison of LCST-transitions of homopolymer mixture，diblock and statistical copolymers of NIPAM and VCL in water [J]. Soft Matter，11：2771 - 2781.

[6] Yuan W Z，Zou H，Guo W，et al. 2012. Supramolecular amphiphilic star-branched copolymer：from LCST-UCST transition to temperature-fluorescence responses [J]. Journal of Materials Chemistry，22：24783 - 24791.

2D-clay Mineral as Drug Delivery Vector for Nanomedicine: Challenges and Chances

Jin-Ho Choy*

(Center for Intelligent Nano-Bio Materials (CINBM), Department of Chemistry and Nano Science, Ewha Womans University, Seoul, 03760, Republic of Korea)

Some challenges have been made to realize new inorganic-clay, organic-clay and bio-clay nanohybrids with two different functions, one from clay moiety and the other from inorganic-, organic- or biomolecular one[1]. Recently we were quite successful in demonstrating that a two-dimensional clay minerals like hydrotalcite (layered double hydroxide), montmorillonite and other natural and synthetic clays can be used as gene or drug delivery vectors. To the best of our knowledge, such clay vectors are completely different from conventionally developed drug delivery ones such as viral-based DNA or naked ones, and other nonviral carriers including liposomes, polymersomes, dendrimers and etc., those which have been limited in certain cases of applications due to their toxicity, immunogenecity, poor integration and etc.[2,3]. However, the present clay drug delivery system (CDDS) becomes more and more important in the future, since it allows tissue targeting function and enhanced cellular permeation efficacy of drug and gene molecules[4,5].

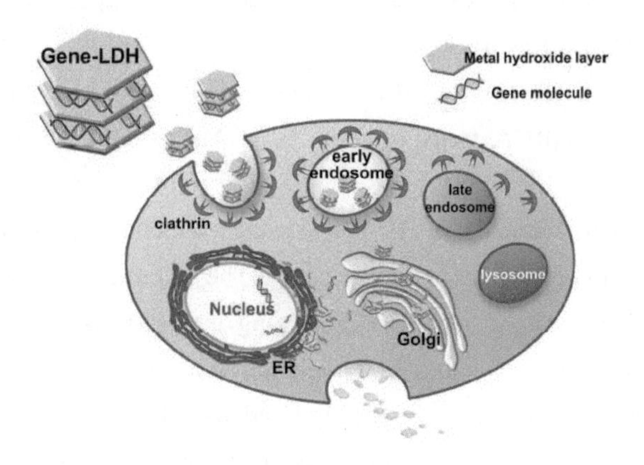

Fig. 1　Biocompatible and multifunctional gene - LDH nano hybrid system[11]

* Corresponding author: Jin-Ho Choy; E-mail: jhchoy@ewha. ac. kr.

In this presentation, a novel nanohybrid concept of delivery nanovehicle with drug or bioactive molecules is proposed to get breakthroughs in CDDSs. And at the same time, some experimental findings of new bio-clay nanohybrids are demonstrated on the basis of CDDSs, corresponding to 2-D clay nanocarriers hybridized with genes or anticancer drugs, for gene-, chemo-and radiation therapies[6-13] (Fig. 1).

References

[1] Choy J H, Kwon S J, Park G S. 1998. High - T_c superconductors in the two-dimensional limit: [$(Py-C_nH_{2n+1})_2HgI_4$] - $Bi_2Sr_2Ca_{m-1}Cu_mO_y$ (m=1 and 2) [J]. Science, 280: 1589 - 1592.

[2] Choy J H, Kwak S Y, Park J S, et al. 1999. Intercalative nanohybrids of nucleoside monophosphates and DNA in layered metal hydroxide [J]. Journal of the American Chemical Society, 121: 1399 - 1400.

[3] Choy J H, Kwak S Y, Jeong Y J, et al. 2000. Inorganic layered double hydroxides as nonviral vectors [J]. Angewandte Chemie-International Edition, 39: 4041 - 4045.

[4] Oh J M, Choi S J, Kim S T, et al. 2006. Cellular uptake mechanism of an inorganic nanovehicle and its drug conjugates: Enhanced efficacy due to clathrin-mediated endocytosis [J]. Bioconjugate Chemistry, 17: 1411 - 1417.

[5] Park D H, Kim J E, Oh J M, et al. 2010. DNA core@inorganic shell [J]. Journal of the American Chemical Society, 132: 16735 - 16736.

[6] Oh J M, Park D H, Choy J H. 2011. Integrated bio-inorganic hybrid systems for nano-forensics [J]. Chemical Society Reviews, 40: 583 - 595.

[7] Choi S J, Choy J H. 2011. Layered double hydroxide nanoparticles as target-specific delivery carriers: uptake mechanism and toxicity [J]. Nanomedicine, 6: 803 - 814.

[8] Park D H, Hwang S J, Oh J M, et al. 2013. Polymer-inorganic supramolecular nanohybrids for red, white, green, and blue applications [J]. Progress in Polymer Science, 38: 1442 - 1486.

[9] Choi G, Kwon O J, Oh Y, et al. 2014. Inorganic nanovehicle targets tumor in an orthotopic breast cancer model [J]. Scientific Reports, 4: 4430.

[10] Park D H, Cho J, Kwon O J, et al. 2016. Biodegradable inorganic nanovector: passive versus active tumor targeting in siRNA transportation [J]. Angewandte Chemie-International Edition, 55: 4582 - 4586.

[11] Choi G, Eom S, Vinu A, et al. 2018. 2D nanostructured metal hydroxides with gene delivery and theranostic functions; a comprehensive review [J]. The Chemical Record, 18: 1 - 22.

[12] Choi G, Kim T H, Oh J M, et al. 2018. Emerging nanomaterials with advanced drug delivery functions; focused on methotrexate delivery [J]. Coordination Chemistry Reviews, 359: 32 - 51.

[13] Choi G, Jeon I R, Piao H, et al. 2018. Highly condensed boron cage cluster anions in 2D carrier and its enhanced antitumor efficiency for boron neutron capture therapy [J]. Advanced Functional Materials, 28: 1704470.

Boron Neutron Capture Therapy with Clay Drug Delivery System

Goeun Choi　Jin-Ho Choy*

(Center for Intelligent Nano-Bio Materials (CINBM),

Department of Chemistry and Nano Science, Ewha

Womans University, Seoul, 03760, Republic of Korea)

The clay drug delivery system (CDDS) becomes more and more important in the near future, since it allows tissue targeting function and enhanced cellular permeation efficacy of drug molecules[1-3]. In the present study, a novel concept of nanohybridization clay and drug or bioactive molecules is proposed to get breakthroughs in drug delivery system[4-7]. And at the same time, drug delivery efficacy of CDDS are demonstrated on the basis of clay nanohybrids like layered double hydroxides (LDHs) hybridized with genes or anticancer drugs for gene-and chemo-therapy. More recently, we attempted to apply the present nanohybrid technology to the boron neutron capture therapy (BNCT), since it has long been needed to develop a boron carrier to deliver boron (^{10}B) -containing molecules, such as boranes, carboranes or borates, into cells sufficiently. The present ^{10}B - LDHs does not show any cytotoxicity *in-vitro* according to the bioassay studies in glioma cell culture lines, and eventually any influence on cell proliferation and viability upto the concentration of 250 μg/mL. In addition, we found that the higher cellular boron concentration upon ^{10}B - LDH treatments could result in high neutron capture efficiency in *in-vitro*.

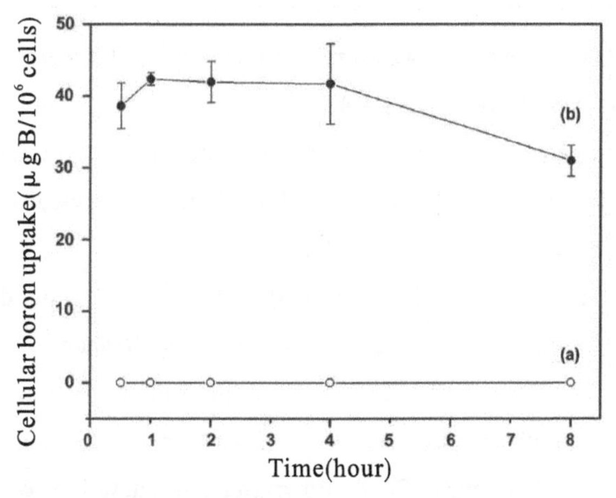

Fig. 1　Cellular boron uptake of BSH (sodium mercaptoundecahydro-closo-dodecaborate, $Na_2B_{12}H_{11}SH$) (a) and BSH - LDH (b) in U87 glioblastoma cell line[8]

* Corresponding author: Jin -Ho Choy; E-mail: jhchoy@ewha. ac. kr.

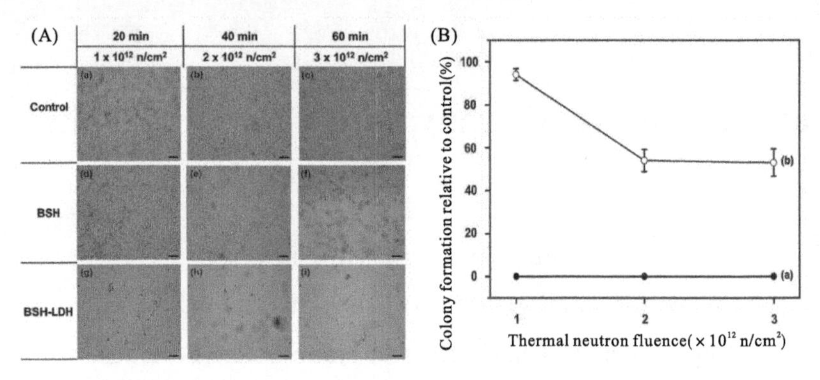

Fig. 2 (A) Microscopic images of U87 cells after thermal neutron irradiation for 20, 40 and 60 min (representing the neutron flux of 1×10^{12} n/cm^2, 2×10^{12} n/cm^2 and 3×10^{12} n/cm^2): control a - c; BSH treated with the boron concentration of 30 μg/mL d - f; BSH - LDH treated with the boron concentration of 30 μg/mL g - i. (Scale bar: 50 μm); (B) Colony formation ability for BSH - LDH - treated U87 cells (a) and BSH treated one after thermal neutron irradiation (b)[8]

It is, therefore, concluded that the present BNCT combined drug delivery system (BNCT - DDS) could provide a promising integrative therapeutic action in chemo-and radiation therapy[8] (Fig. 1 and Fig. 2).

References

[1] Choi S J, Choi G, Oh J M, et al. 2010. Anticancer drug encapsulated in inorganic lattice can overcome drug resistance [J]. Journal of Materials Chemistry, 20: 9463 - 9469.

[2] Choi G, Kim S Y, Oh J M, et al. 2012. Drug - ceramic 2 - dimensional nanoassemblies for drug delivery system in physiological condition [J]. Journal of the American Ceramic Society, 95: 2758 - 2765.

[3] Choi G, Kwon O J, Oh Y, et al. 2014. Inorganic nanovehicle targets tumor in an orthotopic breast cancer model [J]. Scientific Reports, 4: 4430.

[4] Choi G, Piao H, Alothman Z A, et al. 2016. Anionic clay as the drug delivery vehicle: tumor targeting function of LDH - MTX nanohybrid in C33A orthotopic cervical cancer model [J]. International Journal of Nanomedicine, 11: 337 - 348.

[5] Choi G, PiaoH, Kim M H, et al. 2016. Enabling nanohybrid drug discovery through the soft chemistry telescope [J]. Industrial & Engineering Chemistry Research, 55: 11211 - 11224.

[6] Choi G, Eom S, Vinu A, et al. 2018. 2D nanostructured metal hydroxides with gene delivery and theranostic functions; a comprehensive review [J]. The Chemical Record, 18: 1 - 22.

[7] Choi G, Kim T H, Oh J M, et al. 2018. Emerging nanomaterials with advanced drug delivery functions; focused on methotrexate delivery [J]. Coordination Chemistry Reviews, 359: 32 - 51.

[8] Choi G, Jeon I R, Piao H, et al. 2018. Highly condensed boron cage cluster anions in 2D carrier and its enhanced antitumor efficiency for boron neutron capture therapy [J]. Advanced Functional Materials, 28: 1704470.

Graphene-doped Bio-ink in 3D Printed Myocardium

Tian Xiao Mei　Hao Cao　Wen Jun Le*　Dong Lu Shi　Zhong Min Liu
(Shanghai East Hospital, The Institute for Biomedical Engineering & Nano Science, Tongji University School of Medicine, Shanghai 200092, China)

In recent years, three-dimensional (3D) printing artificial technology has become a one of the most promising technologies in the field of regenerative medicine. 3D bioprinting, an extension of 3D printing, is based on additive manufacturing techniques and provides controlled manufacturing of 3D structures in all X, Y and Z directions[1,2]. However, bio-ink is the key to limit the application of this technology. In particular, there is a lack of biocompatible media with bio-electrical signalling. Here we try to develop a 3D printable graphene composite ink, due to the exceptional properties of graphene enable applications in electronics, energy storage, and structural composites[3].

Methods and materials: Graphene powder (38 atomic layers thick, 520 μm long and wide; Graphene Laboratories Inc., USA). Other reagents purchased from Sigma. All printed structures were fabricated using a 3D Bio X (Fig. 1A; Cellink, Sweden). All scaffolds for *in vitro* studies were created by 3D printing contoured 1.0 cm×1.0 cm squares using a 200 μm tip; 300 μm spacing between deposited bio-ink; 180° orientation with every other layer being offset 300 μm in X and Y relative to two layers prior (Fig. 1B and 1C).

Results and discussion: These inks can be utilized by extrusion-based 3D printing (Fig. 1A) under ambient conditions to create sheet structures having features as small as 100 μm, from as few as two layers (<300 μm m thick objects) or hundreds of layers (>10 cm thick object). *In vitro* experiments in simple growth media showed that the ink supports myocardial precursor cell adhesion, viability, proliferation. Although 3D printing technology has been successfully applied to mechanical manufacturing, there are still many difficulties in the application of tissue engineering. For example, the structure is unstable, mainly in the absence of electrical conduction channels, lack of blood vessels to supply nutrients. Graphene has received extensive attention as a promising new biomaterial in recent years. Although its long-term biocompatibility has not been fully evaluated, it is one of the few potential biocompatible materials with conductivity that enhances cell-cell signaling, cell differentiation and cells in a variety of

---

* Corresponding author: Wen Jun Le; E-mail: Wenjunle@tongji. edu. cn; Mobile Phone No.: ++86-15821531253.
Project funded by National Postdoctoral Program for Innovation Talents (BX201700173) and China Postdoctoral Science Foundation (2017M621534).

cell types. Features include those including muscle, heart and nerve tissue.

Conclusion and implications: If 3D bioprinting solves the problem of nerve conductance and vascularization, artificial tissue can obtain oxygen and nutrients from the host body and may survive easily after transplantation.

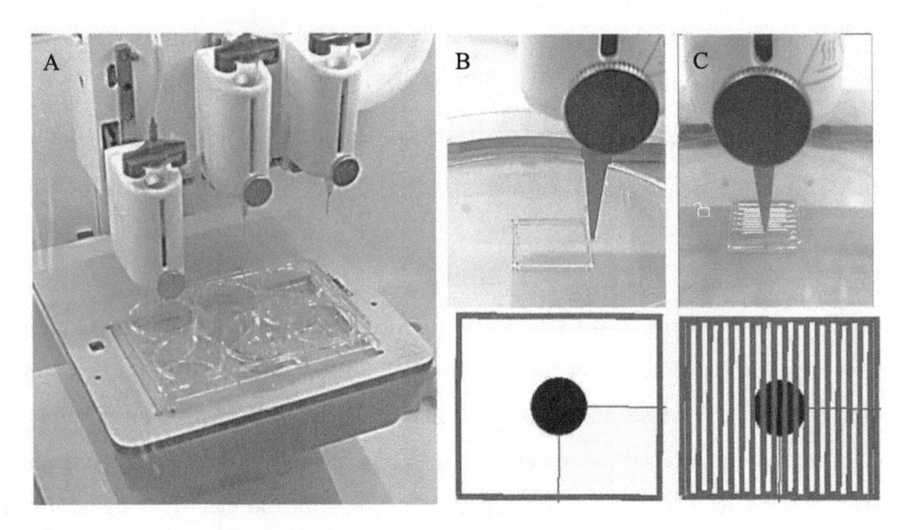

Fig. 1 3D bio-printing operation platform and sheet model

Note: BIO X system are used in the 3D printing platform (A). Use PCL to build a peripheral skeleton (B) and then fill the bio-ink (C)

References

[1] Murphy S V, Atala A. 2014. 3D bioprinting of tissues and organs [J]. Nature biotechnology, 32 (8): 773.

[2] Ma X, Liu J, Zhu W, et al. 2018. 3D bioprinting of functional tissue models for personalized drug screening and in vitro disease modeling [J]. Advanced Drug Delivery Reviews, doi. org/10. 1016/j. addr. 2018. 06. 011.

[3] Jakus A E, Secor E B, Rutz A L, et al. 2015. Three-dimensional printing of high-content graphene scaffolds for electronic and biomedical applications [J]. ACS nano, 9 (4): 4636 - 4648.

Man-made Mineral Fibers Effects on the Expression of Anti-oncogenes and Oncogenes in Lung Tissues of Rats

Yan Cui[1]　Liu Wen Huang[1]　Fa Qin Dong[2]　Qing Bi Zhang[1*]
(1. Southwest Medical University, Luzhou 646000, China;
2. Southwest University of Science and Technology,
Mianyang 621010, China)

Man-made mineral fibers (MMMFs), as a substitute for asbestos, refers to non-crystalline fibrous inorganic substances (silicates) made from several types of minerals. Since the harms associated with the use of asbestos were fully realized, people tend to use MMMFs more frequently than asbestos. Most MMMFs are used for soundproof, heat insulation, fire protection, reinforcement and filtering applications in the world. The fibrous form of MMMFs, which share with the asbestos minerals, also attracted a great concern regarding their possible health effects. Whether MMMFs could change the level of tumor-associated gene expression in rats, and its mechanism was associated with anti-oncogene (p53 and p16) and proto-oncogene (c-jun and c-fos) in vivo has not been fully elucidated. For this reason, this study was focused on exploring the effect of MMMFs on the expression of cancer-related genes in lung tissues of rats and comparing the different toxicity of MMMFs based on physicochemical properties. The three types of MMMFs, namely, ceramic fibers (CF), glass fibers (GF), and rock wool (RW) were prepared into inhalable particles. Then the particle size, morphology and chemical composition were analyzed by laser particle analyzer, scanning electron microscope (PW1404, Philips) and X-ray Fluorescence Spectrometer (TM-100, Hitach) respectively. The Wistar rats were administered by intratracheal instillation of those three MMMFs for 1, 3 and 6 months. Then, several parameters (e. g. Body mass, lung mass, and lung histology) were measured. After that, levels of P53, P16, C-JUN, and C-FOS mRNA and protein were detected by Western blotting and quantitative real-time RT-PCR. In this study, we found that the chronic exposure to three MMMFs (CF, GF and RW) by multiple non-exposure intratracheal instillation in rats could slow body mass growth, increase lung mass, and induce pathological injury of rats' lung tissues (Fig. 1). General conditions showed white nodules and irregular atrophy. In addition, Hematoxylin-eosin (HE) staining revealed inflammatory infiltration, injury of alveolar struc-

* Corresponding author: Qing Bi Zhang; E-mail: qingbizhang@126. com; Mobile Phone No.: +86-18113513199. This work was supported by the National Natural Fund Project of China (No. 41472046).

tures, and fibrosis. Moreover, MMMFs could inactivate anti-oncogene P16 but activate proto-oncogenes (C-JUN and C-FOS) in the mRNA and protein levels (Fig. 2). The effects of CF and GF on the mRNA and proteins levels of cancer-related genes were more robust than the one of RW. Maybe the different physical and chemical characteristics of different MMMFs could be responsible for MMMFs' different effects. In conclusion, MMMFs (CF, GF and RW) could disrupt the balance in expression of cancer-related genes in lung tissues of rats. While an understanding of the determinants of toxicity and carcinogenicity has provided a scientific basis for developing and introducing new safer MMMFs products, the present study explain some carcinogenic mechanisms of MMMFs.

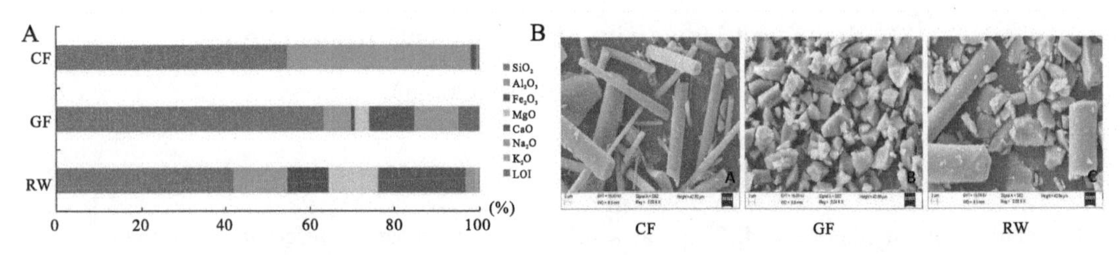

Fig. 1 Chemical composition analysis (A) and morphology of MMMFs (B)

Note: LOI. loss on ignition; CF. ceramic fiber; GF. glass fiber; RW. rock wool. Mag=×2.00K

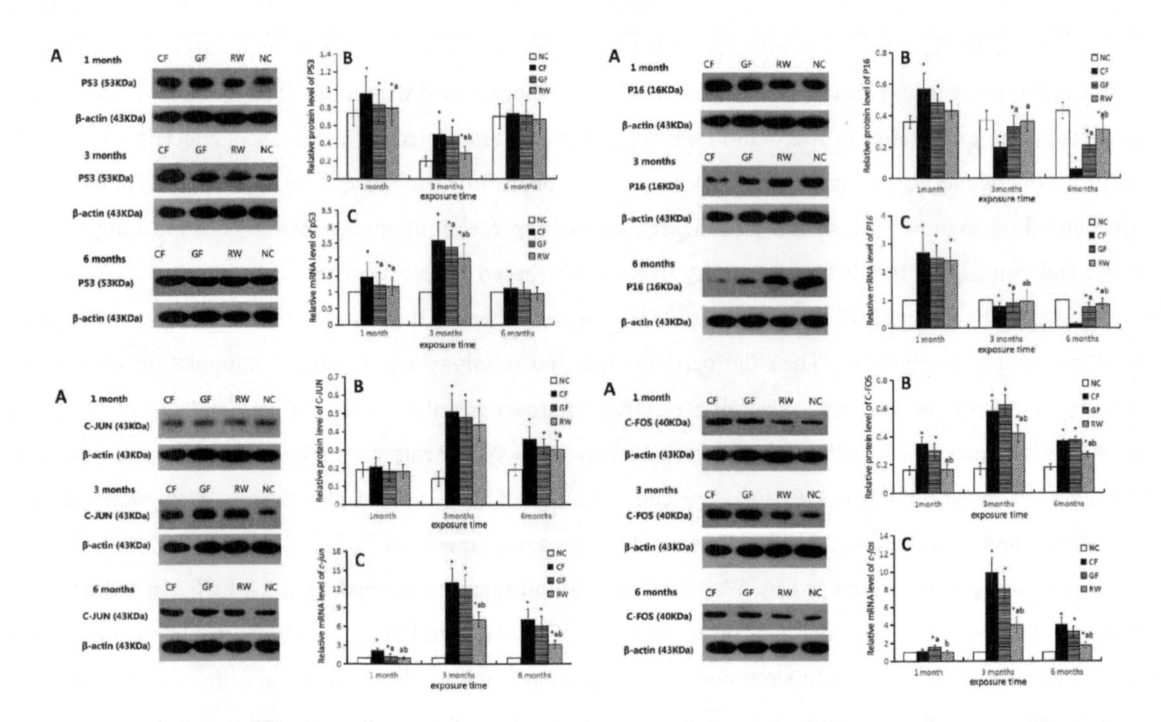

Fig. 2 Effects of MMMFs exposure on anti-oncogene P53 and P16 and oncogene C-JUN and C-FOS expression

Note: A. Western blot analysis; B. Relative protein levels; C. Quantitation of mRNA

Indirectly Electrochemical Detection of Ribavirin Based on Boronic Acid-diol Recognition Using the Platform of 3-aminophenylbornic Acid-electrochemically Reduced Grapheneoxide Modified Electrode

Xiao Lei Zhang　Xiao Yan Wang　Si Yan Guo　Ye Gao
Xiao Tong Li　Gong Jun Yang*
（School of Pharmacy，China Pharmaceutical University，
Nanjing 210009，P. R. China）

Ribavirin，as an antiviral drug，is usually used to treat the patients with chronic hepatitis C virus (HCV) infection. Up to now，several works，mainly including chromatography method[1]，were studied for quantification of ribavirin. In this paper，a novel method for indirectly electrochemical detection of ribavirin based on boronic acid-diol recognition was developed using the sensing element of the platform，which was constructed by 3-aminophenylboronic acid (APBA) -electrochemically reduced graphene oxide (ERGO) modified electrode. The fabrication of the proposed electrode and the principle of electrochemical analysis for ribavirin were sketched in Fig. 1. When the electrode was immersed in solution of ribavirin，complexation of boronic acid groups of APBA with ribavirin would occur at the surface of the proposed electrode and simultaneously cause the effect of steric hindrance，resulting in current decrease due to hindering the redox probe of ferricyanide to access to the surface of the proposed electrode. Under the optimized conditions，the good linear relationship between the relative change in current ($\%\Delta i$) of $[Fe(CN)_6]^{3-/4-}$ and the concentration of ribavirin was obtained in the concentration range of 10.0 to 7.50×10^2 ng/mL. The proposed electrochemical sensor performed the acceptable sensitivity and reproducibility. It was successfully used to determine the content of ribavirin in its injection with satisfactory results.

* E-mail：gjyang@cpu. edu. cn.

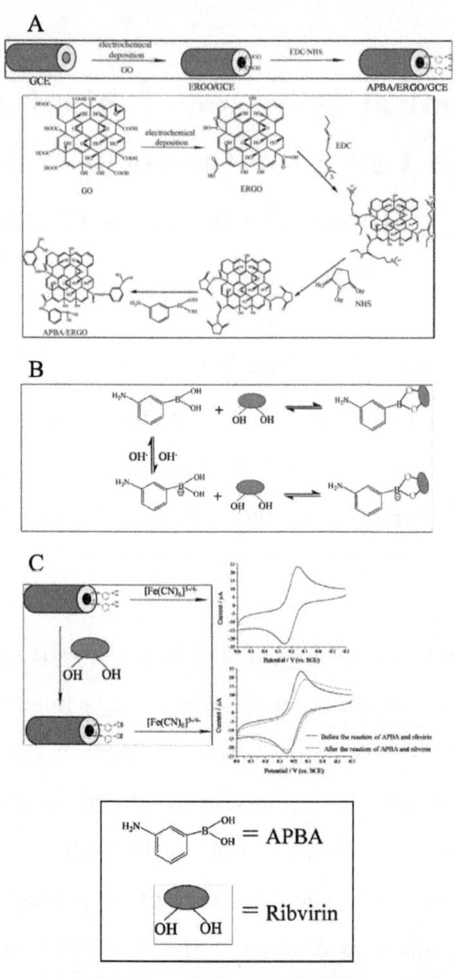

Fig. 1 （A）The process of the preparation of APBA/ERGO/GCE;（B）The binding between APBA and ribavirin;（C）The mechanism of electrochemical detection of ribavirin

Reference

[1] Agnesod D，Nicolo A D，Simiele M，et al. 2014. Development and validation of a useful UPLC-MS/MS method for quantification of total and phosphorylated-ribavirin in peripheralblood mononuclear cells of HCV+ patients [J]. J Pharmaceut Biomed Anal，90：119 - 126.

In Situ Synthesis of Graphene Oxide/gold Nanorods Theranostic Hybrids for Efficient Tumor Computed Tomography Imaging and Photothermal Therapy

Dan Li Bing Mei Sun Bing Di Chen*
(School of Materials Science and Engineering,
Tongji University, Shanghai 200092)

Photothermal therapy (PTT) is one of the most attractive methods for tumor ablation and regression. GNRs is a good material of photothermal conversion. But the aggregation of GNRs presents challenges regarding the improvement of photo-conversion and CT imaging. Graphene oxide (GO) is a two-dimensional (2D) nanostructures. GNRs can be conjugated onto GO for various functionalities and utilized in PTT[1] and drug delivery.

The in situ synthesis of GO/GNRs addressed the issue of the aggregation of the GNRs before their attachment onto the GO, which is straightforward and environment-friendly. The temperature of the GO/GNR nanohybrids increased from 25 to 49.9 ℃ at a concentration of 50 μg/mL after irradiation with an 808-nm laser (0.4 W/cm^2) for 6 min (Fig. 2). Additionally, the GO/GNRs exhibited good optical and morphological stability and photothermal properties after six cycles of laser irradiation. And the GO/GNR nanohybrids have great potential for precise CT-image-guided tumor photothermal treatment.

The process of synthesizing GO/GNRs includes two steps (Fig. 1): first, preparation and noncovalent modification of GO, second, preparation and modification of GO/GNR nanohybrids. GO/GNR hybrids were synthesized using a GO-and gold-seed-mediated in situ growth method at room temperature. This in situ synthesis is highly adaptable for developing other types of GO/NPs.

The GO/GNRs were successfully used as a versatile carrier of the small molecule iohexol for enhanced CT tumor imaging. Moreover, theGO/GNR nanohybrids demonstrated excellent photostability over six laser irradiation cycles and localized tumor killing ability and have therefore great potential for application in precise CT-image-guided tumor PTT.

* Corresponding author: Bing Di Chen; E-mail: aihyao@126. com; Mobile Phone No.: 18721796963. National Natural ScienceFoundation of China (81772285), Shanghai Natural Science Foundation (No. 16ZR1400700) and Shanghai Science and Technology Commission Project (No. 17411968700).

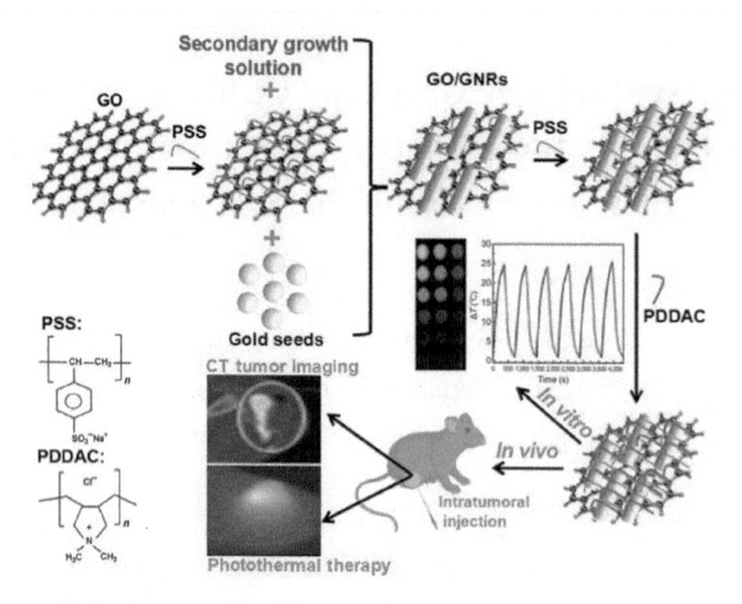

Fig. 1 Schematic illustration of the preparation and application of the GO/GNR hybrids

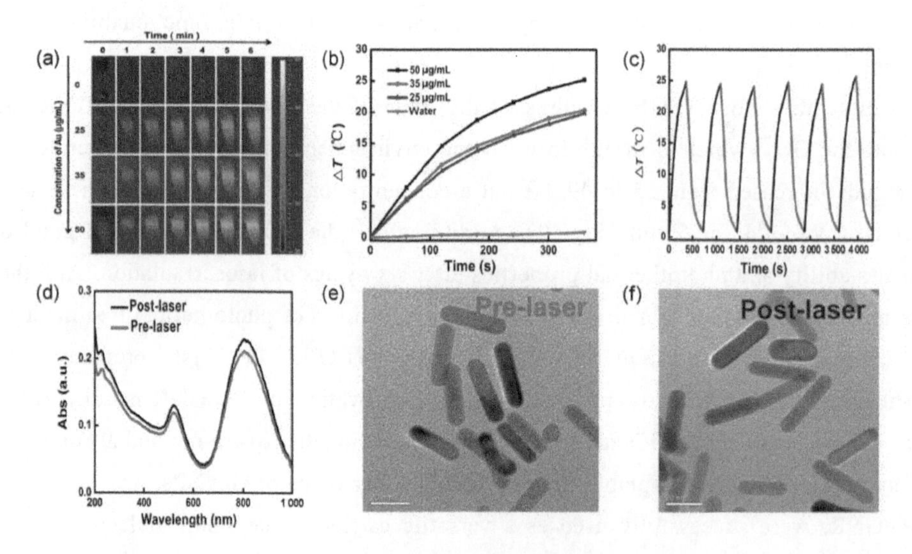

Fig. 2 *In vitro* photothermal experiments

Note: Real-time thermal imaging (a) and temperature profiles for different concentrations of GO/GNRs (b). Temperature variations of GO/GNRs (50 μg/mL) under continuous irradiation with an 808-nm laser for six cycles (c). UV-Vis spectra obtained before and after six cycles of irradiation (d). TEM images acquired before (e) and after (f) six cycles of laser irradiation (0.4 W/cm^2)

References

[1] Zhang W, Guo Z Y, Huang D Q, et al. 2011. Synergistic effect of chemo-photothermal therapy using PEGylated graphene oxide [J]. Biomaterials, 32: 8555 - 8561.

Oxidative Effects on Lungs in Wistar Rats Caused by Long-term Exposure to Four Kinds of China Representative Chrysotile

Yan Cui[1]　Yu Xin Zha[1]　Jian Jun Deng[2]　Qing Bi Zhang[1*]
(1. Southwest Medical University, Luzhou 646000, China;
2. 404 Hospital of Mianyang, Mianyang 621000, China)

Chrysotile, accounting for approximately 90% of all the asbestos, has now been used worldwide. Accumulating evidence has shown that increasing of lung cancer, asbestosis as well as mesothelioma was associated with asbestos (including chrysotile) exposure. However, the molecular mechanisms underlying the toxic effects of chrysotile have not yet been entirely understood. Therefore, four kinds of China representative chrysotile [Shaanxi Shannan (SSX), Gansu Akesai (AKS), Sichuan XinKang (XK) and Qinghai Mangnai (MN)] were administered to Wistar rats to study whether chrysotile could result in pulmonary pathology and oxidative damage, and provide a comprehensive assessment of natural chrysotile minerals from four regions as well.

The chemical composition of all fibers was detected by an X-ray Fluorescence Spectrometer (PW1404, Philips) and the morphology was assessed under a scanning electron microscope (TM - 100, Hitachi). Then the fiber dispersion was administered to 120 Wistar rats by multiple intratracheal instillation. Rats were sacrificed to observe pathological changes in lungs at 1, 3, 6 and 12 months. AKP, LDH and total protein were measured by BALF using commercial kits (Nanjing Jiancheng Bioengineering Institute, Jiangsu, China). ROS, MDA and SOD of lung tissues were examined by commercial colorimetric assay kit (Beijing Chenglin and Nanjing Jiancheng Bioengineering Institute, Beijing, China). Real-time RT-PCR detected the mRNA levels of ho - 1 and hsp70. These results (Fig. 1 and Fig. 2) indicate that chrysotile exposure lead to a significant increase in lung mass and slow the growth of body mass. Inflammatory lesions, destruction of alveolar structures, and pulmonary fibrosis appeared in all chrysotile groups, and the damage would be aggravated with the prolonged exposure time. Biochemical analyses performed on BALF revealed significant increase of cytotoxicity markers (LDH, AKP) and permeability markers (TP). Exposure to chrysotile significantly increased the accumulation of reactive oxygen species (ROS) and the level of lipid peroxidation, but decreased antioxidant capacity in lung tissues. Furthermore, the expression levels of heme oxygenase - 1 (HO - 1) and heat shock protein 70 (HSP70) increased with 1 month of chrysotile exposure and remained elevated for 6 months, whereas 12 months' exposure showed an obvious decrease in expression levels of two factors in XK and MN groups

* Corresponding author: Qing Bi Zhang; E-mail: qingbizhang@126. com; Mobile Phone No.: +86-18113513199. This work was supported by the National Natural Fund Project of China (No. 41472046).

when compared with negative control. Therefore, our results suggested that chronic chrysotile pulmonary injury in Wistar rats was triggered by oxidative damage. Meanwhile, the oxidative damage of MN and XK is stronger than SSX and AKS, the reasons for the difference of oxidative damage between four chrysotile could be attributed to its properties, morphology, chemical composition and particle size. These results will provide useful experimental data for further research on the toxicity and mechanism of chrysotile.

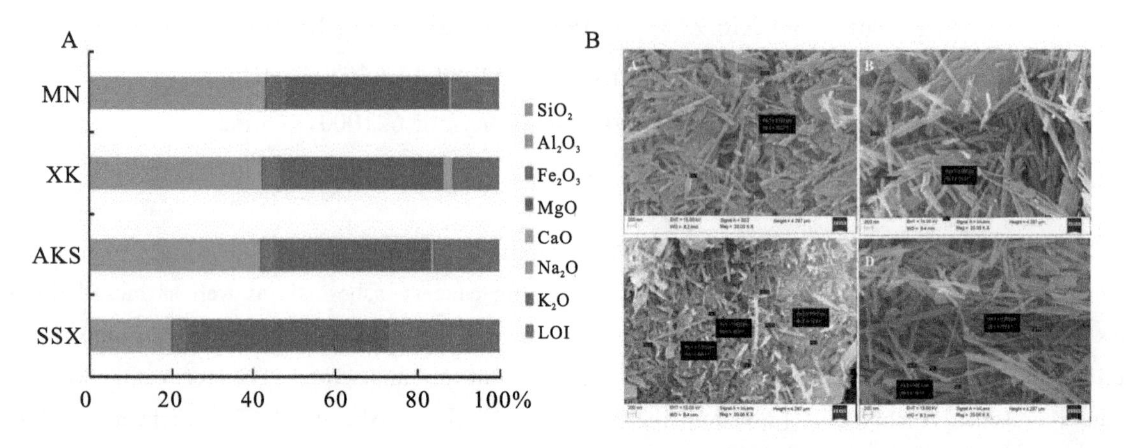

Fig. 1 Chemical composition analysis (A) and scanning electron microscope images (B) of China representative chrysotile mineral samples

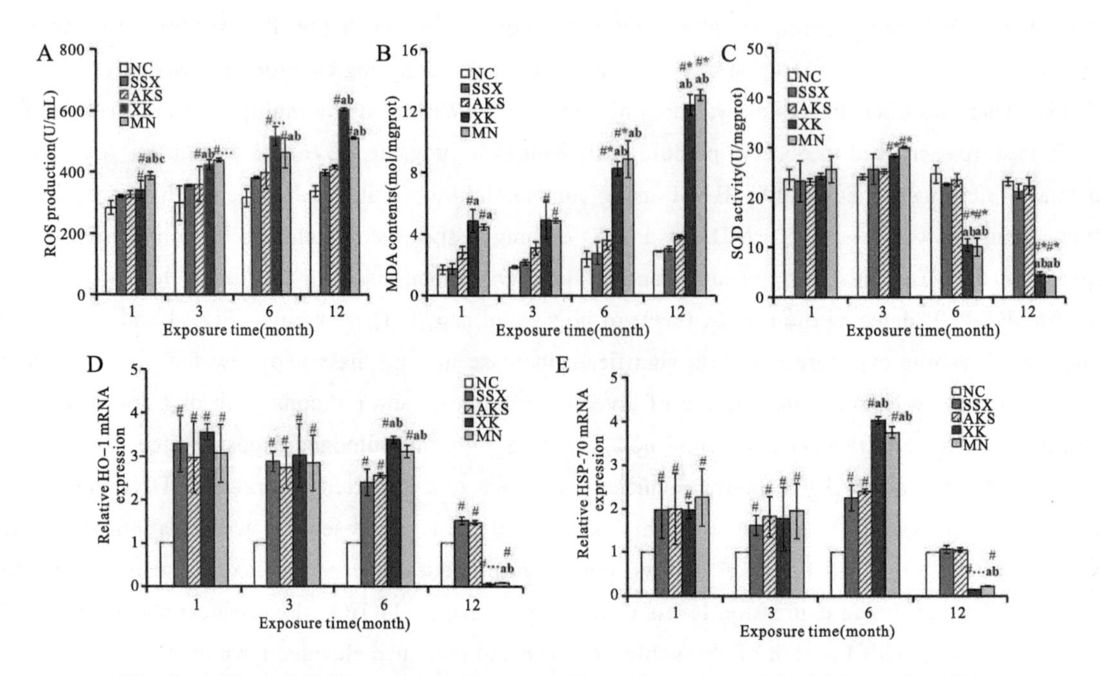

Fig. 2 Effects of China representative chrysotile on the content of ROS (A), MDA (B), the SOD activity (C), HO-1 (D) and HSP70 (E) expression in rat lung

Enzymatic Biofuel Cells Enabled with Carbon Materials

Guo Zhi Wu　Mei Zhao　Dan Zhao　Yue Gao　Zhen Yao　Feng Gao*
(College of Chemistry and Materials Science,
Anhui Normal University, Wuhu 241002, China)

Enzymatic biofuel cells, a special kind of fuel cells that use enzymes rather than conventional noble metals as catalysts through the bioelectrocatalytical means, provide a versatile means to generate electrical power from environmentally friendly biomass or biofuels. Biofuel cells are low-cost and active in moderate conditions (e. g. near-neutral pH and room temperature) and are therefore viewed as a potential green energy technology compared to conventional fuel cells. However, the development of enzymatic biofuel cells is still in its early days. Compared to conventional fuel cells, research on enzymatic biofuel cells are still in the fundamental research stage because of their low stability and power output. Electrodes biocatalytically modified with enzymes are the key for performance of biofuel cells. Rational choice of electrode materials and tailoring are critical in circumventing such challenges[1-5].

Carbon-based materialshave many technological advantages such as facile modification by functional groups, high chemical stability, good biocompatibility, robust mechanical strength, and feasibility of incorporating both hydrophilic and hydrophobic substances[6,7]. These characteristics make carbon materials promising for the development of enzyme electrodes and therefore for the construction of biofuel cells.

In our group, we have devoted to the construction of biofuel cells with different carbon materials[8-12]. For example, Figure 1 shows a direct bioelectrocatalysis-type membrane-less lactate/oxygen biofuel cell using carbon double-shelled hollow spheres (CDSs)[9]. Taking advantage of their excellent catalytic activity to NADH oxidation at lower overpotential at ca. −0.10 V (vs. Ag/AgCl, pH=7.0), and facilitated direct electron transfer (DET) behavior of bilirubin oxidase (BOD) at CS, a membrane-less lactate/oxygen biofuel cell is assembled. Figure 2 shows carbon nanodots (CNDs) -based glucose/air biofuel cell[10]. On the basis of the DET of GOx and BOD at CNDs, a mediator-free DET-type glucose/air enzymatic biofuel cell was assembled by using GOx as biocatalyst for glucose oxidation at bioanode and BOD as biocatalyst for oxygen reduction at biocathode, with a high open-circuit voltage of 0.93 V and maximum power density of 40.8 $\mu W \cdot cm^{-2}$ at 0.41 V. These newly constructured biofuel cells oper-

* Corresponding author: Feng Gao; E-mail: fgao@mail. ahnu. edu. cn; Mobile Phone No.: +86-13955306629. This work was supported by the Natural Science Foundation of China (No. 21575004), Program for New Century Excellent Talents in University (NCET-12-0599), the project sponsored by SRF for ROCS, SEM, and the Foundation for Innovation Team of Bioanalytical Chemistry of Anhui Province.

ating in physiological conditions exhibits enhanced power and stability. In this report, we will show different biofuel cells constructed with different enzymes and carbon-based electrode materials.

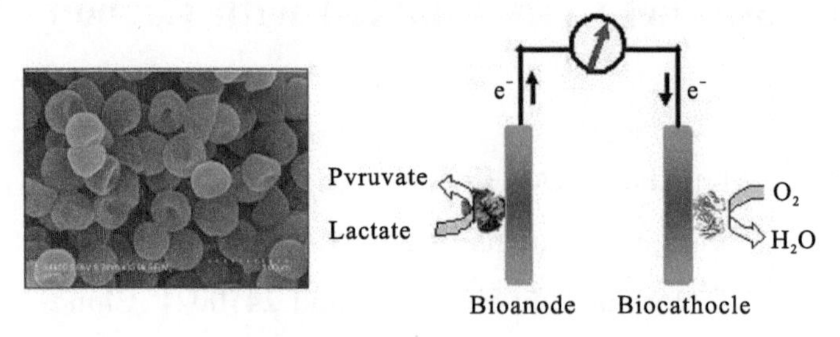

Fig. 1 Lactate/oxygen biofuel cell using carbon double-shelled hollow spheres

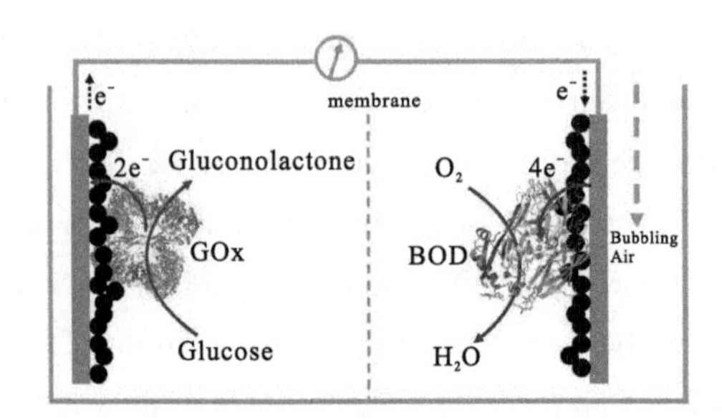

Fig. 2 DET-type glucose/air biofuel cell based on carbon nanodots

References

[1] Catalano P N, Wolosiuk A, Soler-Illia GJA A, et al. 2015. Wired enzymes in mesoporous materials: A benchmark for fabricating biofuel cells [J]. Bioelectrochemistry, 106: 14 - 21.

[2] Nöll T, Nöll, G. 2011. Strategies for "wiring" redox-active proteins to electrodes and applications in biosensors, biofuel cells, and nanotechnology [J]. Chem. Soc. Rev., 40: 3564 - 3576.

[3] Cooney M J, Svoboda V, Lau C, et al. 2008. Enzyme catalysed biofuel cells [J]. Energy Environ. Sci., 1: 320 - 337.

[4] Gao F, Viry L, Maugey M, et al. 2010. Engineering hybrid nanotube wires for high power biofuel cells [J]. Nat. Commun., 1.

[5] Zhou M, Dong S. 2011. Bioelectrochemical interface engineering: toward the fabrication of electrochemical biosensors, biofuel cells, and self-powered logic biosensors [J]. Acc. Chem. Res., 44: 1232 - 1243.

[6] McCreery R L. 2008. Advanced carbon electrode materials for molecular electrochemistry [J]. Chem. Rev., 108: 2646 - 2687.

[7] Yan Y, Zheng W, Su L, et al. 2006. Carbon-nanotube-based glucose/O_2 biofuel cells [J]. Adv. Mater., 18: 2639 - 2643.

[8] Gao F, Yan Y, Su L, et al. 2007. An enzymatic glucose/O_2 biofuel cell: Preparation, characterization and performance in serum [J]. Electrochem. Commun., 9: 989 - 996.

[9] Gao F，Guo X，Yin J，et al. 2011. Electrocatalytic activity of carbon spheres towards NADH oxidation at low overpotential and its applications in biosensors and biofuel cells [J]. RSC Adv., 1：1301 - 1309.

[10] Zhao M，Gao Y，Sun J，et al. 2015. Mediatorless glucose biosensor and direct electron transfer type glucose/air biofuel cell enabled with carbon nanodots [J]. Anal. Chem.，87：2615 - 2622.

[11] Wu G，Gao Y，Zhao D，et al. 2017. Methanol/oxygen enzymatic biofuel cell using laccase and NAD+-dependent dehydrogenase cascades as biocatalysts on carbon nanodots electrodes [J]. ACS Appl. Mater. Interfaces，9：40978 - 40986.

[12] Wu G，Yao Z，Fei B，et al. 2017. An enzymatic ethanol biosensor and ethanol/air biofuel cell using liquid-crystalline cubic phases as hosting matrices to co-entrap enzymes and mediators [J]. J. Electrochem. Soc.，164 (7)：82 - 86.

Near-infrared Laser-triggered Drug-loaded Graphitic Carbon Nanocages for Cancer Therapy

Yu Liang Guo[1⊥] Yang Chen[1,2,3⊥] Pomchol Han[1⊥] Dan Li[1] Xin Gui[1]
Ze Fei Zhang[1] Kai Fu[1] Mao Quan Chu[1*]
(1. Research Center for Translational Medicine at Shanghai East Hospital, School of Life Sciences and Technology, Tongji University, Shanghai 200092, P. R. China;
2. Institute of Biophysics, Chinese Academy of Science, Beijing 100101, P. R. China;
3. University of Chinese Academy of Sciences, Beijing 100049, P. R. China)

In recent years, nanomaterial-induced cancer photothermal therapy (PTT) has attracted great attention all over the world. These nanomaterials include carbon-based nanoparticles, gold nanostructures, magnetic nanoparticles, semiconductor nanocrystals, palladium nanoplates, RuO_2 nanoparticles, copper-based nanomaterials, polymer nanoparticles, cyanine dye-containing nanoparticles, natural chlorophyll, and melanin nanoparticles. Among these photothermal conversion materials, carbon-based nanoparticles especially graphene-based nanomaterials including graphene quantum dots, carbon nanohorns, carbon nanotubes, graphene oxide and reduced graphene oxide (rGO) nanosheets, have unique physical and chemical properties besides excellent photothermal conversions in red or NIR wavelength region, such as high photothermal and chemical stability, large surface-area-to-volume ratio, and aromatic molecular structure.

Graphitic carbon nanocage (GCNC) is also an important member of the graphene-based nanomaterials family. These cage-like nanoparticles have a large internal space and graphitic shells, which have been used as hydrogen-storage materials, electrode materials, catalyst supports and super capacitors. However, few reports have focused on drug delivery and in vivo cancer PTT under 980-nm laser irradiation using these GCNCs.

Herein, we synthesized GCNCs using ferrous oxalate and ethanol as precursors, and found that the as-synthesized GCNCs have great potential for future clinical translation due to their good bio-safety, highly drug-loading capacity and highly efficient cancer therapy (Fig. 1)[1]. The GCNCs had a narrow size distribution [average diameter: (77.6 ± 11.7) nm] and were in hollow structure with a shell that may be composed of tens of graphitic layers.

⊥ These authors contributed equally to this work. * Corresponding author: Mao Quan Chu, E-mail: mqchu98@tongji.edu.cn.

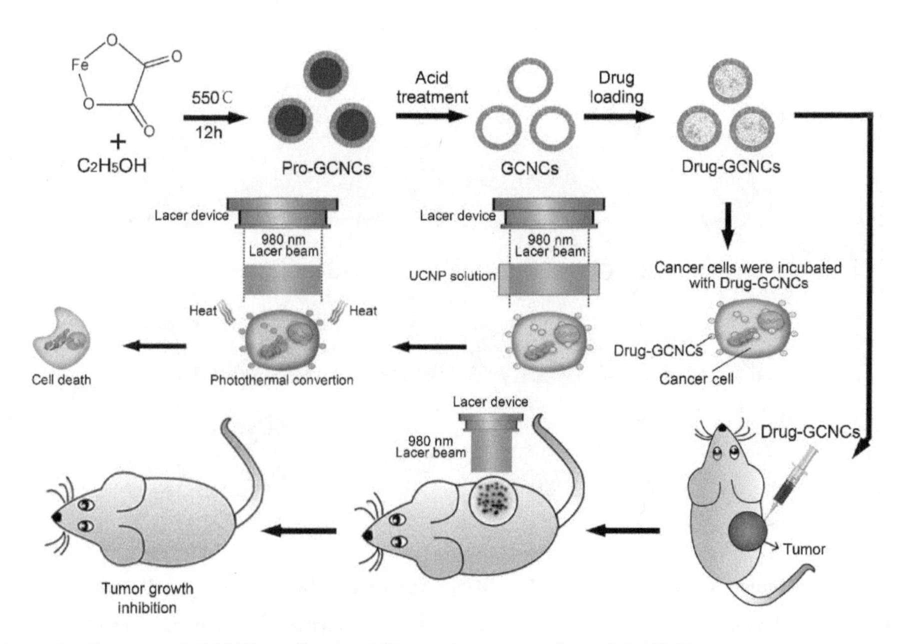

Fig. 1　Schematic diagram of GCNC synthesis, killing of cancer cells and inhibiting mouse tumor growth through synergism between the photothermal effect of GCNCs and cytotoxicity of drugs loaded in the GCNCs

Note: The red upconversion nanoparticles (UCNPs) solution was used to confirm irradiation of the samples (e. g., cells) by the 980-nm laser beam

The cytotoxicity of GCNCs may be lower than that of multi-walled carbon nanotubes (MWCNTs), as when human nasopharyngeal carcinoma cells (CNE cells) were incubated with GCNCs or MWCNTs for 24 h, a significant increase in the cell population in the G0/G1 phase and a significant decrease in those in the S andG2/M phases were observed for the MWCNT-treated cells, but not for the GCNC-treated cells. The photothermal conversion efficiency of the GCNCs on 980-nm laser irradiation (powder density: 0.28 W/cm^2) was 19.2%, which is similar with that of MWCNTs and reduced graphene oxide under the same conditions. We showed that GCNCs could be used for loading various types of drugs, including water-soluble anticancer drug doxorubicin hydrochloride (DOX) and photosensitizers methylene blue (MB) and sodium copper chlorophyllin (SCC), slightly water-soluble cis-dichlorodiammine-platinum (Ⅱ) (DDP) and water-insoluble curcumin (CUR). The growth of mouse tumors intratumorally injected with GCNCs (2 mg/mL, 100 μL) and irradiated with a 980-nm laser was significantly suppressed compared with that in tumor-bearing mice injected with PBS without irradiation, and the tumors finally disappeared at 6 days after irradiation. The cell-killing and tumor-inhibiting efficiencies of the laser-triggered drug-loaded GCNCs were significantly higher than those achieved using PTT with GCNCs and cytotoxic drugs (Fig. 2), which was due to a synergistic effect between PTT and drug toxicity. To improve the biocompatibility and cancer therapy efficiency, GCNCs were further encapsulated in chitosan (CS) polymer nanospheres. After coating with CS polymer, the changes of cell cycle distribution of CNE cells influenced by the GCNCs/CS system were reduced compared with those influenced by the obtained GCNCs, and the final cytotoxicity of the GCNCs/CS was significantly reduced compared with that of the GCNCs; in addition, the cancer therapy efficiency was obviously enhanced, as more drugs had been incorporated into the GCNCs/CS nanocomposites.

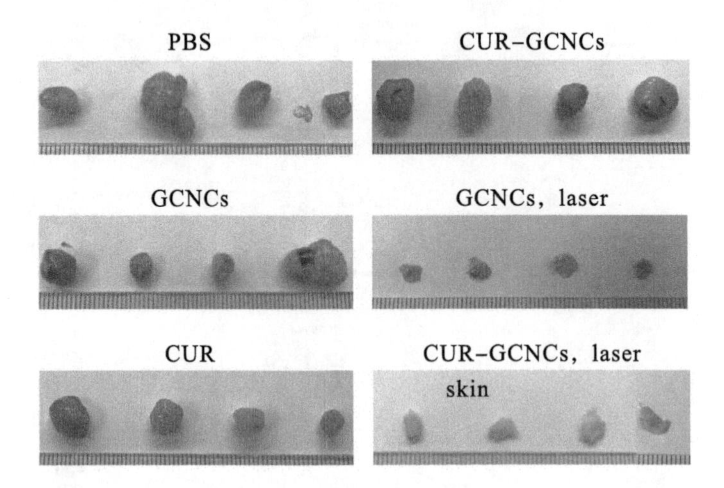

Fig. 2 *In vivo* tumor therapy efficiency of CUR-GCNCs under 980-nm laser irradiation, and controls

Note: The size of tumors at 18 days post initial irradiation (or injection). In laser irradiation groups, the tumor-bearing mice were intratumorally injected with PBS-dispersed GCNCs (2 mg/mL, 100 μL) or CUR-GCNCs (GCNCs: 2 mg/mL, 100 μL) and then irradiated with the 980-nm laser 10 h post-injection for 20 min every other day (when the tumors could no longer be detected, irradiation was stopped). n=4 per group

Reference

[1] Chen Y, Guo Y L, Han P, et al. 2018. Graphitic carbon nanocages as new photothermal agent and drug Carrier for 980 - nm - laser - driven cancer therapy [J]. Carbon, 136: 234 - 247.

第八章

非金属矿与地球化学、生态环境

基于钛柱撑蒙脱石的环境净化修复系统构筑

吴宏海*
（华南师范大学化学与环境学院，广州 510006）

当前全球环境污染问题十分严峻，人类除了暴露于难降解污染物中外，又面临药物和个人护理品（PPCPs）的环境暴露风险。蒙脱石（Mnt）具有巨大的比表面积和优越的阳离子交换性能，但是在光催化降解有机物或表面催化增强 Fe（Ⅱ）还原能力等方面仍存在一些不足。于是，在蒙脱石层间域插入纳米 TiO_2 以设计出一种生态功能材料——钛柱撑蒙脱石（TPMnt）。TPMnt 不仅能够有效地促进有机污染物如邻硝基苯酚（2－NP）的还原转化或光催化降解，而且还可以有效地利用地下的 Fe（Ⅱ）和地上的太阳光。基于上述认识，本文构筑基于 TPMnt 的低碳环境净化修复系统。

钛柱撑蒙脱石的制备参照 Del Castillo 等[1]的制备方法：取 5.0 g 蒙脱石样品放入 1.0 L 水中浸泡 24 h；将钛酸丁酯缓慢加入 2.0 mol/L 盐酸中并磁力搅拌 1.0 h 形成柱化剂；后将柱化剂加入上述蒙脱石悬浊液，搅拌 1.0 h，经离心、洗涤，并在 80 ℃下干燥。干燥样品在不同温度下煅烧 2.0 h 便获得煅烧钛柱撑蒙脱石的系列样品 TPMnt－x ℃，经研磨、过 200 目筛备用。

TPMnt 催化 Fe（Ⅱ）对 2－NP 的还原实验：依次加入 0.8 g 钛柱撑蒙脱石和 200 mL 混合溶液（含 0.022 mmol/L 2－NP、3.0 mmol/L $FeSO_4$、28 mmol/L 缓冲剂溶液以及 200 mmol/L NaCl）到 250 mL 锥形瓶中，盖上橡胶塞，持续通氮气，避光置于 25 ℃恒温磁力搅拌水浴锅中，并间隔取样。取样时需加入 60 μL 2 mol/L 盐酸，以防止反应进一步发生。4 000 r/min 离心分离 15 min，采用针管取上层液体 1 mL，过 0.45 μm 滤膜，后保存至棕色小瓶中供液相色谱分析用。

光催化 TPMnt 对 2－NP 的降解实验：将 0.4 g TPMnt 倒入 100 mL 石英管中，再加入 1 mL 浓度为 2.2 mmol/L 的邻硝基苯酚溶液，把石英管置于光反应仪内部，先避光磁力搅拌 20 min，然后在 300 W 汞灯的紫外照射下启动 2－NP 的光催化降解反应，再继续不断地搅拌并按上述方法间隔 1 h 取样和分析。

未加入 Fe（Ⅱ）的蒙脱石和钛柱撑蒙脱石两个系统分别有 7％和 12％的 2－NP 去除率，且没有 2－AP 还原产物的产生；仅有 Fe（Ⅱ）的均相反应体系中，2－NP 的还原转化缓慢，其 4 h 的去除率为 26％；然而，添加了 Fe（Ⅱ）的蒙脱石非均相系统对 2－NP 4 h 的去除率约为 45％，而添加了 Fe（Ⅱ）的钛柱撑蒙脱石非均相系统对 2－NP 的还原转化快速。这些结果说明，表面配位吸附态 Fe（Ⅱ）为 2－NP 上述还原转化中的关键性物种。这与之前的相关报道一致[2]。例如，

* 通信作者：吴宏海；E-mail：wuhonghai@scnu. edu. cn；手机号：13352871983。国家自然科学基金（41372050）、广东省自然科学基金-重点（2018B030311021）资助项目。

与蒙脱石相比，钛柱撑蒙脱石对 Fe（Ⅱ）的还原性能显著增强，故明显促进了 Fe（Ⅱ）对 2-NP 的还原转化。实际上，蒙脱石和钛柱撑蒙脱石对 2-NP 的吸附去除作用均很弱（图 1）。

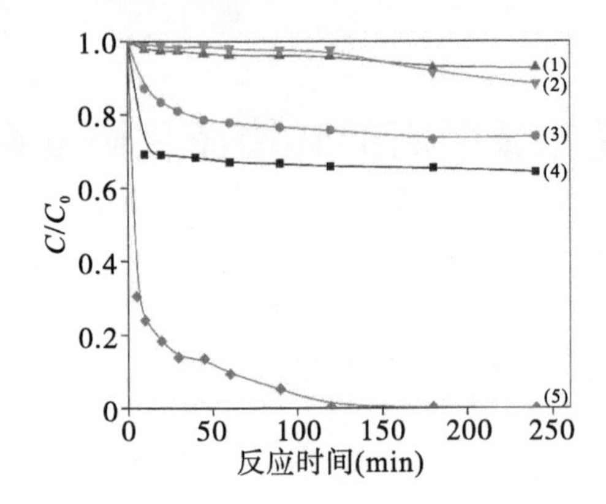

图 1　5 种不同反应系统对 2-NP 去除效果的对比

注：初始条件：22 μM/L 2-NP，pH=6.0，25 ℃，0.2 M NaCl，28 mM MES 或 MOPs 和 4.0 g/L 矿物。5 种反应系统：（1）Mnt；（2）TPMnt200；（3）3.0 mM/L Fe（Ⅱ）；（4）Mnt + 3.0 mM/L Fe（Ⅱ）；（5）TPMnt200 + 3.0 mM/L Fe（Ⅱ）

光催化降解实验的初步结果表明，500 和 600 ℃煅烧的 TPMnt 对 2-NP 的吸附量最大，300 和 400 ℃煅烧的样品次之，而 200 ℃煅烧的样品最小。这些 TPMnt 样品的光催化性能不尽相同，其中 TPMnt400 对 2-NP 的去除率为 86.3%，去除效果最佳。TPMnt600 对 2-NP 的吸附最强，但其对 2-NP 的去除率最低，仅为 60.4%。

参考文献

[1] Del Castillo H L，Grange P. 1993. Preparation and catalytic activity of titanium pillared montmorillonite [J]. Appl. Catal.，A：General，103：23-34.

[2] Wu H H，Song Z H，Lv M X，et al. 2018. Iron-pillared montmorillonite as an inexpensive catalyst for 2-nitrophenol reduction [J]. Clays and Clay Minerals.

nZVI－CNT光催化与希瓦氏菌协同去除U（Ⅵ）的机理

项书宏[1]　程文财[1,2]　丁聪聪[1,2]　刘明学[3]　边　亮[4]　董发勤[4]　聂小琴[1,2*]
（1. 西南科技大学国防科技学院，绵阳 621010；
2. 西南科技大学核废物与环境安全国防重点学科实验室，绵阳 621010；
3. 西南科技大学生命科学与工程学院，绵阳 621010；
4. 西南科技大学固体废物处理与资源化教育部重点实验室，绵阳 621010）

铀在废水中主要以高溶解度的六价铀酰［U（Ⅵ）］形式存在，而四价铀［U（Ⅳ）］的溶解度较低，因此可以通过将六价铀酰还原为四价的难溶沉淀，进而分离去除铀酰。前人研究表明，纳米零价铁具有很高的还原活性，可用于U（Ⅵ）的还原去除，但纳米零价铁易团聚和被氧化，氧化产物覆盖在团聚物表面阻止进一步反应，从而使其失去还原活性，降低利用效率[1]，然而纳米零价铁在反应后转化为纳米铁氧化物，纳米铁氧化物与某些导体材料的结合能够表现出光催化活性，增强还原效果[2]。与此同时，希瓦氏菌属于产电微生物并且能够提供较多的活性位点，在一定条件下能够自主产生生物电子供吸附于菌体表面的其他材料利用，因此本研究结合上述各材料的优势，克服纳米零价铁单独应用的缺陷，提高U（Ⅵ）的去除效率[3]。实验基于纳米零价铁（nZVI）的高还原性能以及高电导率的蜂巢晶格碳材料——石墨烯（RGO）、多壁纳米碳管（CNT）、富勒烯（C60），制备3种复合材料（nZVI－RGO、nZVI－CNT、nZVI－C60），同时结合光照和产电微生物希瓦氏菌用于高效处理含铀废水，系统考察了不同光照条件下几种复合材料对50 mg/L铀酰去除效果、适用条件及铁的溶出情况；同时采用介观和光谱手段，系统探究了反应前后nZVI－CNT材料结构和形貌；利用CASTEP计算铁原子在纳米碳管表面的吸附能以及FeOOH在纳米碳管表面的吸附能。

结果表明，光照条件下的nZVI－CNT材料去除效率最高，能够将纳米零价铁与纳米碳管铀酰去除效率之和提高300%，此外与希瓦氏菌结合后，在nZVI－CNT基础上，总去除效率进一步提高132%。XRD结果表明，nZVI－CNT与U（VI）反应过后，其中Fe^0主要转化为羟基氧化铁（FeOOH），而U（Ⅵ）则转化为难溶的UO_2沉淀。对反应前后的nZVI－CNT材料的XPS和SEM－EDS分析表明，反应过后大量的Fe^0转化为Fe（Ⅲ）或者Fe（Ⅱ），同时U（Ⅳ）元素出现在反应过后的产物中，侧面证明nZVI－CNT对U（Ⅵ）的去除主要通过还原和物理吸附作用（图1）。

本研究体系首先利用纳米零价铁将六价铀酰还原成四价并沉淀，而纳米零价铁转化为纳米铁氧化物；纳米铁氧化物在光照条件下能够产生大量的光电子和空穴，与纳米碳管的结合使得光电子被纳米碳管快速传导给吸附于材料表面六价铀酰用于铀还原，而空穴则与产电微生物表面的生

*通信作者：聂小琴；E-mail：xiaoqin_ nie@163. com；手机号：13908112161。国家重点基础研究发展计划（973）项目（2014CB846003）、国家自然科学基金（41502316、2170110、41703118）、中国博士后科学基金（2017M612991）和中国博士后基金特别资助计划项目（2018T110994）资助。

物电子复合；接收生物电子后的铁氧化物在光照条件下再次表现出光催化效应，重复发生前面铀还原的过程，使得在光照条件下铀去除效果得以明显提升。本研究旨在通过采用环境友好材料，综合利用纳米半导体材料光催化效应，纳米零价铁的高还原性能以及腐败希瓦氏菌的生物产电效应，实现高效去除U（Ⅵ），达到放射性水体绿色修复的目的。

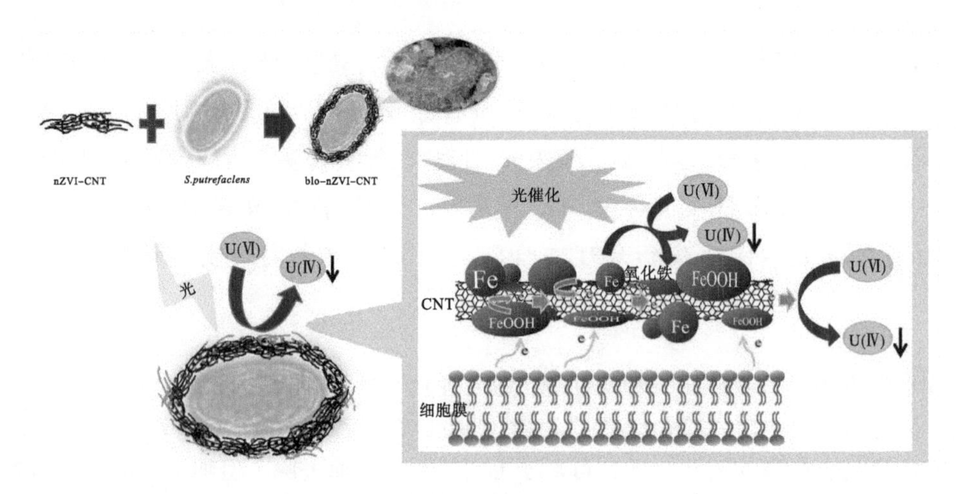

图1 反应机理

参考文献

［1］Fu F，Dionysiou D D，Liu H. 2014. The use of zero-valent iron for groundwater remediation and wastewater treatment：a review［J］. Journal of Hazardous Materials，267：194－205.

［2］Wang Z，Dong X，Li J. 2008. An inlaying ultra-thin carbon paste electrode modified with functional single-wall carbon nanotubes for simultaneous determination of three purine derivatives［J］. Sensors & Actuators B Chemical，131：411－416.

［3］Huang W，Nie X，Dong F，et al. 2017. Kinetics and pH-dependent uranium bioprecipitation by Shewanella putrefaciens under aerobic conditions［J］. Journal of Radioanalytical & Nuclear Chemistry，312：531－541.

蒙脱石层板表面接枝改性及催化纤维素水解性能

杨　淼　周　扬　房　凯　杨海燕　童东绅* 周春晖*

（浙江工业大学化学工程学院，杭州 310032）

生物质能以其清洁性和高储量受到广泛关注。生物质中最具价值的是纤维素，通过纤维素生产的葡萄糖等还原糖，对缓解能源日益枯竭有着巨大的优势，由纤维素生产的生物质清洁燃料[1]被认为是目前化石燃料的最佳代替品。因此，如何催化纤维素水解为葡萄糖等还原糖成为研究人员关注的热点。

本文采用不同浓度的硫酸、盐酸和磷酸分别对蒙脱石进行酸化处理[2]，并逐一测试了其水解纤维素制备还原糖的能力，我们选取 5wt%硫酸、10wt%盐酸和 10wt%磷酸处理的蒙脱石进一步接枝磺酸基团[3,4]，制备出固体酸催化剂（MMT－SO_3H），随后将三者投入到纤维素水解制备还原糖的反应中去，对应的还原糖收率分别为 20.52%、21.76%、24.03%，相比酸化的蒙脱石，其催化性能明显提高（图 1）。

在此基础上，我们探究了不同的巯基接枝时间所制备的催化剂对还原糖收率的影响，结果发现巯基的接枝时间在 2.5 h 所制备的催化剂对应的还原糖收率最高，达到 26.02%（图 2）。原因是巯基的接枝时间越长，蒙脱石上的磺酸根含量也会相应增加，在一定的磺酸根密度下，还原糖收率随着酸度的增加而增加，当酸度过高时，还原糖收率又会逐渐下降。

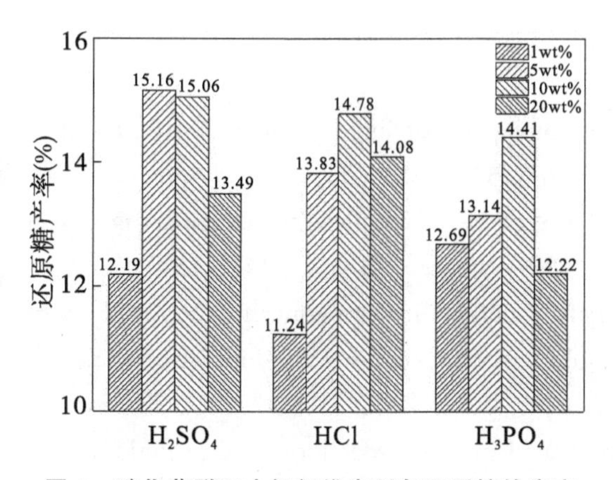

图 1　酸化蒙脱石水解纤维素制备还原糖的产率

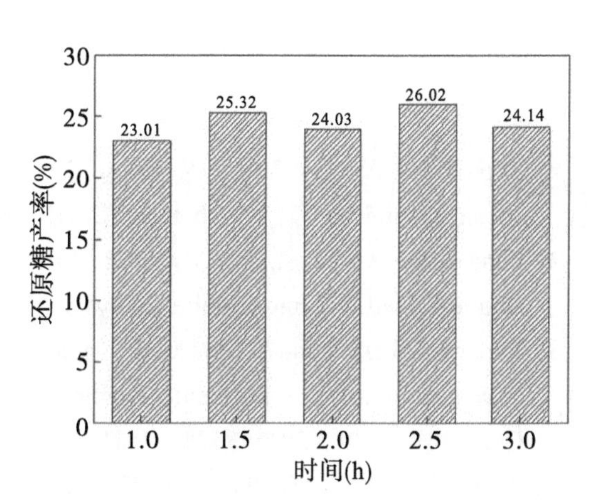

图 2　巯基接枝时间对应的还原糖产率

* 通信作者：童东绅；E-mail：tong980480@sina. com；手机号：13656675377。周春晖；E-mail：amsc_ group@126. com；手机号：13588066098。国家自然科学基金（21506188，41672033）、浙江省自然科学基金（LY16B030010）资助项目。

从图 3 可以看出，分别在 $2\theta=6.5°$、$19.7°$、$35.0°$左右出现了蒙脱石的特征衍射峰。对于 $2\theta=6.5°$左右的峰的出现，经硫酸和盐酸酸化的蒙脱石的特征峰强度出现了减弱，而经磷酸酸化的蒙脱石的特征峰强度未出现多大变化，可能是因为蒙脱石经过酸化处理，其结晶度降低，由于磷酸为中强酸，酸强度不如盐酸和硫酸，致使对蒙脱石内部结构影响小。而层间距变大可能是因为氢离子与层间离子进行置换，且氢离子半径小，导致层间吸引力降低，层间距变大。

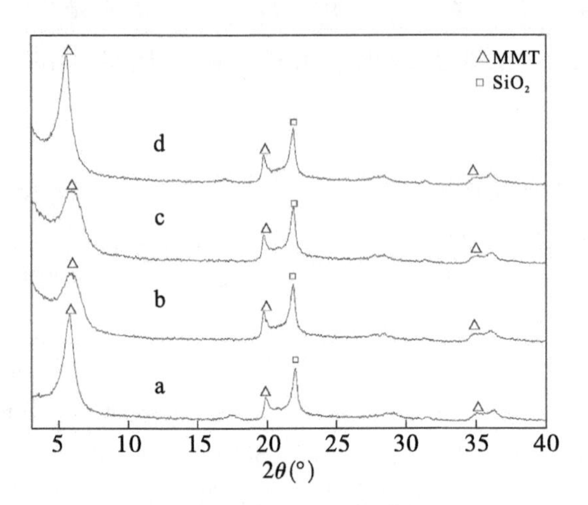

图 3　XRD 衍射图谱

注：a. MMT；b. 5wt% H_2SO_4-MMT-SO_3H；c. 10wt% HCl-MMT-SO_3H；d. 10wt% H_3PO_4-MMT-SO_3H

将蒙脱石先进行酸化再接枝磺酸根所制备的固体酸催化剂对水解纤维素制备还原糖有良好的催化作用，其中，用 10wt%的磷酸处理的蒙脱石所制备的催化剂效果最佳，还原糖的转化率达到 26.02%。

参考文献

[1] Zhou C H，Xia X，Lin C X，et al. 2011. Catalytic conversion of lignocellulosic biomass to fine chemicals and fuels [J]. Chemical Society Reviews，40 (11)：5588-5617.

[2] Tong D S，Xia X，Luo X P，et al. 2013. Catalytic hydrolysis of cellulose to reducing sugar over acid-activated montmorillonite catalysts [J]. Applied Clay Science，74 (1)：147-153.

[3] Varadwaj G B，Rana S，Parida K，et al. 2014. A multi-functionalized montmorillonite for co-operative catalysis in one-pot henry reaction and water pollution remediation [J]. Journal of Materials Chemistry A，2 (20)：7526-7534.

[4] Zhang C，Fu Z，Dai B，et al. 2014. Biochar sulfonic acid immobilized chlorozincate ionic liquid：an efficiently biomimetic and reusable catalyst for hydrolysis of cellulose and bamboo under microwave irradiation [J]. Cellulose，21 (3)：1227-1237.

Ni－TiO_2/凹凸棒石催化剂催化 CO_2 甲烷化

顾　委[1,2]　张　毅[1,2]　杨华明[1,2]　欧阳静[1,2*]

（1. 中南大学资源加工与生物工程学院，长沙 410083；

2. 矿物材料及其应用湖南省重点实验室，长沙 410083）

“能源是工业生产的血脉”，进入 20 世纪以来，气候变化也从一个之前无人问津的话题演变成为人们讨论未来时的核心问题[1]。全球对于能源的研究最开始主要集中于太阳能、风能、氢能的研发、利用。但随着全球气候变暖的出现，人们发现以变废为宝为思考点是开采新能源更加行之有效的方法。CO_2一方面是主要的温室气体，但同时也是一种极为丰富的自然资源，取之不尽，用之不竭[2,3]。Sabatier 于 1902 年第一次提出 CO_2 甲烷化技术，被认为是目前 CO_2循环再利用过程中非常实用有效的技术之一。

目前 CO_2甲烷化技术的不断改进包括对反应器旳改进、反应条件的优化以及反应机理的探索，目的是指导催化剂的研发，发挥催化剂的优势，进一步提高 CO_2甲烷化在工业上的应用，配合全球 CO_2减排。因此，CO_2甲烷化研究的重中之重仍然是研发出在较低温度下活性与选择性更高、稳定性出色并且廉价易回收的催化剂。本研究重点是在前人已有的研究成果中，针对传统加氢催化剂低温催化性能低、易失活和成本高等缺点，并结合金属的高分散性有利于催化性能提高，以 TiO_2为助催化剂，凹凸棒石作为载体，制备了新型 Ni 基催化剂，并系统地研究了 CO_2甲烷化的催化性能。

凹凸棒石原矿来自江苏盱眙；蜂窝陶瓷购自萍乡市三元公司；药剂钛酸四丁酯和六水硝酸镍购自阿拉丁试剂有限公司，均为分析纯。实验主要仪器：实验室纯水系统型号 Eco－S15Q，购自上海和泰仪器有限公司；管式炉和马弗炉型号分别为 OTF－1200X 和 KSL－1100X，均购自合肥科晶材料技术有限公司；气体的成分及含量通过装备有氢焰检测器和热导检测器的鲁南瑞虹化工仪器有限公司生产的 SP－7890 型高性能气相色谱仪检测。

通过提拉浸渍法将提纯凹凸棒石负载在蜂窝陶瓷的孔道表面，采用化学气相沉积法一步合成分散均匀、晶型稳定的 TiO_2纳米颗粒，最后用浸渍法将不同浓度 Ni（NO_3）$_2$负载在凹凸棒石和 TiO_2体系上，通过煅烧得到催化剂。CO_2加氢甲烷化反应用于催化剂的性能评价在常压连续流动固定床微反应器内进行，通过质量流量计控制气体流量。

各样品形貌分析如图 1 所示。从图 1 可以看出，蜂窝陶瓷具有均匀分布的三角状的多孔结构，表面积较大。MOCVD 负载的 TiO_2小球均匀地分布在凹凸棒石的表面，且表面光滑。进一步负载 NiO 后 TiO_2分子表面可以明显地看到一些分布较为均匀絮状物，表面光滑度下降。

---

* 通信作者：欧阳静；E-mail：lhitxu@126.com；手机号：15116338598。国家自然科学基金（51774331，51304242，51374250）、湖南省自然科学基金（2017JJ0351）资助项目。

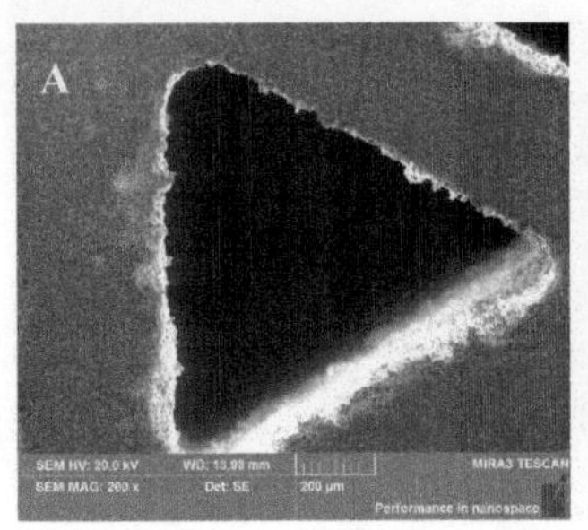

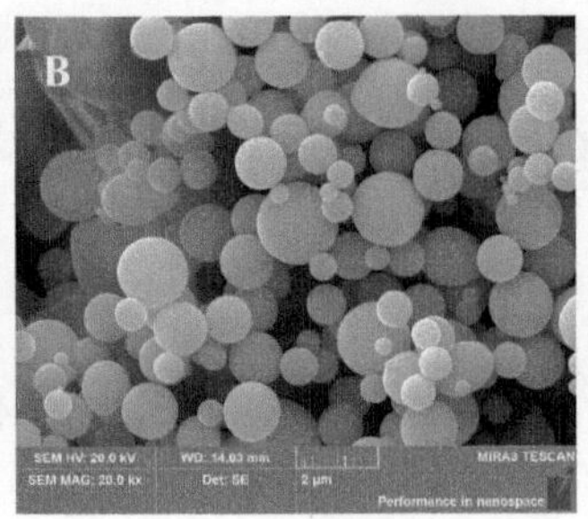

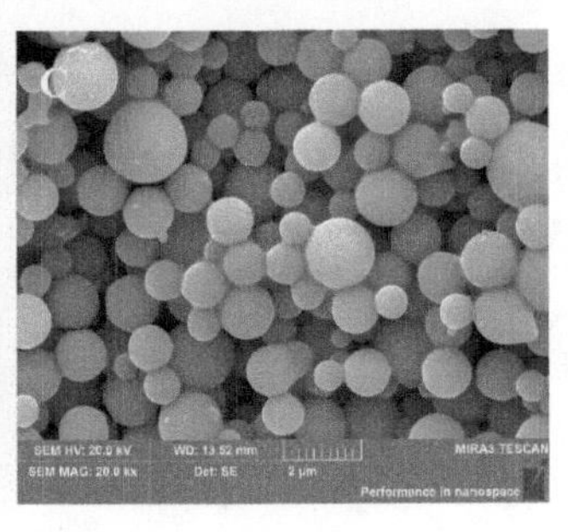

图 1　样品 SEM 图

注：A. 蜂窝陶瓷；B. 负载 TiO_2/ATP；C. 负载 NiO－TiO_2/ATP

催化剂用于 CO_2 甲烷化的催化性能测试，CO_2 转化率相关数据汇总于表 1。从表 1 可以看出，最大的 CO_2 转化率为 93.49%，来自于 Ni（10%）－TiO_2/ATP 催化剂在 400 ℃的反应。

表 1　不同镍负载量的 Ni－TiO_2/ATP 催化剂的 CO_2 转化率（%）

催化剂	实际 Ni 负载量	250 ℃	300 ℃	350 ℃	400 ℃	450 ℃
Ni（5%）－TiO_2/ATP	5.65	26.95	55.75	79.00	86.27	88.71
Ni（10%）－TiO_2/ATP	9.84	29.83	64.01	87.42	93.49	91.73
Ni（15%）－TiO_2/ATP	14.69	27.93	45.69	69.68	82.46	82.81
Ni（20%）－TiO_2/ATP	21.5	31.45	54.93	70.92	79.01	83.20

参考文献

［1］张玉新. 2003. 试谈能源危机和解决的方法［J］. 应用能源技术，4.

［2］Wang J，Huang L，Yang R，et al. 2014. Recent advances in solid sorbents for CO_2 capture and new development trends［J］. Energy Environ. Sci，7：3478－3518.

［3］Dutcher B，Fan M，Russell A G. 2015. Amine-based CO_2 capture technology development from the beginning of 2013：A review［J］. ACS Appl. Mater. Interfaces，7：2137－2148.

熔盐法制备减电荷蒙脱石

何秋芝[1,2]　朱润良[1*]　陈情泽[1,2]　何宏平[1,2]

（1. 中国科学院广州地球化学研究所中科院矿物学与成矿重点实验室/
广东省矿物物理与材料研究开发重点实验室，广州 510640；
2. 中国科学院，北京 100049）

目前，我国黏土矿物矿产资源储量丰富、廉价易得、环境友好。其中，蒙脱石作为一种典型的2∶1型层状硅酸盐黏土矿物，具有较高阳离子交换容量、大比表面积等优点，受到大量研究者的关注。但蒙脱石层间因电荷密度高，层间作用力较强，进而限制了其在环保、材料、医药等领域的应用。而减电荷蒙脱石因其电荷量适中、片层相互作用力较弱，比表面积增大，经改性后，具有优异的吸附性能等优点受到众多研究者的青睐[1]。目前研究中，实现蒙脱石减电荷的方法是利用传统减电荷的方法，即霍夫曼效应[2]，利用具有可交换性的小半径阳离子（例如 Li^{+}、Cu^{2+}、Ni^{2+}），在加热条件下迁移至蒙脱石片层结构，从而得到减电荷蒙脱石。然而该方法过程繁琐，需要耗费大量时间和水资源，所以找到一种快速、简单的方法是十分必要的。目前熔盐制备方法（采用熔融的无机盐作为介质）作为一项对传统液相制备的重要补充方法受到了研究者的广泛关注。熔盐在使用过程中，能为反应物质提供一个稳定、易于控制、非水的液体介质环境，并且使用后可回收利用，不会对环境造成污染。因此，本文以蒙脱石为原料，以含 Li 元素且熔点较低的硝酸锂为熔盐，系统地研究了硝酸锂熔盐对蒙脱石减电荷的效果影响。

实验原料：内蒙赤峰钙基蒙脱石（Mt）。实验试剂：硝酸锂（$LiNO_3$）购自上海阿拉丁生化科技股份有限公司。实验仪器：电热恒温鼓风干燥箱，Bruker D8 Advance X 射线衍射仪。实验方法：称取 0.1 mol $LiNO_3$置于 100 mL 玻璃烧杯中，将烧杯置于 260 ℃烘箱中保温 30 min，待$LiNO_3$完全熔融后，将 1 g 钙基蒙脱石均匀分散于熔融的$LiNO_3$中，最后将混合物置于 260 ℃烘箱中保温不同时间，经洗涤、离心、干燥，即可获得不同电荷的蒙脱石。

图 1 为不同加热时间所得产物的 XRD 图谱。经硝酸锂熔盐调控后，所得产物在～7.2°（2θ）处出现蒙脱石（001）晶面特征衍射峰，对应的 d 值为 1.25 nm，表明经硝酸锂熔盐加热处理后，Li^{+}被成功交换至蒙脱石层间。此外，随着加热时间的延长，产物在～9.1°（2θ）出现了一个新的衍射峰，其对应的 d 值为 0.96 nm，约为单个蒙脱石片层的厚度，表明蒙脱石层间的 Li^{+} 迁移至其片层结构，从而引起了蒙脱石层间域的闭合。并且随着加热时间的增加，该衍射峰的强度逐渐增加，当加热时间为 20 h，蒙脱石层间域完全闭合。

---

* 通信作者：朱润良；E-mail：zhurl@gig. ac. cn；手机号：18666629445。牛顿高级学者基金（NA150190）资助项目。

表1是不同加热时间所得产物的阳离子交换量（CEC）。从表1可以看出，随着加热时间的增加，所得产物的CEC不断减小，表明利用硝酸锂熔盐能够制备不同电荷的蒙脱石。并且加热1 h后所得产物的CEC值只有初始CEC值的59%，表明利用硝酸锂熔盐减电荷的速度是十分快速的。

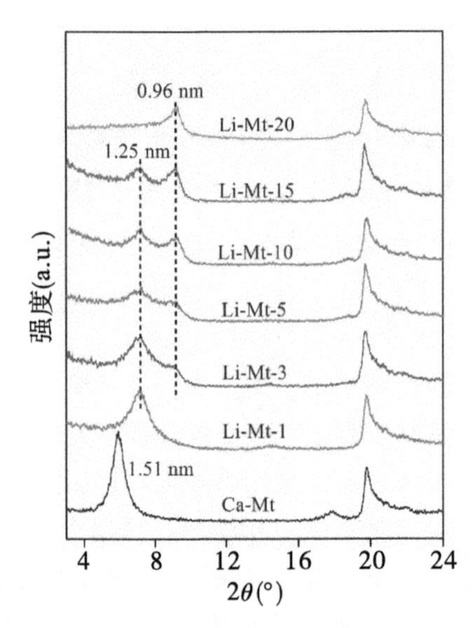

图1　不同加热时间所得产物的XRD图谱

表1　不同加热时间所得产物对应的阳离子交换量

加热时间（h）	CEC（mmol/100g）
0	114
1	68
3	57
5	51
10	34
15	25
20	21

本研究利用硝酸锂熔盐成功制备了不同电荷的蒙脱石，为制备减电荷蒙脱石提供了一个简单快速的方法。

参考文献

[1] Zhu R，Zhao J，Ge F，et al. 2014. Restricting layer collapse enhances the adsorption capacity of reduced-charge organoclays [J]. Applied clay science，88 - 89：73 - 77.

[2] Jana H，Jana M，Peter K. 2001. Effect of heating temperature on Li-fixation，layer charge and properties of fine fractions of bentonites [J]. J. Mater. Chem，11：1452 - 1457.

MgFe 类水滑石及其焙烧产物对磷酸根的吸附机制

刘婷娇 张 盈 张思思 王邵鸿 许 银*
（湘潭大学环境与资源学院环境系，湘潭 411105）

随着社会经济的发展，磷的过度排放造成了水体富营养化等严重的环境问题，给人类的生产、生活带来极大危害，同时也造成巨大的资源浪费[1]。吸附法是利用具有多孔和大比表面积的固体物质吸附水体中的磷，从而达到除磷目的[2]。类水滑石作为一种常见的吸附剂，主要利用类水滑石的比表面积大、层间阴离子的可交换性及类水滑石焙烧后“记忆效应”等作用机制。类水滑石除磷的主要作用机制包括配位络合与离子交换形式的化学吸附及静电引力引发的物理吸附，吸附过程划分为单层吸附、多层吸附等几种简单的类型。类水滑石焙烧前后结构差异显著及可变性大，且吸附性能差异较大，吸附机制尚未明确。

本课题组采用共沉淀法制备 MgFe - LDH 及焙烧后产物 MgFe - CLDH，探究了 MgFe - LDH 和 MgFe - CLDH 对低、中、高浓度磷酸根的吸附机制，并开发了对磷资源回收的方法。

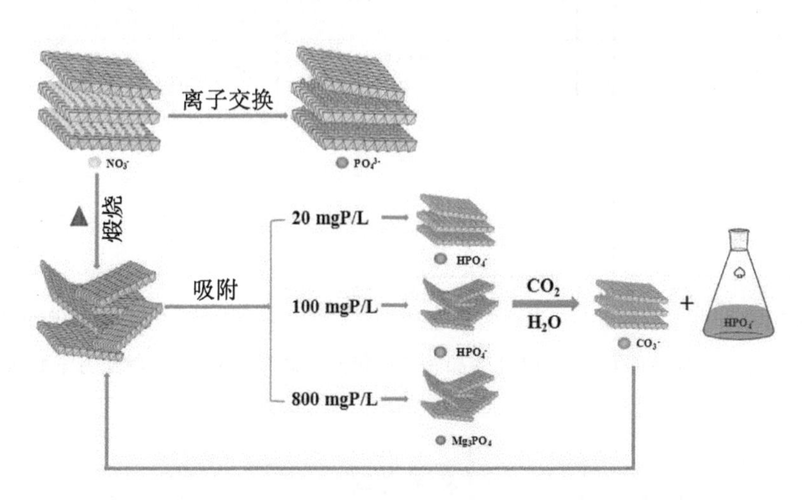

图 1 MgFe - LDH 和 MgFe - CLDH 对 P 的吸附机制示意

MgFe - CLDH 在吸附不同浓度的磷酸根后样品呈现不同的晶体结构（图 1）；MgFe - CLDH 在吸附磷酸根浓度为 0、20 mg/L 后样品都出现类水滑石层状结构的特征衍射峰，这是由于类水滑石本身具有“记忆效应”性质所致，而随着吸附磷酸根浓度的增大，层状结构逐渐消失，且在

---

*通信作者：许银；E-mail：xuyin @ xtu. edu. cn；手机号：18674359286。国家自然科学基金面上项目（51678511）、国家自然科学基金青年项目（51308484）、中国科学院矿物学和成矿学重点实验室开放基金（KLMM20150104）、湘潭大学重大人才计划培育基金项目（16PYZ09）。

磷酸根浓度为 800 mg/L 开始出现 $Mg_3(PO_4)_2$ 沉淀。通过对 MgFe－LDH 和 MgFe－CLDH 的吸附等温线的拟合（图 2）可知：MgFe－LDH 对磷酸根的吸附等温线更符合 Freundich 模型，说明该吸附机制为多层、多种吸附点同时作用的吸附。MgFe－CLD 吸附等温线分为低浓度和高浓度两个阶段；低浓度阶段符合 Langmuir 模型，属于单分子层吸附，高浓度阶段符合 Langmuir－Freundich 模型。分别对 MgFe－LDH 和 MgFe－CLDH 对磷酸根的吸附动力学拟合可得到：MgFe－LDH 和 MgFe－CLDH 对磷酸根的吸附动力学都更符合二级动力学（图 3）。

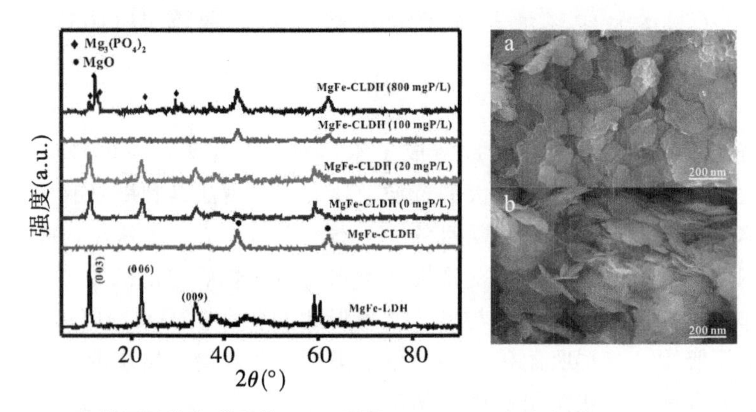

图 2　MgFe－CLDH 吸附不同浓度磷后的 XRD 图及 MgFe－LDH（a）和 MgFe－CLDH（b）的 SEM 图

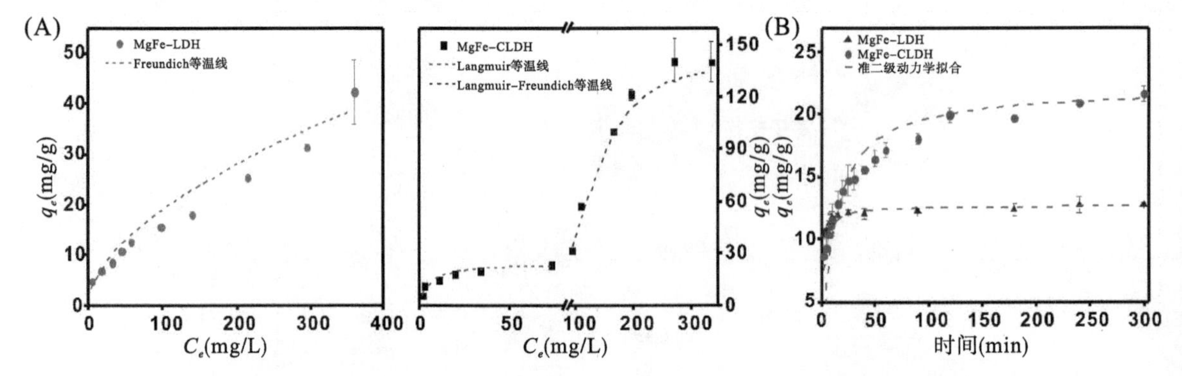

图 3　MgFe－LDH 和 MgFe－CLDH 对磷酸根的吸附等温线（A）及吸附动力学曲线（B）

参考文献

［1］Li R，Wang J J，Zhou B，et al. 2016. Enhancing phosphate adsorption by Mg/Al layered double hydroxide functionalized biochar with different Mg/Al ratios［J］. Science of the Total Environment，559：121－129.

［2］Yu Z，Zhang C，Zheng Z，et al. 2017. Enhancing phosphate adsorption capacity of SDS－based magnetite by surface modification of citric acid［J］. Applied Surface Science，403：413－425.

坡缕石制备硅/硅氧化物多孔材料及其苯吸附性能

朱润良* 陈情泽 朱建喜 何宏平

（中国科学院矿物学与成矿学重点实验室，广东省矿物物理与材料研究开发重点实验室，广州地球化学研究所，广州 510640）

随着我国工业的飞速发展以及城镇化建设的大步迈进，大气环境污染问题日益凸显，尤其是挥发性有机化合物（volatile organic compounds，VOCs）引发的环境问题和人类健康问题，已引起人们的广泛关注。目前，研究者们已经开发了多种 VOCs 处理技术，比如吸附法、膜分离法、燃烧法和生物降解法等。其中，吸附法具有成本低、工艺简单、净化效率高等优势而成为 VOCs 治理中应用最广的方法。吸附法的关键在于选择性能优异的吸附剂。活性炭是当前应用最广的 VOCs 吸附剂，具有大比表面积、高吸附容量等优点，但同时存在热稳定性低、孔结构单一、吸湿性强及回收利用难等不足。另一种常用吸附剂沸石分子筛具有良好的热稳定性和丰富的微孔结构，对 VOCs 分子吸附量高，但其复杂的制备和高昂的成本极大地限制了其工业生产及使用。

黏土矿物具有天然微纳米结构、良好热稳定性、丰富的储量等优点[1,2]，作为一种具有前景的 VOCs 吸附剂前驱体受到了越来越多的关注。直接将原始黏土矿物用于吸附 VOCs 还存在一些问题，比如其表面通常含有丰富羟基，容易吸水，对非极性和弱极性 VOCs 分子的吸附能力有限，而且有限的比表面积和孔结构也限制了最终的吸附容量。进一步提高黏土矿物基 VOCs 吸附剂的吸附性能是当前研究者们努力探索的重要方向。Seliem 等[3]发现使用十六烷基三甲基溴化铵改性的蒙脱石具有较好的疏水性，表现出远优于原始蒙脱石的甲苯吸附性能。但有机黏土矿物的层间域空间和孔隙常被表面活性剂覆盖，难以提供足够的吸附空间和位点，且热稳定性低，因此对 VOCs 吸附效果有限。Wang 等[4]先制备了大比表面积的多孔异构蒙脱石，再通过硫酸处理将模板剂原位转化为碳质涂层，增强材料疏水性，所得样品对甲苯具有较高的静态吸附量（～257 mg/g）。但其制备过程繁琐，且需要使用有机胺等有毒试剂，不利于大规模制备和使用。因此，基于黏土矿物开发吸附性能好、稳定性高且成本低的 VOCs 吸附剂仍然十分重要。

本研究以典型链层状黏土矿物坡缕石为前驱体，采用熔盐助镁热还原法，制备了兼具大比表面积和疏水表面的硅/硅氧化物多孔材料。将所得多孔材料用作 VOCs 吸附剂，以典型 VOCs 分子苯作为探针分子，通过静态和动态苯吸附实验，探讨硅/硅氧化物多孔材料对苯的吸附性能及吸附机理，并结合原始坡缕石和以坡缕石为前驱体制备的无定型二氧化硅及硅纳米颗粒对比样，进一步证实硅/硅氧化物多孔材料在 VOCs 吸附领域的优势。

实验结果表明，硅/硅氧化物多孔材料兼具无定型二氧化硅和硅纳米晶的特征，呈现表面粗糙的纳米颗粒形貌，具有大比表面积（307 m^2/g）、大总孔容（0.95 cm^3/g）以及多级孔结构（即

*通信作者：朱润良；E-mail：zhurl@gig. ac. cn；手机号：18666629445。国家自然科学基金（41572031，21177104，41322014）、广东省“科技创新青年拔尖人才”计划项目（2014TQ01Z249）资助项目。

微孔、介孔和大孔)；对苯分子表现出较高的静态和动态吸附容量（分别为 585.7 和 316.2 mg/g)、较快的扩散/传质速率以及良好的抗水性。其良好的苯吸附性能归因于以下几点：样品中的硅纳米晶增强了多孔性特征，防止无定型二氧化硅的团聚，增大比表面积，有利吸附更多的苯分子；多级孔结构提高了苯分子在孔道内的扩散和传质性能；疏水性较强的硅纳米晶表面增强了样品的抗水性，使样品在一定湿度条件下仍保持较高的苯吸附容量，对实际环境中的 VOCs 吸附具有重要意义。

参考文献

［1］Zhu R L，Chen Q Z，Zhou Q，et al. 2016. Adsorbents based on montmorillonite for contaminant removal from water：A review［J］. Appl. Clay Sci.，123：239－258.

［2］Chen Q Z，Zhu R L，Ma L Y，et al. 2017. Influence of interlayer species on the thermal characteristics of montmorillonite［J］. Appl. Clay Sci.，135：129－135.

［3］Seliem M K，Komarneni S，Cho Y，et al. 2011. Organosilicas and organo-clay minerals as sorbents for toluene［J］. Appl. Clay Sci.，52（1)：184－189.

［4］Wang Y B，Su X L，Xu Z，et al. 2016. Preparation of surface-functionalized porous clay heterostructures via carbonization of soft-template and their adsorption performance for toluene［J］. Appl. Surf. Sci.，363：113－121.

微米级黄铁矿氧化行为的差异性

杜润香[1,2]　鲜海洋[1,2]　魏景明[1]　朱建喜[1*]　何宏平[1]

（1. 中国科学院矿物学与成矿学重点实验室/广东省矿物物理与材料研究开发重点实验室，广州 510640；
2. 中国科学院大学，北京 100049）

黄铁矿是一种常见的硫化物矿物。当黄铁矿被暴露到大气环境中时将发生氧化。黄铁矿的氧化是地球 Fe、S 循环的重要分支，也是形成酸性矿山废水的主要原因[1,2]。目前，黄铁矿氧化的研究主要以具有各种高能面的粉末为对象，探究其氧化机理和氧化速率[3,4]。然而，自然界中的黄铁矿大多以低能面的形式稳定存在，当以具有高表面活性的粉末作为研究对象时，研究结果可能与自然界中矿物实际的表面反应性存在偏差。

本研究将天然五角十二面体黄铁矿和立方体黄铁矿在手套箱研磨成微米级粉末，并结合合成的微米级 {100} 单形晶，考察了它们在湿度分别为 47%、77%、98%的空气中的氧化行为。利用 X 射线光电子能谱（XPS）检测不同时间黄铁矿表面氧化产物与未氧化产物的相对含量。结果表明，黄铁矿表面上氧化物种和未被氧化的物种之间的比值（$C_{oxidized}/C_{unoxidized}$）与氧化时间（$t$）呈现明显的正相关关系。以斜率 k 来表征黄铁矿的平均氧化速度，对比不同湿度条件下黄铁矿的氧化差异（表 1），进一步证明当以黄铁矿粉末为研究对象时，可能会过高估算自然界黄铁矿的氧化速率，而不能精确地反映其在自然环境中的真实地球化学行为。

前期的研究结果表明[5]，黄铁矿 {100} 和 {210} 晶面在不同湿度空气中氧化行为存在明显差异性。然而，将天然五角十二面体和立方体黄铁矿研磨后，它们在空气中的氧化行为差异不再显著。这说明研磨后的矿物颗粒难以准确反映研磨前矿物晶面反应性的信息。在湿度为 47%和 98%的空气中，研磨后的立方体黄铁矿粉末的平均氧化速率和合成的粒度更小的 {100} 单形晶

表 1　黄铁矿破碎粉晶（P210 和 P100）与合成单形晶（S100）的平均粒径及不同湿度下用于表征平均氧化速度的斜率 *k*

湿度（%）	P210	P100	S100
47	0.002 1	0.004 0	0.000 6
77	0.004 4	0.004 4	0.002 4
98	0.001 3	0.001 5	0.000 9
平均粒径（μm）	1.91	0.77	0.40

* 通信作者：朱建喜；E-mail：zhujx@gig. ac. cn；手机号：13428862003。国家自然科学基金（41573112）。

的平均氧化速率相差1个数量级。这种氧化行为的差异表明，即使排除粒度因素，研磨后的矿物颗粒反应性也难以完全代替自然界矿物颗粒的反应性。

本研究通过比较研磨后的天然黄铁矿粉末和合成的{100}单形晶的表面反应差异为例，说明了从矿物的晶面角度出发考察其表面反应性，可以更好地诠释自然界中发生在天然矿物表面反应性信息，为地球化学模型的构建奠定更好的理论基础。

参考文献

［1］Gillozano C，Davila A F，Losaadams E，et al. 2017. Quantifying fenton reaction pathways driven by self-generated H_2O_2 on pyrite surfaces［J］. Scientific Reports，7：43703.

［2］Murphy R，Strongin D R. 2009. Surface reactivity of pyrite and related sulfides［J］. Surface Science Reports，64：1－45.

［3］Rimstidt J D，Vaughan D J. 2003. Pyrite oxidation：a state-of-the-art assessment of the reaction mechanism［J］. Geochimica Et Cosmochimica Acta，67：873－880.

［4］Moses C O，Nordstrom D K，Herman J S，et al. 1987. Aqueous pyrite oxidation by dissolved oxygen and by ferric iron［J］. Geochimica Et Cosmochimica Acta，51：1561－1571.

［5］Zhu J X，Xian H Y，Lin X J，et al. 2018. Surface structure-dependent pyrite oxidation in relatively dry and moist air：implications for the reaction mechanism and sulfur evolution［J］. Geochimica Et Cosmochimica Acta，228.

碳化钢渣建筑材料的研究进展

王爱国　何懋灿　孙道胜*　徐海燕　刘开伟　经　验
（安徽建筑大学安徽省先进建筑材料重点实验室，合肥 230022）

钢铁是我国工业发展中的重要建筑材料，而钢渣是炼钢过程中的工业副产品，占粗钢产量的8%～15%。由于钢渣水硬性和火山灰性相对较差，硬度大且存在大量游离氧化钙和方镁石，所以当其用作建筑材料时存在用量小且安定性不良的缺陷[1]。但是有学者研究发现钢渣对二氧化碳表现出高的碳酸化反应性[2]。众所周知，我国在工业发展中不可避免地产生了大量的CO_2，CO_2引起的温室效应也在不断加剧，若能利用CO_2碳化钢渣，不仅能吸收消耗温室气体，并且能减少钢渣堆积，同时碳化后的钢渣制品性能较好。

本文从热力学[3]和动力学[4]两方面综述钢渣碳化机理，着重分析钢渣碳化过程的影响因素，阐明不同碳化条件对碳化钢渣建筑材料强度和体积稳定性的影响，从微观结构与宏观性能上分析碳化制品，并提出了钢渣碳酸化在研究和应用中存在的一些问题。

钢渣碳化的主要影响因素有胶凝材料组成、CO_2浓度和压力、碳化温度和时间、预养护和初始压力、水分及外加剂。对于胶凝材料组成，有研究发现在相同碳化条件下，掺5% CaO+15% MgO组分的钢渣混凝土强度相对较高，随着反应性MgO比例的增加，碳酸盐固化砂浆的抗压强度明显高于大气固化砂浆，这可能是由于Mg^{2+}的存在促进了镁方解石的生长、碳酸盐的凝聚，从而改善了微结构，使抗压强度提高。除此之外，掺加矿物掺合料亦能提高其早期强度及碳化深度；对于CO_2的浓度和压力，有研究发现提高CO_2的浓度和压力均有利于提高碳化程度，这是因为CO_2在钢渣建筑材料中的扩散速率会直接影响制品碳化进程，CO_2浓度和压力值越大，越容易溶解进入钢渣建筑材料孔隙中，碳酸化养护程度也越高，但达到一定压力值后，碳化程度不再增加；对于碳化温度和时间，研究发现提高碳化温度和延长碳化时间均会增加钢渣建筑材料碳化深度，且温度对钢渣建筑材料强度提高的影响更为显著；对于预养护和初始压力，研究发现初始水养护有利于钢渣建筑材料碳化的进行，且一定的初始压力不仅保障了钢渣建筑材料的成型，也对其早期强度做出了一定贡献，但初始压力过大，会影响钢渣建筑材料碳化进程；对于水分及外加剂，研究发现水分对钢渣建筑材料碳化的效果起着重要影响，且液态外加剂的加入也有利于碳化进行，这是因为气—液—固三相反应速度更快。经CO_2养护前后的SEM见图1。经过CO_2养护后，钢渣建筑材料微观结构更致密，界面更紧密，孔隙率更低（图1）。其体积稳定性更好（图2）。

随着社会发展，钢渣等废弃物利用成为人们越来越关注的一个焦点，钢渣碳酸化后的早期强度和体积稳定性都得到提高，然而对于其耐久性性能的影响还需要进一步探索。目前研究与应用

---

*通信作者：孙道胜；E-mail：sundaosheng@163. com；手机号：13805510929。国家自然科学基金（51778003）、高校优秀中青年骨干人才国内外访学研修项目（gxfxZD2016134）、安徽省高等教育人才项目（皖教高［2014］11号文）。

的碳酸化手段大多适用于实验室条件或者半成型制品的加工，亟待应用于实际工业化阶段。这对于废弃物高效利用、环保、高性能建筑材料等方面，将具有巨大的环保和经济效益。

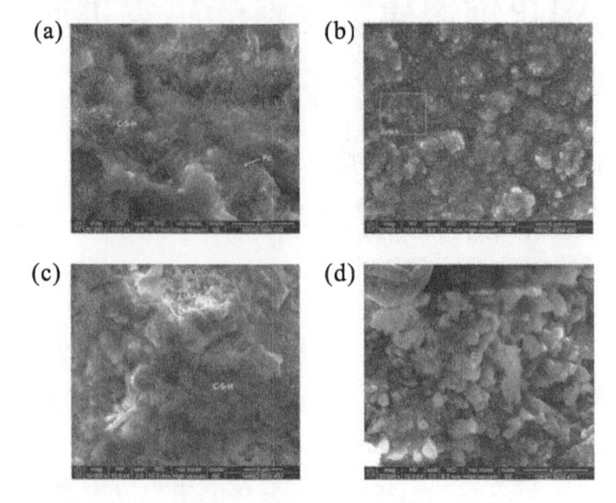

图 1　钢渣浆体碳化前后的 SEM 图[5]

注：a. S 浆体，CO_2 养护前；b. S 浆体，CO_2 养护 14 d 后；c. SC 浆体，CO_2 养护前；d. SC 浆体，CO_2 养护 14 d 后

图 2　混凝土试件先潮湿养护或 CO_2 养护后再在 60 ℃水中养护 60 d 的对比[1]

注：a. SCM－S，先潮湿养护；b. SCM－S，先 CO_2 养护；c. SCLM－S，先潮湿养护；d. SCLM－S，先 CO_2 养护

参考文献

[1] Mo L，Zhang F，Deng M，et al. 2017. Accelerated carbonation and performance of concrete made with steel slag as binding materials and aggregates [J]. Cement & Concrete Composites，83：138－145.

[2] Mahoutian M，Ghouleh Z，Shao Y. 2014. Carbon dioxide activated ladle slag binder [J]. Construction & Building Materials，66（1）：214－221.

[3] 常钧，吴昊泽. 2010. 钢渣碳化机理研究 [J]. 硅酸盐学报，38（7）：1185－1190.

[4] Wu J C S，Sheen J D，Shyanyeh Chen A，et al. 2001. Feasibility of CO_2 fixation via artificial rock weathering [J]. Industrial & Engineering Chemistry Research，40（18）：3902－3905.

[5] Mo L，Zhang F，Deng M. 2016. Mechanical performance and microstructure of the calcium carbonate binders produced by carbonating steel slag paste under CO_2 curing [J]. Cement & Concrete Research，88：217－226.

膨胀石墨制备方法及应用研究

张晚佳[1,2]　高志勇[1,2*]
（1. 中南大学资源加工与生物工程学院，长沙 410083；
2. 战略含钙矿物资源清洁高效利用湖南省重点实验室，
中南大学，长沙 410083）

石墨是碳质元素结晶矿物，其工艺特性主要决定于它的结晶形态。根据结晶形态不同，天然石墨分为三类，即块状石墨、鳞片石墨和隐晶质石墨。其中鳞片石墨的性能最优越，工业价值最大。膨胀石墨（EG）是天然鳞片石墨经强酸和强氧化剂的插层处理、高温膨化得到的一种疏松多孔的蠕虫状物质[1,2]。膨胀石墨同时也沿袭了天然鳞片石墨的性能，具有极强的电导率、耐高温、抗腐蚀、抗辐射特性。与天然鳞片石墨相比，膨胀石墨的结构松散、多孔且弯曲、密度降低，导致体积和表面积扩大、表面能提高。膨胀石墨松散、多孔、弯曲的结构特点决定了其具有极强的抗震性、抗扭曲性、耐压性、吸附性。

由于膨胀石墨的独特性质，国内外的科研工作者尝试了多种手段制备膨胀石墨，主要为Hummers法[1]或改进的Hummers法[3]，具体为强氧化化学法、电化学法[4]等，但后续制备过程仍需要高温处理。电化学法仍然是目前最成熟的方法，包括强酸和弱酸电解质两种方法，具有耗酸量小、酸液循环使用率高、层间插入物均匀稳定等优点，但也存在污染大、能耗高等缺点。比较温和的制备方法，如高压釜法、微波法[5]、超声氧化法和室温一步法[6]等仍处于起步阶段。其中高压釜法和微波法制备的膨胀石墨质量高、结构完整，随着进一步对制备参数和方法的改进，有望实现膨胀石墨的高效、低能耗和低污染化的大规模生产。

膨胀石墨独特的结构与性质决定了其应用的广泛性，如图1所示。膨胀石墨具有良好的可压缩性、回弹性、自黏结性、低密度等优异性能，被广泛用于密封领域。膨胀石墨结构以大孔为主，对油脂类有机大分子吸附性能优越且化学稳定性好，可用于水污染处理。膨胀石墨具有抗高温性，而且电导率高、导热性好，兼具高表面活性及非极性表面，被用于高能电池领域。膨胀石墨在高温下可以快速膨胀，隔绝热能辐射，并促进基体炭化，被广泛应用于阻燃领域。由于独特的层状结构、较大的比表面积和较低的密度等特点，膨胀石墨被应用于医学、储能储电、化工等多个领域。需要特别指出，热膨胀石墨（TEG）还被用于特殊领域，如核工业领域[7]，用于核工业生产的放射性废油的固化，提高固化的水泥化合物的力学性能。

---

*通信作者：高志勇，副教授，博士生导师，E-mail：zhiyong. gao@csu. edu. cn；手机号：13574892751。国家自然科学基金（51774328，51404300）、中国科协青年人才托举工程项目（2017QNRC001）、湖南省自然科学基金（2018JJ2520）项目资助。

目前膨胀石墨制备和应用取得了很大的进展，但膨胀石墨的制备方法和改性手段仍需继续探索和改进。探索更加高效、低耗、环保的常温常压制备方法和改性手段，从而实现膨胀石墨的高效、环保、低能耗规模化工业生产，进一步拓宽其应用范围和领域，将是未来膨胀石墨领域的主要研究方向。

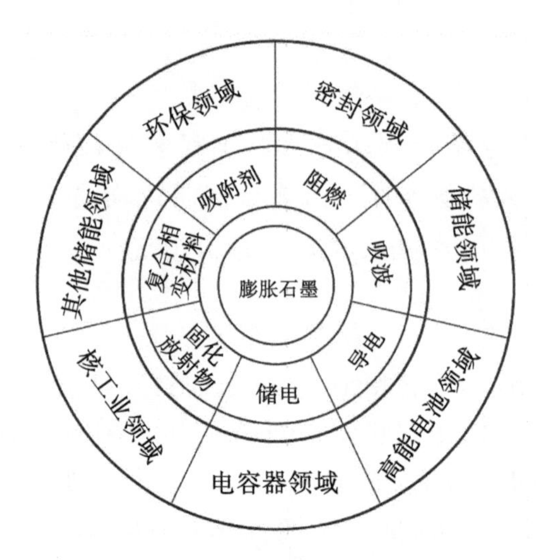

图 1　膨胀石墨的特点及应用总结

参考文献

[1] Falcao E H L, Blair R G, Mack J J, et al. 2007. Microwave exfoliationof a graphite intercalation compound [J]. Carbon, 45: 1364 - 1369.

[2] Nuno G, Luis C C, Vitor A, et al. 2017. Insights into the physical properties of biobased polyurethane/expanded graphite composite foams [J]. Composites Science and Technology, 138: 24 - 31.

[3] Liu T, Zhang R J, Zhang X S, et al. 2017. One-step room-temperature preparation of expanded graphite [J]. Carbon, 119: 544 - 547.

[4] 于仁光，乔小晶，刘伟华. 2003. 影响电化学法制备的膨胀石墨的膨胀容积因素研究 [J]. 精细石油化工进展，10：8 - 10.

[5] 赵正平. 2002. 混酸系（HNO_3- H_3PO_4）制备无硫可膨胀石墨 [J]. 非金属矿，25：26 - 28.

[6] 蒋述兴，李光桥. 2013. 二次插层制备可膨胀石墨的实验研究 [J]. 非金属矿，36：13 - 30.

[7] Tyupina E A, Sazonov A B, Sergeecheva Y V, et al. 2016. Application of thermally expanded graphite for the cementation of cesium-and tritium-containing waste oils [J]. Materials of Power Engineering and Radiation-Resistant Materials, 7: 196 - 203.

黏土矿物吸附磷酸根和Cd（Ⅱ）

杨奕煊[1,2]　朱润良[1*]　傅浩洋[1,2]　何宏平[1,2]

（1. 中国科学院广州地球化学研究所中科院矿物学与成矿重点实验室/
广东省矿物物理与材料研究开发重点实验室，广州 510640；
2. 中国科学院大学，北京 100049）

黏土矿物是天然的纳米材料和地质吸附剂，具有成本低廉、储量丰富和环境友好的优点，在重金属污染控制领域有良好应用前景，相关研究工作得到了广泛关注[1]。不同黏土矿物由于结构不同，对重金属离子的吸附性能和机制存在差异；另一方面，环境中的共存阴离子对黏土矿物吸附阳离子的影响显著，其机制同样需要探明。因此，有必要研究黏土矿物表面结构和共存离子对重金属离子吸附行为的影响。

蒙脱石和海泡石是两种典型黏土矿物的代表。蒙脱石为层状 2∶1 结构，结构天然带负电荷，并将由层间域中的 K^+、Na^+、Ca^{2+} 等阳离子来平衡，从而具有阳离子交换功能，并作为其主要的阳离子吸附机理。海泡石为层链状 2∶1 型结构，结构天然带负电荷，具有一定的离子交换性；但其表面具有丰富的－OH 基团，推测对重金属离子的吸附有较大贡献，但具体吸附机制不够明确。有必要对比研究这两种结构不同、表面基团差异较大的黏土矿物对阴、阳离子的吸附行为。磷酸根是一种环境中广泛存在的含氧阴离子，对重金属离子的吸附存在重要影响，例如 pH＝5 时磷酸根能促进水中 Cd 在水铁矿表面 Fe－OH 上的吸附[2]。因此，本文选用磷酸根和 Cd（Ⅱ）分别作为含氧阴离子和重金属离子的代表，研究了蒙脱石、海泡石两种不同结构的典型黏土矿物对磷酸根和 Cd（Ⅱ）单一及共同吸附的特点。

实验中使用的 KOH、HNO_3、KH_2PO_4、$Cd(NO_3)_2 \cdot 4H_2O$ 等分析纯试剂均购自上海化学试剂厂。钙基蒙脱石产自内蒙古自治区，海泡石购自 Sigma－Aldrich 试剂网。本实验涉及两种不同的吸附体系：黏土矿物对磷酸根或 Cd（Ⅱ）的单一吸附、对磷酸根和 Cd（Ⅱ）的同时吸附。单一吸附体系中，磷酸根和 Cd（Ⅱ）的初始浓度均为 0.05～1.0 mmol/L。共同吸附体系中，将磷酸根浓度固定，其两组次实验浓度分别 0.1 和 0.6 mmol/L，Cd（Ⅱ）浓度选为 0.05～1.0 mmol/L。所有实验吸附剂加入量为 2.5 g/L，控制吸附 pH 值为 5。

磷酸根对蒙脱石和海泡石吸附 Cd（Ⅱ）的影响分别如图 1a 和图 1b 所示。结果表明，磷酸根抑制了 Cd（Ⅱ）在蒙脱石上的吸附，促进了 Cd（Ⅱ）在海泡石上的吸附；随着磷酸根浓度上升，抑制和促进作用都得到加强。单一吸附 Cd（Ⅱ）体系中，蒙脱石对 Cd（Ⅱ）的饱和吸附量为 0.159 mmol/L，高于海泡石的 0.106 mmol/L；但随着磷酸根浓度提升至 0.6 mmol/L，蒙脱石的饱和吸附量降至 0.112 mmol/L，低于此时海泡石的 0.131 mmol/L。这是由于该 pH 值下 Cd^{2+} 与 PO_4^{3-} 可

* 通信作者：朱润良；E-mail：zhurl@gig. ac. cn；手机号：18666629445。国家重点研发计划（2016YFD0800704）、牛顿高级学者基金（NA150190）资助项目。

能形成[Cd（HPO_4)$_2$]$^{2-}$、[Cd（HPO_4)$_3$]$^{4-}$等阴离子复合物，不利于蒙脱石的离子交换；但磷酸根能与海泡石表面大量的－OH基团发生配体交换，带来更多吸附位点。另外，在较低Cd平衡浓度（小于0.1 mmol/L）时，海泡石对Cd（Ⅱ）的吸附量高于蒙脱石，这是由于此时Cd（Ⅱ）与海泡石表面－OH的配体交换吸附比蒙脱石的离子交换吸附更为强烈；随着Cd平衡浓度增大，海泡石表面的－OH位点基本被完全占据，但蒙脱石依旧能提供吸附位点。

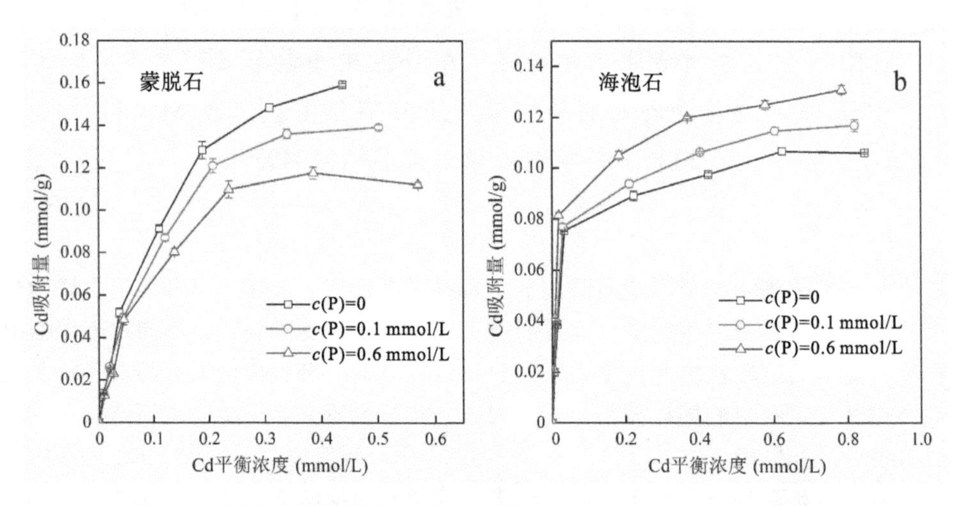

图1　不同浓度P时蒙脱石（a）和海泡石（b）吸附Cd（Ⅱ）等温线

本研究探究了蒙脱石和海泡石对磷酸根和Cd（Ⅱ）的单一、共同吸附行为，为黏土矿物在重金属治理以及在地球化学过程中对阴、阳离子的共吸附行为提供理论参考。

参考文献

[1] Sen Gupta S，Bhattacharyya K G. 2016. Adsorption of metal ions by clays and inorganic solids [J]. RSC Advance，4：28537－28586.

[2] Liu J，Zhu R L，Liang X L，et al. 2018. Synergistic adsorption of Cd（Ⅱ）with sulfate/phosphate on ferrihydrite：An in situ ATR－FTIR/2D－COS study [J]. Chemical Geology，477：12－21.

水化水泥同时吸附磷酸根和 Cd（Ⅱ）

傅浩洋[1,2]　朱润良[1*]　陈情泽[1,2]　何宏平[1,2]

（1. 中国科学院广州地球化学研究所中科院矿物学与成矿重点实验室/
广东省矿物物理与材料研究开发重点实验室，广州 510640；
2. 中国科学院大学，北京 100049）

人类活动作为地球上一项巨大的地质营力，时刻改变着周边的环境系统，如影响元素的地球化学行为等。而水泥作为人类活动中生产和使用最多的工程材料，其自 1930 年到 2013 年以来利用量已达到约 762 亿吨[1]。可以说，地球都市人口是生活在一个由混凝土为主所构建的人造环境中。此外，重金属离子和含氧阴离子在人类环境中普遍存在，因此有必要探究混凝土对这些元素的迁移、转化、归趋等地球化学行为。

水泥是一种粉状水硬性无机胶凝材料，遇水会发生水化反应，并生成以水化硅酸钙（C－S－H）、Ca（$OH)_2$（CH）、钙矾石（AFt）为主的水化产物。其中 C－S－H 凝胶具有低结晶度、较高比表面等特点，可通过吸附、共生和层间位置的化学置换等方式固化外来离子。CH 提供了大量的 OH^-，可促使重金属产生界面沉淀。同时钙矾石也可通过化学置换在晶体柱间和通道内容纳许多外来离子。先前有报道称，溶解态的阳离子会以 M（$H_2O)_n$的形式结合于水泥基材料的表面或者孔隙间，这也使重金属元素替换 M（$H_2O)_n$中的阳离子成为可能[1]。对此，本文选用磷酸根和 Cd（Ⅱ）作为含氧阴离子和重金属离子的代表，研究了水化水泥对磷酸根和 Cd（Ⅱ）单一和共同吸附的特点。

实验中使用的 NaOH、HNO_3、NaH_2PO_4、Cd（$NO_3)_2$·$4H_2O$ 等分析纯试剂均购自上海化学试剂厂。普通硅酸水泥与水以 2∶1 的比例混合并老化 28 d 制得水化水泥，并以此作为后续实验吸附材料。本实验涉及两种不同的吸附体系：单一污染物在水化水泥上的吸附；两种污染物在水化水泥上的同时吸附体系。在单一吸附体系中，磷酸根和 Cd（Ⅱ）的初始浓度分别选为 0.2～2.5 mg/L和 5～30 mg/L。在共同吸附体系中，两种污染物的初始浓度设置与单一体系一致。磷酸根和 Cd（Ⅱ）作为共存离子的浓度皆为 5、15 和 25 mg/L，并控制 Cd（Ⅱ）或磷酸根为 25 mg/L，所有实验吸附剂加入量为 0.4 g/L，控制吸附 pH 值为 5。

Cd（Ⅱ）对水化水泥吸附磷酸根的影响如图 1a 所示。结果表明，在没有 Cd（Ⅱ）存在的条件下，水化水泥也能有效地吸附磷酸根，且吸附量高达 7.9 mg/g 以上。这是由于水化水泥中含有大量与磷酸根具有较高的亲和能力钙元素。随着 Cd（Ⅱ）的加入（5、15 和 25 mg/L），水化水泥吸附磷酸根的能力增强，且随着 Cd（Ⅱ）浓度的增大水化水泥吸附磷酸根的吸附量增大。当 Cd（Ⅱ）添加量为 25 mg/L 时，水化水泥对磷酸根吸附能力达到 16 mg/g，远高于单独吸附磷酸根的

---

* 通信作者：朱润良；E-mail：zhurl@gig. ac. cn；手机号：18666629445。国家重点研发计划（2016YFD0800704）、牛顿高级学者基金（NA150190）资助项目。

量（10 mg/g）。另一方面，磷酸根也能够促进水化水泥对 Cd（Ⅱ）的吸附，如图 1b 所示。在没有磷酸根存在的情况下，水化水泥吸附 Cd（Ⅱ）的效果较差。然而，磷酸根能明显增强水化水泥对 Cd（Ⅱ）的吸附效果。当磷酸根的浓度为 15 mg/L 时，水化水泥对 Cd（Ⅱ）的吸附量显著增大，当磷酸根的浓度增大到 25 mg/L 时，水化水泥对 Cd（Ⅱ）的最大吸附量（12 mg/g）约为对单一的 Cd（Ⅱ）的最大吸附量（1.8 mg/g）的 6.7 倍。综上可知，Cd（Ⅱ）和磷酸根在水化水泥上存在协同吸附的效应。这种协同作用可能是由于磷酸根和 Cd（Ⅱ）在水化水泥表面形成了三元配合物或是由于它们之间的静电作用和表面沉淀导致的[2]。

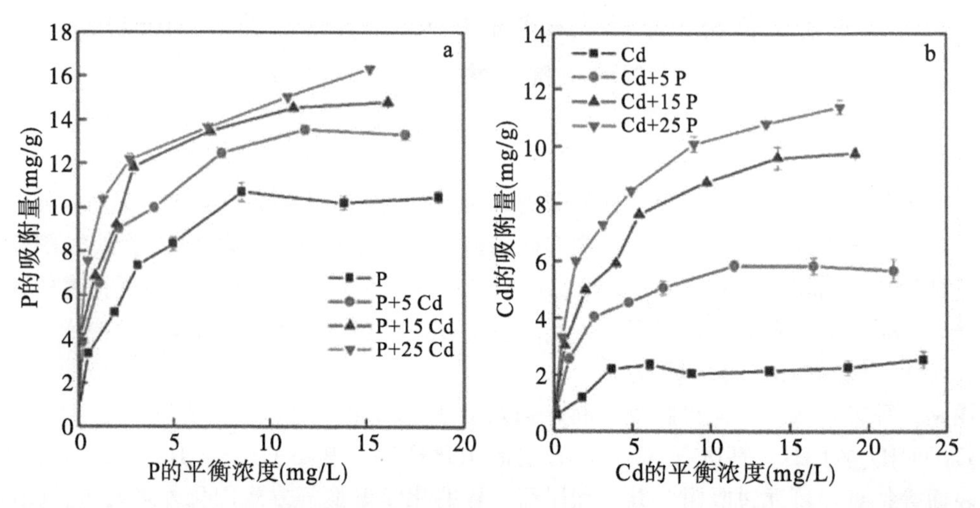

图 1　不同浓度的 Cd（Ⅱ）对水化水泥吸附磷酸根的影响（a）和不同浓度 P 对水化水泥吸附 Cd（Ⅱ）的影响（b）

本研究探究了水化水泥在重金属离子和含氧阴离子单一及同时存在条件下的吸附行为，可为水化水泥对环境中普遍存在的阴阳离子的地球化学过程提供重要的理论指导。

参考文献

[1] Kundu S, Kavalakatt S S, Pal A, et al. 2004. Removal of arsenic using hardened paste of Portland cement: batch adsorption and column study [J]. Water Research, 38: 3780 - 3790.

[2] Liu J, Zhu R, Xu T, et al. 2016. Co-adsorption of phosphate and zinc (ii) on the surface of ferrihydrite [J]. Chemosphere, 144: 1148 - 1155.

聚乙烯亚胺改性蒙脱石高效吸附富勒醇

陈情泽　朱润良*　朱建喜　何宏平
（中国科学院矿物学与成矿学重点实验室，广东省矿物物理与材料研究开发重点实验室，广州地球化学研究所，广州 510640）

富勒烯及其衍生物作为一类重要的碳纳米材料，在环境修复、生物医药和材料科学等领域具有重要的应用。当前，随着富勒烯基碳纳米材料的扩大生产及应用，它们将不可避免地进入环境中，势必影响自然环境和人体健康。虽然富勒烯具有高度疏水性，但是水溶性富勒烯［如水溶性富勒烯聚集体和富勒醇（PHF）］均能通过多种方法（如超声溶解、溶剂交换等）方便制备。水溶性的增强拓展了富勒烯的应用范围，但同时也增加其对人体健康和环境的危害。已有研究表明，水溶性富勒烯对细菌和人体细胞具有生物毒性[1]。因此，关于水体中水溶性富勒烯高效去除的前瞻性研究一方面有利于控制其潜在风险，另一方面对于碳纳米材料的可持续发展具有重要意义。其中，PHF 作为富勒烯在环境中的主要产物，其污染控制尤其需要被重视。

蒙脱石具有天然二维纳米结构、高阳离子交换容量和大比表面积，作为一类廉价高效、环境友好的吸附剂，在环境修复领域具有广阔的应用前景[2]。然而，蒙脱石片层的负电性导致蒙脱石对阴离子污染物的亲和力较差。研究表明，通过阳离子改性剂修饰蒙脱石能够制备得到多种功能吸附剂，如阳离子表面活性剂改性蒙脱石（吸附有机污染物）和聚合羟基金属离子改性蒙脱石(吸附含氧酸根)，大大扩展其应用范围。其中，阳离子聚合物具有较高的荷质比，容易饱和蒙脱石的阳离子交换量，改变蒙脱石表面电性，从而可以为阴离子污染物提供吸附位点。考虑到 PHF 在水体中通常表现为负电性，阳离子聚合物改性蒙脱石也许能作为 PHF 的高效吸附剂。由此，本研究选择典型阳离子聚合物聚乙烯亚胺（PEI）作为改性剂来修饰蒙脱石，考察 PEI 改性蒙脱石对水体中 PHF 的吸附行为，并探讨可能的吸附机制[3]。图 1 为实验流程示意图。

吸附实验结果表明，PEI 改性蒙脱石在较广 pH 值范围内均对 PHF 具有较好的吸附能力。Zeta 电位和红外光谱分析结果表明，这主要是由静电引力和氢键吸附共同作用导致。此外，PEI 改性蒙脱石对 PHF 的吸附量随着 PEI 负载量的增加而增大，随溶液 pH 值的减小而增大（图 2)。PHF 在所有吸附剂上的吸附等温线数据均能较好符合 Langmuir 模型，PEI 改性蒙脱石对 PHF 的最大吸附量高达 210 mg/g，远超过未改性蒙脱石（15 mg/g)。XRD 和 TEM 结果显示，PHF 主要吸附在改性蒙脱石外表面。综上所述，阳离子聚合物 PEI 改性蒙脱石对 PHF 具有较好的亲和力，能够作为水体中 PHF 的高效吸附剂。

* 通信作者：朱润良；E-mail：zhurl@gig. ac. cn；手机号：18666629445。国家自然科学基金（41572031，21177104，41322014）、广东省“科技创新青年拔尖人才”计划项目（2014TQ01Z249）资助项目。

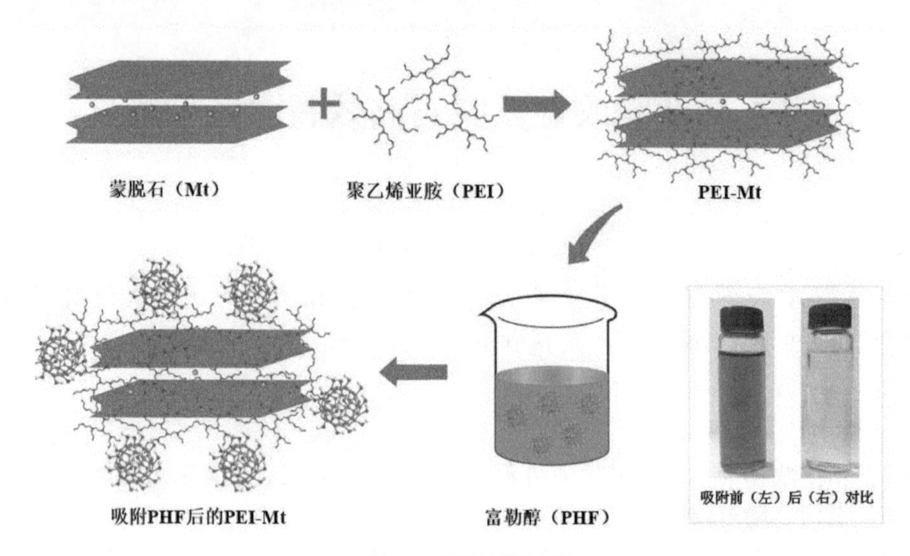

图1 实验流程示意

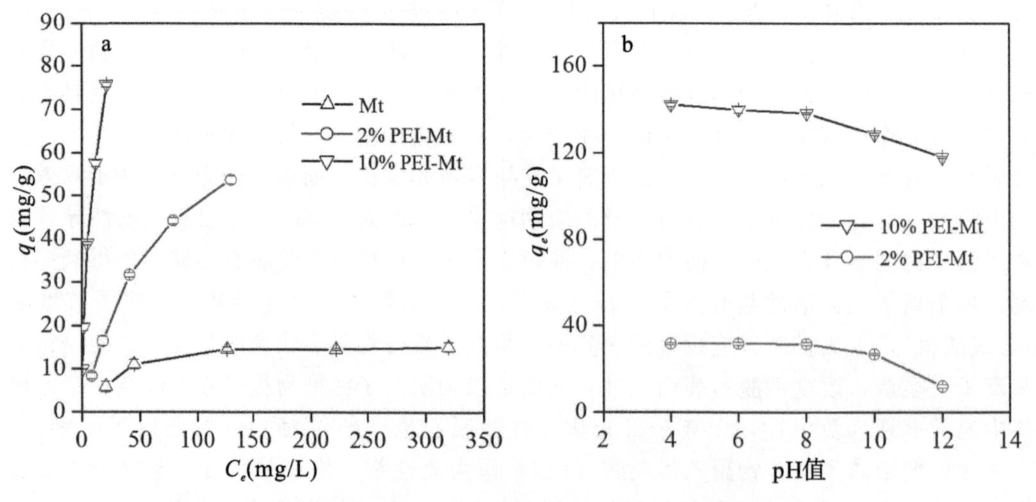

图2 未改性蒙脱石和PEI改性蒙脱石吸附PHF的吸附等温线（a）及PEI改性蒙脱石在不同pH值吸附PHF的吸附量（b）

参考文献

[1] Wielgus A R，Zhao B Z，Chignell C F，et al. 2010. Phototoxicity and cytotoxicity of fullerol in human retinal pigment epithelial cells [J]. Toxicol. Appl. Pharm.，242：79－90.

[2] Zhu R L，Chen Q Z，Zhou Q，et al. 2016. Adsorbents based on montmorillonite for contaminant removal from water：A review [J]. Appl. Clay Sci.，123：239－258.

[3] Chen Q Z，Zhu R L，Zhu Y P，et al. 2016b. Adsorption of polyhydroxy fullerene on polyethylenimine-modified montmorillonite [J]. Appl. Clay Sci.，132：412－418.

环境矿物材料在地下水修复中的应用

张思思　张　盈　刘婷娇　王邵鸿　许　银*
（湘潭大学环境与资源学院环境系，湘潭 411105）

随着工农业的发展和城市化进程的加快，由于城市生活垃圾和工业“三废”等的不合理处置，农业生产中农药、化肥的大量使用，污染物经地表水和土壤迁移进入地下水，造成全国地下水污染状况日趋加重[1]。其中药物及个人护理品（PPCPs）是一种新兴的污染物，其具有微量高毒的特点，对人类健康造成了极大的威胁。目前对 PPCPs 的降解主要有化学氧化、活性炭吸附、膜处理等方法。其中双酚 A 是一种典型的 PPCPs，使用环境矿物复合材料吸附地下水中双酚 A，该方法高效绿色廉价[2]。渗透性反应墙（PRB）是在自然水力梯度下，地下水污染羽渗流通过反应介质，污染物与介质发生物理、化学或生物作用得到阻截或去除，这个过程中材料的渗透性能尤为重要。本课题将类水滑石、海泡石、活性炭 3 种材料通过交联作用制备成凝胶小球作为吸附介质，该材料渗水性好，不会造成装置的堵塞，且持续运行时间长。

将水滑石、海泡石、活性炭、海藻酸钠混合均匀后在氯化钙溶液中交联成球，如图 1 所示。由其 XRD 谱图（图 2）可知，在 $2\theta=7.29°$、$9.34°$ 处出现海泡石的特征峰，在 $11.35°$处出现了水滑石的 003 特征峰，在 $28.54°$时出现活性炭的特征峰，在凝胶小球中，成功地将各材料结合在一起。

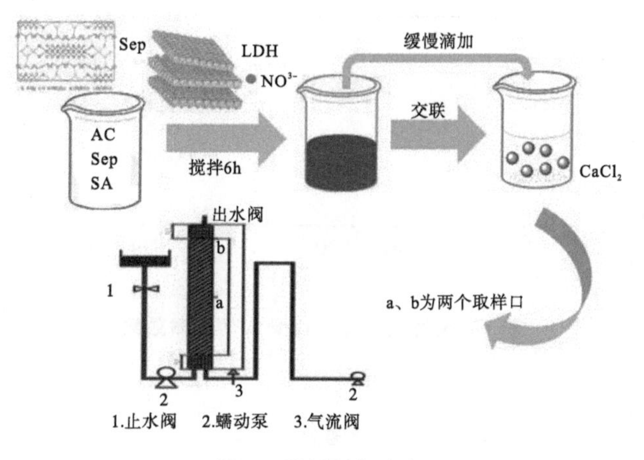

图 1　材料制备流程

*通信作者：许银；E-mail：xuyin@xtu. edu. cn；手机号：18674359286。国家自然科学基金青年项目（51308484）、国家自然科学基金面上项目（51678511）、湘潭大学重大人才计划培育基金项目（16PYZ09）、中国科学院矿物学和成矿学重点实验室开放基金（KLMM20150104）。

对吸附柱进行 9 d 的检测，结果见图 3。双酚 A 经蠕动泵从有机玻璃柱底进入处理，通水后约 58 min 顶部出水口即开始出水，出水速率稳定，在出水实测中，出水速率为 1.048 mL/min，反应进行 9 d（处理水量约为 13.6 L）时，吸附柱的处理效率为 44%。

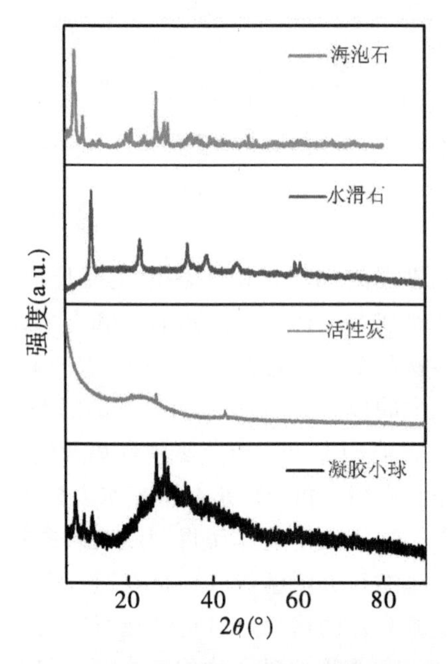

图 2　凝胶小球及其制备材料的 XRD 图

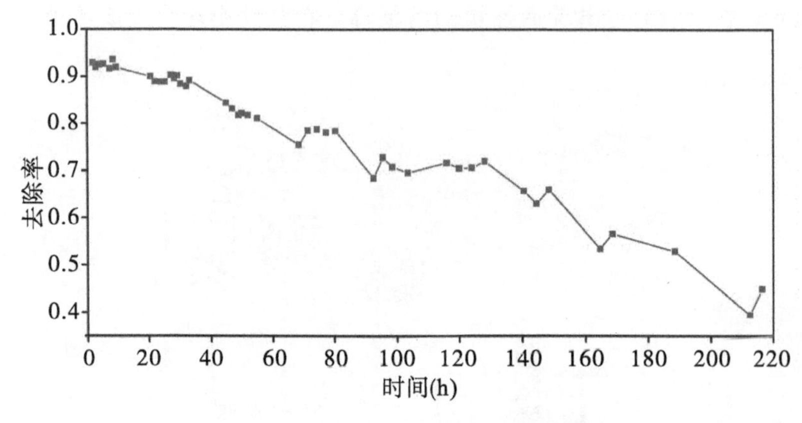

图 3　吸附柱实验对双酚 A 的吸附效果

参考文献

［1］Tesoriero A J，Terziotti S，Abrams D B. 2015. Predicting redox conditions in groundwater at a regional scale［J］. Environmental Science & Technology，49（16）：9657－9664.

［2］Huang G X，Wang C Y，Yang C W，et al. 2017. Degradation of bisphenol A by peroxymonosulfate catalytically activated with $Mn_{1.8}Fe_{1.2}O_4$ nanospheres：Synergism between Mn and Fe［J］. Environmental Science & Technology，51（21）.

CuMgFe 类水滑石构建及湿式催化氧化硝基苯的性能

王劭鸿 刘婷娇 张思思 许 银*

（湘潭大学环境与资源学院环境系，湘潭 411105）

硝基苯作为重要的化学原料，广泛应用于国防、印染、塑料、农药和医药等行业[1]。目前，吸附、超声降解、光降解、高级氧化过程等水处理方法已经用于硝基苯废水的去除。其中，高级氧化过程中的常温常压湿式空气催化氧化作为一种新型技术，由于其处理效率高、低成本，具有广阔的应用前景[2,3]。本课题组前期已经研发多种基于类水滑石的钼基催化剂，能有效在常温常压下湿式空气催化氧化染料废水。基于以上的研究背景，本课题通过对类水滑石的结构调控，研发在常温常压下对硝基苯废水去除的催化剂。

本课题组采用共沉淀法制备 CuMgFe－LDH、CuFe－LDH 和 MgFe－LDH，研究其常温常压湿式空气催化氧化降解硝基苯的性能。

如图 1 所示，在常温常压下，曝气量为 3.0 L/min 时，CuMgFe－LDH 比 MgFe－LDH 和 CuFe－LDH 在 60 min 前有着更快的去除能力，且在 60 min 的去除率分别是 58.37％、45.78％、25.11％。

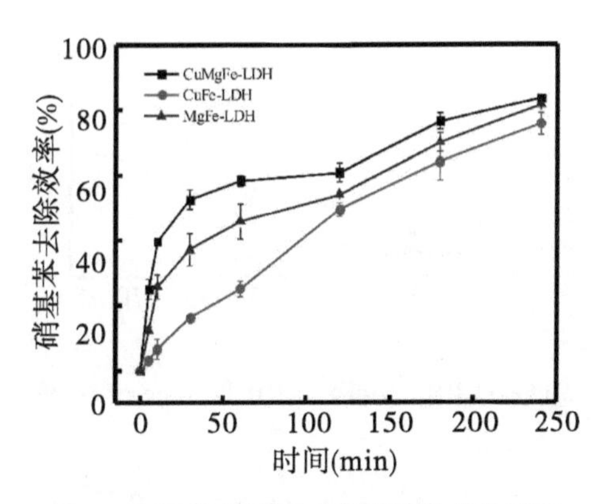

图 1 不同类水滑石对硝基苯的去除效率

通过对 CuMgFe－LDH 在有无曝气条件下比较分析（图 2）可看出：在常温常压下，曝气量为3.0 L/min 时，未曝气条件下，CuMgFe－LDH 在 240 min 对硝基苯的去除率为 29.88％，而曝气条件下，CuMgFe－LDH 在 240 min 对硝基苯的去除率为 83.18％，其去除效率将近是未曝气条件

*通信作者：许银；E-mail：xuyin @ xtu. edu. cn；手机号：18674359286。国家自然科学基金面上项目（51678511）、国家自然科学基金青年项目（51308484）、中国科学院矿物学和成矿学重点实验室开放基金（KLMM20150104）、湘潭大学重大人才计划培育基金项目（16PYZ09）。

下的 2.8 倍。同时在 60 min 前，曝气条件下的 CuMgFe－LDH 比未曝气条件下的 CuMgFe－LDH 有着明显的更大的去除速率，故 CuMgFe－LDH 对去除硝基苯有着良好的催化性能。

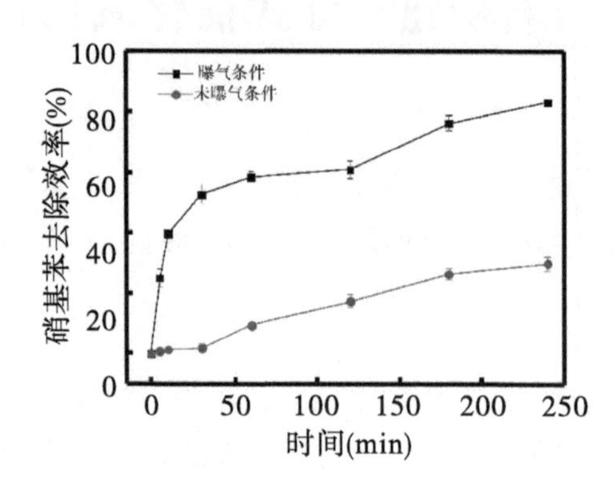

图 2　CuMgFe－LDH 在曝气与未曝气条件下对硝基苯的去除效率

从 XRD 分析（图 3）可看出：CuMgFe－LDH、MgFe－LDH 和 CuFe－LDH 均有明显的（003）、（006）、（009）、（110）等特征衍射峰的存在，说明其晶体生长良好。CuMgFe－LDH 对比 MgFe－LDH 可以看出，CuMgFe－LDH 衍射峰位置向右偏移但偏移程度极小，说明其层间距相差无几，其 d 值为 7.79。而 CuMgFe－LDH 对比 CuFe－LDH 可以看出，CuFe－LDH 衍射峰位置明显右移，说明层间距变小，其 d 值为 6.93。

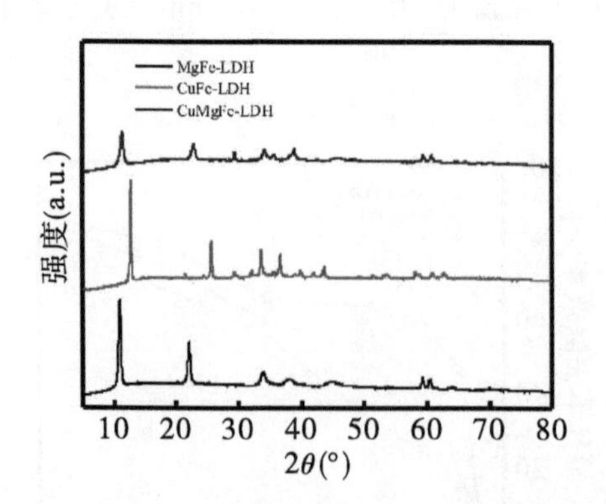

图 3　CuMgFe－LDH、MgFe－LDH 和 CuFe－LDH 的 XRD 图

参考文献

［1］Fu H Y，Zhu D Q. 2016. Graphene oxide-facilitated reduction of nitrobenzene in sulfide-containing aqueous solutions［J］. Environmental Science. & Technology，47：4204－4210.

［2］Li G，Wang B D，Xu W Q，et al. 2018. Enhanced nitrobenzene adsorption in aqueous solution by surface-silanized fly-ash-derived SBA－15［J］. ACS Earth Space Chemistry，2：246－255.

［3］Chi Z，Lei J，Ding L，et al. 2018. Mechanism on emulsified vegetable oil stimulating nitrobenzene degradation coupled with dissimilatory iron reduction in aquifer media［J］. Bioresource Technology，260：38－42.

氧化亚铜/蒙脱石催化甘油脱水氧化制丙烯酸

付超鹏[1]　张　浩[1]　吴书涛[1]　周春晖[1,2*]
（1. 浙江工业大学化学工程学院，杭州 310014；
2. 青阳非金属矿研究院，青阳 242800）

蒙脱石（montmorillonite，MMT）是一种 2∶1 的天然黏土矿物，理论化学式为（$Mg_y \cdot nH_2O$）（Al_2-yMg_y）Si_4O_{10}（OH）$_2$，即单位晶胞的蒙脱石是由两层硅氧四面体和一层铝氧八面体构成，具有阳离子交换性；且 MMT 本身兼具 B 酸位和 L 酸位，所以常将 MMT 作为脱水催化剂和催化剂载体。铜是一种过渡金属元素，具有 0、＋1、＋2 等多种氧化态，单质和氧化物均具有良好的氧化还原活性，所以铜基材料是很好的氧化还原催化剂，且对于甘油的转化具有很好的催化作用[1]。丙三醇，俗称甘油，是最简单的三元醇，在自然界中以甘油三酯的形式存在于动物和植物体内。在工业上，皂化制皂和生产柴油均会副产甘油。目前工业上生产生物柴油常用的方法是酯交换法（图 1），使用该方法生产生物柴油，每生产 1 吨生物柴油约副产 0.1 吨甘油[2]，这也成为甘油的主要来源。能否在较温和的条件下将甘油转化为高附加值的产品，是当前研究的热点[3]。

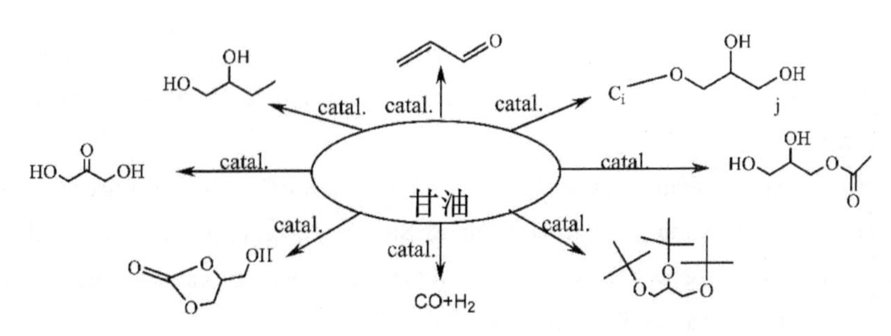

图 1　甘油衍生物的合成途径[4]

本研究用酸化的蒙脱石（MMT－H）和醋酸铜在水热条件下反应，生成 Cu_2O/MMT－H 催化剂。用甘油作为反应物，Cu_2O/MMT－H 作为脱水氧化催化剂，H_2O_2 作为氧化剂，在水浴条件下，将甘油经脱水氧化制备了丙烯酸。

醋酸铜在高温条件下会分解，生成 Cu_2O，利用水热法，将生成的 Cu_2O 负载到 MMT－H 上，成功制备了兼具酸性和氧化还原活性位的 Cu_2O/MMT－H 催化剂。从 XRD 图（图 2）可看出，该方法不仅制备出了 Cu_2O，且 Cu_2O 结晶性较好。

*通信作者：周春晖，E-mail：clay@ zjut. edu. cn；手机号：13588066098。国家自然科学基金（21373185，41672033）资助项目。

将该催化剂用于催化甘油脱水氧化的反应中，利用 H_2O_2 做氧化剂，将甘油经脱水氧化一步转化为丙烯酸。

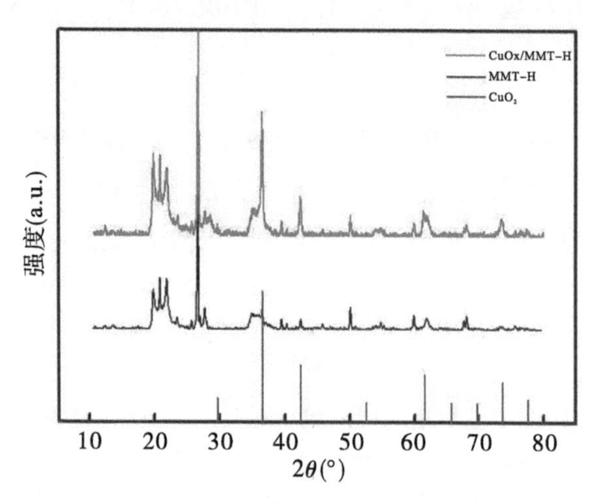

图 2　MMT 表面 Cu 的物相鉴定

本研究通过水热法，利用醋酸铜高温分解的性质，简单高效地制备出了兼具酸性位和氧化还原活性位的 Cu_2O/MMT－H 催化剂；并将该催化剂用于甘油的脱水氧化反应，在水浴条件下，成功地将甘油转化为丙烯酸。

参考文献

[1] Zhou C H，Beltramini J N，Lin C X，et al. 2011. Selective oxidation of biorenewable glycerol with molecular oxygen over Cu-containing layered double hydroxide-based catalysts [J]. Catalysis Science & Technology，1：111－122.

[2] Kong P S，Aroua M K，Wan M A W D. 2016. Conversion of crude and pure glycerol into derivatives：A feasibility evaluation [J]. Renewable & Sustainable Energy Reviews，63：533－555.

[3] Zhou C H，Zhao H，Tong D S，et al. 2013. Recent advances in catalytic conversion of glycerol [J]. Catalysis Reviews，55：369－453.

[4] Zhou C H，Beltramini J N，Fan Y X，et al. 2008. Cheminform abstract：chemoselective catalytic conversion of glycerol as a biorenewable source to valuable commodity chemicals [J]. Chemical Society Reviews，37：527－549.

铜锰氧化物/坡缕石的结构、分散性能及其热催化氧化甲醛的转化机制

刘 鹏[1,2] 梁晓亮[1,2*] 何宏平[1,2] 陈汉林[1,2] 朱建喜[1,2]
（1. 中国科学院广州地球化学研究所，广州 510640；
2. 中国科学院大学，北京 100049）

坡缕石是一种层链状含水镁铝硅酸盐矿物，也是一类在我国安徽、江苏等地储量丰富的非金属矿产资源。坡缕石具有纳米棒束状形貌，比表面积巨大，孔隙结构发达，酸性位点丰富，热和化学稳定性良好。当前坡缕石基初级材料主要用于吸附、填充材料、石油钻井、载体等领域。开发环境催化剂产品是坡缕石高附加值利用的一种重要途径。甲醛作为基础化工原料广泛应用于装饰装修、石化、医药等生产领域，同时是一种常见的大气污染物，容易诱发白血病和呼吸道疾病。热催化氧化是处理甲醛等有机废气的一种高效、绿色、经济的末端控制方法。本研究选用安徽明光的坡缕石原矿作为载体，通过硝酸盐浸渍—焙烧法制备坡缕石负载型铜锰氧化物，应用于高浓度甲醛的热催化氧化处理，并运用现代谱学技术探明铜锰氧化物在坡缕石上物相组成、分散状态及其对甲醛的热催化氧化性能的制约机制。

坡缕石加入不同配比的硝酸铜和硝酸锰溶液中，搅拌分散，滴加氨水至完全沉淀。烘干后，研磨筛分前驱体，并在 450 ℃焙烧 2 h。依据铜和锰的负载量，所制备催化剂命名为 1Cu5Mn、2.5Cu5Mn、5Cu5Mn、10Cu5Mn。催化性能实验在固定床反应器进行，在上海光源 BL14W1 线站进行 XAFS 实验，*in situ* DRIFT 表征在配备有高温反应池的 Bruker Vertex－70 上完成。

图 1A 对比坡缕石载体（PG－400）、单一负载铜或锰氧化物/坡缕石（10Cu0Mn 和 0Cu10Mn）和机械混合铜锰氧化物/坡缕石（mixed 5Cu5Mn）与复合负载铜锰氧化物/坡缕石（5Cu5Mn）的甲醛催化性能。结果表明，5Cu5Mn 甲醛催化性能突出，在 221 ℃（T90）实现 90%的 CO_2生成率。图 1B 反映随着铜含量增加，锰氧化物催化甲醛性能提升，反应抗水性能也提高，H_2O 逐渐呈现促进作用。在 5% H_2O 的条件下，5Cu5Mn 和 10Cu5Mn 的 T90 分别从 201 和 197 ℃降低至 185 和 181 ℃。催化性能结果表明铜锰之间存在强烈的协同作用，促进甲醛向 CO_2转变。

10Cu0Mn 和 0Cu10Mn 的 XRD 谱图中存在相应氧化物的衍射峰（CuO、bixbyite Mn_2O_3 和 pyrolusite MnO_2）。而 1Cu5Mn、2.5Cu5Mn、5Cu5Mn 和 10Cu5Mn 的 XRD 谱图中只有坡缕石的衍射峰，说明复合负载能够提高氧化物分散性能。SEM 图中显示，坡缕石均匀分散铜锰氧化物，负载颗粒粒径范围为 10～50 nm。

* 通信作者：梁晓亮；E-mail：liangxl@ gig. ac. cn；Tel：020-85290075。本研究受到广东省自然科学基金研究团队项目（S2013030014241）、广东省科技计划协同创新与平台环境建设项目（2017A050501048）、广州市珠江科技新星项目（201806010069）的资助。

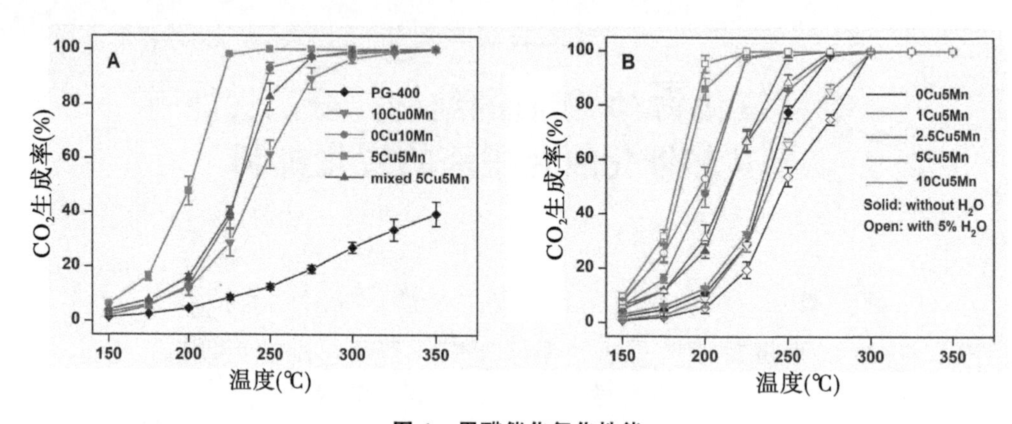

图 1　甲醛催化氧化性能

注：A. 单一与复合负载铜锰氧化物催化性能；B. 水对催化性能的影响

XAFS 谱表明，1Cu5Mn、2.5Cu5Mn、5Cu5Mn 和 10Cu5Mn 中 Cu 以＋2 价为主，位于四面体位，结构占位与 Cu 在 CuO 中不同（畸变八面体位）。Mn 以＋3 和＋4 价为主，位于八面体位，结构占位与 Mn 在 Mn_2O_3和 MnO_2中不同。扩展边 XAFS 发现，坡缕石负载的铜锰氧化物其局域微结构与尖晶石 $CuMn_2O_4$相同。Cu_2MnO_4是一类活性氧化物，具有优异的氧化还原性能，故而能够显著提高催化剂催化氧化甲醛的效果。

In situ DRIFT 分析结果表明，甲醛吸附在催化剂表面，首先转变成为甲酸盐，然后分解成 CO_2和 H_2O。反应路径中有两个速率控制步骤，一是吸附在催化剂表面的甲醛向甲酸盐的转变；二是转变而来的甲酸盐分解为 CO_2和 H_2O。实验结果发现，$CuMn_2O_4$中的含 Cu 催化位点能够加速甲酸盐在催化剂表面的积累，而含 Mn 催化位点具有更快的甲酸盐分解速率（图 2）。

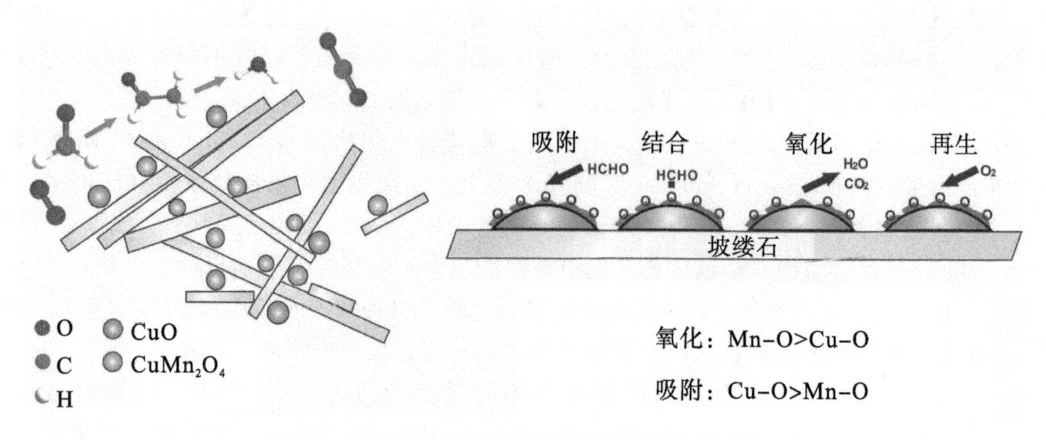

图 2　铜锰氧化物/坡缕石热催化氧化甲醛机制

An Organically Modified Bentonite with Stability to Hypersaline Brines

Andras Fehervari[1]　Usma Shaheen[2]　Will Gates[1*]

Abdelmalek Bouazza[2]　Tony Patti[2]　Terry Turney[2]

(1. Deakin University, Burwood, 3125, Australia;

2. Monash University, Melbourne, 3800, Australia)

Coal seam gas exploration and production presents risks to surface and ground water quality due to handling, intermittent storage and disposal of hypersaline brines which can have severe environmental consequences. Highly saline leachates can also adversely impact on the hydraulic performance of clay-based barrier systems because the high ionic strength (I) causes shrinkage of the swollen smectite micro-fabric and increases the meso and macro-porosity of the bentonite. Thus, improvements in bentonite-based barriers, such as geosynthetic clay liners, are urgently needed enhance their ability to retain hypersaline brines.

Table 1　Interlayer spacing of glycerol carbonate (GC) modified sodium smectite as determined by X - ray diffraction[5]

Form of smectite	d - value (nm)
Na - Smectite	1.29±0.03
Ca - Smectite	1.58±0.03
GC - Na - Smectite	1.89±0.03
washed with I=1 M NaCl	1.86±0.04
washed with I=1.5 M $CaCl_2$	1.61±0.04

Glycerol carbonate is a cyclic organic carbonate that can interact strong with the hydrated interlayer sodium cation in smectite even under conditions of high ionic strength, similar to propylene carbonate[1,2]. Glycerol-carbonate modified sodium bentonite[3,4] was evaluated for its chemical and geotechnical properties and hydraulic performance to hypersaline brines[5,6]. Glycerol carbonate is also useful as a modifying agent, in that it can be functionalised[3] and polymerised[4]. X - ray diffraction (Table 1) and infrared spectroscopy analysis indicated the glycerol carbonate modified smectite (extracted from benton-

* Will P. Gates; E-mail: will. gates@deakin. edu. au; +61 3 9246 8373. The Australian Research Council's Discovery Project Grants program (DP1095129).

ite) was strongly resistant to NaCl brines, but still had some resistance to $CaCl_2$ brines[5].

Glycerol carbonate bentonites had greater swelling and solution retention, as well as lower fluid loss and hydraulic conductivity than unmodified Na - bentonite to as high as I=3 M $CaCl_2$ and I=5 M NaCl leachates[4]. In particular, the saturated hydraulic conductivity of the glycerol carbonate modified Na - bentonite remained $<1.0\times10^{-11}$ m/s (Figure 1). Under the same conditions, the unmodified Na - bentonite had a hydraulic conductivity of $>1.9\times10^{-10}$ m/s. Furthermore, glycerol carbonate remain intercalated when the interlayer cations were only partially hydrated.

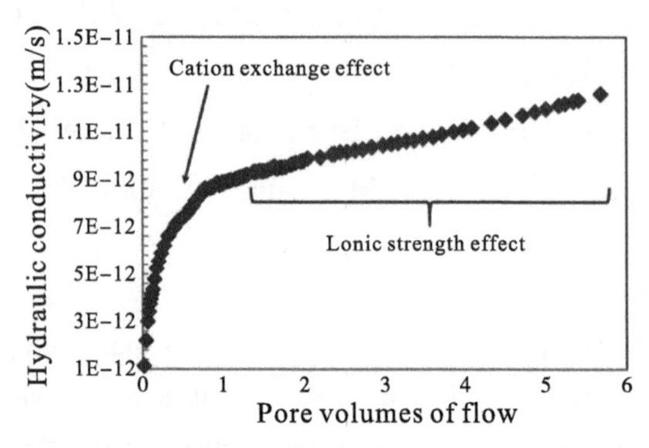

Fig. 1 Hydraulic conductivity of non-prehydrated glycerol carbonate modified sodium bentonite in 1 M (I=3 M) $CaCl_2$[6]

References

[1] Onikata M, Kondo M, Hayashi N, et al. 1999. Complex formation of cation-exchanged montmorillonites with propylene carbonate: osmotic swelling in aqueous electrolyte solutions [J]. Clays and Clay Minerals, 47: 672.

[2] Katsumi T, Ishimori H, Onikata M, et al. 2008. Long-term barrier performance of modified bentonite materials against sodium and calcium permeant solutions [J]. Geotextiles and Geomembranes, 26: 14 - 30.

[3] Gates W P, Shaheen U, Turney T W, et al. 2016. Cyclic carbonate-sodium smectite intercalates [J]. Applied Clay Science, 124 - 125: 94 - 101.

[4] Shaheen U, Turney T W, Saito K, et al. 2016. Pendant cyclic carbonate-polymer/Na-smectite nanocomposites via in situ intercalative polymerization and solution intercalation [J]. Journal of Polymer Science, Part A: Polymer Chemistry, 54: 2421 - 2429.

[5] Fehervari A, Gates W P, Turney, T W, et al. 2016. Cyclic organic carbonate modification of sodium bentonite for enhanced containment of hyper saline leachates [J]. Applied Clay Science, 134: 2 - 12.

[6] Fehervari A, Gates W P, Patti A F, et al. 2016. Potential hydraulic barrier performance of cyclic organic carbonate modified bentonite complexes against hyper-salinity [J]. Geotextiles and Geomembranes, 44: 748 - 760.

Clay Mineralogical Constraints on Weathering in Response to Early Eocene Hyperthermal Events in the Bighorn Basin, Wyoming (Western Interior, USA)

Chao Wen Wang[1,2*] Rieko Adrianes[3] Han Lie Hong[4] Jan Elsen[3]
Noël Vandenberghe[3] Lucas J. Lourens[2] Philip D. Gingerich[5]
Hemmo A. Abels[6*]

(1. Gemmological Institute, China University of Geosciences, Wuhan 430074, P. R. China;
2. Department of Earth Sciences, Utrecht University, Heidelberglaan 2, 3584 CS, Utrecht, Netherlands;
3. Department Earth and Environmental Sciences, KU Leuven, Celestijnenlaan 200E, B - 3001 Leuven, Belgium;
4. State Key Laboratory of Biogeology and Environmental Geology, China University of Geosciences, Wuhan 430074, P. R. China;
5. Department Earth and Environmental Sciences, University of Michigan, Ann Arbor, Michigan 48109, USA;
6. Department Geosciences and Engineering, Delft University of Technology, Stevinweg 1, 2628 CN, Delft, Netherlands)

Series of transient greenhouse warming intervals in the early Eocene provide an opportunity to study the response of rock weathering and erosion to changes in temperature and precipitation. During greenhouse warming, chemical weathering is thought to increase the uptake of carbon from the atmosphere, while physical weathering and erosion control sediment supply[1]. A large ancient greenhouse warming event is the Paleocene-Eocene Thermal Maximum at 56 Ma. In many coastal sites, an increase in the abundance of kaolinite clay during the Paleocene-Eocene Thermal Maximum is interpreted as the result of

* Corresponding author: Chao Wen Wang; E-mail: cwwang_cug@cug. edu. cn; Mobile Phone No.: +86-13297062002. Hemmo A. Abels; E-mail: h. a. abels@tudelft. nl. Wang is grateful for grants from the National Natural Science Youth Foundation of China (grant 41602037), Natural Science Youth Foundation of Hubei (grant 2016CFB183), the Postdoctoral Science Foundation of China (grant 2015M582301), and Fundamental Research Funds for the Central Universities, China University of Geosciences (Wuhan, CUG160848). Abels acknowledges the Netherlands Organization for Scientific Research (NWO) Earth and Life Sciences (ALW) for a VENI grant (863. 11. 006).

reworking from terrestrial strata due to enhanced runoff caused by increased seasonal precipitation and storminess during a time of decreased vegetation cover[2]. In the continental interior of North America, Paleocene-Eocene Thermal Maximum paleosols show more intense pedogenesis and drying, which are indicated by deeply weathered and strongly oxidized soil profiles[3]. The weathering and oxidation could be related to temperature and precipitation changes, but also to increased time available for weathering and increased soil permeability in coarser sediment. Here, we provide evidence for enhanced climate seasonality, increased erosion of proximal laterites and intrabasinal floodplain soils, and a potential slight increase in chemical weathering during the smaller early Eocene hyperthermals (Eocene Thermal Maximum 2 and H2) postdating the Paleocene-Eocene Thermal Maximum, for which no previous clay mineral data were available. Hyperthermal soil formation at the site of floodplain deposition causes a similar, insignificant clay mineralogical change as occurred during the background climates of the early Eocene by showing small increases in smectite and decreases in illite-smectite and illite. Remarkably, the detrital sediments during the hyperthermals show a similar pedogenic-like increase of smectite and decreases of mixed-layer illite-smectite and illite, while the kaolinite and chlorite proportions remained low and unchanged. Since sedimentation rates and provenance were similar during the events, enhanced smectite neoformation during soil formation in more proximal settings, and associated reworking, is the likely process causing this clay mineralogical change. The hundreds to thousands of year time scales at which individual paleosols were formed were probably too short for significant alteration of the rocks by in situ chemical weathering despite changing climates during the two post-Paleocene-Eocene Thermal Maximum greenhouse warming episodes. The relatively small signal, however, raises the question of whether increased chemical weathering can indeed be a strong negative feedback mechanism to enhanced greenhouse gas warming over the time scales at which these processes act.

References

[1] Kump L R, Brantley S L, Arthur M A. 2000. Chemical weathering, atmospheric CO_2, and climate [J]. Annual Review of Earth and Planetary Sciences, 28: 611-667.

[2] McInerney F A, Wing S L. 2011. The paleocene-eocene thermal maximum: A perturbation of carbon cycle, climate, and biosphere with implications for the futurc [J]. Annual Review of Earth and Planetary Sciences, 39: 489-516.

[3] Kraus M J, Riggins S. 2007. Transient drying during the Paleocene-Eocene Thermal Maximum (PETM): Analysis of paleosols in the Bighorn Basin, Wyoming [J]. Palaeogeography, Palaeoclimatology, Palaeoecology, 245: 444-461.

Acid-alkali Treated Natural Mordenite Supported Platinum Nanoparticles for Efficient Catalytic Oxidation of Formaldehyde at Room Temperature

Qian Guo　Xiao Ya Gao*
(Faculty of Environmental Science and Engineering, Kunming University of Science and Technology, Kunming 650500, P. R. China)

Natural zeolite is a valuable natural resource. Compared with synthetic zeolite, natural zeolite is rich in resources and low in cost. The cost of extracting 'industrial material' from natural zeolite is only 1%—5% of synthetic zeolite. If natural zeolite can be used as a catalyst carrier to achieve catalytic oxidation of formaldehyde at low temperature, its practical application prospects will be enormous. In the catalytic oxidation of formaldehyde, supported Pt catalysts exhibit excellent catalytic performance and can completely oxidize formaldehyde at room temperature[1]. And it is the most widely used catalyst in the catalytic oxidation of formaldehyde. Wei *et al*. studied the acid-treated TiO_2 nanobelt supported platinum nanoparticles for the catalytic oxidation of formaldehyde, and they found Pt/TiNB-ac exhibits much higher catalytic activity, which can efficiently convert formaldehyde to CO_2 and H_2O at ambient temperature[2]. In contrast, natural zeolites are more valuable to use. However, natural zeolite also needs a certain pretreatment to regulate the silicon-aluminum ratio and specific surface area of the carrier to better exert its catalytic effect on formaldehyde.

In this work, Pt/MORn - H - OH compositecatalyst was prepared by a facile two-step method, namely, acid-alkali treated natural mordenite and the load of platinum (Pt) nanoparticles. The structure and morphology of Pt/MORn - H - OH were characterized by XRD, TEM and XPS. The crystalline phase of the zeolite after acid and alkali treatment did not change significantly and the average particle diameters of Pt particles of 1% Pt/MORn - H - OH were the smallest, 2.8 nm (Fig. 1). Results showed that 1% Pt/MORn - H - OH exhibited the best catalytic activity and oxidized HCHO completely at room temperature (Fig. 2). The enhanced catalytic activity was benefited from the stronger capability of loading Pt nanoparticles.

* Corresponding author: Xiao Ya Gao; Mobile Phone No.: 18187529369.

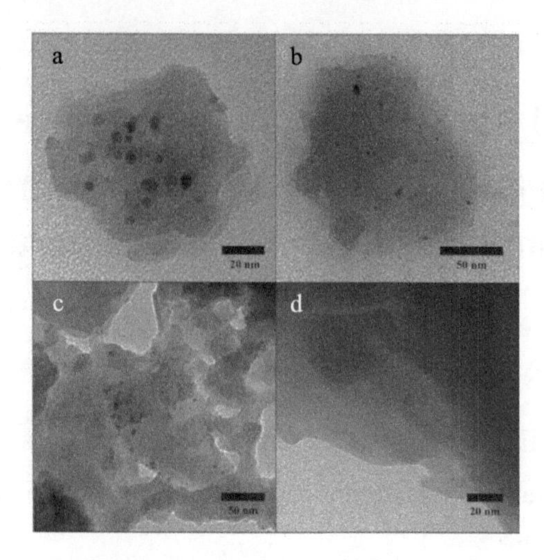

Fig. 1 TEM images of the series of 1% Pt/MORn catalysts

Note: a. 1% Pt/MORn; b. 1% Pt/MORn - H; c. 1% Pt/MORn - OH; d. 1% Pt/MORn - H - OH

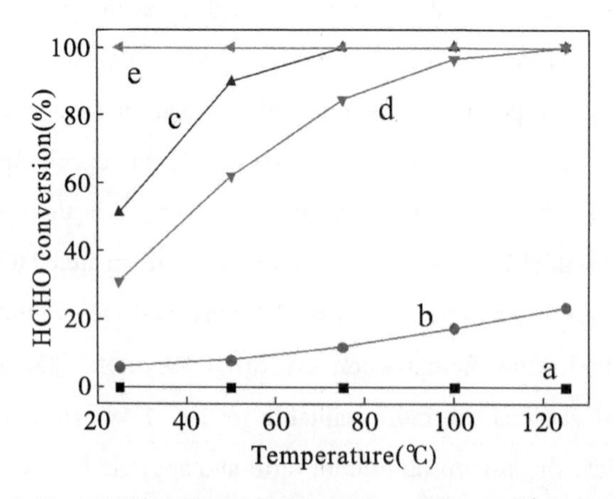

Fig. 2 Catalytic activities of the MORn and the series of 1% Pt/MORn catalysts

Note: a. MORn; b. 1% Pt/MORn; c. 1% Pt/MORn - H; d. 1% Pt/MORn - OH; e. 1% Pt/MORn - H - OH

References

[1] Zhang Y, Zhang M, Cai Z, et al. 2012. A novel electrochemical sensor for formaldehyde based on palladium nanowire arrays electrode in alkaline media [J]. Electrochimica Acta, 68 (5): 172 - 177.

[2] Cui W, Xue D, Yuan X, et al. 2017. Acid-treated TiO_2 nanobelt supported platinum nanoparticles for the catalytic oxidation of formaldehyde at ambient conditions [J]. Applied Surface Science, 411: 105 - 112.

Influence of Acid Leaching of Coal Gangue on the Performance of Cordierite Porous Ceramic

Hai Yan Xu* Jie Li Jing Chen Ai Guo Wang
Guo Tian Wu Dao Sheng Sun
(School of Materials & Chemical Engineering,
Anhui JianZhu University, Hefei 230601, China)

Coal gangue is a complex industrial solid waste discharged when coal is excavated and washed in the production course. The amount of coal gangue accumulated in China has already reached 3.8 billion tons; moreover, the stockpile of gangue is increasing at a rate of 0.2 billion tons per year[1,2]. The disposal of such a large quantity of this solid waste requires a lot of land and has caused many serious environmental problems, especially the atmosphere pollution caused from spontaneous combustion and water pollution from leaching and release of heave metal ions.

In the study, porous cordierite ceramics have been fabricated using 66.2wt.% coal gangue (from Huainan Coal clayey material), 22.7wt.% basic magnesium carbonate and 11.1wt.% bauxite by a polymeric sponge infiltration method[3]. The gangue was pre-treated with the mixed acid (25% hydrochloric acid and 15% sulfuric acid 1 : 1 mixture) at 80 ℃. X-ray diffractometry (XRD, Bruker Advance D8 diffractometer), and electron universal testing machines (RGGER-3010, Shenzhen) have been used to investigate the crystal structure and compressive strength. The porosity has been obtained by the method based on Archimedes' principle.

Porosity and comprehensive strength of the porous sample sintered at 1 150—1 250 ℃ using the coal gangue leached at different volume of acid have been shown in Fig. 1. The porosity decreases with sintering temperature as well as the ratio of acid to gangue at each point of sintering temperature. The compressive strength improves with sintering temperature up to 1 200 ℃ but reduces greatly at 1 250 ℃. However, it increases with the ratio of acid to gangue at every point of sintering temperature. The porosity of the sample sintered at 1 200 ℃ decreases from 83.09% with the ratio of gangue to acid of 4 : 1 to 76.40% with that of 16 : 1, while compressive strength improves from 2.47 MPa to 3.16 MPa.

* Corresponding author: Hai Yan Xu; E-mail: xuhaiyan@ahjzu.edu.cn; Mobile Phone No.: +86-13514971927. Financial support of the research and development project in science and technology of Anhui Province (No. 1301042127) is gratefully acknowledged.

The samples prepared with the gangue leached with acid show higher compressive strength than these prepared with the gangue without acid leaching. The samples sintered at 1 250 ℃ show lower porosity and compressive strength than these sintered at low temperature (1 150—1 200 ℃), which mainly due to the samples distortion and the large number of bubbles on the sample surface.

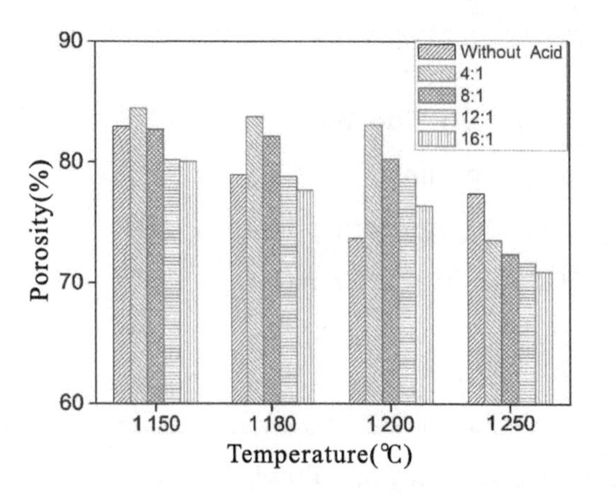

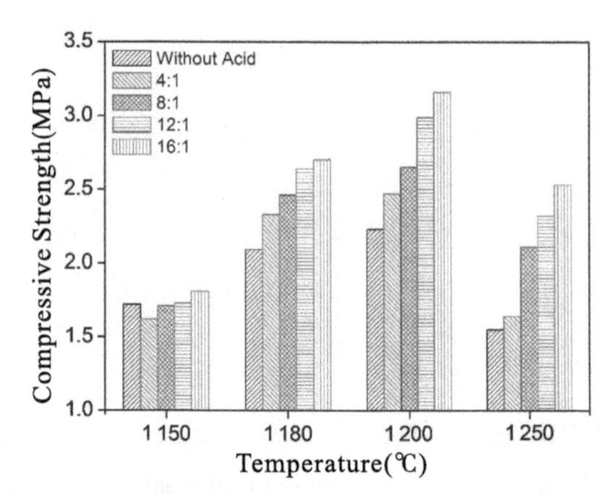

Fig. 1 Porosity and comprehensive strength of the porous sample sintered at 1 150—1 250 ℃ using the coal gangue leached with the varied ratio of gangue to acid

In conclusion, the samples prepared with gangue after acid leaching not only maintain high porosity, and also increase the compressive strength significantly. The cordierite porous ceramics with a porosity of 73.69% prepared with raw coal gangue exhibited average compressive strength of 2.23 MPa. While the ceramics prepared with coal gangue after acid leaching for 16 : 1 showed the increased porosity of 76.40% and the improved compressive strength of 3.16 MPa. Therefore, the acid leaching can control the content of impurities in coal gangue to promote sintering process and result in the improved performance of the obtained porous ceramics.

References

[1] Ji H P, Fang M H, Huang Z H, et al. 2014. Phase transformation of coal gangue by aluminothermic reduction nitridation: Influence of sintering temperature and aluminum content [J]. Applied Clay Science, 101: 94 - 99.

[2] Zhou J E, Dong Y C, Stuart H, et al. 2011. Utilization of sepiolite in the synthesis of porous cordierite ceramics [J]. Applied Clay Science, 52: 328 - 332.

[3] Lü Q, Dong Xinfa Z Z. 2014. Environment-oriented low-cost porous mullite ceramic membrane supports fabricated from coal gangue and bauxite [J]. J Hazard Mater, 273: 136 - 145.

Mechanochemical Activation of Phlogopite for Fixation of Copper and Zinc Ions

Ahmed Said[1]　Qi Wu Zhang[1]　Qu Jun[2]　Yan Chu Liu[1]
(1. School of Resources and Environmental Engineering, Wuhan University of Technology, Wuhan 430070, China;
2. Key Laboratory of Catalysis and Materials Science of the State Ethnic Affairs Commission & Ministry of Education, Hubei Province, South-Central University for Nationalities, Wuhan 430074, China)

Phlogopite was activated by ball milling to be directly used as heavy metal adsorbent with the optimal milling speed at 300 r/min. Before the final breakdown of the crystal structure, phase transformation of trigonal phlogopite into monoclinic by milling at 300 r/min was observed accompanying with obvious dissolution of K^+, Mg^{2+} and OH^-. Theadsorption data confirmed that the heavy was fixated by the surface precipitation and cation exchange mechanism. The proposed craft to produce heavy metal adsorbent was facile and environmental friendly with outstanding heavy metal fixation efficiency which has a great potentiality for practical application.

Phlogopite ($KMg_3[Si_3AlO_{10}](OH, F)_2$) can be widely found all over the world as an important industrial silicate mineral. Excessive comminution of mica is a common phenomenon in the mica processing industry which produces massive waste mica. To recycle the waste mica and remedy the heavy metal pollution in the water, the exploitation of waste mica seems to be an alternative option. Mechanochemistry has been proved to be an efficient way to reuse the useless clays[1,2]. This work the focus was put on the mechanochemical activation of phlogopite to adsorb heavy metal ions from wastewater.

A phlogopite sample from Bazhou, Xinjiang, China was used in this experiment. 2.0 g of phlogopite powder was put in a planetary ball milling (Pulverisset-7, Fritsch, Germany) which was equipped with two pots of zirconia (45 cm^3 of internal volume) with 7 zirconia balls of 15 mm in diameter. The rotational speed was varied from 0 to 600 rpm while the milling time was fixed at 120 minutes. 0.05, 0.1, 0.2 and 0.5 g of the milled sample were stirred in 100 mL 100 ppm Cu^{2+} or Zn^{2+} solution for two hours following by centrifugation to remove the solid. After that, the solution was tested of the remaining concentration of Cu^{2+} or Zn^{2+} by the Atomic Absorption Spectrometry (AA 6880, Shimadzu, Japan) for the calculation of removal efficiency.

Fig. 1 displays the phase transformation of phlogopite with milling speed varied at 0, 300 and 600 r/min. The natural phlogopite without milling was nearly a pure phase without other observable impurities. It is very interesting to note that, before the complete breakdown of the crystal structure, phase transformation of phlogopite from trigonal to monoclinic happened at relatively mild milling condition at 300 r/min. This is different from the commonly observed phenomena with other clay minerals such as kaolinite, serpentine, talc or pyrophyllite, of which gradual amorphization is usually observed without appearances of other intermediate phases.

Fig. 2 shows the Cu^{2+} removal efficiency by phlogopite milled at different speed with different dosage in which the 300 r/min sample gained the optimal results. Combining with the XRD results in Fig. 1 and the K^{+}, Mg^{2+} and OH^{-} dissolution results, it could be concluded that the heavy metal was fixated by the surface precipitation and cation exchange mechanism.

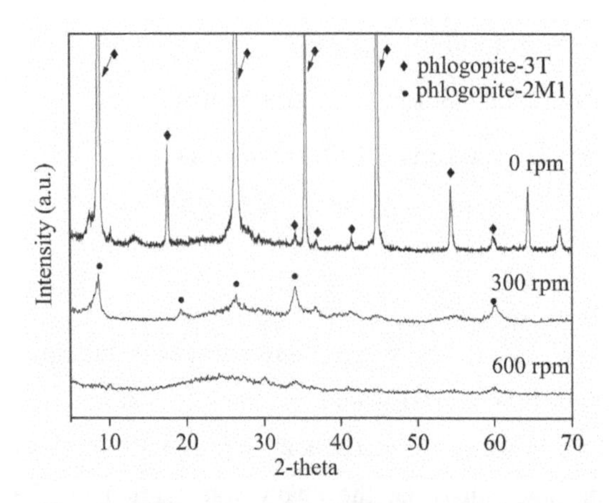

Fig. 1 XRD patterns of phlogopite milled at different speeds

Fig. 2 Cu^{2+} removal efficiency by phlogopite milled at different speed with different dosage

The proposed craft to produce heavy metal adsorbent was facile and environmental friendly without polluted water or gas emission. The prepared products possessed high removal efficiency toward heavy metal ions. Above all, the craft and the products could be widely used for remediation of heavy metal water pollution.

References

[1] Huang P, Li Z, Chen M, et al. 2017. Mechanochemical activation of serpentine for recovering Cu (Ⅱ) from wastewater [J]. Appl. Clay Sci., 149: 1 - 7.

[2] Qu J, Zhang Q, Li X, et al. 2016. Mechanochemical approaches to synthesize layered double hydroxides: a review [J]. Appl. Clay Sci., 119: 185 - 192.

Spinel-type Cobalt-manganese Oxide Catalyst for Degradation of Orange Ⅱ Using a Novel Heterogeneous Photo-chemical Catalysis System

Qing Zhuo Ni　Jian Feng Ma*　Bo Yuan Zhu
(School of Environmental and Safety Engineering, Changzhou University, Jiangsu 213164, China)

Recently, advanced oxidation processes (S－AOPs) based on sulfate radical (SO_4^-) and hydroxyl radical have attracted intensive attention for rapidly oxidizing organic compounds. Similar to the method of Fenton activation to produce ·OH, SO_4^- can be generated using various transition metal ions (Co^{2+}, Cu^{2+}, Fe^{2+}, Mn^{2+} etc.) to catalyze sulfite, bisulfite (BS), persulfate (PS) etc[1,2].

Herein, we have designed and synthesized a spinel-type cobalt-manganese oxide. Firstly, 8 mL ammonia water (25wt%) was dropped into 10 mL 0.2 mol/L Co $(NO_3)_2$ solution. Then, 20 mL of 0.2 mol/L Mn $(NO_3)_2$ solution was added to the mixture followed by stirring for 120 min in air. Afterwards, the mixtures were dried by heating at 180 ℃ for 6 h, yielding the Co-Mn oxide catalyst.

XRD analysis showed that the catalyst had a cubic $MnCo_2O_{4.5}$ spinel structure. The catalyst was formed with nanoparticles (25－65 nm) and high specific surface area about 85.7 m^2/g as confirmed by SEM and BET analyses.

As seen through a series of controlled tests, the degradation rates of Orange Ⅱ varied under different conditions (Fig. 1). With only Co-Mn catalyst, almost 22% of Orange Ⅱ was removed, due to its favorable adsorption ability. For the experiment with Co-Mn catalyst but under visible light irradiation, a significant enhancement of Orange Ⅱ removal (about 56%) was obtained, suggesting that Co-Mn catalyst can be used as photocatalyst. However, for the Co-Mn catalyst/BS system, the degradation dramatically increased with a removal of 87%. Furthermore, a higher rate of about 94% of the degradation was achieved in the Co-Mn catalyst/BS/Vis system. The possible mechanism is illustrated in Eqs. (1) to (10).

$$2HSO_3^- - + O_2 \rightarrow SO_4^- + SO_4^{2-} + 2H^+ \quad (1)$$

$$catalyst + hv \rightarrow e- + h^+ \quad (2)$$

$$e- + O_2 \rightarrow O_2^- \quad (3)$$

$$h+ + H_2O \rightarrow \cdot OH + H^+ \quad (4)$$

$$catalyst + HSO_3^- \rightarrow SO_3^- \quad (5)$$

* Corresponding author: Jian Feng Ma; E-mail: jma@zju. edu. cn; Mobile Phone No.: 814-865-1542.

$$SO_3^- + O_2 \rightarrow SO_5^- \quad (6)$$

$$SO_5^- + HSO_3^- \rightarrow SO_4^- + SO_4^{2-} \quad (7)$$

$$Co(\text{III}) + Mn(\text{II}) \rightarrow Co(\text{II}) + Mn(\text{III}) \quad (8)$$

$$Co(\text{III}) + Mn(\text{III}) \rightarrow Co(\text{II}) + Mn(\text{IV}) \quad (9)$$

$$h^+ / O_2^- / SO_4^- / \cdot OH / Mn(\text{III}) + dyes \rightarrow products \quad (10)$$

Cycling experiments were conducted with the Co-Mn catalyst/BS/Vis system for three cycles. As summarized in Fig. 2, there was surprisingly little difference between the removals using fresh and recycled Co-Mn catalyst.

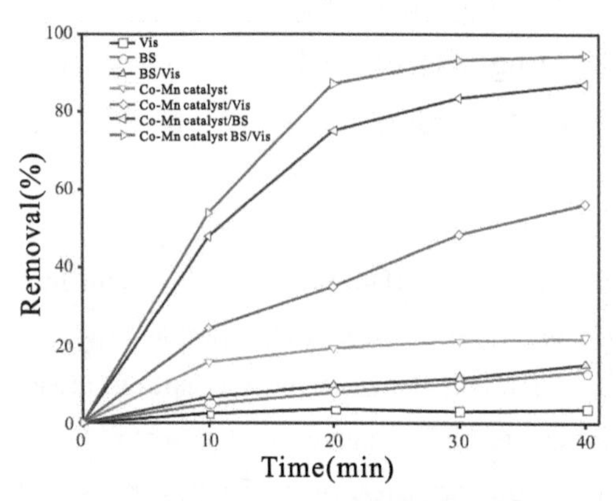

Fig. 1 Orange Ⅱ degradation with different systems

Conditions: Orange Ⅱ 60 mg/L, Co-Mn catalyst 0.6 g/L, $NaHSO_3$ 2 g/L

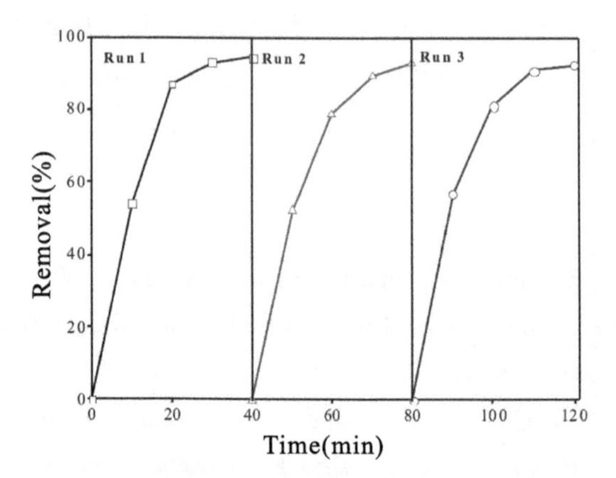

Fig. 2 Cycling runs for the degradation of Orange Ⅱ in Co-Mn catalyst/BS/Vis system

Conditions: Orange Ⅱ 60 mg/L, Co-Mn catalyst 0.6 g/L, $NaHSO_3$ 2 g/L

Co-Mn catalyst exhibited much better Orange Ⅱ degradation capability in the heterogeneous photochemical catalysis system with the assistance of $NaHSO_3$ and supplemented by visible light irradiation than that in either photocatalysis only in visible light or chemocatalysis only in the presence of bisulfite. Results demonstrated that Co-Mn catalyst showed not only excellent catalytic activity but also stability and satisfactory recyclability. Some active species such as Mn (Ⅲ), holes and superoxide and sulfate and hydroxyl radicals were found to be responsible for the outstanding degradation of Orange Ⅱ dye by the Co-Mn catalyst synthesized here.

References

[1] Yang S, Yang X, Shao X, et al. 2011. Activated carbon catalyzed persulfate oxidation of Azo dye acid orange 7 at ambient temperature [J]. Journal of Hazardous Materials, 186: 659－666.

[2] Ranguelova K, Chatterjee S, Ehrenshaft M, et al. 2010. Protein radical formation resulting from eosinophil peroxidase-catalyzed oxidation of sulfite [J]. Journal of Biological Chemistry, 285: 24195.

Using Visible and Near-infrared Reflectance Spectroscopy to Decipher Clay-mineral and Paleoclimate Information of a Loess/paleosol Sequence

Kai Peng Ji[1]　Qian Fang[1,2]　Lu Lu Zhao[1,2]　Han Lie Hong[1*]

（1. School of Earth Sciences，China University of Geosciences，Wuhan 430074，China；

2. Department of Soil，Water and Environmental Science，University of Arizona，AZ 85721，USA）

The properties of clay minerals in soils can respond sensitively to changes in pedogenic weathering intensity and past climate. Increasing efforts have been devoted to visible and near-infrared reflectance （VNIR） spectroscopy to monitor and assess soil attributes in the past few decades[1]. Nevertheless，rare research has been conducted to test whether the clay-mineral sensitive region alone，i. e.，the short-wave infrared（SWIR；1 000—2 500 nm）region，has potential for predicting clay properties and reconstructing palaoclimate. In addition，the spectral parameters extracted from the entire VNIR spectra were suggested to be useful proxies in highly weathered soils[2]，yet their utility in less weathered soils remains unclear. Here we investigated in detail a weakly-moderately weathered soil sequence（i. e. Shangbaichuan-SBC loess/paleosol sequence）in central China. The partial least squares regression（PLSR）approach was used to establish direct quantitative relationships between spectral signals and clay properties.

Good predictive models were established for I，（Kao+Sm）/I，and Kao/I with RPD values exceeding 1.4. The actual validation results for these properties are listed in Fig. 1，where the results are slightly better than the cross-validation results. Compared with the clay minerals，the clay-mineral ratios，i. e.，（Kao+Sm）/I and Kao/I，are better predicted using chemometric models.

The spectral parameters exhibit similar trends with geochemical and mineralogic weathering proxies，revealing four distinct intervals（Units 1—4）within the SBC sequence（Fig. 2）. The D_{2200}/D_{1900} and AS_{2200} parameters are suggested as reliable weathering proxies. Compared with the AS_{1400} parameter，they perform better in correlation with clay mineralogic（$|r| \geq 0.34$，$p<0.001$，$n=197$）and chemical proxies（$r\geq 0.66$，$p<0.05$，$n=18$）. The AS_{1400} and D_{900} parameters are rather variable throughout the profile，thus a seven-point moving average method is preferred in interpreting paleoclimate evolution.

---

* Corresponding author：Han Lie Hong；E-mail：honghl8311@aliyun. com；Mobile Phone No.：+8613627265008. This study was supported by the NSF of China（41772032 and 41472041），and Special Funding for Soil Mineralogy（CUG170106）.

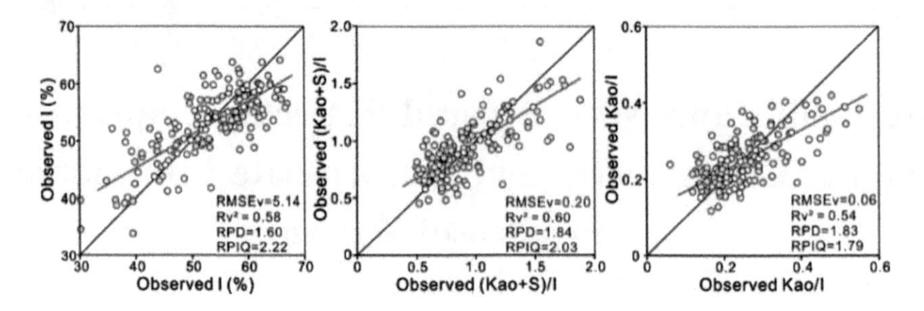

Fig. 1　Actual validation plots for SBC clay minerals

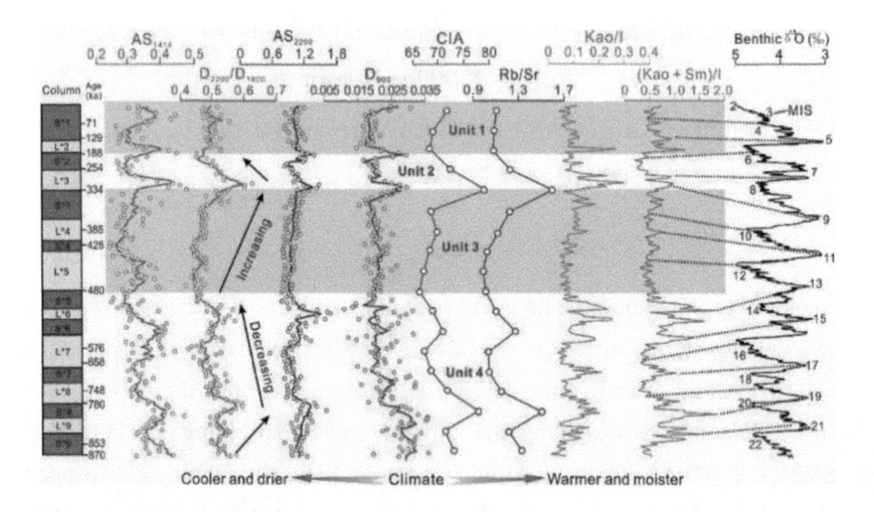

Fig. 2　Profiles of the spectral parameters, geochemical weathering proxies, and clay-mineral ratios in the SBC section

Note: For the spectral parameters, seven-point moving averages were calculated

The VNIR spectral parameters are demonstrated to have the potential to provide high-resolution information for predicting clay composition and reconstructing past paleoclimate evolution in weakly-moderately weathered soils. Compared with conventional weathering proxies and previous studies[1,2], the spectral parameters are suitable in various soil types. At least four qualified parameters (AS_{1400}, D_{2200}/D_{1900}, AS_{2200} and D_{900}) can be extracted from the VNIR spectrum, and paleoclimate evolution can be reconstructed with multiple proxies.

References

[1] Zhao L, Hong H, Liu J, et al. 2018. Assessing the utility of visible-to-shortwave infrared reflectance spectroscopy for analysis of soil weathering intensity and paleoclimate reconstruction. *Palaeogeography, Palaeoclimatology, Palaeoecology* in press, doi. org/10. 1016/j. palaeo. 2017. 07. 007.

[2] Fang Q, Hong H, Zhao L, et al. 2017. Tectonic uplift-influenced monsoonal changes promoted hominin occupation of the Luonan Basin: Insights from a loess-paleosol sequence, eastern Qinling Mountains, central China [J]. Quaternary Science Reviews, 169: 312－329.

Degradation of Methylene Blue in a Heterogeneous Fenton-like Reaction Catalyzed by Fe_3O_4@Expanded Graphite

Yun Shan Bai* Dan Liu Qing Hua Mao Hong Zhu Ma*
(School of Chemistry and Chemical Engineering, Shaanxi Normal University, Xi' an 710119, China)

With the rapid development of industry, the discharge of industrial wastewater increased year by year and environment pollution problem caught more and more attention. Dyes and pigments are the main organic pollutants in wastewater released from various industries, including textile, food, paper, printing, leather and cosmetic[1-3]. The presence of dyes in the surroundings, even at low concentrations, is very harmful to human health and the environment[4]. Therefore, it is necessary to find appropriate treatment strategies for efficient removal of dyes from wastewater system before discharge. The heterogeneous Fenton-like reaction is an advanced oxidation process which has gained wide spread acceptance for high removal efficiency of recalcitrant organic contaminants. In this work, Fe_3O_4 anchored to expanded graphite (EG) (Fe_3O_4@EG) was synthesized through a solvothermal method[5]. The catalytic properties and possible mechanism of Fe_3O_4@EG toward methylene blue (MB) were investigated (Fig. 1). The influences of various experimental parameters, such as contact time, temperature, initial concentration of dye, initial pH, adsorbent dosages and H_2O_2 dosage on methylene blue (MB) removal were studied. At the optimal condition: 50 min, 303 K, 200 mg/L MB at pH 2.50, 0.025 g/10 mL Fe_3O_4@EG, 0.7 mL/L H_2O_2, 95.55% MB was degraded. Langmuir isotherm and the pseudo-second-order kinetics provided the best correlation with the experimental data. Intra-particle diffusion model showed that adsorption process affected by external mass transfer and diffusion. The reusability of the Fe_3O_4@EG was tested and still over 79% MB removal was obtained after five cycles (Fig. 2). All the results revealed that the Fe_3O_4@EG composites could act as a promising candidate for high-efficient removal of contaminants from wastewater in future practical use.

* Corresponding author: Yun Shan Bai & Hong Zhu Ma; E-mail: ysbai@snnu. edu. cn; hzmachem@snnu. edu. cn; Phone No. 13636819611, 18049007258.

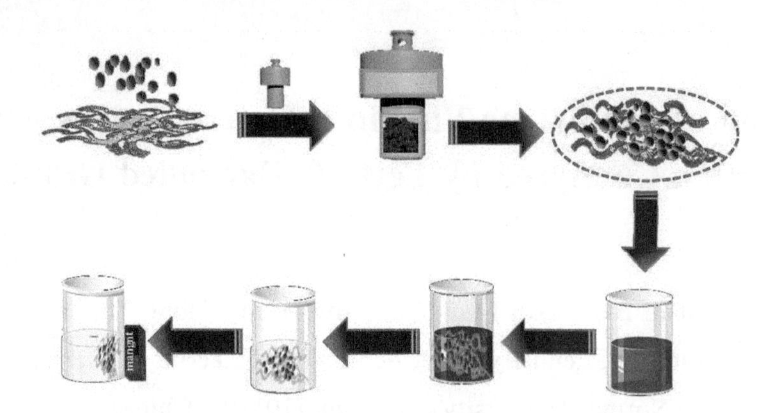

Fig. 1 Illustration of Fe_3O_4@EG synthesis and its application in MB degradation

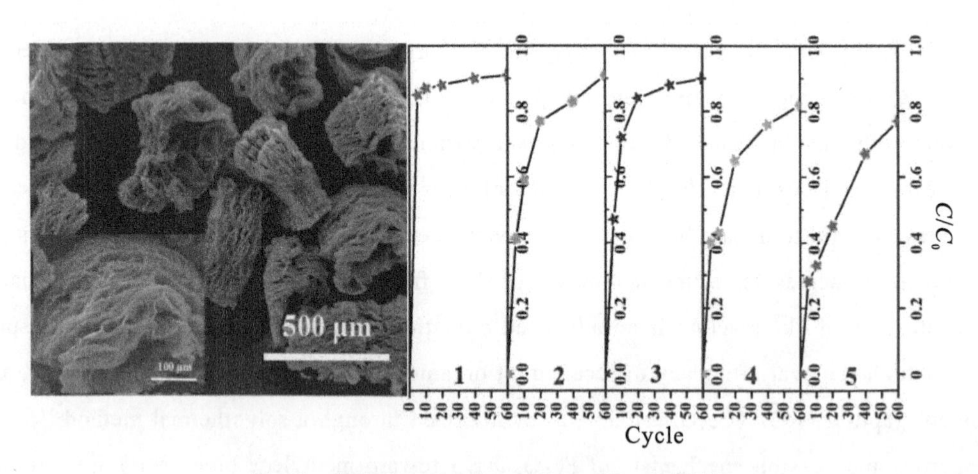

Fig. 2 SEM image and recycling of Fe_3O_4@EG

References

[1] Sabbaghan M, Adhami F, Aminnezhad M. 2018. Mesoporous Jarosite/MnO_2, and Goethite/MnO_2, nanocomposites synthesis and application for oxidation of methylene blue [J]. Journal of Structural Chemistry, 59: 463 - 473.

[2] Banerjee S, Benjwal P, Singh M, et al. 2018. Graphene oxide (rGO) -metal oxide (TiO_2/Fe_3O_4) based nanocomposites for the removal of methylene blue [J]. Applied Surface Science, 439: 560 - 568.

[3] Du X D, Wang C C, Liu J G. 2017. Extensive and selective adsorption of ZIF - 67 towards organic dyes: Performance and mechanism [J]. Journal of Colloid Interface Science, 506: 437 - 441.

[4] Zhou X P. 2017. Toxic effects of simulated printing and dyeing wastewater on ruditapes philippinarum [J]. Advances in Marine Sciences, 4: 23 - 29.

[5] Qian W C, Luo X P, Wang X, et al. 2018. Removalof methylene blue from aqueous solution by modified bamboo hydrochar [J]. Ecotoxicology & Environmental Safety, 157: 300 - 306.

Degradation Process and Mechanism of Orange Ⅱ by Using CuS Coupline Persulfate under Visible Light

Bo Yuan Zhu Jian Feng Ma* Qing Zhuo Ni

(School of Environmental and Safety Engineering, Changzhou University, Jiangsu 213164, China)

The rapid development of industry has caused serious environmental pollution problems, such as organic wastewater. Advanced oxidation processes are effective for removing many organic contaminants from water because they generate strong radical oxidants, such as hydroxyl radical (OH^-) and sulfate radical (SO_4^-). Persulfate activation is a new advanced oxidation technology, which can produce strong oxidizing and highly active SO_4^- under the excitation of heat, light and transition metal[1].

Chemical agents including $CuCl_2 \cdot 2H_2O$, $Na_2S \cdot 9H_2O$ and $K_2S_2O_8$ were of analytical grade and used without further purification. The experimental instruments used so far are UV - vis spectrometer and X - ray diffractometer.

The preparation of CuS was carried out as follows: 12 mmol of $CuCl_2 \cdot 2H_2O$ and 12 mmol of $Na_2S \cdot 9H_2O$ were dissolved in a beaker containing 28 mL of deionized water and 14 mL of ethanol, and labeled A and B, respectively. Then, A was slowly added to B under a vigorous stirring condition. Then, the black suspension was transferred to a 100 mL Teflon-lined stainless steel autoclave and reacted at 140 ℃ for 10 h. At the end, the product was washed for several times with deionized water and absolute ethanol and dried at 60 ℃ in the oven.

The catalytic activity of CuS were monitored by degradation of Orange Ⅱ under irradiation with visible light (120 W LED lamp). 15 mg of CuS catalyst was added to Orange Ⅱ (50 mL, 60 mg/L) and stirred for 60 min to reach adsorption-desorption equilibrium. Then, 100 mg $K_2S_2O_8$ was added to the solution with the visible light turning on. 2 mL solution were sampled every 20 min and filtrated to remove the solid particles. Then the absorbance of the sampled solution was determined by UV-vis spectrophotometer at 484 nm.

The degradation of Orange Ⅱ under different reaction conditions is shown in the Fig. 1: $CuS/K_2S_2O_8$/Vis (98.88%) >$CuS/K_2S_2O_8$ (35.67%) >$K_2S_2O_8$/Vis (27.35%) >CuS/Vis (23.42%) >$K_2S_2O_8$(7.76%). The results suggested that $CuS/K_2S_2O_8$/Vis system is faster than any other systems. As shown in Fig. 2, the catalytic activity of CuS didn’t decrease significantly, which could degrade over 98% of Orange Ⅱ after three cycles of degradation within 120 min.

* Corresponding author: Jian Feng Ma; E-mail: jma@zju. edu. cn; Mobile Phone No.: 13775020169.

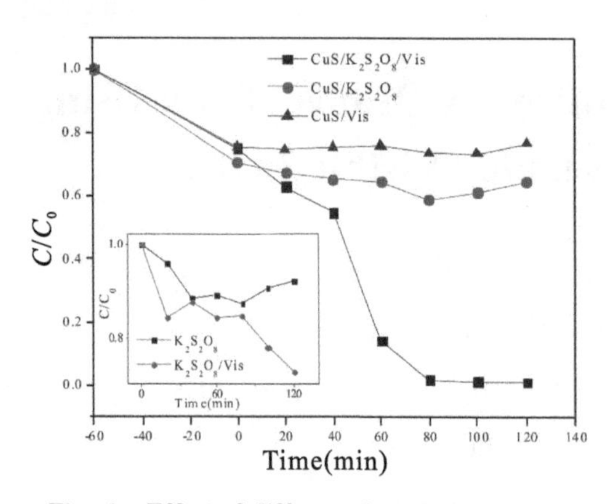

Fig. 1 Effect of different degradation systems on the degradation of Orange Ⅱ

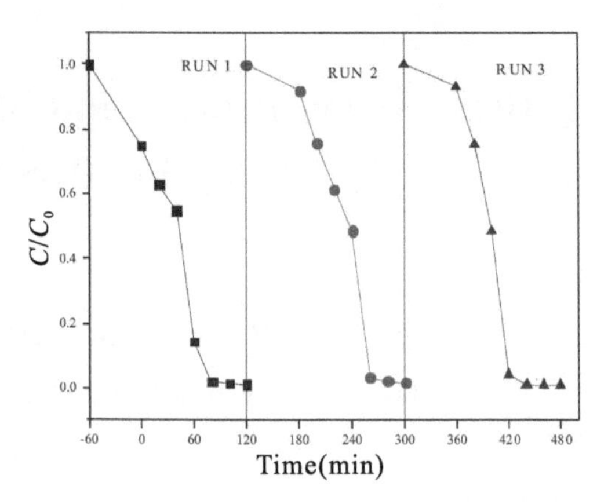

Fig. 2 Effects of different scavengers on the degradation of Orange Ⅱ in $CuS/K_2S_2O_8$/Vis system

Methanol and tert-butanol were used as free radical quenching agents[2]. Based on quenching experiment, a possible mechanism of the degradation of Orange Ⅱ by $CuS/K_2S_2O_8$/Vis system was proposed by Eqs. (1) — (6).

$$S_2O_8^{2-} + hv \rightarrow 2SO_4^- \quad (1)$$

$$\text{Orange II} + hv \rightarrow \text{Orange II}^* \quad (2)$$

$$CuS + H^+ \rightarrow Cu^{2+} + H_2O \quad (3)$$

$$Cu^{2+} + hv \rightarrow Cu^+ \quad (4)$$

$$Cu^+ + S_2O_8^{2-} \rightarrow Cu^{2+} + SO_4^- \ (\text{or } OH^-) \quad (5)$$

$$SO_4^- \text{ or } OH- + \text{Orange II}^* \rightarrow CO_2 + H_2O + \cdots \quad (6)$$

CuS showed remarkable reactivity, excellent reusability in the degradation of Orange Ⅱ dye with presence of $K_2S_2O_8$ and visible light, which was attributed to the large production of SO_4^- and OH^-. The wastewater treated in this study is only a laboratory simulated wastewater. Further research can be made to use more complex wastewater or actual wastewater, providing technical support and data support for industrial applications and promoting the development of water treatment technology.

References

[1] Tan C, Gao N, Deng Y, et al. 2012. Heat-activated persulfate oxidation of diuron in water [J]. Chemical Engineering Journal, 203: 294 - 300.

[2] Zeng T, Zhang X, Wang S H. et al. 2015. Spatial confinement of a Co_3O_4 catalyst in hollow metal-organic frameworks as a nanoreactor for improved degradation of organic pollutants [J]. Environmental Science & Technology, 49: 2350.

Clay Mineral Adsorbents for Effective Environmental Remediation

Sudipta Ramola[1,2,3*]　Chun Hui Zhou[2,3*]

(1. DBS School of Agriculture and Allied Sciences，Dehradun，Uttarakhand，India；

2. Research Group for Advanced Materials & Sustainable Catalysis（AMSC），State Key Laboratory Breeding Base of Green Chemistry-Synthesis Technology，College of Chemical Engineering，Zhejiang University of Technology，Hangzhou 310014，China；

3. Qingyang Institute for Industrial Minerals，Qingyang 242804，China)

Industrial minerals play an important role in day to day life of modern human beings. These are generally non-metallic，non-fuel based minerals having some extraordinary chemical or physical properties which make them irreplaceable in many everyday products. Clays，bentonite，limestone，gravel，kaolin，silica，diatomite，barite，gypsum and talc are some of the examples of widely used industrial minerals having a very wide spectrum of application such as construction，ceramics，paints，electronics，filtration，plastics，glass，detergents and paper. The term clay applies to the materials having a particle size of less than 2 μm. Utilization of clays have been done in wide spectrum such as in agriculture，construction，pharmaceuticals，food processing and many other industrial applications. Apart from many commercial applications，clay minerals (both in natural as well as modified form) are extensively used for environmental remediation especially waste water treatment.

Biochar is a black carbonaceous product of pyrolysis. A broad spectrum of feedstock can be used to prepare the biochar such as agrowaste，wood based waste as well as terrestrial and aquatic weed species. Biochar has multi-dimensional uses and environmental remediation has been recently recognized as a promising area where biochar can be successfully applied (Fig. 1). The preparation of biochar is done at varied temperature and residence time which greatly influence its physico-chemical properties and hence adsorption efficiencies of biochar for different inorganic and organic contaminants. Biochar has similar porous structure as that of activated carbon but is much cheaper，often equally or at times more efficient adsorbent as well as more versatile in terms of environmental and agronomic applications[1].

---

* Corresponding author：Sudipta Ramola，E-mail：ramola. sudipta@gmail. com，Phone No. +91 8279822989；Chun Hui Zhou，E-mail：chc. zhou@aliyun. com. The financial support from the National Natural Scientific Foundation of China (41672033).

Biochar can increase soil productivity by improving physico-chemical properties of soil. In addition to this, biochar also enhances nutrients such as N, P, K, Ca, Mg to soil and hence alter nutrient cycles also. Mixing of biochar in soil also stimulate the microbial population and activate dormant soil microorganisms and hence increases microbial respiration. After biochar addition, a number of processes occur that increases CO_2 fluxes in soil. These increased fluxes may be due to factors such as i) Some of the components of biochar is biologically consumed by micro, meso and macro-organisms ii) some of the carbon from biochar may get released abiotically in the form of carbonates or chemically sorbed CO_2 iii) there may be interaction between biochar and native soil organic matter. The fact that biochar is of recalcitrant nature limited the possibility of its use as a food source by soil microorganisms. However, in contrast to this, there is microbial assimilation of biochar basically because of many direct and indirect physico-chemical changes induced by biochar in soil such as presence of reactive surfaces, change in soil pH, porosity, soil tensile strength, water holding capacity and bulk density. The conversion of lignocellulosic material by pyrolysis imparts many changes at micro and macro level with basically homogenization of wood cell walls and disappearance of middle lamella. This results in increasing the porosity and internal surface area (because of dehyroxylation). This increased porous surface coupled with sorption of organic compounds, dissolved organic matter and ammonium ions by either varied functional groups of biochar and/or increased CEC at biochar surface promote habitats to microbes. Biochar also has positive effects on soil pH, especially acidic soil where addition of biochar causes increased alkanity in soil and thus increased microbial activity. Biochar addition also increases organic carbon, calcium content, soil water holding capacity and thus enhances soil suitability to host microbes. However, the other aspect of effect of biochar on microbial growth also lies in fact that biochar is a good adsorbent of toxic compounds including heavy metals and pesticides which may ultimately increase the microbial growth in bulk soil but may adversely influence the direct use of biochar by them[2,3].

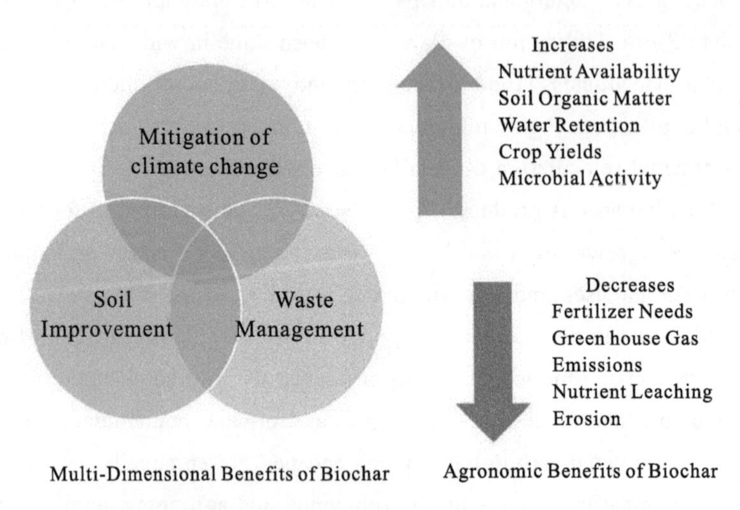

Fig. 1 Multi-dimensional and agronomic benefits of biochar

References

[1] Xue Y，Gao B，Yao Y，et al. 2012. Hydrogen peroxide modification enhances the ability of biochar（hydrochar）produced from hydrothermal carbonization of peanut hull to remove aqueous heavy metals：batch and column tests [J]. Chem. Eng. J.，200：673 - 680.
[2] Ameloot N，Graber E R，Verheijen F G A，et al. 2013. Interactions between biochar stability and soil organisms：review and research needs [J]. European Journal of Soil Science，64：379 - 390.
[3] Chen L，Chen X L，Zhou C H，et al. 2017. Environmental-friendly montmorillonite-biochar composites：facile production and tunable adsorption-release of ammonium and phosphate [J]. Journal of Cleaner Production，156：648 - 659.

附　录

第二届世界非金属矿科技和产业论坛主办单位

WFIM－2 Organizers and Sponsors

中国·安徽·青阳

Qingyang · Anhui · China

主题：非金属矿科学前沿和绿色高新技术

Theme：Cutting-edge Science and Green High-tech Industry

2018 年 10 月 20 日（星期六）—23 日（星期二）

- 青阳非金属矿研究院
- 青阳县经济和信息化委员会
- 青阳县科学技术局
- 青阳县西华镇人民政府
- 青阳县经济开发区西华工业园（筹）
- 池州非金属矿产业技术创新战略联盟
- 浙江工业大学化学工程学院工业催化学科
- 绿色化学合成技术国家重点实验室培育基地

第二届世界非金属矿科技和产业论坛国际科学咨询委员会

WFIM－2 International Scientific Advisory Committee

Cyril Aymonier (France)

Georgios Christidis (Greece)

Saverio Fiore (Italy)

Emilia García-Romero (Spain)

Jaime Gómez-Morales (Spain)

Suryadi Ismadji (Indonesia)

Josef Breu (Germany)

John Keeling (Australia)

John Mungai Kinuthia (UK)

Sridhar Komarneni (USA)

Victoria Krupskaya (Russia)

Mercedes Suárez (Spain)

Riccardo Tesser (Italy)

Jan J. Weigand (Germany)

Chengzhong Yu (Australia)

Shinya Hayami (Japan)

Ru-Shi Liu (China)

Yang Kim (Japan)

Gang Wei (Australia)

Martino Di Serio (Italy)

Godwin Ayoko (Australia)

Sabine Petit (France)

Reiner Dohrmann (Germany)

Maguy Jaber (France)

Jin-Ho Choy (Korea)

Huijun Zhao (Australia)

Peter Crowley Ryan (USA)

Joseph William Stucki (USA)

第二届世界非金属矿科技和产业论坛学术委员会

WFIM－2 Scientific Committee

鲁安怀（北京大学）

董发勤（西南科技大学）

何宏平（中国科学院广州地球化学研究所）

陈天虎（合肥工业大学）

周根陶（中国科技大学）

陆现彩（南京大学）

蔡元峰（南京大学）

冯雄汉（华中农业大学）

洪汉烈（中国地质大学（武汉））

黄　菲（东北大学）

刘　羽（福州大学）

刘钦甫（中国矿业大学）

吕国诚（中国地质大学（北京））

秦　善（北京大学）

孙红娟（西南科技大学）

万　泉（中国科学院地球化学研究所）

王爱勤（中国科学院兰州化学物理研究所）

汪立今（新疆大学）

王林江（桂林理工大学）

吴宏海（华南师范大学）

吴平霄（华南理工大学）

杨华明（中南大学）

杨　燕（浙江大学）

杨志军（中山大学）

朱建喜（中国科学院广州地球化学研究所）

赵红挺（杭州电子科技大学）

袁　鹏（中国科学院广州地球化学研究所）

申宝剑（中国石油大学（北京））

季生福（北京化工大学）
范　杰（浙江大学）
罗永明（昆明理工大学）
郑水林（中国矿业大学）
严春杰（中国地质大学（武汉））
梁金生（河北工业大学）
张以河（中国地质大学（北京））
鲍康德（温州医科大学）
边　亮（西南科技大学）
储茂泉（同济大学）
刘海波（合肥工业大学）
刘明贤（暨南大学）
欧阳静（中南大学）
王文波（中国科学院兰州化学物理研究所）
周岩民（南京农业大学）
朱润良（中国科学院广州地球化学研究所）
周春晖（浙江工业大学）

第二届世界非金属矿科技和产业论坛主席和会务组织

WFIM－2 Chairpersons and Secretariats

论坛名誉主席（Honorary Chairman）：

George E. Christidis（Greece）

Professor of Economic Geology Industrial Mineralogy

论坛主席： 周春晖

共同主席：

边　亮　储茂泉　刘海波　刘明贤

欧阳静　王文波　周岩民　朱润良

论坛地方组织委员会：

章双宏　章晓山　王效锋　汪兴华　王朝虎　周春晖

庞经龙　鲍康德　孙正力　刘初旺　周　磊　张　静

夏淑婷　杨海燕　沈程程　李贵黎　吴琦琦　童东绅

Freeman B. Kabwe　Sudipta Ramola

青阳非金属矿研究院第一届国际科技咨询委员会

International Scientific and Technical Advisory Committee for Qing Yang Institute for Industrial Minerals

Cyril Aymonier（France）

Georgios Christidis（Greece）

Saverio Fiore（Italy）

Emilia García-Romero（Spain）

Jaime Gómez-Morales（Spain）

Suryadi Ismadji（Indonesia）

Josef Breu（Germany）

John Keeling（Australia）

John Mungai Kinuthia（UK）

Sridhar Komarneni（USA）

Victoria Krupskaya（Russia）

Mercedes Suárez（Spain）

Riccardo Tesser（Italy）

Jan J. Weigand（Germany）

青阳非金属矿研究院第一届学术和技术委员会

Academic and Technical Committee for Qing Yang Institute for Industrial Minerals

鲁安怀（北京大学）

董发勤（西南科技大学）

何宏平（中国科学院广州地球化学研究所）

陈天虎（合肥工业大学）

周根陶（中国科技大学）

陆现彩（南京大学）

蔡元峰（南京大学）

冯雄汉（华中农业大学）

洪汉烈（中国地质大学（武汉））

黄　菲（东北大学）

刘　羽（福州大学）

刘钦甫（中国矿业大学）

吕国诚（中国地质大学（北京））

秦　善（北京大学）

孙红娟（西南科技大学）

万　泉（中国科学院地球化学研究所）

王爱勤（中国科学院兰州化学物理研究所）

汪立今（新疆大学）

王林江（桂林理工大学）

吴宏海（华南师范大学）

吴平霄（华南理工大学）

杨华明（中南大学）

杨　燕（浙江大学）

杨志军（中山大学）

朱建喜（中国科学院广州地球化学研究所）

赵红挺（杭州电子科技大学）

鲍康德（温州医科大学）

周春晖（浙江工业大学）